Carbohydrates in Solution

Carbohydrates in Solution

A symposium sponsored by the
Division of Carbohydrate Chemistry
at the 162nd Meeting of the
American Chemical Society,
Washington, D. C.,
Sept. 14-15, 1971.

Horace S. Isbell
Symposium Chairman

ADVANCES IN CHEMISTRY SERIES **117**

AMERICAN CHEMICAL SOCIETY

WASHINGTON, D. C. 1973

ADCSAJ 117 1–421 (1973)

Library of Congress Catalog Card 73–81038
ISBN 8412–0178–1
PRINTED IN THE UNITED STATES OF AMERICA

Advances in Chemistry Series

Robert F. Gould, *Editor*

FOREWORD

ADVANCES IN CHEMISTRY SERIES was founded in 1949 by the American Chemical Society as an outlet for symposia and collections of data in special areas of topical interest that could not be accommodated in the Society's journals. It provides a medium for symposia that would otherwise be fragmented, their papers distributed among several journals or not published at all. Papers are referred critically according to ACS editorial standards and receive the careful attention and processing characteristic of ACS publications. Papers published in ADVANCES IN CHEMISTRY SERIES are original contributions not published elsewhere in whole or major part and include reports of research as well as reviews since symposia may embrace both types of presentation.

CONTENTS

PREFACE

Recognition that the properties of sugars and related substances depend on conformational as well as structural factors opened a new chapter of chemistry, which in the past 45 years has engulfed not only carbohydrate but organic and biological chemistry as well. Carbohydrate molecules have, at definite points in space, groups capable of interaction with substances in the environment in stereospecific manner. In many instances the intermolecular forces are sufficiently strong to alter conformations and thereby influence equilibria and the course of ensuing reactions. Striking effects of solvent and other molecules in the solution environment on the properties of carbohydrates were noted long ago but could not be studied satisfactorily for lack of suitable methods. Now newly developed methods make detailed studies possible, and many investigations of the behavior of carbohydrates in solution are in progress. This volume covers briefly four main areas of carbohydrate chemistry.

The first six chapters of the text deal primarily with tautomeric equilibria and the reactions of monosaccharides. Bentley and Campbell discuss determination of anomeric equilibria by gas chromatography of trimethyl silyl ethers, a technique which has greatly extended our knowledge of the composition of sugar solutions. Anderson and Garver use computer techniques to study mutarotation reactions. Perlin and coworkers apply carbon-13 and proton NMR spectra to study the equilibrium composition of certain ketones in various media. A paper by Rendleman deals with the interaction of reducing sugars with metal hydroxides and with ionization, epimerization, and rearrangement reactions. Isbell rationalizes the reactions of sugars with oxygen and hydrogen peroxide under alkaline conditions by formation of intermediate hydroperoxide adducts which decompose by oxidative cleavage and rupture of the carbon chain. Fatiadi applies electron spin resonance to study phenylhydrazones in alkaline solutions.

Chapters 7 to 12 deal with factors affecting conformational equilibria and complex formation. Angyal explains the effect of calcium chloride on the anomeric equilibria of certain sugars by formation of especially stable complexes. These involve an axial–equatorial–axial sequence of three hydroxyl groups on a six-membered ring or a cis–cis sequence on a five-membered ring. Lemieux and Brewer use model compounds to study solvation effects on the orientation of the hydroxymethyl groups at the 5-position of hexopyranose structures. Horton and co-workers report the use of low-temperature NMR spectroscopy and the method of averaging of spin couplings for determining the relative proportions of conformers in solutions of various carbohydrate derivatives. Jeffrey uses x-ray techniques to study crystal structures and relates these to the conformation of molecules in solution. A paper by Montgomery discusses potentiometric methods for studying stabilities of some nickel and borate complexes. Acree reports recent studies of boric acid complexes of polyols.

Chapters 13 through 16 relate to biological materials and physiological phenomena. Pigman reviews and explains certain structural changes arising from treatment of glycoproteins with alkalies. Whistler reports that the shape of a polysaccharide in solution depends in part on the environmental solvent and solute molecules. Shallenberger ascribes the relative sweetness of various sugars to specific stereochemical features of the glycol groups. He finds that a gauche (staggered) conformation results in maximum sweetness. Bailey reviews transport systems for sugars in living organisms, a complex subject now being actively investigated in several laboratories.

Chapters 17 through 21 deal with carbohydrate-enzyme systems. Hehre presents some new ideas on the action of amylases. Kabat presents some new immunochemical studies on the carbohydrate moiety of certain water-soluble blood-group substances and their precursor antigens. Hassid reviews the role of sugar phosphates in the biosynthesis of complex saccharides. Pazur and co-workers present information obtained by isotopic techniques on the nature of enzyme-substrate complexes in the hydrolysis of polysaccharides. Gabriel presents a common mechanism for the production of 6-deoxyhexoses. An intermediate nucleoside-5′-(6-deoxyhexose-4-ulose pyrophosphate) is formed in each of the syntheses.

These papers show the broad scope of the symposium and the intense interest of chemists in the changes which take place in compounds prior to and during chemical reactions of carbohydrates in solution.

Horace S. Isbell

The American University
Washington, D. C.
July 1972

1

Analytical Methods for the Study of Equilibria

IAIN M. CAMPBELL and RONALD BENTLEY

Department of Biochemistry, Faculty of Arts and Sciences,
University of Pittsburgh, Pittsburgh, Pa. 15213

The majority of the many methods used to study the composition of equilibrium solutions of carbohydrates examine the mixture without separating the individual components. With the discovery that the anomeric forms of sugars could be readily separated by gas chromatography of their trimethylsilyl ethers, a new approach to the problem was found. A protocol was developed for the direct gas chromatographic analysis of the amount of each anomer present in an aqueous solution. The protocol can be used on the micro scale and can be used in enzyme assays such as that for mutarotase. The method has been made more effective by combining gas chromatography with mass spectrometry. It is shown how mass spectral intensity ratios can be used to discriminate anomers one from another. The application of these methods to the study of complex mutarotations is discussed.

What's past is prologue (*1*). The year 1971 marked the one hundred and twenty-fifth anniversary of the discovery by Dubrunfaut of the changes in optical rotation which occur when crystalline glucose is dissolved in water (*2*). Some 50 years after Dubrunfaut's discovery, the general phenomenon of change of optical rotation of a solution (which could be often a change in direction as well as in magnitude) was termed mutarotation by Lowry (*3*). Although the reason for these and similar changes remained unclear for a long time, it was recognized by 1899 that the stable product obtained when the optical rotation was constant was not a separate isomer of glucose but rather an equilibrium mixture essentially of two isomeric forms (*3*).

As knowledge of the isomeric structures available to a sugar such as glucose became ever more detailed, there was an increase in the number of isomers postulated to exist in aqueous solution. In his classical work on optical rotatory power, Lowry described the situation for glucose in terms of the two α- and β-pyranose forms and noted that

"... the presence of an open-chain aldehyde or aldehydrol in the equilibrium mixture must also be postulated in order to provide a mechanism for the inversion of the terminal >CHOH radical. The presence of 5-ring (furanose) sugars in the mixture must also be admitted. . . . The final equilibrium must therefore include, in addition to the open-chain aldehydic sugar . . . four glucosidic sugars (α- and β-pyranose, α- and β-furanose) corresponding with the four ethyl glucosides which have now been isolated (*4*)."

The question of the detailed composition of sugar solutions in various solvents together with the chemistry of the mutarotation reaction has continued to attract attention through the years. In their classical review, "Mutarotation of Sugars in Solution," completed in 1969, Isbell and Pigman have cited references to more than 320 papers and books (*5, 6*). The appearance of this volume justifies Isbell's faith that carbohydrate chemistry continues to have a significant impact "on the development of new concepts permeating all branches of chemistry and biochemistry (*7*)."

General Methods

In the earliest experiments optical rotation, and to a lesser extent, optical rotatory dispersion were used to study mutarotation reactions and to calculate the composition of equilibrium solutions. (The presence of many chiral elements and the accompanying optical properties have dominated the whole area of carbohydrate chemistry. Thus, in the 1942 preface to a National Bureau of Standards circular, Lyman Briggs remarked ". . . that carbohydrate chemistry, and carbohydrate industry dependent upon carbohydrate chemistry, could hardly have developed to the magnitude it has attained in recent years without the aid and the guidance of the polariscope (*8*)." The publication of this circular might be said to end the first epoch of carbohydrate chemistry; since glucose was isolated from starch in 1792, this era covers a span of one and one-half centuries.) Such methods are relatively poor, however, since the information they give is composite—individual isomers in an equilibrium solution cannot be examined independently. This was realized in 1942 when it was concluded, "there is no really satisfactory method for determining the proportions of the labile components in complex mixtures (*9*)."

Since 1942 some new analytical techniques have become available for the study of sugar solutions, notably nuclear magnetic resonance spectroscopy (NMR) and gas–liquid chromatography (GLC). The NMR method is discussed elsewhere in this volume; GLC-based techniques are described here.

Gas–Liquid Chromatography of Carbohydrates

While the application of GLC to problems of microanalysis has had far reaching results in all areas of chemistry and biochemistry, the development of methods for use with carbohydrates and other polyhydroxy compounds was slower than with other classes of compounds. The major problem was the general lack of volatility of the polyhydroxy compounds and the fact that derivatives known to be volatile could not readily be prepared in quantitative yields by simple micro procedures.

The first application of GLC in carbohydrate chemistry was to separate fully methylated methyl glycopyranosides of simple pentoses and hexoses (*10*), and much literature now covers these derivatives (*11, 12, 13, 14*). Shortly after this work, acetate derivatives were examined by various workers (*15, 16*), and a little later another development was the use of these acetates with thin-film columns containing liquid phases of high thermal stability such as silicone polymers (*e.g.*, SE-30) and fluoroalkyl silicone polymers (*e.g.*, QF-1) (*17*).

Although such derivatives are extremely useful in many applications, the synthetic steps to prepare them quantitatively are a little too vigorous and time consuming to permit facile use in sugar equilibria studies. Such is not the case with the trimethylsilyl (TMSi) derivatives. Following initial studies on the use of these derivatives in the carbohydrate field (*18, 19, 20*), the development of a simple, rapid, quantitative method for the preparation of TMSi derivatives of bile acids (*21*) suggested the extension of this derivatization method to carbohydrates. Thus, beginning in 1963 we began a systematic examination of the preparation and analysis of TMSi derivatives of numerous carbohydrates (*22, 23, 24*). The compounds, either dissolved or suspended in pyridine, reacted with hexamethyldisilazane in the presence of trimethylchlorosilane as a catalyst. The use of pyridine was advantageous in dealing with polyhydroxy compounds, and even at room temperature, silylation was complete within minutes. Isolation of the derivatives was unnecessary; an aliquot of the reaction mixture could be injected directly onto the column.

An example of the scope of the method is the tetrasaccharide, stachyose. Even with the equipment limitations of 1963, this material with 14 protected hydroxyl groups and a molecular weight for the derivative of 1676 was successfully chromatographed.

The results obtained with a single, pure anomer of a sugar such as glucose were particularly interesting. When a sample of α-D-glucose was treated with a mixture of pyridine:hexamethyldisilazane:trimethylchlorosilane (10:2:1, *v/v*) the gas chromatogram showed a single major peak, with at most, a minor second component which was well separated from the major component (*see* Figure 1). A similar peak distribution was obtained from β-D-glucose. However, the position of the minor peak from

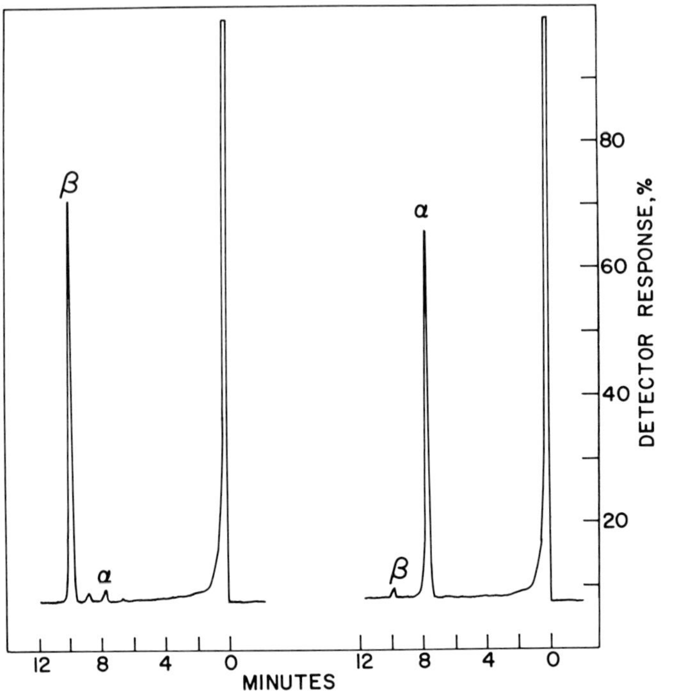

Figure 1. Gas chromatography of the TMSi derivatives of the glucose anomers.

The solid anomers were derivatized with a premixed cocktail and chromatographed on a column of OV-1 (6 feet by 2 mm). The column temperature was held at 175°C for the first three minutes and then was increased at the rate of 4°C per minute.

the α anomer was the same as that of the major peak from the β anomer while the minor peak from the β anomer corresponded exactly with the major peak from the α anomer. (The minor peaks in this early work accounted at most for 5% of total. In this work, the anomers were first dissolved in pyridine then subjected to trimethylsilylation. By direct reaction of the solid anomers with a premixed "cocktail," this anomerization can be kept at a much lower level.)

Since each anomer yielded essentially a single peak, it appeared that all of the free OH groups were reacting rapidly with the reagent, even the anomeric one, and a potential method of freezing out components of an equilibrium (or other) solution was at hand. This proved to be the case when methyl α-D-glucopyranoside, silylated in the usual way, gave a chromatographic peak indistinguishable from that obtained with a pure sample of methyl tetra-*O*-trimethylsilyl-α-D-glucopyranoside, prepared as described by Hedgley and Overend (*18*). Final proof of the complete trimethylsilylation of all OH groups is given by mass spectrometric evidence, discussed below.

When the residue, obtained on evaporation of an aqueous equilibrium solution of glucose, was derivatized with the standard reagent, two peaks were obtained on gas chromatography, corresponding to the peaks obtained with the two separate α and β anomers; the ratio of the peak areas was $\alpha:\beta = 39.8:60.2$, agreeing well with the ratio 36.2:63.8 determined by the optical method or 37.4:62.6 determined by the bromine oxidation technique (*25*).

These observations have been extended to many sugars other than glucose. As before when pure anomers were available, a single major peak was obtained; whereas with dried mutarotation mixtures, two or more peaks resulted. The TMSi derivatives of the anomeric forms were generally well separated with the β anomers usually having the longer retention times (more precisely, the anomer having the anomeric OH group in an equatorial position). Thus, when an equilibrium solution containing idosan and the seven other aldohexoses was chromatographed on an EGS polyester column, 13 peaks were separated, some containing more than one component.

Application of GLC Method to Sugar Equilibrium Studies

Since the GLC–TMSi technique offered a new approach to the study of the composition of aqueous solutions, efforts were made to develop a method for direct trimethylsilylation in the presence of water. It was necessary that the trimethylsilylation procedure did not change the anomeric composition of the solution. This problem was solved by rapid dilution of a small volume (*ca.* 5 μliters) of the aqueous solution into 0.1 ml of dimethylformamide, followed by rapid freezing at liquid air temperature. The frozen mixture was then treated with a silylation reagent in which the proportions of hexamethyldisilazane and trimethylchlorosilane were increased relative to the pyridine (pyridine:hexamethyldisilazane:trimethylchlorosilane, 4:1:1, *v/v*). After standing for about 30 minutes at room temperature, samples were ready for gas chromatographic analysis (*26*). (A similar type of GLC based mutarotation assay was developed independently by Semenza and his colleagues (*27*).) Although a heavy precipitate formed as a result of the reaction of the water with the reagent, it was still possible to inject the solution directly into the column since the precipitate settled rapidly.

When samples of a pure glucose anomer were dissolved in water and a sample removed for silylation within about 30 seconds, only one major peak was obtained, and the chromatogram was indistinguishable from that obtained by treatment of the solid anomer with the standard silylation reagent. Occasionally, minor peaks were present, preceding the peak of the α anomer; these may have represented partial trimethylsilyl derivatives (*28*).

As a solution of one anomer was allowed to mutarotate, an increasing amount of the second anomer became evident until finally the usual equilibrium mixture was obtained (*see* Figure 2). By determining the peak areas, the percent of each anomer present at any given time interval could be calculated. Since the equation for the mutarotation coefficient, k, can be expressed in the form of Equation 1, the coefficient can be determined

$$k = \frac{1}{t} \log_{10} \frac{x_\infty}{x_\infty - x_t} \tag{1}$$

$$x_\infty = \%\ \beta \text{ at equilibrium; } x_t = \%\ \beta \text{ at time, } t$$

from the slope of the plot of $\log_{10} \frac{x_\infty}{x_\infty - x_t}$ against time, t. In Figure 3 are

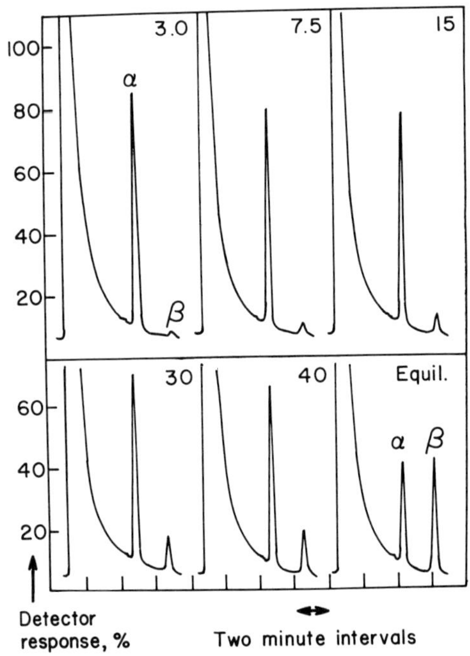

Figure 2. The mutarotation of α-D-glucose followed by GLC.

The mutarotation conditions were those described previously (26): the figures in the upper right hand portions of the GLC traces are time in minutes after solution of the solid α anomer.

shown such plots for the mutarotation of α-D-glucose in the standard pH 4.7 phthalate buffer used in many polarimetric determinations and in phosphate buffer where mutarotation is about four times as fast (29).

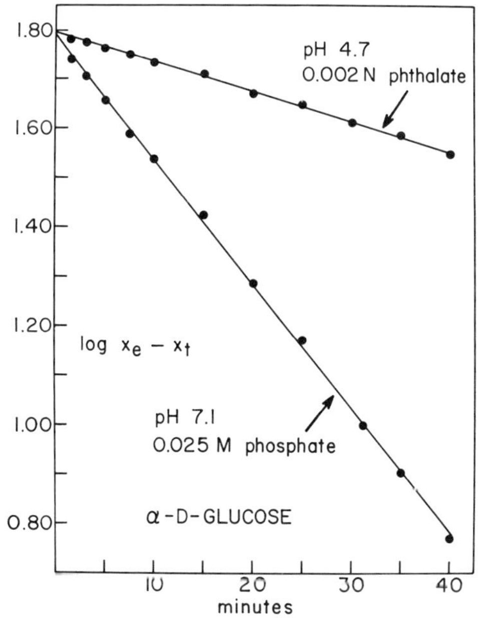

Figure 3. Determination of mutarotation coefficients by the GLC–TMSi method. The conditions were those described previously (26).

Straight line plots are obtained by this method; furthermore, the mutarotation coefficients for several sugars, determined under the same conditions as used in polarimetric work, give values identical to those obtained optically (*see* Table I).

In this work we have identified the anomers based on comparison of retention times with known standards whenever they were available.

Table I. Mutarotation Kinetics by Gas Chromatography of TMSi Derivatives[a]

Carbohydrate	*Mutarotation Coefficient (logarithms to base 10, reciprocal minutes)* *Gas Chromatography*		*Polarimetry*
α-D-Glucose	0.00630	(4)	0.00629
β-D-Glucose	0.00626	(2)	0.00625
α-D-Galactose	0.00795	(1)	0.00803
β-D-Arabinose	0.0299	(3)	0.0300
α-D-Xylose	0.0204	(2)	0.0203

[a] All values determined in pH 4.7, 0.002*N* potassium phthalate buffer at 20°. The gas-chromatographic values are averages with the number of determinations indicated by the number in parentheses.

Lee, Acree, and Shallenberger have in some cases carried out actual isolations of the anomeric trimethylsilyl per-*O*-trimethylsilyl glycosides by preparative gas chromatography and have then characterized the separated derivatives by optical rotation, elementary analysis, and NMR spectroscopy (*30*). These characterizations, which have been done for glucose, mannose, and galactose (*see* below) confirm the usefulness of the GLC technique.

The situation just discussed for glucose is described as a simple mutarotation. In such cases, as already indicated (*see* Figure 3 for example), the plot of $\log_{10} \frac{x_\infty}{x_\infty - x_t}$ against time is linear. It is only necessary to invoke the interconversion of the α- and β-pyranose forms, and no significant amounts of other isomers can be detected in the aqueous equilibrium solution by the gas chromatographic method. More complex situations are known, however, where two separate linear portions of the plot are found. Thus, with galactose there is an initial fast mutarotation, followed by a slow mutarotation. Separate mutarotation coefficients may be calculated for each of the phases. The slow process was considered to involve the usual α-pyranose ⇌ β-pyranose change while furanoid intermediates were postulated to account for the initial, rapid isomerization (*5, 6*).

It was interesting to see if gas chromatography could contribute any understanding to this problem. We noted at an early date that the usual equilibrium solution of galactose contains two major and one minor components when examined by gas chromatography (*see* Figure 4) (*23*). The proportion of this third component, γ-galactose, was considerably increased by first warming a pyridine solution of galactose, followed by immediate derivatization (*see* Table II). By this same technique a small amount of a third component was also detected for glucose. Using preparative gas–liquid chromatography, Shallenberger and Acree (*31*) actually isolated the TMSi derivative of the γ-galactose and tentatively identified it as a furanose since it lacked a free carbonyl absorption in the infrared spectrum. Later, with different chromatographic conditions a total of four components was found in pyridine solutions of galactose, and at equilibrium the following composition was determined:

α-Furanose	13.7%
β-Furanose	23.4
α-Pyranose	31.7
β-Pyranose	31.2

In this work the isomeric trimethylsilyl galactosides were separated by preparative gas–liquid chromatography as before, and identification of the various anomers was based primarily on NMR measurements (*32*).

We found for the aqueous equilibrium solution of galactose 5.4% of

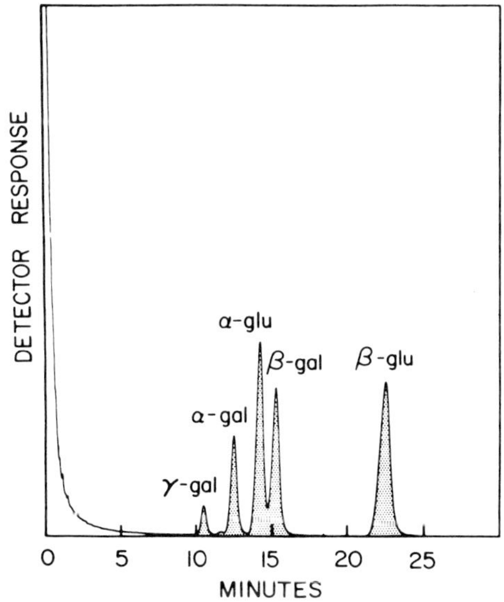

Figure 4. Gas chromatogram of TMSi derivatives prepared from an aqueous equilibrium solution of glucose and galactose. The column was 2.5% SE-30, isothermal at 160°. Gal = galactose; glu = glucose.

Table II. Percentage Composition of Equilibrium Solutions

	Aqueous				*Pyridine*		
Carbohydrates	*Gas Chrom.*		*Polarim.*		*Gas Chrom.*		
	α	β	α	β	α	β	γ
D-Glucose	39.8	60.2	36.2	63.8	47.3	49.5	3.2
D-Galactose	31.9	62.6 [a]	29.6	70.4	29.9	45.8	24.3
D-Mannose	72.0	28.0	68.8	31.2	78.3	21.7	—

[a] And 5.4% of γ.

γ-isomer (*see* Table II) (*23*). This amount did not seem to change significantly during mutarotation (*26*). More recently, Acree *et al.* have determined the following composition for the aqueous equilibrium solution of galactose (*33*).

α-Furanose	1.0%
β-Furanose	3.1
α-Pyranose	32.0
β-Pyranose	63.9

The total furanose component of 4.1% determined by these authors agrees well with our earlier value. (The pseudo-first-order rate constant for the formation of β-D-galactopyranose was the same as the first-order mutarotation coefficient for the slow phase, and also the rapid phase mutarotation coefficient was identical with the pseudo-first-order rate constant for the formation of β-D-galactofuranose. Thus, for the first time direct evidence for the formation of furanose forms in the fast mutarotation reaction was obtained. It was also concluded from thermodynamic considerations that the transition state for the various tautomerization reactions is unlike the structure of any of the pyranose or furanose isomers.) By similar GLC methods, mutarotated solutions of fructose have been shown to contain 33% of furanoside components (*34*) and ~0.5 to 0.6% of penta-*O*-trimethylsilyl-keto-D-fructose (*35*). Work by Anderson on the galactose problem is described in this volume.

Mutarotase Assay

A particularly interesting situation where the speed associated with the GLC-TMSi method proves important is in the study of mutarotase. In 1949 during a study of the oxidation of glucose by the enzyme, notatin (glucose oxidase), evidence suggesting that this enzyme also catalyzed the mutarotation of glucose was obtained (*36*). In a detailed study of notatin preparations, Keilin and Hartree showed that some of them con-

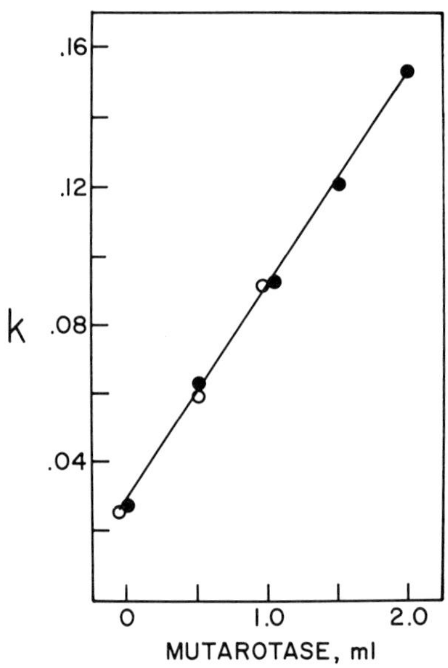

Figure 5. Mutarotase assay by the GLC–TMSi method. The enzyme preparation was obtained from hog kidney, and the conditions are those described previously (26).

tained a second enzyme, named mutarotase, which catalyzed the mutarotation reaction (but not the oxidation) (*37*). This enzyme is the most active catalyst known for the mutarotation reaction, the catalytic coefficient being at least 2×10^5 min^{-1}. The early assays for this enzyme were based on polarimetry; for samples also containing glucose oxidase activity, anaerobic polarimetry was necessary to prevent changes in optical rotation as a result of the formation of gluconolactone and gluconic acid. Although other analytical methods such as automatic polarimetry were developed later, none of them have been particularly convenient. The gas chromatographic technique, however, proved well suited to assay this enzyme activity. For example, a preparation of hog kidney mutarotase was used over a fourfold concentration range; mutarotation coefficients were determined by the gas chromatographic method, and for part of the range by polarimetry. As can be seen from Figure 5 the same straight line plot of mutarotation coefficient against enzyme volume was obtained by either method. Values for the mutarotation coefficient could be obtained by gas chromatography in solutions that were too turbid for assay by the polarimeter.

Determination of Anomer Configuration

Many enzymes are known which liberate a free sugar, and it is frequently important to know in which anomeric form the sugar is produced. Polarimetric methods have proved far from ideal for this purpose. This is another area in which the GLC-TMSi method is being increasingly valuable. For example, the sucrase–isomaltase enzyme from rabbit intestine hydrolyzed either sucrose or palatinose with the liberation of α-D-glucose; thus, these enzymatic hydrolyses proceed with retention of configuration at the carbonyl position (*27*). This method was more sensitive than the polarimetric analysis, which is in any case complicated by the known transglucosidation activity of sucrase producing small amounts of oligosaccharides of unknown optical rotation.

Parrish and Reese also used the GLC-TMSi method to determine the anomeric configuration of carbohydrates released by enzyme action (*38*). With α and β exo-glucanases the stereochemical course of the reaction was inversion of configuration; glucosidases, α and β, gave retention of configuration. More recently, similar results were obtained by Barnett, who used α- and β-D-glucopyranosyl fluoride and α-D-galactopyranosyl fluoride as substrates for the enzymatic hydrolyses (*39*). Other results involving retention and inversion of configuration have been summarized (*40*).

Critique of the GLC–TMSi Method

The GLC–TMSi method alone is an excellent technique to use in the detailed analysis of sugar solutions where speed, small quantities of material, or complex mixtures are involved. It has the added advantages that solutions whose composition is changing can be frozen for analysis, and the technique can also be applied to turbid or pigmented solution-areas wherein polarimetric assay would be extremely difficult. Despite this versatility the GLC–TMSi method suffers from one major disadvantage. It gives no information which directly and absolutely defines the chemical constitution of the transient or equilibrium components; at best, a measure of relative definition is available through the judicious use of standards. Thus, in our mutarotation studies our confidence that the first GLC peak was the TMSi derivative of α-D-glucopyranose, the second, the TMSi derivative of β-D-glucopyranose, was based solely on the fact that the derivative of pure α-D-glucopyranose had the same retention time as the first and that of pure β-D-glucopyranose as the second. This disadvantage although tolerable in many straightforward studies proves a major impediment to more detailed work. While, as described above, this problem can be resolved by isolating sufficient quantities of each of the components by preparative GLC and subjecting them to classical structure determining methods, such action cancels any saving of time use of the analytical method confers. A technique must be devised, therefore, to give structure determination in more routine and rapid fashion.

Low Resolution Mass Spectrometry in Combination with the GLC–TMSi Method

To meet the need mentioned above, resort was made to combined GLC–mass spectrometry (MS) and the fact that, at least in principle, structural and geometrical isomerism influence mass spectral cracking patterns (*41, 42*). Before our studies were begun in 1969, it was not clear from the literature whether this influence was as easily discernible in the spectra of sugar anomers and their derivatives as it was in those of sugar ring isomers and their derivatives (*43, 44, 45*). Much evidence indicated it was discernible. Thus in their pioneering work, Reed and co-workers (*46, 47*) had shown that ion appearance potential differences were present in the isomeric pair, methyl α-D-glucopyranoside and methyl β-D-glucopyranoside. Biemann *et al.* later demonstrated that the spectral characteristics of the anomeric pentaacetates of D-glucopyranose and D-mannopyranose were such as to distinguish between them (*48, 49*). That the anomeric forms of permethylated D-glucose, D-galactose, and D-mannose were distinguishable mass spectrometrically has been established by the work of Kochetkov and Vol'fson and their co-workers (*50,*

51, 52, 53). Intensity ratio differences were also reported by Heyns and Scharmann to exist between the isomeric methyl α- and β-tri-*O*-methyl-D-lyxopyranosides (*54*).

In the disaccharide field intensity difference attributable to the configuration at the anomeric center had also been reported. Data presented by Kochetkov (*55*) indicated that such was the case with pertrimethylsilylated α and β, 1 → 4 and 1 → 6 glucosylglucoses. Vink *et al.* (*56*) concluded that the configuration at the anomeric center of the reducing unit was the most important factor influencing a set of ten peaks common to the spectra of pertrimethylsilylated α- and β-lactose and β-cellobiose. Moreover, these workers used ratios of these ten peaks to distinguish α-lactose from its anomer.

By contrast, however, in their classic paper DeJongh *et al.* claimed that the mass spectra of the anomeric forms of penta-*O*-trimethylsilylated glucopyranose were identical, as were those of the anomers of the corresponding galactopyranose (*57*). No differences were noted in the spectra of the corresponding pertrimethylsilylated methyl glycosides. Thus, the possibility arose that with the trimethylsilylated monsaccharides no easily discernible influence of anomeric configuration on fragmentation pattern

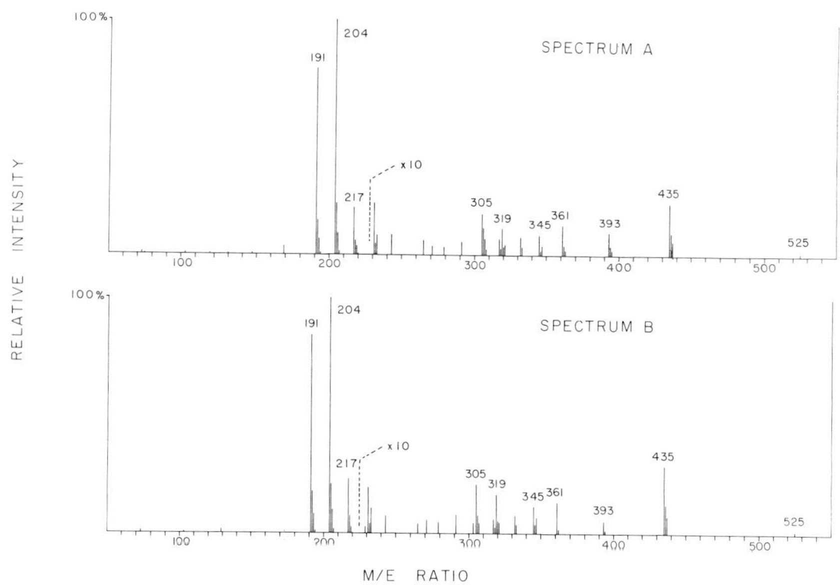

Figure 6. Mass spectra of trimethylsilyl tetra-O-trimethylsilyl-D-glucopyranosides; Spectrum A is α anomer, Spectrum B is β anomer.

Recording conditions: LKB 9000 single focusing spectrometer operating at 17 ev with an ion source temperature of 270°C and a molecular separator temperature of 250°C. The materials were eluted from a 3% OV 101 column before mass spectrometry.

occurred and consequently that mass spectrometry could not be used to give the structural information that was needed to make the GLC–TMSi method of examining sugar equilibria really workable.

Before abandoning the idea, however, the work of DeJongh *et al.* was repeated. A lower ionization voltage was used to limit secondary fragmentation processes and hopefully to accentuate stereochemical differences. The 17 ev mass spectra of the pertrimethylsilylated derivatives of α- and β-D-glucopyranose are shown in Figure 6. The two spectra are very similar; however, intensity differences do occur. In particular, attention has been focused on the ratio of the two peaks with m/e values of 435 and 393. The m/e 435 peak corresponds to the parent molecular ion (PMI) with loss of CH_3 and TMSiOH, the 393 peak to the shard A minus CH_3 (57).

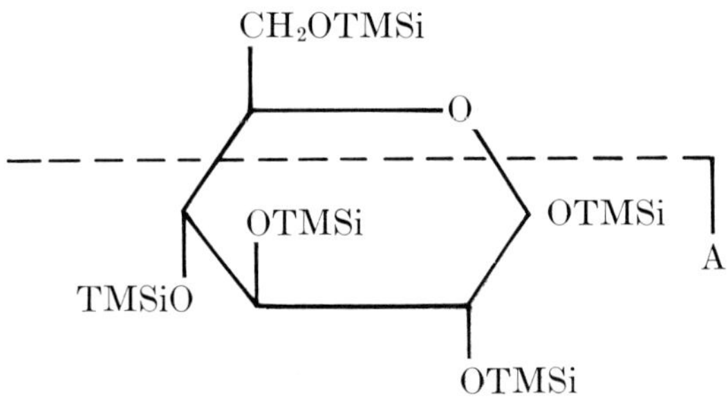

The ratio 435/393 is larger in the β anomer spectrum than in that of the α anomer. The numerical value of this ratio is shown in Table III together with corresponding values for the pertrimethylsilylated anomers of galactopyranose and mannopyranose. Each value is the average of at least 10 independent measurements and is subject to a ±15% error. Within this range the ratio value is constant provided the ion source temperature is not varied. Therefore, the ratio can be used to afford structure determining information in sugar equilibria studies. At present, the range of this information is being extended by considering the other pertrimethylsilylated aldopyranoses and by selecting, as was done in the fatty acid field (58), a more extensive set of peaks to allow each of the 16 pertrimethylsilylated aldopyranoses to be differentiated, one from another.

Similar results have been obtained with the tetra-*O*-trimethylsilyl derivatives of the anomers of methyl glucopyranoside, methyl galactopyranoside, and methyl mannopyranoside. The 17 ev mass spectrum of the tetra-*O*-trimethylsilylated form of the anomers of methyl glucoside are shown in Figure 7. Again the traces are similar, but intensity differences are discernible. The ratio $(377/361)_{17\,ev}$ has proved most useful.

Table III. Sugar Intensity Ratios

	$(435{:}393)_{17ev}$ *in persilylated sugar*	$(377{:}361)_{17ev}$ *in persilylated methyl glycoside*
α-D-Glucose	100:37	100:73
β-D-Glucose	100:18	100:22
α-D-Galactose	100:75	100:190
β-D-Galactose	100:40	100:28
α-D-Mannose	100:58	100:75
β-D-Mannose	100:28	100:31

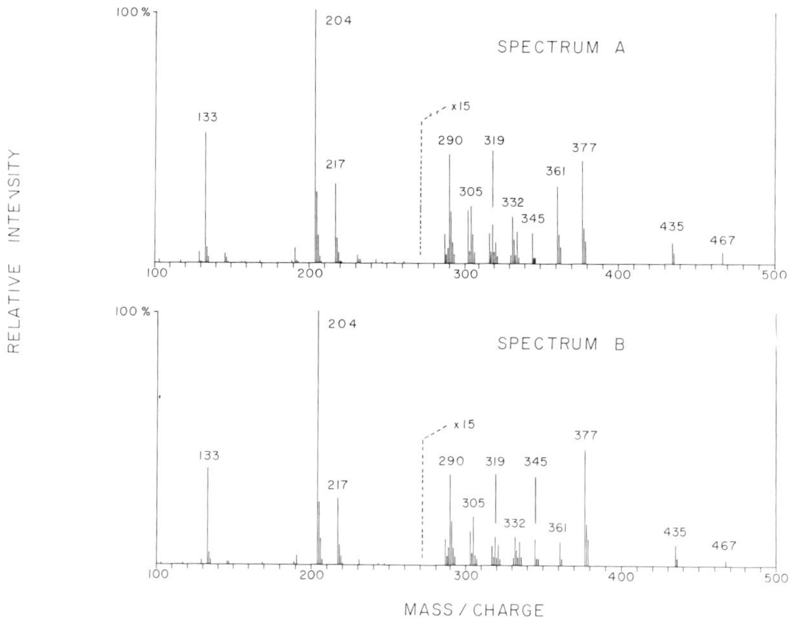

Figure 7. Mass spectra (17 ev) of methyl tetra-O-trimethylsilyl-D-glucopyranosides; Spectrum A is the α anomer, Spectrum B is β anomer (recording conditions as in Figure 6)

The first of these peaks corresponds to PMI minus CH_3 and TMSiOH, the second to PMI minus CH_3O and TMSiOH (*55*). The average values of the ratio for at least ten runs are shown in Table III. Again the ratio is consistently and reproducibly different for a given anomeric pair.

Although such ratios have proved to be valuable, the whole approach to structure definition through their use is still a relative one. A large volume of standard data is required to set up the appropriate intensity ratio and to prescribe its bounds. To realize our goal of using mass spectrometry as a more absolute structure determining probe, we must determine the physical basis for the intensity ratio differences. This will not be easy since present-day mass spectrometric theory is not adequate to treat rigorously a molecule as complex as a derivatized sugar. Recently, however, some empirical results have been obtained, and they are mentioned here. In Table IV are contained the intensities of the *m/e* 377 and *m/e* 361 ions in the spectra of the anomers of the methyl tetra-*O*-trimethylsilyl pyranosides of glucose, galactose, and mannose, expressed as a proportion of the sum (Σ') of the eight largest ions in the spectrum.

Table IV. Ratios $(377:\Sigma')_{17ev}$ and $(361:\Sigma')_{17ev}$ in Persilylated Methyl Glycosides ($\Sigma' = 133 + 134 + 191 + 204 + 205 + 206 + 217 + 218$)

	$(377:\Sigma')_{17ev}$	$(361:\Sigma')_{17ev}$	$(377:361)_{17ev}$
α-D-Glucose	0.0108	0.0081	100:75
β-D-Glucose	0.0130	0.0023	100:18
α-D-Galactose	0.0040	0.0078	100:195
β-D-Galactose	0.0064	0.0017	100:27
α-D-Mannose	0.0066	0.0053	100:80
β-D-Mannose	0.0106	0.0029	100:27

This sum approximates total ion current. In the three sugars the $377:\Sigma'$ value for an α anomer is always less than the corresponding value for the β anomer. For the $361:\Sigma'$ ratio the situation is reversed; the value for the α anomer is always larger than that of the β anomer. Considering this latter finding and recalling that the formation of the *m/e* 361 ion involves a loss of OCH_3—*i.e.*, the group whose configuration differentiates the α from the β form, it seems possible that differences in its abundance between the two forms represent a difference in the rate of its formation more than in the rate of its destruction. If this is the case, the finding that the OCH_3 group in the α configuration leaves more rapidly than one in the β configuration is easily understood on energetic grounds and could be of considerable structure determining value.

From what has been discussed above, it will be clear that the combined GLC–TMSi–MS approach has considerable potential. We have already begun to use it to analyze the composition of the equilibrium of glucose in boiling pyridine. When the silylating reagent was added to such a refluxing solution, subsequent GLC revealed at least four peaks (*see* Figure 8). (In Figure 8 solid α-D-glucose (5 mg) in reagent grade but not dried pyridine (2 ml) was refluxed for 35 minutes. While still at reflux, a mixture of pyridine (0.5 ml), hexamethyldisilazane (0.5 ml), and trimethylchlorosilane (0.25 ml) was added through the condenser. The mixture was then allowed to cool to room temperature before chromatography on OV-1 under the conditions described in Figure 1). The

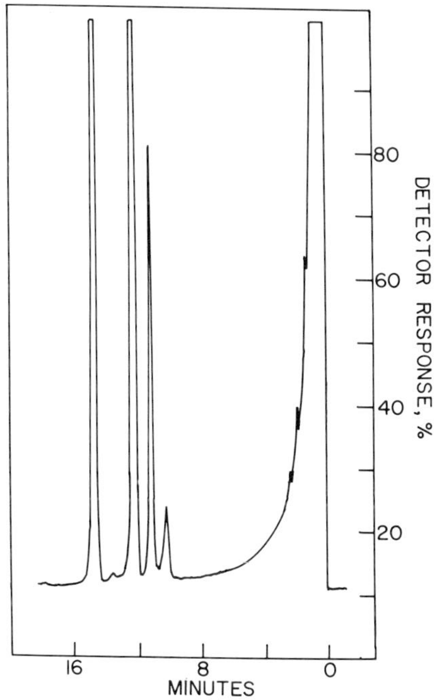

Figure 8. The equilibrium solution of glucose in boiling pyridine

two major peaks were clearly the α- and β-pyranose anomers as would be expected. Both minor peaks contain a furanose material as evidenced by mass spectrometry. There is, however, considerable evidence that more than one component is present in the second peak. With improved separation and data processing capability we hope positively to identify these minor constituents of the equilibrium mixture.

Scheuer and his co-workers have recently further illuminated the relationships that exist between stereochemistry and fragmentation pattern in pertrimethylsilylated monosaccharides (59). They noted that the *m/e* 147 ion in trimethylsilylated glycols resulted from a cyclic elimination of the unit $TMSiO^+ = Si(CH_3)_2$. Arguing that the intensity of this ion could be related to the physical ease with which six, five, and seven membered cyclic transition states could be constructed in ordinary molecular models of α and β glucose, galactose, and mannose, these workers deduced that the intensity of the *m/e* 147 ion should drop off in the order: β-glucose, β-mannose, β-galactose, α-glucose, α-mannose, α-galactose. The experimentally determined order was: β-glucose, β mannose, α-glucose, β-galactose, α-mannose, α-galactose. It is clear that some fundamentally significant concept has been encountered and is ready for use in structural determination.

Acknowledgment

The authors are pleased to acknowledge the collaboration of E. Farish and J. Naworal in obtaining the mass spectrometric data reported here. That portion of the work was supported by the U. S. Public Health Service (RR-00273).

Literature Cited

1. Shakespeare, W., "The Tempest," Act II, Scene i, First Folio Edition, William Jaggard, London, 1623.
2. Dubrunfaut, A. P., *Compt. Rend.* (1846) **23,** 38.
3. Lowry, T. M., *J. Chem. Soc.* (1899) **75,** 211.
4. Lowry, T. M., "Optical Rotatory Power," p. 272, Dover, New York, 1964.
5. Pigman, W., Isbell, H. S., *Advan. Carbohyd. Chem.* (1968) **23,** 11.
6. Isbell, H. S., Pigman, W., *Advan. Carbohyd. Chem. Biochem.* (1969) **24,** 13.
7. Isbell, H. S., Letter to all participants in Symposium "Chemistry of Carbohydrates in Solution."
8. Briggs, L. J., "Polarimetry, Saccharimetry and the Sugars," Preface, United States Government Printing Office, Washington, D. C., 1942.
9. Bates, F. J. *et al.*, "Polarimetry, Saccharimetry and the Sugars," p. 441, United States Government Printing Office, Washington, D. C., 1942.
10. McInnes, A. G., Ball, D. H., Cooper, F. P., Bishop, C. T., *J. Chromatogr.* (1958) **1,** 556.
11. Bishop, C. T., *Advan. Carbohyd. Chem.* (1964) **19,** 95.
12. Wells, W. W., Sweeley, C. C., Bentley, R., "Biomedical Applications of Gas Chromatography," p. 169, H. A. Szymanski, Ed., Plenum, New York, 1964.
13. Berry, J. W., *Advan. Chromatogr.* (1966) **2,** 271.
14. Clamp, J. R., Bhatti, T., Chambers, R. E., *Methods Biochem. Anal.* (1971) **19,** 229.
15. Bishop, C. T., Cooper, F. P., *Can. J. Chem.* (1960) **38,** 388.
16. Kircher, H. W., *Anal. Chem.* (1960) **32,** 1103.
17. VandenHeuvel, W. J. A., Horning, E. C., *Biochem. Biophys. Res. Commun.* (1961) **4,** 399.
18. Hedgley, E. J., Overend, W. G., *Chem. Ind. (London)* (1960) 378.
19. Ferrier, R. J., Singleton, M. F., *Tetrahedron* (1962) **18,** 1143.
20. Ferrier, R. J., *Tetrahedron* (1962) **18,** 1149.
21. Makita, M., Wells, W. W., *Anal. Biochem.* (1963) **5,** 523.
22. Bentley, R., Sweeley, C. C., Makita, M., Wells, W. W., *Biochem. Biophys. Res. Commun.* (1963) **11,** 14.
23. Sweeley, C. C., Bentley, R., Makita, M., Wells, W. W., *J. Amer. Chem. Soc.* (1963) **85,** 2497.
24. Sweeley, C. C., Wells, W. W., Bentley, R., "Methods in Enzymology," p. 95, Vol. 8, E. F. Neufeld and V. Ginsburg, Eds., Academic, New York, 1966.
25. Bates, F. J. *et al.*, "Polarimetry, Saccharimetry and the Sugars," p. 455, United States Government Printing Office, Washington, D. C., 1942.
26. Bentley, R., Botlock, N., *Anal. Biochem.* (1967) **20,** 312.
27. Semenza, G., Curtius, C.-H., Kolinska, J., Muller, M., *Biochim. Biophys. Acta* (1967) **146,** 196.
28. Kim, S. M., Bentley, R., Sweeley, C. C., *Carbohyd. Res.* (1967) **5,** 373.
29. Keilin, D., Hartree, E. F., *Biochem. J.* (1952) **50,** 331.
30. Lee, C. Y., Acree, T. E., Shallenberger, R. S., *Carbohyd. Res.* (1969) **9,** 356.
31. Shallenberger, R. S., Acree, T. E., *Carbohyd. Res.* (1966) **1,** 495.

32. Acree, T. E., Shallenberger, R. S., Mattick, L. R., *Carbohyd. Res.* (1968) **6**, 498.
33. Acree, T. E., Shallenberger, R. S., Lee, C. Y., Einset, J. W., *Carbohyd. Res.* (1969) **10**, 355.
34. Curtius, H.-C., Muller, M., Vollmin, J. A., *J. Chromatogr.* (1968) **37**, 216.
35. Sherman, W. R., Goodwin, S. L., *J. Chromatogr. Sci.* (1969) **7**, 167.
36. Bentley, R., Neuberger, A., *Biochem. J.* (1949) **45**, 584.
37. Keilin, D., Hartree, E. F., *Biochem. J.* (1952) **50**, 341.
38. Parrish, F. W., Reese, E. T., *Carbohyd. Res.* (1967) **3**, 424.
39. Barnett, J. E. G., *Biochem. J.* (1971) **123**, 607.
40. Bentley, R., "Molecular Asymmetry in Biology," p. 216, Vol. II, Academic, New York, 1970.
41. Meyerson, S., Weitkamp, A. W., *Organ. Mass Spectrom.* (1968) **1**, 659.
42. Natalis, P., "Mass Spectrometry," p. 379, NATO Advanced Study Inst., Glasgow, 1964.
43. Kochetkov, N. K., Chizhov, O. S., *Advan. Carbohyd. Chem.* (1966) **21**, 39.
44. Hanessian, S., "Methods of Biochemical Analysis," p. 105, Vol. 19, D. Glick, Ed., Interscience, 1971.
45. Cooks, R. G., Johnson, G. S., "Specialist Periodical Reports—Mass Spectrometry," p. 164, Vol. 1, Chemical Society, London.
46. Finan, P. A., Reed, R. I., Sneddon, W., *Chem. Ind. (London)* (1958) 1722.
47. Finan, P. A., Reed, R. I., Wilson, J. M., *J. Chem. Soc.* (1963) 5945.
48. Biemann, K., Schnoes, H. K., McCloskey, J. A., *Chem. Ind. (London)* (1963) 448.
49. Biemann, K., DeJongh, D. C., Schnoes, H. K., *J. Amer. Chem. Soc.* (1963) **85**, 1763.
50. Kochetkov, N. K., Vul'fson, N. S., Chizhov, O. S., Zolotarev, B. M., *Dokl. Akad. Nauk. SSSR* (1963) **151**, 336.
51. Kochetkov, N. K., *J. Sci. Ind. Res.* (1964) **23**, 468.
52. Vul'fson, N. S., Chizhov, O. S., Zolotarev, B. M., *Materialy Vses. Konf. po Probl. "Khim. i Obmen Uglevodov,"* 3rd Moscow (1963) 51; *Chem. Abs.* (1966) **65**, 14605f.
53. Vul'fson, N. S., Chizhov, O. S., Zolotarev, B. M., *Khim. Obmen Uglevodov* (1965) 51; *Chem. Abs.* (1967) **67**, 117187r.
54. Heyns, K., Scharmann, H., *Tetrahedron* (1965) **21**, 507.
55. Kochetkov, N. K., Chizhov, O. S., Molodtsov, N. V., *Tetrahedron* (1968) **24**, 5587.
56. Vink, J., deRidder, J. J., Kamerling, J. P., Vliegenthart, J. F. G., *Biochem. Biophys. Res. Commun.* (1971) **42**, 1050.
57. DeJongh, D. C., Radford, T., Hribar, J. D., Hanessian, S., Bieber, M., Dawson, G., Sweeley, C. C., *J. Amer. Chem. Soc.* (1969) **91**, 1728.
58. Campbell, I. M., Naworal, J., *J. Lipid Research* (1969) **10**, 589.
59. Havlicek, S. C., Brennan, M. R., Scheuer, P. J., *Organ. Mass Spectrom.* (1971) **5**, 1273.

RECEIVED December 2, 1971.

2

Computer Modeling of the Kinetics of Tautomerization (Mutarotation) of Aldoses: Implications for the Mechanism of the Process

LAURENS ANDERSON and JOHN C. GARVER

Department of Biochemistry, College of Agricultural and Life Sciences, University of Wisconsin, Madison, Wis. 53706

Simulation by computer has been used to examine certain features of the kinetics of tautomerization of α-galactopyranose and β-arabinofuranose, which have complex mutarotations. In both cases the calculations indicate a tautomer which approaches or exceeds its equilibrium level early in the process. Such tautomers have not yet been characterized in studies made on the two sugars. The computations also suggest that the formation rates of furanoses from aldehydo *forms may be no greater than the formation rates of pyranoses. Pyranose–pyranose interconversion* via *a cyclic oxycarbonium ion is eliminated for galactose by the slow rate of ^{18}O exchange between sugar and water. 2,3-Di-*O*-methylarabinose, generated by the periodate oxidation of 3,4-di-*O*-methylmannitol, can be used to study relative rates of ring closure to pyranoses and furanoses.*

Sugars with complex mutarotation—*i.e.*, mutarotation which deviates from simple first order kinetics—are especially interesting in studies of the mutarotation phenomenon. Well known examples of such sugars are galactose, arabinose, and talose. As shown for galactose in Figure 1, plots of log $[(r_t\text{-}r_{eq})/(r_0\text{-}r_{eq})]$ for these sugars are typically biphasic (r_0, r_t, and r_{eq} = rotation at time zero, time t, and equilibrium, respectively). For many years the presence of significant amounts of more than two tautomers in mutarotating or equilibrated solutions of sugars of this type had been postulated (*1*), but direct confirmation of this hy-

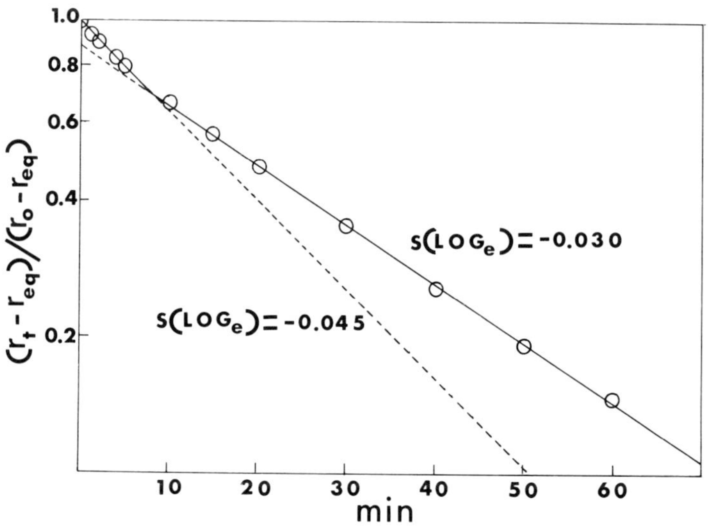

Figure 1. Mutarotation of α-D-galactopyranose in water at 25°C (6) (s = slope)

pothesis and detailed study of the progress of the compositional changes had to await the development of two modern methods, proton magnetic resonance spectroscopy (PMR) and gas–liquid chromatography (GLC). Lemieux and Stevens (*2*) and Sweeley *et al.* (*3*), pioneering applications of these respective techniques, showed that solutions of ribose, galactose, arabinose, and several other sugars contained three or four components. Later with Lemieux and Conner (*4*), Anderson studied the kinetics of tautomerization of 2-deoxy-D-*erythro*-pentose (deoxyribose) by PMR. Similarly GLC was used by Acree *et al.* (*5*) to study D-galactose and by Conner in Anderson's laboratory (*6, 7*) to study D-galactose and L-arabinose.

The progress curves obtained by Conner for α-D-galactopyranose and β-L-arabinopyranose are shown in Figures 2 and 3. Although tautomerizing solutions of galactose in water contain a small amount of α-furanose (*5*), we measured for this sugar only the two pyranose forms and what is probably total furanose. What is important about these curves, because of earlier postulations about the nature of the two phases of complex mutarotations (*8*), is the lack of any evidence that furanose forms accumulate, then decrease, or even that these forms approach their equilibrium levels especially rapidly. All the curves seem to represent simple exponential processes, a conclusion which is substantially verified by plotting the data in logarithmic form. This is done for the galactose curves in Figure 4. From the slopes of the log plots, a half-time for the approach of each tautomer to its equilibrium level can be calculated.

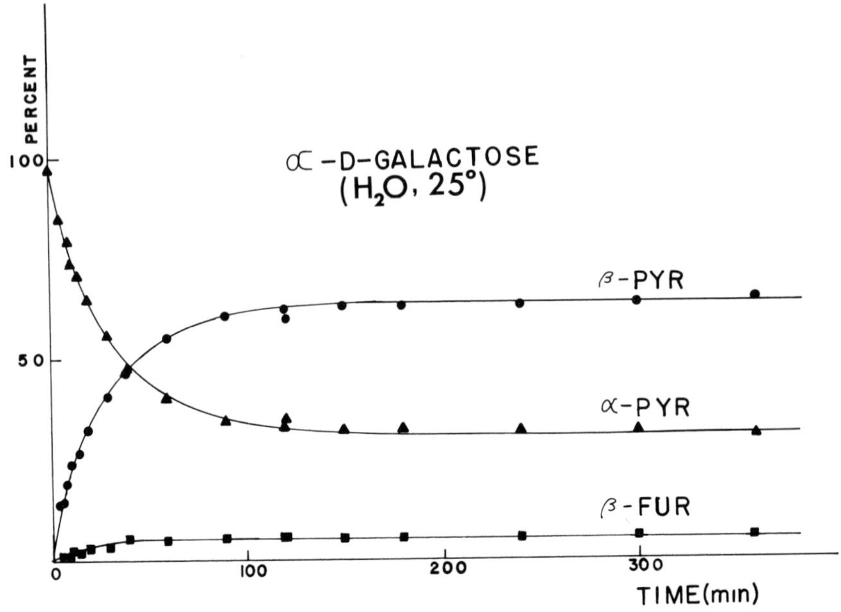

Figure 2. Composition of tautomerizing solutions of α-D-galactopyranose in water at 25°C (6), as observed by GLC over SE 52 on Anakrom A

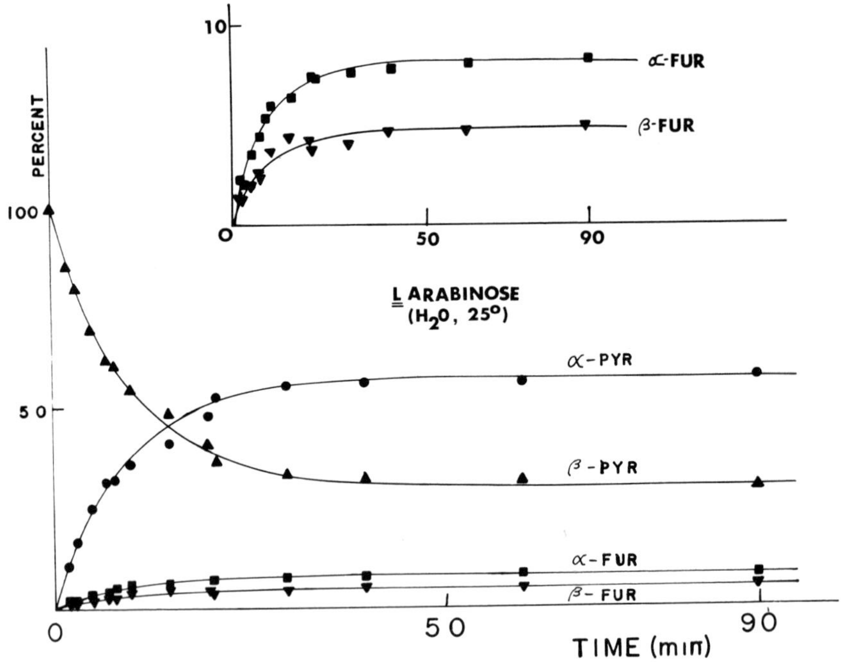

Figure 3. Composition of tautomerizing solutions of β-L-arabinopyranose in water at 25°C (6)

For α-D-galactopyranose, under the conditions used, these were α-pyranose, 19.9 minutes, β-pyranose, 20.3 minutes, and furanose, 16.2 minutes. Thus, the rate at which the furanoses approach equilibrium is somewhat, but not markedly greater than the rates for the pyranose tautomers. A similar treatment of the data for β-L-arabinopyranose gave a set of four straight lines from which it could be seen that the rates of approach to equilibrium were nearly identical for all the tautomers. The values were α-pyranose, 6.8 minutes; β-pyranose, 6.7 minutes; α-furanose, 6.4 minutes; and β-furanose, 7.2 minutes.

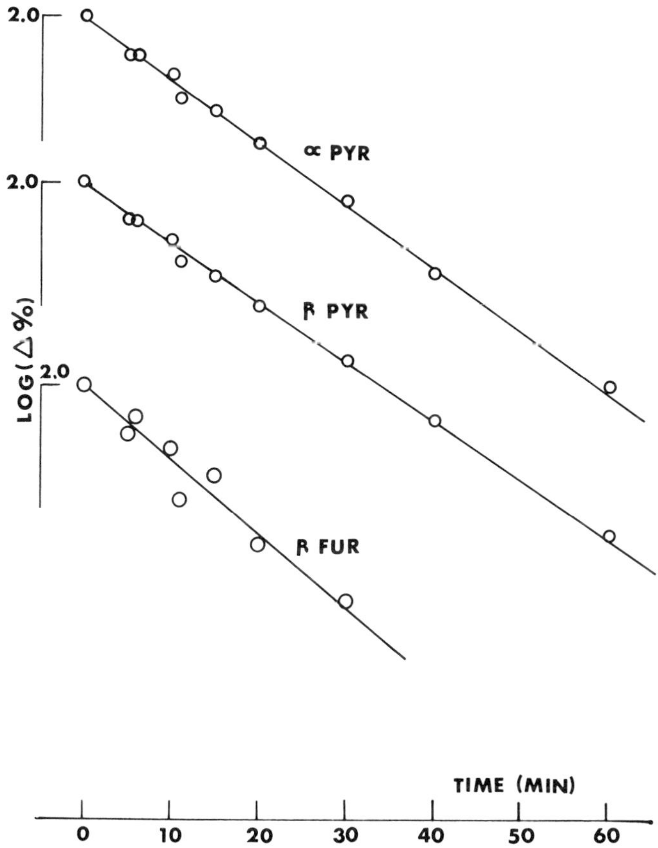

Figure 4. Logarithmic replots of the data of Figure 2 (percent unaccomplished change vs. time)

A characteristic feature of the optical data for complex mutarotations is that they show a rapid initial change in rotation followed by a longer, slower phase of approach to the equilibrium value (*e.g.*, Figure 1). The question arose as to whether these biphasic plots are quantitatively accounted for by the changes in the composition of the solution observed by GLC techniques. More specifically the question was if one uses a known value or reasonable estimate of the specific rotation of each tautomer and sums the contributions of all tautomers at successive points in time, would the observed optical rotation curve be reproduced? Initial efforts to answer this question for galactose and arabinose by manual calculation suggested a negative answer, but the result was uncertain because of uncertainty as to how to adjust the composition curves to a constant sum of 100%. We, therefore, studied the possibility of modeling complex mutarotations by computer.

Computer Modeling of the Tautomerization Process

As a basis for the modeling, we have used the formulation of mutarotation in which the ring forms of the sugar are interconverted *via* a central intermediate, taken to be the free aldehyde form (*1*, *9*). (In Reference 9 mechanisms with more than one ring-opened, aldehyde-like intermediate are considered. However, in the absence of definite evidence for multiple intermediates, we have assumed the simple mechanism involving a single, central intermediate because it is amenable to mathematical manipulation. Even if the true mechanism is more complex, the computations are meaningful if they duplicate within experimental error the progress curves for the measurable components of the system. The computer runs described here seem to meet this criterion.) The case for three significant ring forms is illustrated by the following scheme.

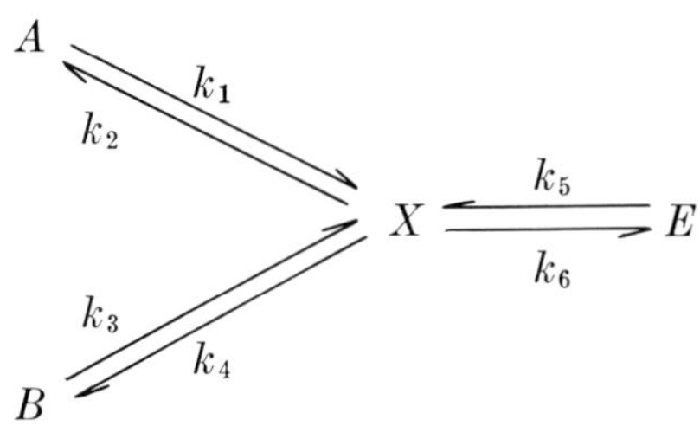

For this mechanism, a differential equation may be written for the rate of change of each ring form, letting the symbols *A*, *B*, and *E* stand for concentrations:

$$\frac{dA}{dt} = -k_1A + k_2X$$

$$\frac{dB}{dt} = -k_3B + k_4X \qquad (1)$$

$$\frac{dE}{dt} = -k_5E + k_6X$$

These with the equation for material balance

$$C = A + B + E + X \qquad (2)$$

where C is the total concentration of sugar describe sufficiently the course of the reaction. The method for solving this system of equations is presented in the Appendix. Solution gives

$$\begin{aligned} A &= a_1e^{-m_1t} + a_2e^{-m_2t} + a_3e^{-m_3t} + A^* \\ B &= b_1e^{-m_1t} + b_2e^{-m_2t} + b_3e^{-m_3t} + B^* \\ E &= d_1e^{-m_1t} + d_2e^{-m_2t} + d_3e^{-m_3t} + E^* \\ X &= C - A - B - E \end{aligned} \qquad (3)$$

where A^*, B^*, and E^* are equilibrium concentrations, the a_i, b_i, and d_i are integration constants and m_1, m_2, and m_3 are functions of the rate constants.

The use of Equation 3 for simulation requires that values be assigned to the initial concentration of each component, the equilibrium concentration of each component, the specific rotation of each component, and the k's with odd subscripts or the k's with even subscripts (the second k for each arm of the scheme is calculated from the given k and the equilibrium constant for that arm). Fortran programs have been developed which, given the boundary conditions and parameters listed above, direct the computation of the initial and final specific rotations, the k's not specified, and the integration and exponential constants for Equation 3. Next are calculated for specific points in time the percentage of each component present and the percent unaccomplished change for that component, the specific rotation, the percent unaccomplished change in specific rotation, and for the latter quantity the slope of a plot of its natural log *vs.* time. Table I shows the printout of a typical computation with some of the foregoing items omitted.

For exploratory simulations the values used for the specific rotations and equilibrium levels of the components were those pertaining to galactose, which was studied as a three-component system. As the effect of varying the rate constants was examined, the behavior of the system could be related to the values assigned to the ring opening rates (k's with odd subscripts), particularly those associated with the components present in major amounts at equilibrium. For galactose these are the α-pyranose (k_1) and β-pyranose (k_3). The value assigned to the amount of central intermediate at equilibrium (X^*) can be varied over wide limits—from vanishingly small to over 2%—with little effect on the progress of the ring forms to equilibrium. For given values of the ring opening rates, the ring-closure rates (k's with even subscripts) vary

Table I. Computer Simulation of the

```
PROGRAM CHC-4CB,    RUN NO.

BOUNDARY CONDITIONS-

                    AI  =   1.0000
                    BI  =  98.0000
                    EI  =     .5000
                    FI  =     .5000
                    XI  =     .0000

                    SRIN  =     .18457+03

PARAMETERS-

   K1  =    .17500+01        K2  =    .35333+03
   K3  =    .70000-01        K4  =    .76615+01
   K5  =    .37000+02        K6  =    .10441+04
   K7  =    .18000+02        K8  =    .29829+03

INTEGRATION CONSTANTS-

   R1   =   -.17297+04       R2   =   -.99464-01
   QA1  =    .42621-02       QA2  =   -.55451+02
   QB1  =    .92328-04       QB2  =    .67355+02
   QE1  =    .12856-01       QE2  =   -.73290+01
   QF1  =    .36323-02       QF2  =   -.43164+01
   Q1   =   -.15125-02       Q2   =    .82274+02
```

TIME	FA	FB	FE	FF
.00	100.00	100.00	100.00	100.00
1.00	90.40	90.53	89.65	94.39
2.00	81.84	81.96	81.16	85.45
3.00	74.09	74.20	73.48	77.36
4.00	67.08	67.18	66.52	70.04
5.00	60.73	60.82	60.22	63.41
6.00	54.98	55.06	54.52	57.40
7.00	49.77	49.85	49.36	51.97
8.00	45.06	45.13	44.69	47.05
9.00	40.79	40.85	40.46	42.59
10.00	36.93	36.99	36.63	38.56
15.00	22.46	22.49	22.27	23.45
20.00	13.66	13.68	13.55	14.26
25.00	8.31	8.32	8.24	8.67

```
HALF-LIFE A  =    6.954
HALF-LIFE B  =    6.969
```

Mutarotation of β-L-Arabinopyranose

```
AEQ  =    .56533+02
BEQ  =    .30646+02
EEQ  =    .79010+01
FEQ  =    .46400+01
XEQ  =    .28000+00

SREQ =    .10223+03
```

```
SRA  =    .82500+02
SRB  =    .18750+03
SRE  =   -.56000+02
SRF  =    .55000+02
SRX  =    .00000
```

```
R3   =   -.23058+02        R4   =   -.73452+01
QA3  =    .33608-01        QA4  =   -.12002+00
QB3  =    .67524-03        QB4  =   -.19991-02
QE3  =   -.15178+00        QE4  =    .66912-01
QF3  =    .11952+00        QF4  =    .53206-01
Q3   =    .17972+00        Q4   =   -.11097+00
```

X	POT	FR	SLOPE
.000	184.570	100.000	-.108+00
.455-01	176.713	90.458	-.995-01
.677-01	169.661	81.894	-.995-01
.878-01	163.277	74.140	-.995-01
.106+00	157.497	67.121	-.995-01
.122+00	152.264	60.766	-.995-01
.137+00	147.527	55.013	-.995-01
.151+00	143.238	49.805	-.995-01
.163+00	139.356	45.089	-.995-01
.174+00	135.840	40.820	-.995-01
.184+00	132.658	36.955	-.995-01
.222+00	120.735	22.475	-.995-01
.245+00	113.483	13.668	-.995-01
.258+00	109.073	8.312	-.995-01

```
HALF-LIFE E  =    6.871
HALF-LIFE F  =    7.388
HALF-LIFE SR =     6.961
```

inversely with the value of X^*. In the computations cited below, the equilibrium percentages of X were taken to be those estimated by polarography (*10*) (0.081% for galactose, 0.28% for arabinose).

With the general mathematical characteristics of the system established, we considered that further manipulation should focus on models in which not only the equilibrium composition but also the progress of the individual components with time paralleled that observed experimentally. It was recognized that there would be an essentially infinite number of sets of rate constants satisfying a given set of experimental data. As a criterion of correspondence between calculation and experiment, the half-time for the approach of each component to its equilibrium level was chosen. The ratio k_1/k_3 was then varied. Trial values having the desired ratio were assigned, a computation was made, and the values were successively adjusted until the average $t_{1/2}$ for the α- and β-pyranoses was within a few hundredths of a minute of the observed figure. Simultaneously, k_5 was adjusted to bring $t_{1/2}$ for the furanoses into correspondence. Final adjustment of k_5 was by machine iteration. Successful calculations were carried out for k_1/k_3 ratios from 50 to 1/16. Table II shows the detail for runs from the middle and the extremes of this range.

Table II. Sets of Rate Constants[a] with a Range of k_1/k_3 Values, Which Simulate the Tautomerization[b] of α-D-Galactopyranose in Water at 25°C

		k_1/k_3		
k_n	*Rate for*	*50*	*2*	*1/16*
k_1	α-*p*, opening	0.560	0.0456	0.0240
k_3	β-*p*, opening	0.0112	0.0228	0.384
k_5	β-*f*, opening	0.0141	0.0283	1220
k_2	α-*p*, closing	221	17.9	9.44
k_4	β-*p*, closing	8.68	17.7	297
k_6	β-*f*, closing	0.923	1.85	80000

[a] For the three-component model.
[b] With each set the computed $t_{1/2}$'s for the approach of the individual components to their equilibrium levels are equal to the observed $t_{1/2}$'s, within experimental error.

To answer the question above, whether the observed optical mutarotation of galactose could be explained by the three-component model constrained to experimental $t_{1/2}$'s, the optical rotatory values from the computations listed in Table II were plotted *vs.* time. The three plots did not differ significantly from each other, in spite of the variation in the ratio k_1/k_3. The plot for $k_1/k_3 = 2$ is compared with the experimental values in Figure 5. The observed rapid initial drop in rotation is not duplicated in the computed plot. No reasonable combination of

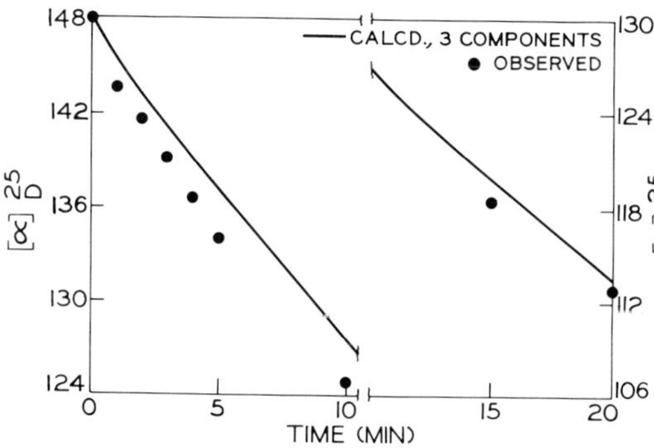

Figure 5. Mutarotation of α-D-galactopyranose, calculated for the three-component system and compared with the observed values

rate constants can account for this initial drop when $t_{1/2}$ for the progress of the furanoses is only moderately shorter than $t_{1/2}$ for the pyranoses.

From the rate constant obtained from the experiments of Acree *et al.* (5), β-galactofuranose approached equilibrium much more rapidly ($t_{1/2}$ 5.3 minutes) than in our work. A simulation using this $t_{1/2}$ gives an optical rotation curve dropping slightly more rapidly than the plot of Figure 5 but still significantly different from the experimental curve. We can therefore conclude that the mutarotation of galactose involves a fourth component, with a low specific rotation, which has the characteristic of increasing in amount very rapidly during the first few minutes.

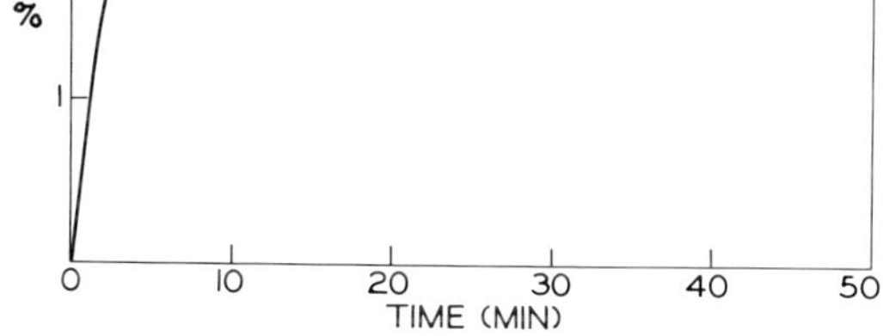

Figure 6. Computed progress curve for a fourth component of the galactose system, assuming rate constants and equilibrium level as shown

Figures 6 and 7 illustrate calculations which include the putative fast component. With $k_1/k_3 = 2$ and appropriate rate constants for the formation (k_8) and reversion to X (k_7) of the fast component, this component would increase to nearly its equilibrium level in 3 minutes, overshoot, then slowly decline to equilibrium. The optical rotation curve computed on this basis (Figure 7) follows the experimental one very closely for the first 20 minutes.

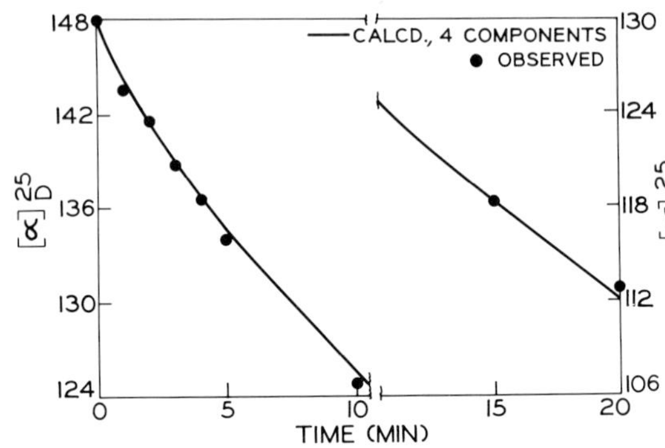

Figure 7. Mutarotation of α-D-galactopyranose, calculated for the four-component system and compared with observed values

A priori the fast component in the galactose system could be either the α-furanose or the aldehydrol (*gem*-diol) form of the sugar. Acree *et al.* (5) have suggested that α-galactofuranose, present to the extent of 1% at equilibrium, probably forms no more rapidly than the β-furanose. Unfortunately, kinetic data are not available for water solutions. However, data on the rate of the galactose–$H_2{}^{18}O$ exchange reaction (*see* below) make the aldehydrol form seem unlikely as a fast component.

Simulation of the mutarotation of arabinose closely followed the work on galactose, except that four components plus a central intermediate were included, and the β-pyranose was the starting anomer. The components were the α-pyranose (k_1, k_2), β-pyranose (k_3, k_4), α-furanose (k_5, k_6), and β-furanose (k_7, k_8). Table III shows sets of values for the k's which yield the experimentally observed half-times for the approach of the components to their equilibrium levels.

For arabinose the half-times for all components are essentially the same. As long as this constraint is obeyed, the calculated optical rotation curves would be expected to show no rapid initial drop, and this is what was found. Therefore, we must postulate a fifth, as yet undetected, fast component in mutarotating arabinose solutions. Here the only reasonable candidate is the aldehydrol form. In future work with arabinose efforts should be made to detect this form and follow quantitatively its progress.

Table III. Sets of Rate Constants[a] with a Range of k_3/k_1 Values Which Simulate the Tautomerization[b] of β-L-Arabinopyranose in Water at 25°C

		k_3/k_1		
k_n	*Rate for*	*50*	*2*	*1/35*
k_1	α-*p*, opening	0.0331	0.0668	2.54
k_3	β-*p*, opening	1.65	0.134	0.0725
k_5	α-*f*, opening	0.0339	0.0695	6360
k_7	β-*f*, opening	0.0297	0.0612	3090
k_2	α-*p*, closing	6.68	13.5	512
k_4	β-*p*, closing	181	14.6	7.94
k_6	α-*f*, closing	0.957	1.96	179000
k_8	β-*f*, closing	0.492	1.01	51200

[a] For the four-component model.
[b] With each set the computed $t_{1/2}$'s for the approach of the individual components to their equilibrium levels are equal to the observed $t_{1/2}$'s within experimental error.

Ring Closure of the Aldehydro *Form to Pyranoses and Furanoses*

In the sets of *k*-values successfully simulating the progress of the galactose and arabinose systems to equilibrium (Tables II and III), the rate constants for the ring closure of the *aldehydo* sugars to furanose forms are usually smaller than those for closure to pyranose forms. (The exception to this statement will be discussed below.) This is contrary to what would be expected from data on ring closures to tetrahydrofuran and tetrahydropyran (*11, 12*) and from much work on carbohydrates. In the carbohydrate field it is well known that methyl glycofuranosides form at much greater rates than methyl glycopyranosides (*13*), and that in reactions giving anhydroalditols the isomers with five-membered rings are formed, almost excluding those with six-membered rings (*14, 15, 16*). Thus, one may question any proposed mechanism of mutarotation featuring rates of pyranose ring formation higher than those of furanose ring formation.

An extension of the simple mechanism in which ring forms are interconverted *via* a central, *aldehydo* intermediate includes direct pathways for the interconversion of the two pyranoses on one hand, and the two furanoses on the other. The existence of such direct pathways would permit a starting pyranose anomer to be converted rapidly to the other pyranose, as observed, even though the rate constant for the closure of the *aldehydo* form to the furanose ring were much greater than that for closure to the pyranose ring.

Galactose-$H_2{}^{18}O$ Exchange. A possible mechanism for the interconversion of the two pyranoses (or furanoses), not involving the *aldehydo* sugar, operates *via* a cyclic oxycarbonium ion (*1*). The participa-

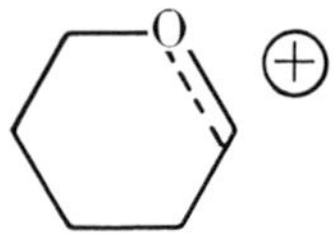

1

tion of such an ion in the mutarotation of glucose has been excluded by the finding that the exchange of O-1 of the sugar with the oxygen of water is a much slower process than the mutarotation (*17, 18*). However, in view of the considerations just raised, we felt that the possibility should be checked for galactose. With the assistance of Irwin Reich the rate of exchange of O-1 of galactose with $H_2{}^{18}O$ was measured. It was necessary to work at temperatures of 50°C or higher to get convenient rates (Figure 8). The rate constant for the exchange in unbuffered water at 25°C, obtained by extrapolation, corresponds to a half-time of 200 hours, orders of magnitude slower than the mutarotation. We conclude that the oxy-

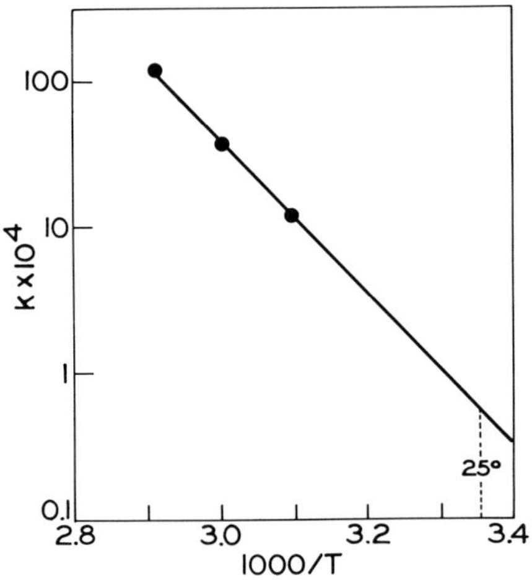

Figure 8. Plot of log k vs. *1/*T *for the galactose-*$H_2{}^{18}O$ *exchange reaction. Each value of the constant* k *is the − slope of a plot of ln* [(P* − P)/P*] vs. *time where* P* *and* P *are the atom percent excess* ^{18}O *in O–1 of the galactose at equilibrium and at time* t, *respectively.*

carbonium ion pathway is not significant in the interconversion of α-galactopyranose and β-galactopyranose.

In the computer simulation there was one set of conditions under which the rates of furanose ring closure would be greater than those for pyranose ring closure. By considering the odd numbered *k*'s for the two pyranoses when the ratio *k* (starting pyranose)/*k* (other pyranose) is made progressively smaller, a point is reached where the *k*'s (for ring opening and closure) associated with the minor components (furanoses) must be made very large to maintain the observed $t_{1/2}$'s for approach to equilibrium. This is illustrated by the last columns in Tables II and III, respectively.

It is possible that similar relationships exist among the rate constants associated with the galactose and arabinose tautomerizations. However, it seems worthwhile to ask whether the ring closure of an *aldehydo* sugar necessarily follows the same rules as methyl glycoside formation or anhydroalditol formation. Is it possible that in the sugar five-membered and six-membered rings are formed at similar rates? An experimental examination of this question requires a system for generating quantities of *aldehydo* sugar in a very short time (seconds) at room temperature or below in the pH range 3–5.

Analysis of Freshly Cyclized 2,3-di-O-methyl-D-arabinose. The system 3,4-di-*O*-methyl-D-mannitol (**2**)—sodium metaperiodate meets the foregoing specifications. 3,4-Di-*O*-methylmannitol was rapidly cleaved by periodate to 2,3-di-*O*-methyl-D-arabinose (**3**), which is resistant to further oxidation (*19*). We prepared the crystalline β-anomer of this sugar (the α-anomer had previously been crystallized (*20*)) and showed that on a suitable GLC column the trimethylsilyl derivatives of the four ring forms could be separated (Figure 9). The individual peaks were characterized by preparative GLC and mass spectrometry. This allowed the analysis of the mixtures of anomeric forms generated in kinetic experiments (*see* p. 34).

The results obtained with the 3,4-di-*O*-methylmannitol—periodate system are preliminary. However, we have found no evidence that the mixture of 2,3-di-*O*-methylarabinose tautomers sampled after 15 seconds of reaction is appreciably richer in furanose forms than the equilibrium mixture. In a control experiment a furanose-rich mixture of 2,3-di-*O*-methylarabinose anomers was generated by heating the sugar in pyridine, and the reversion to the normal equilibrium mixture in water was followed polarimetrically at 5°C. The halftime for the process was 4.4 minutes. Thus, it should be possible to determine whether the furanose forms are the preponderant initial products of the cyclization of aldehydo-2,3-di-*O*-methylarabinose in periodate experiments. Work along this line is being continued.

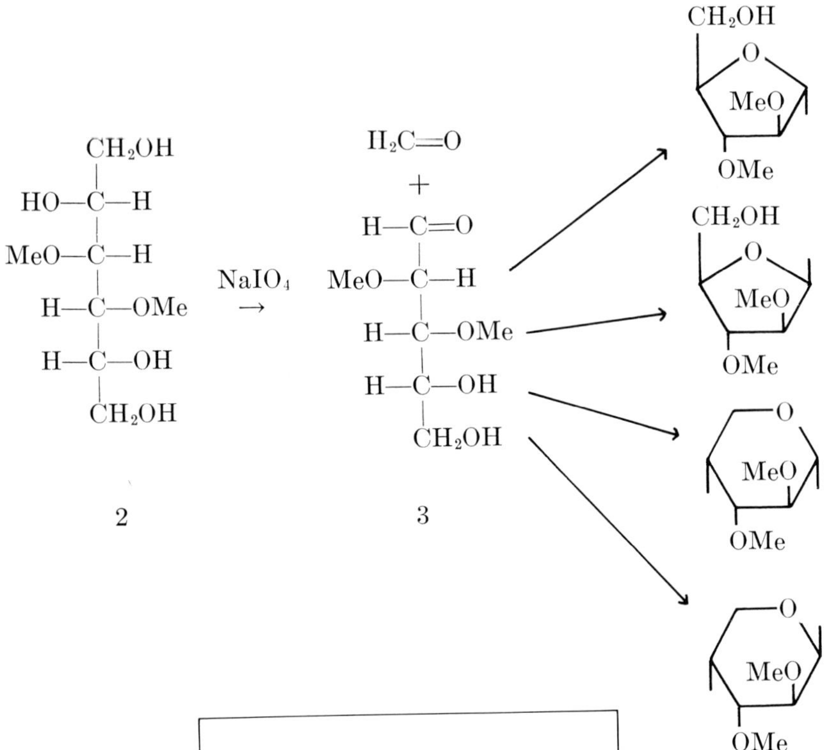

2 3

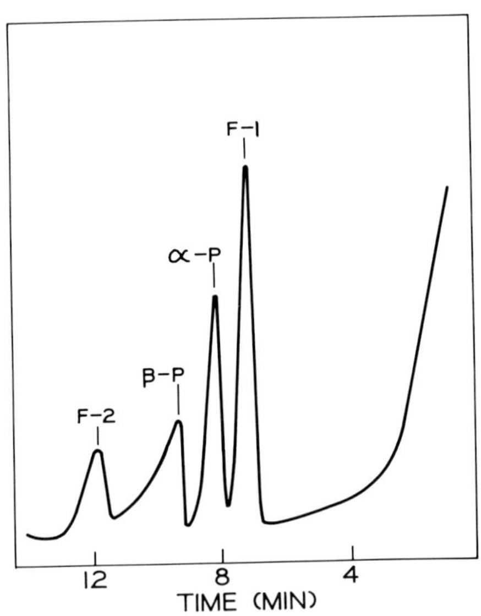

Figure 9. Gas chromatographic separation of the tautomeric forms of 2,3-di-O-methyl-D-arabinose as the bis(trimethylsilyl) derivatives (Column: QF-1 on Anakrom A)

Experimental

Computations were carried out on the IBM 1620 and the Univac 1108 computers. Programs (Fortran) were prepared for the cases where *n*, the number of components in addition to the central intermediate, is 2, 3, 4, and 5. These programs are available on request from the authors.

Galactose–$H_2{}^{18}O$ Exchange. Samples of recrystallized D-galactose (α/β-pyranose, 1 mg) and isotopic water (40 at. % excess ^{18}O, 10 μliters) were incubated in sealed 75 × 8 mm (od) borosilicate glass tubes. At intervals the contents of a tube were freeze dried and trimethylsilylated with hexamethyldisilazane and chlorotrimethylsilane (2:1) in pyridine. The trimethylsilyl ethers were purified with a 6 feet × 6 mm (od) glass column packed with 5% SE-52 on 80–90 mesh Anakrom A in an Aerograph Autoprep model A-700 gas chromatograph. The carrier gas was helium at 200 ml/minute, the temperature 215° C. The β-pyranose fraction was collected and examined in an AEI MS-9 mass spectrometer. Repeated scans were made over the *m/e* range 185–195 since the ion at *m/e* 191 contains O-1 of the sugar (*21*). Values for the atom percent excess ^{18}O in O-1 were calculated (*22*) from measurements of the peaks at *m/e* 191 and 193 in each scan and averaged.

2,3-Di-*O*-methyl-D-arabinose was prepared by a slight modification of the method of Goldstein, Sorger-Domenigg, and Smith (*19*). A portion of the sirupy product, purified by chromatography on silica gel G with methanol-chloroform 1:9 *v/v*, crystallized on seeding. Thereafter, crystalline material was obtained directly from crude preparations by seeding. The crystals were isolated by trituration and filtration with additional crops collected from the mother liquors. The principal fractions had mp 71°–74°C, $[\alpha]_D{}^{20}$ −178° (initial) → −109° (equilibrium) (*c* 1.4, H_2O) and showed single spots in two TLC systems. GLC showed one major peak with a small peak (*ca.* 5% of the total) corresponding to a second anomeric form of the sugar.

A melting point of 82°–83°C and $[\alpha]_D{}^{20}$ −85° → −100° were reported for 2,3-di-*O*-methyl-D-arabinose by Verheijden and Stoffyn (*20*). These authors' preparation must have been largely or wholly an α-anomer, whereas ours is preponderantly β. In view of the rarity of crystalline furanoses, it is presumed to be the β-pyranose, and in the PMR spectrum run at 0°C on a sample freshly dissolved in D_2O the anomeric proton gave a signal, with the appearance of an unresolved doublet, at τ 4.50. This chemical shift and a small value of $J_{1,2}$, would be expected for H-1 of a β-arabinopyranose in the *1C* conformation.

GLC Analysis of 2,3-Di-*O*-methylarabinose Samples. Samples were trimethylsilylated with hexamethyldisilazane and chlorotrimethylsilane (2:1) in pyridine at room temperature and resolved on 6 ft × 3 mm (od) glass columns packed with 5.5% QF-1 on 80–90 mesh Anakrom A. The carrier gas was nitrogen, the temperature 115°C. The sugar, equilibrated in water and then freeze dried, gave the chromatogram shown in Figure 9. The major peak from the crystalline sugar corresponded to the third peak (in order of increasing retention time) of this chromatogram.

Fractions corresponding to each of the peaks of Figure 9 were isolated by preparative GLC, which required two types of columns, QF-1 on Anakrom A and Carbowax 20M on Anakrom A. The relative intensity pattern in the mass spectrum of the second fraction (2nd peak, Figure 9) was nearly identical with that of the β-pyranose fraction, which identifies it as the α-pyranose derivative. The relative intensity patterns for the first and fourth fractions were similar to each other but markedly different from the other two. Thus these fractions are considered to be the furanose fractions.

Appendix

Kinetic analysis of reaction schemes of the type discussed in this paper requires the simultaneous solution of the system of differential equations in Equation 1. A formal procedure for the solution of this set is presented below.

With the material balance Equation 2, it is possible to eliminate X from the differential equations as follows:

$$\begin{aligned} \frac{dA}{dt} + (k_1 + k_2)A + k_2B + k_2E &= k_2C \\ \frac{dB}{dt} + k_4A + (k_3 + k_4)B + k_4E &= k_4C \\ \frac{dE}{dt} + k_6A + k_6B + (k_5 + k_6)E &= k_6C \end{aligned} \tag{4}$$

These are now rewritten using operator notation as

$$\begin{aligned} (D + k_1 + k_2)A + k_2B + k_2E &= k_2C \\ k_4A + (D + k_3 + k_4)B + k_4E &= k_4C \\ k_6A + k_6B + (D + k_5 + k_6)E &= k_6C \end{aligned} \tag{5}$$

The quantities in parentheses are linear differential operators which can, however, be handled as if they were algebraic quantities. The determinant of the coefficients of A, B, and E is thus

$$f(D) = \begin{vmatrix} (D + k_1 + k_2) & k_2 & k_2 \\ k_4 & (D + k_3 + k_4) & k_4 \\ k_6 & k_6 & (D + k_5 + k_6) \end{vmatrix} \tag{6}$$

which may be expanded to yield a cubic equation in D. The three roots of the equation $f(D) = 0$ are all negative and will be denoted as $-m_1$, $-m_2$, and $-m_3$.

The solution may now be written:

$$
\begin{aligned}
A &= a_1e^{-m_1t} + a_2e^{-m_2t} + a_3e^{-m_3t} + A^* \\
B &= b_1e^{-m_1t} + b_2e^{-m_2t} + b_3e^{-m_3t} + B^* \\
E &= d_1e^{-m_1t} + d_2e^{-m_2t} + d_3e^{-m_3t} + E^* \\
X &= C - A - B - E
\end{aligned} \qquad (3)
$$

where, A^*, B^*, and E^* are the equilibrium concentrations, and the a_i, b_i, and d_i are integration constants.

If all rate constants are specified, the equilibrium concentrations of the components can be obtained by setting the derivatives in Equation 4 equal to zero and solving the resulting simultaneous equations together with the material balance Equation 2.

The integration constants are obtained by differentiating the expressions for *A*, *B*, and *E* (Equation 3) twice with respect to time and setting $t = 0$. This gives for the equation for *A*, the following:

$$
\begin{aligned}
a_1 + a_2 + a_3 &= A_0 - A^* \\
-m_1a_1 - m_2a_2 - m_3a_3 &= A_0' \\
m^2{}_1a_1 + m^2{}_2a_2 + m^2{}_3a_3 &= A_0''
\end{aligned} \qquad (7)
$$

where, A_0, A_0', and A_0'' represent the values of *A*, dA/dt, and d^2A/dt^2, respectively, at $t = 0$. Simultaneous solution of 7 thus gives the constants a_1, a_2 and a_3. An identical procedure is used to obtain the b_i and d_i.

The procedure illustrated here may be applied directly to any system of *n* components which are interconverted *via* a single, central intermediate. A similar approach may be used to integrate the differential equations for any network of unimolecular reactions.
plus a central intermediate. Their procedure involves using the Laplace–
the kinetic analysis of reaction networks, including the three components
McLaughlin and Rozett (23) have recently described a method for
Carson transform and gives results identical with ours.

Acknowledgments

This work was supported by the College of Agricultural and Life Sciences, and by Research Grant AM10588 from the National Institute of Arthritis and Metabolic Diseases, NIH and Research Grant RR 00226 from the Division of Research Resources, NIH. Laurens Anderson also acknowledges a Special Fellowship from the Institute of General Medical Sciences, NIH. Irwin Reich assisted with the computations in their later phases, and H. K. Schnoes and students provided mass spectrometric analyses. Merry McCall did the experiments with 2,3-di-*O*-methyl-D-arabinose. We thank Pierre Stoffyn for seed crystals of this sugar.

Literature Cited

1. Pigman, W., Isbell, H. S., *Advan. Carbohyd. Chem.* (1968) **23,** 11.
2. Lemieux, R. U., Stevens, J. D., *Can. J. Chem.* (1966) **44,** 249.
3. Sweeley, C. C., Bentley, R., Makita, M., Wells, W. W., *J. Amer. Chem. Soc.* (1963) **85,** 2497.
4. Lemieux, R. U., Anderson, L., Conner, A. H., *Carbohyd. Res.* (1971) **20,** 59.
5. Acree, T. E., Shallenberger, R. S., Lee, C. Y., Einset, J. W., *Carbohyd. Res.* (1969) **10,** 355.
6. Conner, A. H., M.S. Thesis, University of Wisconsin, Madison, 1970.
7. Conner, A. H., Anderson, L., *Carbohyd. Res.*, in press.
8. Pigman, W., "The Carbohydrates," p. 53, Academic, New York, 1957.
9. Isbell, H. S., Pigman, W., *Advan. Carbohyd. Chem. Biochem.* (1969) **24,** 13.
10. Cantor, S. M., Peniston, Q. P., *J. Amer. Chem. Soc.* (1940) **62,** 2113.
11. Eliel, E. L., "Stereochemistry of Carbon Compounds," p. 198, McGraw–Hill, New York, 1962.
12. Capon, B., *Quart. Rev. (London)* (1964) **18,** 45.
13. Bishop, C. T., Cooper, F. P., *Can. J. Chem.* (1963) **41,** 2743.
14. Hudson, B. G., Barker, R., *J. Org. Chem.* (1967) **32,** 3650.
15. Barker, R., *J. Org. Chem.* (1970) **35,** 461.
16. Heard, D. D., Hudson, B. G., Barker, R., *J. Org. Chem.* (1970) **35,** 464.
17. Rittenberg, D., Graff, C., *J. Amer. Chem. Soc.* (1958) **80,** 3370.
18. Bentley, R., Bhate, D. S., *J. Biol. Chem.* (1960) **235,** 1225.
19. Goldstein, I. J., Sorger-Domenigg, H., Smith, F., *J. Amer. Chem. Soc.* (1959) **81,** 444.
20. Verheijden, J. P., Stoffyn, P. J., *Bull. Soc. Chim. Belges* (1959) **68,** 699.
21. DeJongh, D. C., Radford, T., Hribar, J. D., Hanessian, S., Bieber, M., Dawson, G., Sweeley, C. C., *J. Amer. Chem. Soc.* (1969) **91,** 1728.
22. Thorpe, S. R., Sweeley, C. C., *Biochemistry* (1967) **6,** 887.
23. McLaughlin, E., Rozett, R. W., *Chem. Tech.* (Feb. 1971) 120.

RECEIVED December 2, 1971.

3

Carbon-13 and Hydroxyl Proton NMR Spectra of Ketoses

A Conformational and Compositional Description of *Keto*-hexoses in Solution

A. S. PERLIN, P. HERVE DU PENHOAT, and H. S. ISBELL

McGill University, Montreal, Canada, and The American University, Washington, D. C.

A study of [13]*C NMR spectra of* keto-*hexoses and some derivatives in various solvents and of hydroxyl PMR spectra of these compounds in methyl sulfoxide is presented. Uniformly close similarities are found between the chemical shifts of the* keto-*hexoses and configuration related* aldo-*pentoses and -hexoses, permitting conformational and anomeric configurational designations to be made.* [13]*C spectral characteristics of ketofuranoses are discussed, particularly with reference to the deshielding observed relative to pyranose* [13]*C nuclei. Equilibrium compositions of ketoses and of partially substituted ketoses in different media are reported. A study of the hydroxyl proton spectra of* D-*fructose-containing disaccharides is described, and the earlier designation of crystalline lactulose as a* D-*fructofuranose derivative is confirmed. The results generally agree with information from optical rotatory studies and application of conformational analysis.*

Information about the characteristics of *keto*-hexoses in solution has been derived mainly from optical rotatory data (*1, 2, 3, 4*) and in recent years by application of the principles of conformational analysis (*5, 6*). In the current study an attempt is made to describe the conformation and composition of these sugars in solution by nuclear magnetic resonance (NMR) spectroscopy, a highly sensitive means for examining stereochemistry and for differentiating between isomeric species.

It is most straightforward to begin with α-L-sorbopyranose (**1**) (*see* Figure 1) because this sugar shows virtually no mutarotational change in water (*1, 7*); thus, it remains basically in the α-pyranose form. By analogy with the established (*8*) structure of the crystalline material, this sugar exists in solution in the 1C(L) pyranose chair conformation (**1**) because all substituents are thereby optimally oriented—*i.e.*, the 1-carbinol group and the 3, 4, and 5 hydroxyl groups are all equatorial, and OH-2 is axial. (With the ketohexopyranoses in general it would be expected that the 1-carbinol group maintains almost invariably an equatorial orientation and thus is a dominant factor determining the conformations adopted by these sugars. This simultaneously places the 2-OH group in an axial alignment which, assuming that the anomeric effect in *keto*-hexoses is of the same magnitude as in aldoses (*5*), supplements the conformational impact of the equatorial 1-carbinol group.)

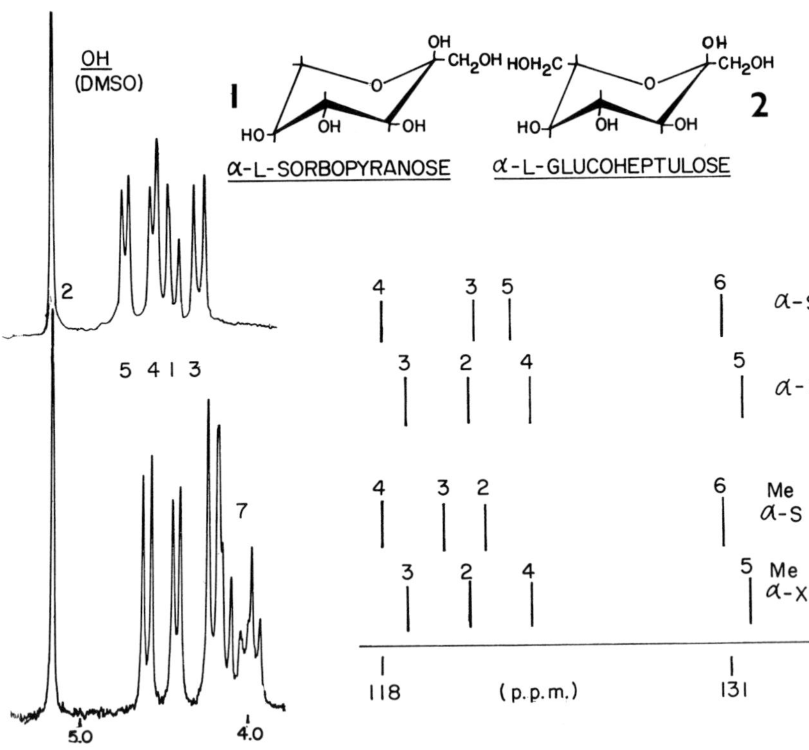

Figure 1. Hydroxyl PMR signals for α-D-sorbopyranose (upper left) and α-L-glucoheptulose (lower left) in methyl sulfoxide-d$_6$. ^{13}C chemical shifts for aqueous solutions of α-D-sorbopyranose (α-S), α-D-xylopyranose (α-X) and their methyl glycosides (Me-α-S and Me-α-X, respectively) (ppm relative to downfield $^{13}CS_2$).

Since **1** seems to be entirely reasonable, there is even greater assurance that this is the conformation of the configurationally related heptulose, α-D-*gluco*-heptulose (**2**), because the latter possesses an additional equatorial primary carbinol group to anchor the molecule securely in this conformation. NMR spectroscopic examination gives experimental evidence to support these theoretical considerations. Thus, the O-H NMR spectra of **1** and **2** in methyl sulfoxide (Figure 1) (*see* Reference *9*), being closely similar, suggest a strong steric affinity between **1** and **2** as well. The chemical shifts of signals OH-2 to -5 in the two spectra are almost coincident, and the spacings of the corresponding doublets are about the same. Furthermore, the ^{13}C NMR (and also O-H NMR) spectrum of α-L-sorbose shows a pattern strikingly analogous to that of the configurationally related *aldo*-pentose, α-D-xylopyranose (Figure 1; spectral data for the methyl glycosides and for α-D-xylopyranose are included). From what is now known about ^{13}C chemical shifts of carbohydrates (*10, 11*), this strongly indicates a close conformational identity between these two sugars.

These spectra have detected only one species of L-sorbose in aqueous solution. However, after prolonged storage in methyl sulfoxide 5–10% of a second component is generated. Although it has not as yet been properly identified, there is no counterpart in spectra of the heptulose, and most likely it is the β-pyranose—*i.e.*, the existence of furanose forms of **1** and **2** should have about equal probability, whereas the β-pyranose form of **2** in a chair conformation should be much less stable than that of **1**.

α-D-Tagatose (**3**) shows a greater, although still small, rotational change in water (*3, 12*). As seen in the ^{13}C and O-H NMR spectra (Figure 2), the minor component accounts for about 10% of the mixture (and its percentage does not change on storage in methyl sulfoxide). This second species likely is β-D-tagatopyranose because its ^{13}C chemical shifts correspond closely to expectation (Table I), the 1C(D) chair conformation being assumed, and they are upfield relative to the shifts of furanoses (*see* below). As to the α-anomer, one may again use the analogy between its spectra and those of the related heptulose and aldopentose (Figure 2)—*i.e.*, α-D-*manno*-heptulopyranose (**4**) (OH spectrum) and α-D-lyxopyranose (^{13}C spectrum) to support the expectation (*5, 6, 9*) that **3** adopts the Cl(D) chair shown. Also suggested, is that relative to **1** the shape of this chair is not altered by introducing an axial hydroxyl group at position-3 in place of an equatorial one. This follows from the close agreement found (Table I) between observed ^{13}C chemical shifts and those calculated from interaction-shift relationships of the kind found earlier with aldopyranoses (*11*).

In the crystalline state β-D-fructopyranose (**5**) possesses the 1C(D) conformation (*13*), and it is clear that it retains this conformation in solu-

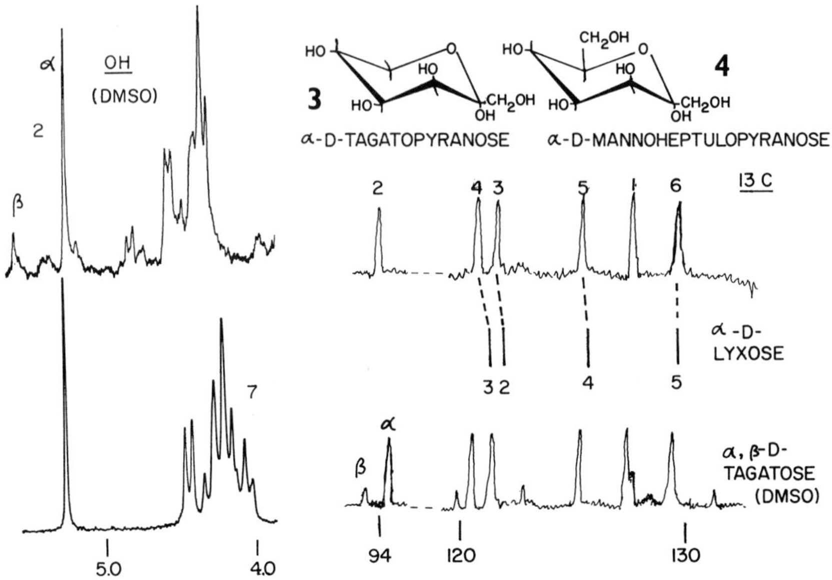

*Figure 2. Hydroxyl PMR signals for α-*D*-tagatopyranose (upper left) and α-*D*-mannoheptulopyranose (lower left) in methyl sulfoxide-*d_6*. *^{13}C* NMR spectra of α,β-*D*-tagatose (3) and α-*D*-lyxose in water and of α,β-*D*-tagatose in methyl sulfoxide-*d_6* (ppm relative to downfield *$^{13}CS_2$*).*

tion (*5, 6*). As with **1** and **3**, the overall close analogy between spectra (Figure 3) of **5** and those of α-L-*gala*-heptulose (**6**), its 7-carbon homolog, and β-D-arabinopyranose favors this 1C(D) conformational assignment. However, the ^{13}C chemical shifts of β-D-fructopyranose are not as readily related to those of **1** as are the chemical shifts of **3** (Table I). This discrepancy parallels earlier findings with respect to D-glucose *vs.* D-galactose or D-xylose *vs.* D-arabinose (*11*). In all three instances, *keto*-hexose, *aldo*-hexose, and *aldo*-pentose, the designation of OH-5 (ketose) or OH-4(aldose) as an axial substituent does not correspond well to its expected impact on shielding of the adjacent carbons. Tentatively, these observations raise the possibility that the exact shape of the β-D-fructopyranose chair differs slightly from that of **1** and **3**. By extrapolation this same possibility applies for the related aldoses and, agreeing with the above comment, for β-D-tagatopyranose (*i.e.*, the latter is correlated with β-D-fructopyranose in Table I).

Mutarotation changes and rates of change for D-fructose show that β-D-fructofuranose is a plentiful component at equilibrium in water and that both of the α-anomers are relatively unimportant (*2*). The recently-described ^{13}C NMR spectrum of D-fructose (*14*) confirms and amplifies this early polarimetric evidence, giving an equilibrium composition of about 6:3:1:trace for β-pyranose:β-furanose:α-furanose:α-pyranose. Figure 4, which shows the anomeric ^{13}C signals detected in our experiments at 25.15 MHz, reproduces essentially these data of Doddrell and Allerhand (*14*).

When the mutarotated solution of D-fructose is concentrated and the water replaced by methyl sulfoxide, the anomeric-OH signals (Figure 4A) present a quantitative relationship close to that shown by the ^{13}C NMR spectrum. This analogy allows the assignments designated and gives an isomeric ratio of 6.5:2.7:1:trace—*i.e.*, in good agreement with the ^{13}C data. However, during storage in methyl sulfoxide, the composition changes slowly so as to favor more strongly the furanoses. At equilibrium the latter comprise 85% of the total (dimethylformamide seems to exert a similar effect; in this solvent D-fructose exists as furanose to the extent of 80% (*15*)), and the ratio is 1:3:1.3:trace (Figure 4B). The corresponding partial spectra for L-*gala*-heptulose (**6**) are virtually superimposable on Figures 4A and 4B.

This pronounced tendency for an increase in the relative stability of the furanose forms in methyl sulfoxide has been observed earlier with L-arabinose, D-galactose, and some other aldoses (*16, 17*). Its origin is obscure, pyranose forms are possibly stabilized preferentially by solvation with water (*18*), or, alternatively, methyl sulfoxide might enhance the relative stability of furanose forms. Figures 4A and 4B, together with related data, indicate that the 1-carbinol group of β-D-fructofuranose (**7**), which produces the doublet of doublets at 5.0 ppm, is involved in hydrogen bonding, probably with OH-4. (This surmise is based on the fact that signal OH-1 of lactulose (Figure 4E) is not shifted downfield as observed with β-D-fructofuranose.) Although this association would favor **7**, it probably is not a major factor because a similar H-bonding arrangement does not seem to apply for the α-furanose, which also becomes more prominent in methyl sulfoxide, nor can it be invoked for the *arabino*- and *galacto*-aldofuranoses.

The disaccharide, leucrose (5-*O*-α-D-glucopyranosyl-β-D-fructopyranose) (**8**) exhibits an OH-2 singlet having the same chemical shift as that of β-D-fructopyranose and two clearly separated groups of four OH signals each (Figure 4C)—the upfield group derives from the D-fructose moiety whereas the other group corresponds more closely in shift to OH signals of α-D-glucosides (*16, 19, 20*). After equilibration in water or in methyl sulfoxide, a small proportion of the α-anomer of **8** becomes detectable—*i.e.*, as with D-fructose. The 1,3-linked analog, turanose, (**9**), produces an OH-2 singlet (Figure 4D) that is intermediate in shift between those of β-D-fructopyranose and the β-furanose (**7**). However, after equilibration in water or in methyl sulfoxide, two additional singlets appear at lower field (as indicated by the arrows), and the relative in-

Table I. ^{13}C Shielding

Inversion[a]	*Carbon*	*Chemical Shift (A)*
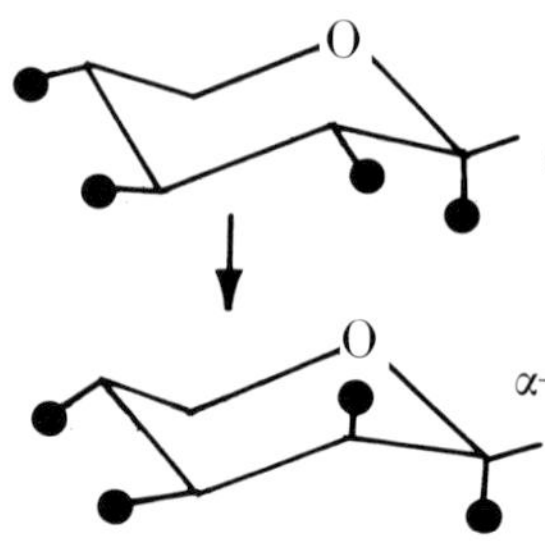 (A) α—D—S → α—D—T (B)	1 2 3 4 5 6	127.9 94.3 121.1 117.7 122.3 130.0
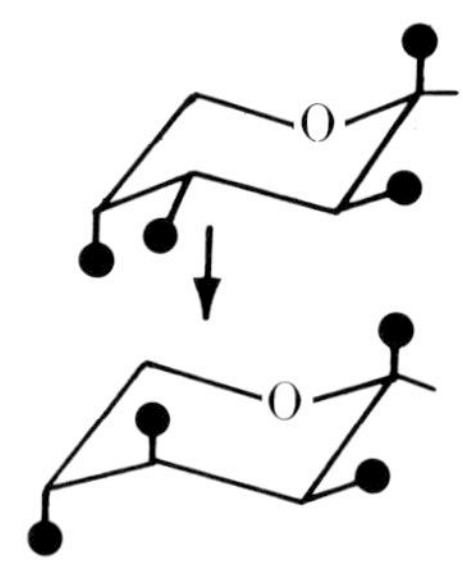 (A) β—D—F → β—D—T (B)	1 2 3 4 5 6	128.0 94.6 123.5 122.8 124.8 129.4
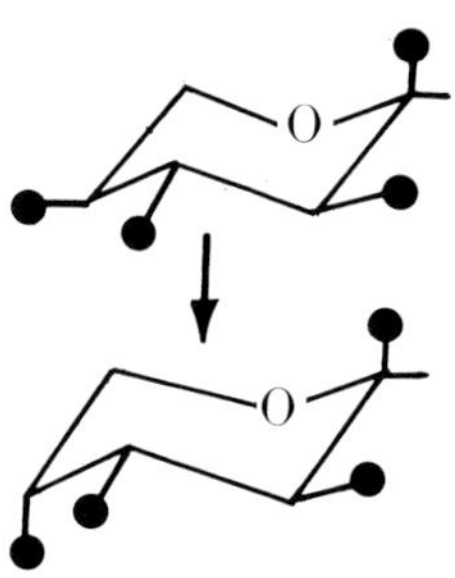 (A) α—L—S → β—D—F (B)	1 2 3 4 5 6	127.9 94.3 121.1 117.7 122.3 130.0

[a] Interconversion of A into B: S, sorbopyranose; T, tagatopyranose; F, fructopyranose.
[b] ax., axial substituent; adj., adjacent to axial substituent; H//O, syn-diaxial; H,O, interaction; O/O, gauche, O,O interaction; O//O, syn-diaxial; O,O interaction.

tensities of these three peaks closely mirror the pattern exhibited by D-fructose. (These similarities are evident also from ^{13}C spectra of turanose and fructose (*14*).) Therefore, crystalline turanose is a β-D-fructopyranose derivative. Figure 4D indicates from the relatively low field

Differences Between Epimers

Interaction [b]	*Change in Shielding*	*Chemical Shift (B) Calc.* [c]	*Found*
O	O	127.5	127.0
adj.3, −O/O—3	+, −	94.3	94.3
ax.O, −O/O—2	+, −	121.1	121.4
adj.3	+	120.7	120.6
H//O—3	+	125.3	125.0
O	O	130.0	129.3
O	O	128.0	127.6
O//O—4, −O//H—4	+ −	93.6	93.5
adj.4	+	126.5	127.4
ax.O,O//O—2,O//H—6	+ + +	120.8	120.6
−H//O—2, −O/O—5	− −		
adj.4, −O/O—4	+ −	124.8	122.7
H//O—4	+	132.4	132.3
O	O	127.9	128.0
O	O	94.3	94.6
H//O—5	+	124.0	123.5
adj.5	+	120.7	122.8
ax.O	+	125.3	124.8
adj.5	+	133.0	129.4

[c] Based on shielding effects given in Ref. *10* and *11*; a value of 2.5 ppm is added to or subtracted from the value for Λ, depending upon whether an increase (+) or decrease (−) in shielding is anticipated for the interaction listed. For an O//O interaction, the value is 2.0 ppm.

position of two doublets (at 5.0 and 5.5 ppm) that two secondary hydroxyl groups are hydrogen bonded, an association analogous to that found by Casu *et al.* (*20*) with maltose. Reference to molecular models suggests that such an association is most likely to involve OH-2 of the D-glucosyl residue and OH-4 of the D-fructosyl residue.

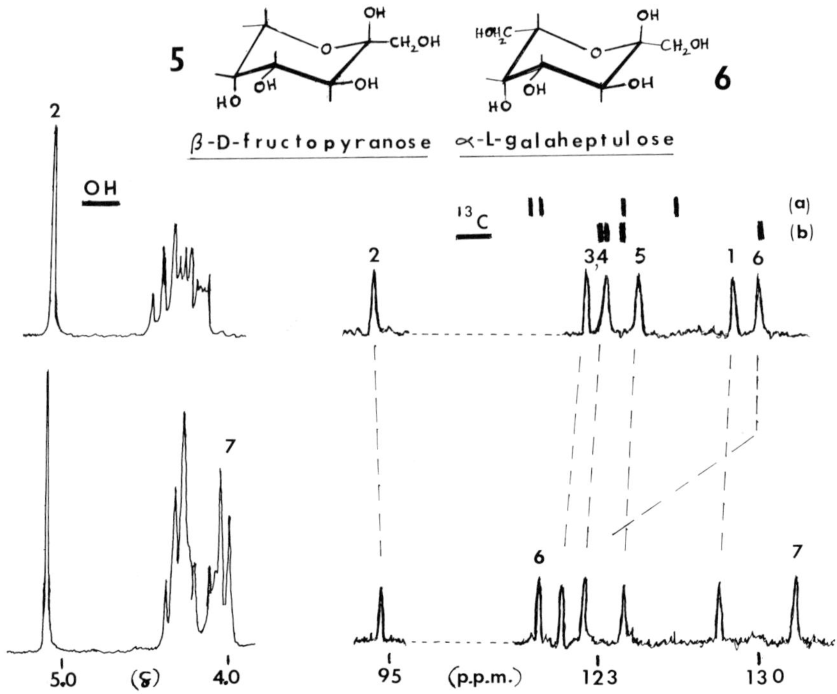

*Figure 3. Hydroxyl PMR signals for β-*D*-fructopyranose* (**5**) *(upper left) and α-*L*-gala-heptulose* (**6**) *(lower left) in methyl sulfoxide-*d_6. *^{13}C NMR spectra of* **5** *and* **6** *(in water). The diagonal line relating the [13]C-6 resonances of* **5** *and* **6** *reflects the large downfield shift attributable to replacement of one H-6 with the 7-carbinol group. [13]C chemical shifts for α- and β-*D*-arabinopyranose (a and b, respectively)* (10, 11) *are inserted to illustrate the close conformational affinity between* **5** *and β-, though not α-,* D*-arabinose (ppm relative to downfield* $^{13}CS_2$*).*

Based on these hydroxyl proton spectral patterns and that of Figure 4E, crystalline lactulose (4-*O*-β-D-galactopyranosyl-D-fructose) can be designated a derivative of β-D-fructofuranose—*i.e.,* **10**. This supports the conclusion derived by Isbell and Pigman (*2*) from the rotatory properties of the compound. Despite repeated purification small proportions of β-fructopyranose and α-fructofuranose forms are retained in the crystal (Figure 4E). Their relative proportions increase in water so that the OH-2 pattern becomes closely similar to that in Figure 4A and resembles that in Figure 4B after equilibration in methyl sulfoxide—all of which agrees with the given assignments.

NMR spectral information about D-psicose (D-allulose) is particularly interesting because little is known of the solution characteristics of this *keto*-hexose. Having not as yet been obtained in any crystalline modification and with an equilibrium specific rotation close to zero, its preferred anomeric configuration and ring size have not been deducible. The O—H and ^{13}C NMR spectra (Figure 5) recorded for this syrupy sugar concur

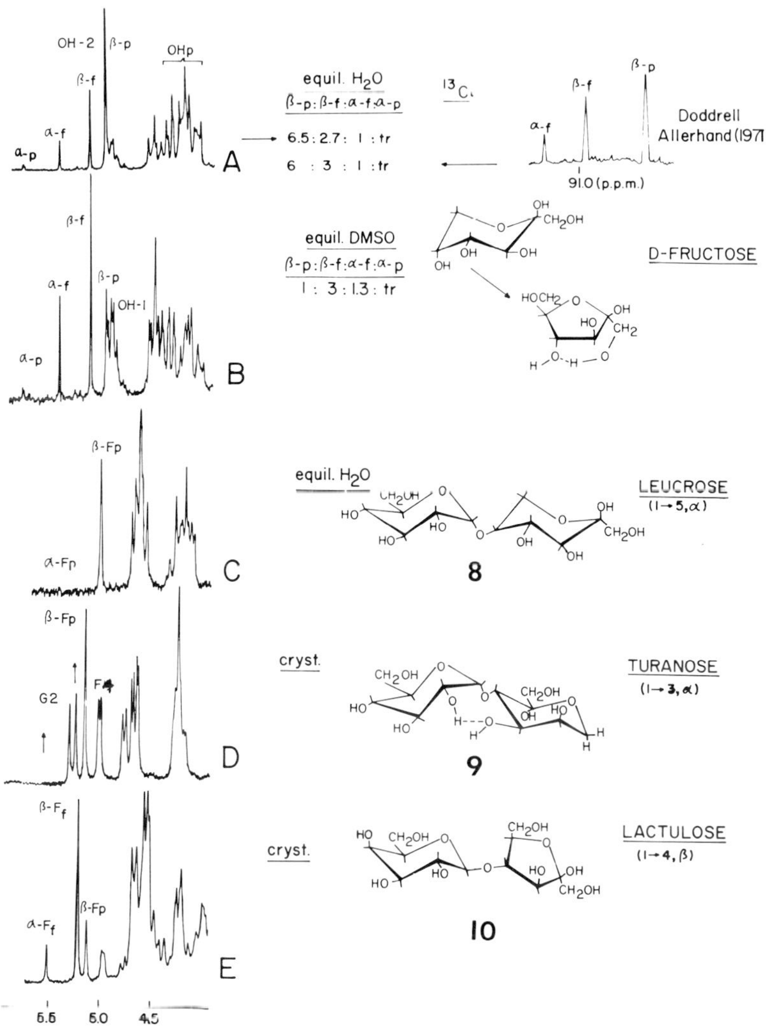

Figure 4. Hydroxyl PMR signals (solvent, methyl sulfoxide-d_6) for A) D-fructose after prior equilibration in water; B) D-fructose at equilibrium in methyl sulfoxide-d_6; C) leucrose, freshly dissolved crystals; D) furanose, freshly dissolved crystals (arrows show the positions of signals that increase in intensity with time) G2-hydroxyl proton-2, F4-hydroxyl proton-4; E) lactulose, freshly dissolved crystals. At the upper right are ^{13}C-2 signals at 25.15 MHz for D-fructose in water (see Reference 14).

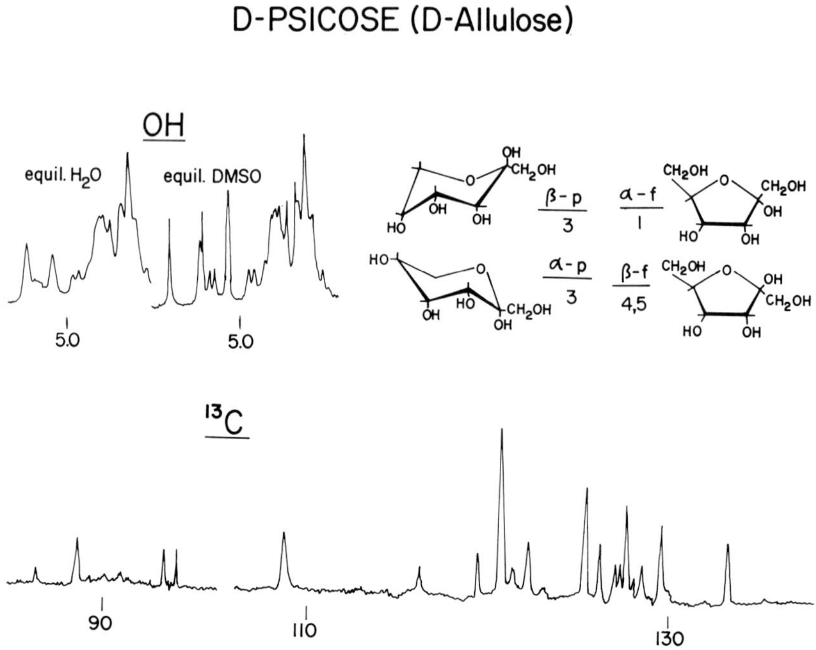

Figure 5. ^{13}C NMR spectrum of D*-psicose in water; the ratios of the anomeric pyranoses and furanoses (upper right) are derived from the relative intensities of the ^{13}C-2 signals in the region of 90 ppm (ppm relative to $^{13}CS_2$). At the upper left are hydroxyl PMR signals for* D*-psicose (solvent, methyl sulfoxide-d_6) after prior equilibration in water and at equilibrium in methyl sulfoxide-d_6.*

in showing that three major components and a minor one are present in solution, the overall pyranose-furanose ratio being about 1:1. A furanose, tentatively the α-anomer, accounts for the strongest ^{13}C signal (by reference to 6-*O*-methyl-D-allulose), whereas the two pyranoses each constitute about one quarter of the mixture. In methyl sulfoxide the relative intensity of the C-2 signal furthest downfield (resulting from the other furanose) increases, and that of the most upfield C-2 signal (probably resulting from the β-pyranose) decreases, each by 40–50%. Thus, energy differences between all four D-allulose species are small, which is to be expected because of the destabilizing interactions that are inherent in each structure.

One further aspect of the ^{13}C NMR data merits comment. In the spectra of D-fructose and D-allulose, the anomeric ^{13}C signals of furanose forms appear downfield from those of pyranose forms. As already noted by Hall and Johnson (*21*), this same kind of distinction is observed with D-ribose. Broader examination of these spectra and also those of various other aldoses shows that stronger deshielding of furanose ^{13}C nuclei is not confined to the anomeric position but is much more general. Thus, total shielding for a given furanose sugar, as reflected in the sum of its ^{13}C chemical shifts, is always substantially less than that of its pyranose counterpart, amounting to 25–65 ppm for the group of compounds examined (*22*). Now the ^{13}C chemical shifts of aldopyranoses and ketopyranoses, as well as those of cyclohexane derivatives (*23, 24*), show consistently that destabilizing non-bonded interactions are associated with increased shielding. Generally, therefore, the less stable the compound in these series, the greater is its total shielding in comparison with isomeric compounds (*25*). Thus the order of shielding in furanose relative to that of pyranoses seems to be inconsistent with the fact that furanoses are the less stable and should accordingly incorporate greater destabilizing (stronger shielding) interactions (except in a few instances).

It is possible that this difficulty can be resolved by a fuller evaluation of the ^{13}C chemical shift data. Basically, it may not be feasible to compare chemical shifts of 5- and 6-membered ring isomers directly. From preliminary studies (*26, 27*) it appears that within a given series of 6-membered ring compounds the contribution of the diamagnetic contribution (σ_d) is highly important in determining shielding differences between isomers and that the paramagnetic term (σ_p) does not obscure these differences. This situation may or may not apply for 5-membered ring compounds. However, limited data now available on furanose sugars (*27*) indicate that the latter should contain overall less shielded ^{13}C nuclei than the corresponding pyranoses, as far as the diamagnetic contribution is concerned. The electronic effect, as expressed in electron density calculations (*28*), may be relatively small. Therefore, as found above with D-fructose and D-allulose and also with a substantial group of aldosides (*22*), the fact that ^{13}C nuclei of furanoses are on the average relatively deshielded does not necessarily alter previous conclusions about chemical shift-interaction relationships in pyranoses and cyclohexanes. Furthermore, it is to be hoped that an extension of this more rigorous treatment of the ^{13}C chemical shift information will help to define the conformations of furanoses in solution.

Acknowledgments

The authors gratefully acknowledge the generous support of the Pulp and Paper Research Institute of Canada and the National Research Council of Canada. Spectra were kindly recorded by H. J. Koch, R. Simoneau, and M. Mazurek.

Literature Cited

1. Pigman, W. W., Isbell, H. S., *J. Res. Nat. Bur. Stand.* (1937) **19**, 443.
2. Isbell, H. S., Pigman, W. W., *J. Res. Nat. Bur. Stand.* (1938) **20**, 773.
3. Karabinos, J. V., *Advan. Carbohyd. Chem.* (1952) **7**, 99.
4. Pigman, W. W., Isbell, H. S., *Advan. Carbohyd. Chem.* (1968) **23**, 11; (1969) **24**, 13.
5. Angyal, S. J., "Conformational Analysis," p. 408, E. L. Eliel, N. L. Allinger, S. J. Angyal, and G. A. Morrison, Eds., Wiley, New York, 1967.
6. Stoddart, J. F., "Stereochemistry of Carbohydrates," p. 90, Wiley, New York, 1971.
7. Lobry de Bruyn, C. A., van Ekenstein, W. A., *Rec. Trav. Chim.* (1897) **16**, 262; (1900) **19**, 1.
8. Kim, S. H., Rosenstein, R. D., *Acta. Crystallogr.* (1967) **22**, 648.
9. Jochims, J., Taigel, G., Seeliger, A., Lutz, D., Driesen, H. E., *Tetrahedron Lett.* (1967) 4363.
10. Dorman, D. E., Roberts, J. D., *J. Amer. Chem. Soc.* (1970) **92**, 1355.
11. Perlin, A. S., Casu, B., Koch, H. J., (1970) **48**, 2596.
12. Khouvine, Y., Arragon, G., Tomoda, Y., *Bull. Soc. Chim.* (1939) [5] **6**, 354.
13. Eiland, P. F., Pepinsky, R., *Acta. Crystallogr.* (1950) **3**, 160.
14. Doddrell, D., Allerhand, A., *J. Amer. Chem. Soc.* (1971) **93**, 2779.
15. Kuhn, R., Grassner, H., *Ann.* (1957) **610**, 122.
16. Perlin, A. S., *Can. J.* (1966) **44**, 539.
17. Mackie, M., Perlin, A. S., *Can. J. Chem.* (1966) **44**, 2039.
18. Kabayama, M. A., Patterson, D., *Can. J. Chem.* (1958) **36**, 563.
19. Casu, B., Reggiani, M., Gallo, G. G., Vigevani, A., *Tetrahedron Lett.* (1964) **39–40**, 2839.
20. Casu, B., Reggiani, M., Gallo, G. G., Vigevani, A., *Tetrahedron* (1966) **22**, 3061.
21. Hall, L. D., Johnson, L. F., *Chem. Commun.* (1969) 509.
22. Koch, H. J., Perlin, A. S., unpublished.
23. Buchanan, G. M., Ross, D. A., Stothers, J. B., *J. Amer. Chem. Soc.* (1966) **88**, 4301.
24. Dalling, D. K., Grant, D. M., *J. Amer. Chem. Soc.* (1967) **89**, 6612.
25. Perlin, A. S., Koch, H. J., *Can. J. Chem.* (1970) **48**, 2639.
26. Cyr, N., Perlin, A. S., Whitehead, M. A., *Can. J. Chem.* (1972) **50**, 814.
27. Cyr, N., Perlin, A. S., Schwarcz, J., *Chem. Inst. Can. Conf., Quebec,* Abstr. No. 160, July 5-7, 1972.
28. Zhdanov, Y. A., Minkin, V. I., Ostroumov, Y. A., Dorofeenko, G. A., *Carbohyd. Res.* (1968) **7**, 156.

RECEIVED January 14, 1972.

4

Ionization of Carbohydrates in the Presence of Metal Hydroxides and Oxides

J. A. RENDLEMAN, JR.

Northern Regional Research Laboratory, Peoria, Ill. 61604

Metal hydroxides interact rapidly and reversibly with carbohydrates to form metal hydroxide–carbohydrate adducts and through hydroxylic proton abstraction to form alcoholates. The formation of an alcoholate, or oxyanion, is an essential step in numerous inter- and intramolecular reactions of polyhydroxy compounds. Bases also attack aldehydo and keto sugars to give carbanions that are precursors for isomerization, epimerization, internal oxidation and reduction, and group migration. Carbohydrate acidity can be explained in terms of statistical factors, electrostatic field effects, polar group interactions, intramolecular hydrogen bonding, steric hindrance, and entropy of ionization.

Studies of the interaction of metal bases with carbohydrates were begun in the early half of the 19th century with chemists such as Berzelius (*50*) and Peligot (*36*). Peligot was perhaps the first to prepare alkali-metal and alkaline-earth metal alcoholates of sugars. He deserves recognition for initiating the investigation of reactions between reducing sugar and alkali that ultimately lead to saccharinic acid formation (*37*). Alcoholates result simply from the loss of a proton from a hydroxyl group whereas saccharinic acid formation and numerous alkali-induced isomerizations, except for anomerization, are initiated by carbanion formation. These carbanions stem from carbon–hydrogen ionization at a carbon atom adjacent to a strongly electron-attracting substituent such as a carbonyl or carboxyl group and are stabilized by delocalization of the negative charge over several atoms. One of the resonance structures contributing to this hybrid anion is an enolate ion.

Ionization of the Hydroxyl Group

Monohydric aliphatic alcohols having no strong electron-withdrawing substituent to facilitate ionization of the hydroxyl group are extremely weak acids in either aqueous or alcoholic media. The dissociation constant in aqueous solution at 25°C for methanol (*4*) is 2.9×10^{-16}, for ethanol (*4*) and possibly other primary and secondary alkyl alcohols, $\sim 1 \times 10^{-16}$. In alcoholic media ionization constants may be 1/10–1/100 as large as those in aqueous media (*22*).

Studies in aqueous solution (*4*) have shown that p*K* values for various substituted methanols correlate well with the inductive effects of the substituent groups as measured by Taft–Ingold σ values.

In contrast to the low acidity of monohydric aliphatic alcohols is the high acidity of polyhydroxy compounds in which the hydroxyl groups are located near one another. The greater the number of hydroxyl groups, the greater the acidity. The ionization constants for carbohydrates generally are in the range of 10^{-12}–10^{-14}. Electrophoretic migration in aqueous 0.1*N* sodium hydroxide solution (*19*) indicates that reducing sugars are the most easily ionized and that straight-chain alditols have on the average about the same acidity as cyclitols and glycosides of similar molecular weight and hydroxyl content. Small differences in electrophoretic rates are not significant as far as acidity is concerned because many factors in addition to proton lability contribute to the observed electrophoretic movement.

High acidity of reducing sugars is caused by high lability of the hydrogen atom of the hemiacetal (anomeric) hydroxyl group, a condition that apparently stems from an electron-withdrawing polar effect exerted upon this group by the ring oxygen. This polar effect, often called an inductive effect, is propagated by (a) successive polarization of σ bonds between substituent and reaction site or by (b) electrostatic field interaction between dipoles (*3*). The former mechanism depends upon the number and nature of the bonds and the number of the paths between substituent (*e.g.*, ring oxygen atom) and reaction site (*e.g.*, C-1 hydroxyl group), whereas the field-effect mechanism presumes that the polar effect originates in bond dipole moments and is propagated according to the classical laws of electrostatics. Distance rather than number of bonds between substituent and reaction site would be important, and the polar effect would show an angular dependence (*20, 51*).

In view of recent work by Neuberger and Wilson (*33*) pertaining to the relative acidities of various methyl glycosides as determined by retention times on a strongly basic ion-exchange resin (OH^- form), the field-effect process may well be the operative mechanism. Neuberger and Wilson offer reasonable explanations for the acidities of these glycosides, based upon the angles of the various dipoles to each other and to the hydroxyl group that ionizes. Methyl α- and β-D-glucopyranoside will illustrate this (Figure 1). In the β-anomer the dipole moment of the ring oxygen atom and the dipole moment of the glycosidic C–O bond largely reinforce each other and cause the ionization constant for the C-2 hydroxyl group to be high, compared with the ionization constant for the same group in the α-anomer. In the α-anomer the dipole moments cancel each other to a significant extent. Evidence that the β-anomer is more acidic than the α-anomer was obtained earlier by Derevitskaya and co-workers (*16*) who titrated electrometrically these glycosides with potassium hydroxide in ethylenediamine solution (*38*). In ethylenediamine simple carbohydrates behave more strongly as polybasic acids than they do in aqueous media where there is only a slight tendency for more than one hydroxyl group to ionize (*38*). Presumably, the most acidic hydroxyl group in an unsubstituted methyl glycoside is the one at position 2 because of the proximity of this hydroxyl group to the electron-withdrawing acetal function. Chemical evidence for this high acidity has been obtained in several rate studies of Williamson-type reactions where carbohydrate oxyanions are attacked by various electrophilic reagents (*34, 38, 40*). Thus, the magnitude of the ionization constant of the 2-hydroxyl group is a major factor determining the overall acidity of a glycoside.

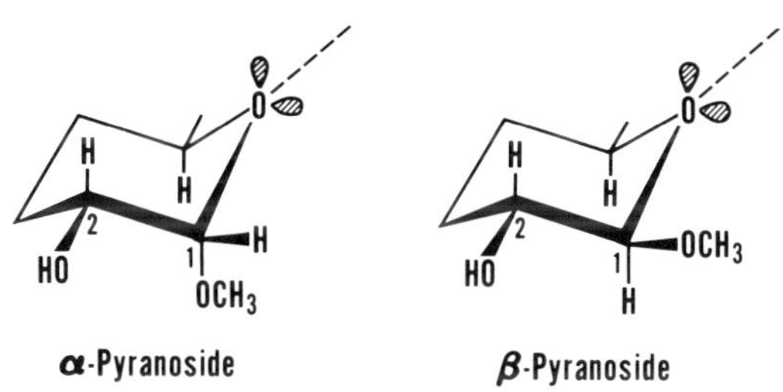

Figure 1. Methyl α- and β-D-glucopyranoside, showing orientation (dotted line) of the resultant negative end of ring oxygen dipole (the hydrogens and hydroxyl groups on C-3, -4, and -6 have been omitted for clarity)

Ionization constants have been determined for numerous simple carbohydrates (*10, 13, 15, 25, 45*), as well as for cellulose (*32, 43*), wheat starch (*43*), and alginate (*43*). Selected carbohydrates with their corresponding pK values are presented in Table I. The analytical methods involved in these determinations include conductimetry, potentiometric titration, thermometric titration, and polarimetry. Polarimetry was used by Smolenski and co-workers (*45*) to calculate a first and a second ionization constant for sucrose at 18°C ($K_1 = 3 \times 10^{-13}$; $K_2 = 3 \times 10^{-14}$).

Table I. Ionization Constants (Hydroxyl Group) of Carbohydrates in Water[a]

	*p*K			
Compound	*0°C*	*10°C*	*25°C*	*40°C*
D-Glucose	12.92	12.72	12.35	
D-Galactose		12.82	12.35	
D-Mannose		12.45	12.08	11.81
D-Arabinose			12.43	
D-Ribose			12.21	
D-Lyxose			12.11	
D-Xylose			12.29	
2-Deoxy-D-glucose		12.89	12.52	12.28
2-Deoxy-D-ribose		12.98	12.65	
Lactose			11.98	
Maltose			11.94	
Raffinose			12.74 (18°C)	
Sucrose			12.51	
D-Fructose		12.53	12.03	
D-Glucitol	14.14		13.57 (18°C)	
D-Mannitol	14.09		13.50 (18°C)	
Glycerol			14.4	

[a] Values from References *10*, *13*, and *15*.

They found that their optical rotation data could best be explained in terms of two ionizations. For simplicity in their calculations, all other workers assumed that only one hydroxyl group per molecule was ionizable. Table I shows that the acidities of the aldoses are similar to one another, except for 2-deoxy-D-ribose whose somewhat lower acidity can perhaps be explained through the concept of intramolecular hydrogen bonding, which will be discussed later. At 25°C glycerol ($pK = 14.4$) has about the same acidity as water ($pK = 14.0$) and is about 10 times more acidic than methanol ($pK = 15.5$). Lactose and maltose, which are reducing disaccharides, seem to be somewhat more acidic than aldopentoses. This higher acidity can be explained by the larger number of hydroxyl groups in the disaccharides. In other words, a statistical factor is important.

On a purely statistical basis where such factors as polar and steric effects are disregarded, a diol should be twice as acidic as a monohydroxy compound, and sucrose should possess a first ionization constant that is as large as that of an alditol containing the same number of hydroxyl groups. Evidence to support this view may be found in the electrophoretic measurements of Frahn and Mills (*19*) who found, for example, that in aqueous sodium hydroxide solution the rate of migration of sucrose (which has eight hydroxyl groups) is roughly the same as that of certain heptitols. Furthermore, a trend toward higher acidity with an increase in the number of hydroxyl groups is evident in a homologous series of sugar alcohols (*19*). Any purely statistical treatment of carbohydrate ionization would be similar to the treatment given to ionization of polycarboxylic acids (*9, 21*).

The second ionization constant K_2 of a dicarboxylic acid should, from statistical considerations alone, equal $K_1/4$. However, experimentally, values of K_2 for all dicarboxylic acids are less than $K_1/4$, but approach $K_1/4$ as the distance between ionized carboxyl group and incipient ionized carboxyl group increases. Presumably the negative charge of the monoanion alters the electrostatic field about the remaining carboxyl group and increases the work required to remove the unit positive charge of the proton. The same relationship between K_1 and K_2 should hold for a diol. Calculation of an approximate value for K_1/K_2 should be possible with Bjerrum's equation (*9*)

$$\ln \frac{K_1}{K_2} = \frac{Ne^2}{RTDr}$$

where N is Avogadro's number, R is the gas constant, T is the absolute temperature, e is the charge on the proton, D is the effective dielectric constant of the region between ionized group and unionized group, and r is the distance separating the two groups. On the one hand for most practical purposes, second ionizations of simple carbohydrates may be ignored because of the low magnitude of the second ionization constant. On the other hand, in polysaccharides the great distances between hydroxyl groups in the same chain facilitate multiple ionization.

In addition to the effect of polar groups (discussed above) and the electrostatic effect of neighboring anionic charge, there are other factors that affect carbohydrate acidity. Among them are steric and entropy effects and intramolecular hydrogen bonding.

Steric effects are probably important because molecular geometry and bulky substituents can influence intramolecular hydrogen bonding. Intramolecular hydrogen bonding between hydroxyl groups is possibly nonexistent in dilute aqueous solutions where carbohydrate hydroxyl groups perhaps hydrogen bond exclusively with the solvent. However, in nonaqueous solvents, particularly aprotic solvents, intramolecular hydrogen bonding between neighboring hydroxyl groups is a well-established phenomenon (*11, 26*) and could have an important influence on hydroxyl acidity. Of perhaps greater importance than intramolecular hydrogen bonding between hydroxyl groups is intramolecular hydrogen bonding between hydroxyl group and anionic oxygen. This type of bonding should occur easily in aqueous and nonaqueous media. The orientation of an anionic oxygen and a neighboring hydroxyl group with respect to each other determines whether hydrogen bonding can occur. As can be seen in Figure 2, only the relationships, *e,e* and *e,a,* allow bonding between vicinal groups on a pyranose ring. However, a 1,3-cis-diaxial relationship would also allow bonding. Such bonding would stabilize the incipient anionic charge and promote ionization of the nonbonded hydroxyl proton.

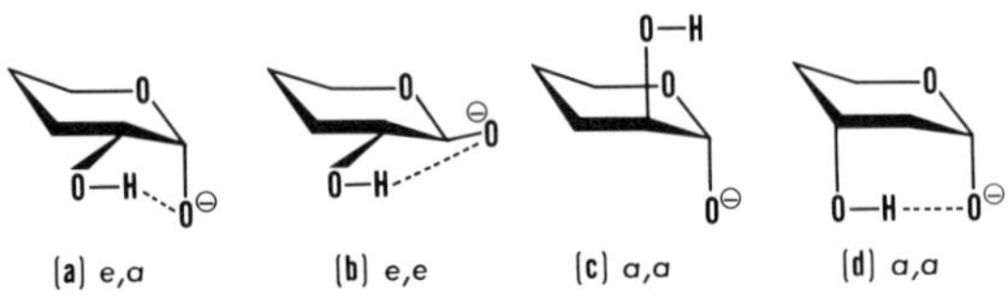

Figure 2. Pyranose structures showing axial–equatorial orientations that favor intramolecular hydrogen bonding between anionic oxygen and hydroxyl group

Brown, McDaniel, and Häfliger (9) pointed out that if there are two acidic groups close to each other in a molecule, the mutual electrostatic influence can substantially increase the p*K* difference between the two groups. This p*K* difference would be especially large if a hydrogen bond could form between the two acidic groups, as shown in Figure 3.

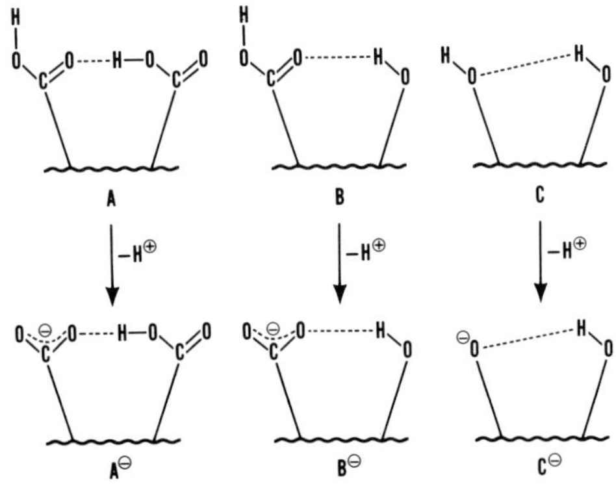

Figure 3. Three types of dibasic acids and their corresponding anionic forms

The energy required to remove the first proton is partly compensated for by the energy gained in the transition of compounds A, B, and C into the corresponding oxyanions A^-, B^-, and C^-. The acidity of the second group (*i.e.*, unionized hydroxyl) is lowered if its proton is held more

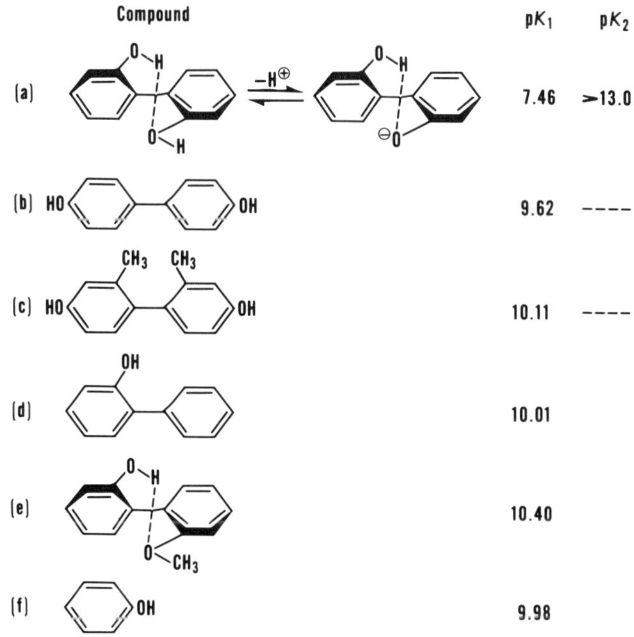

Figure 4. Structures of phenol and various mono- and dihydroxybiphenyls showing steric relationships that permit intramolecular hydrogen bonding. Values of pK are for aqueous solutions at 20°C (31).

firmly in the hydrogen bonding with an anionic oxygen than it is without such bonding.

Study of the literature has revealed only one important publication concerning the effect that intramolecular hydrogen bonding between hydroxyl groups and anionic oxygen has on the acidity of hydroxyl groups. Working with numerous biaryl compounds, Musso and Matthies (*31*) showed by spectrophotometric analysis and potentiometric titration that interactions like that shown in C^- (Figure 3) do occur. Figure 4 shows the results of some of their studies in aqueous solution. Note the great effect that intramolecular hydrogen bonding has on Compound (a); Compound (e) is less acidic than Compound (d). This difference could be caused by intramolecular hydrogen bonding of the lone hydroxyl group in Compound (e) to the oxygen of the methoxyl group. Baker and Shulgin (*2*) have reported that methoxyl groups are better hydrogen-bond formers than hydroxyl groups.

At this stage of our knowledge, the occurrence of hydrogen bonding between hydroxyl groups and anionic oxygen in carbohydrate alcoholates has not been shown conclusively. That it probably does occur is strongly indicated by various published data. Examples of carbohydrates whose behavior in alkaline media can perhaps be explained best by such intramolecular bridging are shown in Figures 5, 6, and 7. When β-D-levoglucosan (Figure 5) is methylated with dimethyl sulfate in 19% sodium hydroxide solution, the hydroxyls at C-2 and C-4 are reactive as compared with the one at C-3 (*34*). Hydroxyls at C-2 and C-4 are capable of intramolecular hydrogen bonding whereas the hydroxyl at C-3 is not. In

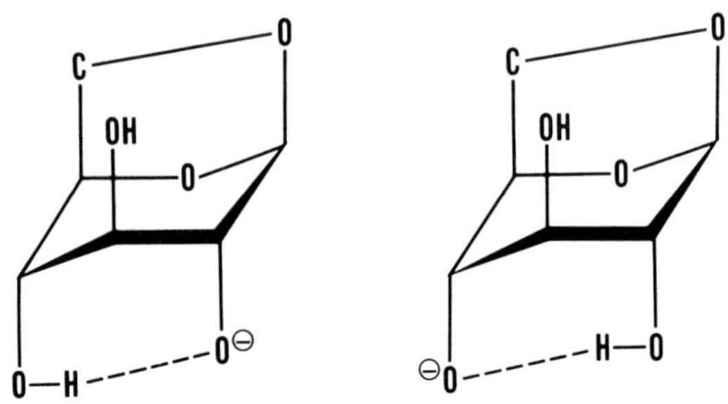

Figure 5. Hypothetical anionic forms of β-D-levoglucosan capable of intramolecular hydrogen bonding (the hydrogens on the carbon atoms have been omitted for clarity)

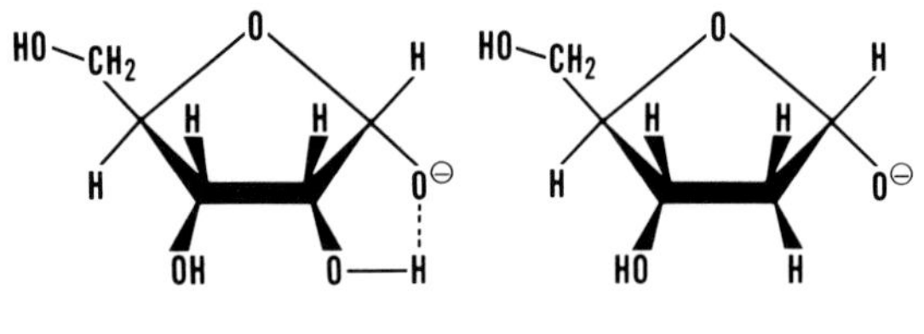

Figure 6. (a) α-D-Ribofuranose oxyanion; (b) 2-deoxy-α-D-ribofuranose oxyanion

Figure 6 are shown D-ribose and 2-deoxy-D-ribose. In D-ribose an anionic oxygen at C-1 is capable of bonding with the adjacent hydroxyl on C-2; whereas in the deoxy compound, such bonding cannot exist. D-Ribose would, therefore, be expected to have the higher acidity, and it does. Adenosine (Figure 7) is acidic with a reported p*K* value of 12.4 (±0.1) (*25*). This acidity is identical with that of D-glucose. Replacing either the 2′- or the 3′-hydroxyl group with either a hydrogen atom or a methoxyl group results in a great loss of acidity (*25*). Because both hydroxyls are essential for the high acidity of adenosine, intramolecular hydrogen bonding between hydroxyl group and anionic oxygen could be involved here also.

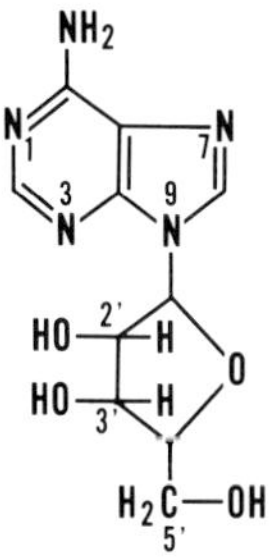

Figure 7. Adenosine

To a large extent, hydrogen bonding could explain relative reactivities and, thus, relative acidities of the different hydroxyl groups within a polyhydroxy compound. That such bonding does in some way influence acidity has been suggested by Roberts, Wade, and Rowland (*39, 40*) and Croon (*14*). It is possible to apply the concept of hydrogen bonding between a hydroxyl group and anionic oxygen to explain the acidities of the hydroxyl groups in methyl D-glucopyranoside. In aqueous media the acidity of the hydroxyls decreases in the order 2-OH >> 6-OH > 3-OH > 4-OH (*38, 40*). Activated by the ring oxygen, the 2-hydroxyl group ionizes most easily, and as shown in Figure 8, the resulting anionic oxygen causes the 3- and 4-hydroxyl groups to become oriented preferentially in its direction. (If there were no alkali present to enhance the negative character of O-2, there would probably be little if any hydrogen bonding between the 3- and 4-hydroxyl, and 3- and 4-hydroxyl would more than likely be bonded preferentially to solvent molecules of the aqueous medium.)

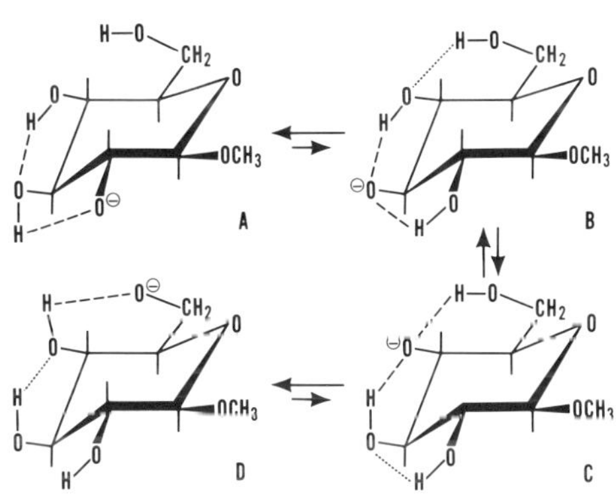

Figure 8. Hypothetical equilibria between oxyanionic forms of methyl β-D-glucopyranoside. The dotted lines represent hydrogen bonds; weaker bonds are indicated by the lighter dots.

Intramolecular hydrogen bonding increases the negative charge on the oxygen atoms of the 3- and 4-hydroxyl groups and lowers their acidities below that which they would normally possess in the absence of an anionic oxygen at C-2. All four oxyanions would exist in equilibrium with one another with anion A (Figure 8) predominating. Reactivity of the 6-hydroxyl group cannot be satisfactorily explained now. Perhaps the preference shown by an attacking reagent for a less sterically hindered hydroxyl (*28*) is sufficient reason for the 6-hydroxyl to be more reactive than the 3- or 4-hydroxyl group. In other words steric hindrance, as well as relative acidity, could be important in determining relative reactivity (*41*). It is possible, however, that there are polar effects that increase the positive character of the hydrogen of the 6-hydroxyl group. Molecular orbital calculations on the aldohexose molecule (*42*) indicate a slightly higher positive charge on the hydrogen of the 6-hydroxyl than on that of either the 3- or 4-hydroxyl group. Substitution of a methoxyl for a hydroxyl group at C-2 should increase the acidities of the 3- or 4-hydroxyl groups because in the absence of intramolecular hydrogen bonding with an anionic oxygen at C-2 there would be a lower negative charge on the oxygens of the 3- and 4-hydroxyls. A low negative charge favors O-H ionization. It should be remembered, however, that superimposed upon this hydrogen-bonding effect is a possible polar effect caused by the proximity of an alkoxyl dipole. Various studies have strongly indicated that the reactivity of one secondary hydroxyl group can be increased by the substitution of an adjacent hydroxyl group (*14, 28*).

The 6-hydroxyl group of methyl 3-*O*-methyl-β-D-glucopyranoside has recently been shown (*39*) to be about one-fourth as reactive as the 6-hydroxyl of methyl 3-*O*-methyl-β-D-glucopyranoside toward substitution in dilute alkaline solution. It was proposed that the higher reactivity of the 3-*O*-methyl derivative is possibly caused by a greater tendency for hydrogen bonding between the C-6 anionic oxygen and the C-4 hydroxyl.

Differences in entropy of ionization are often largely responsible for differences in acidity between weak acids of similar strength (*1*). The anomers of D-glucose give an example of such a situation. From a study in which ionization constants and thermodynamic parameters of ionization were determined, Los and Simpson (*29*) have shown that at 25° α-D-glucose is one-half as acidic as β-D-glucose. Their data indicated that β-D-glucose is the stronger acid because the magnitude of the entropy decrease on ionization is much less for the β-anomer. Furthermore, temperature coefficients indicate that the two anomers are equal at −38°C. Such a phenomenon is not easily explained by polar effects that alter the strength of the O—H bond. However, an interpretation based on entropies is likely since there are examples in the literature (*18*) of pairs of carboxylic acids whose order of strength can actually be reversed by a temperature change. These reversals have been attributed to entropy effects.

At present a complete physical interpretation of entropy change and of the variation in entropy change from one weak acid to another is not possible. The ionization of acids, weak or strong, is accompanied by a decrease in entropy; in other words, the system becomes more highly ordered. To a large extent, differences in entropy of ionization between various weak acids of similar structure probably reflect differences in degree of solvation of the anion; even effects that electron-withdrawing substituents have on ionization can be interpreted in terms of entropy differences arising from differences in solvation of anions (*1*). Internal chelation, or intramolecular hydrogen bonding, in the anion should certainly determine somewhat the magnitude of entropy change. The larger the extent of intramolecular hydrogen bonding, the smaller the decrease in entropy during ionization of the acid.

Free energies, enthalpies, and entropies of ionization for a number of reducing sugars have been determined by Christensen and co-workers (*13*). Their data, however, are for equilibrium mixtures of anomers rather than for individual anomeric forms.

In nonaqueous media intramolecular hydrogen bonding should be more pronounced than in aqueous media, and the formation of cation-containing chelates of the general types shown below becomes a distinct possibility (*38*).

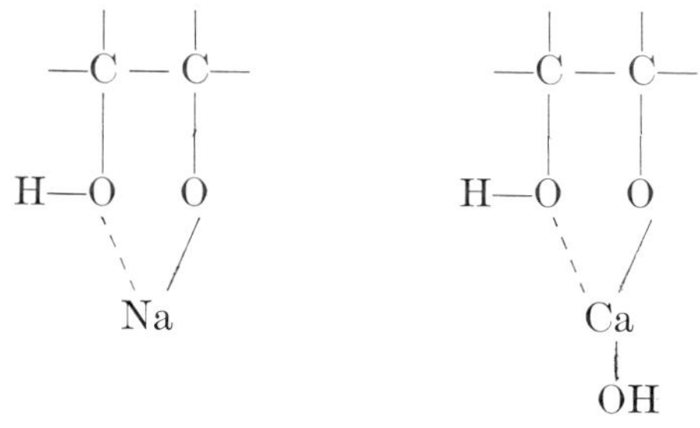

Alcoholates of this type are often capable of associating with an additional molecule of carbohydrate, and reaction stoichiometry is significantly influenced by size of metal cation and concentrations of reactants.

Alcoholates formed from polyvalent cations are more stable than those from monovalent cations. Charley and co-workers (*12*) have presented evidence that sugars and polyols interact with Fe^{3+} in mildly alkaline, aqueous solution to form somewhat stable, soluble chelates. Under similar conditions, D-fructose was shown to react with various polyvalent cations: Fe^{3+}, Fe^{2+}, Mn^{2+}, Co^{2+}, Ni^{2+}, Zn^{2+}, Ca^{2+}, Sr^{2+}, and Ba^{2+}.

By pH titrations when Fe^{2+} interacts with excess D-fructose, two protons are released by the sugar; similarly, Fe^{3+} effects the release of three protons. A sample of ferric alcoholate of D-fructose was shown by elemental analysis to consist of D-fructose, iron, and sodium in the ratio 2:2:1. Alcoholates of Tl^{3+}, Cu^{2+}, Cr^{3+}, Al^{3+}, and Pb^{2+} have also been reported (*49*). Sugars seem to be able to react directly with PbO in aqueous solution; however, nothing is known about the structure of the product or the mechanism of the reaction.

In partially aqueous media the isolated products of reaction between alkali-metal hydroxide and carbohydrate seem to be metal hydroxide adducts rather than alcoholates (*38*). With simple sugars, the ratio of metal hydroxide to carbohydrate in these adducts is greater than 1.0 at high alkali concentration. Rowland (*40*) has found that the relative reactivities of hydroxyl groups in methyl α- or β-D-glucopyranoside vary with sodium hydroxide concentration, and he has invoked the concept of sodium hydroxide adduct formation to explain this phenomenon. At low alkali concentration ($\leqslant 0.1M$), the relative reactivities are those that would normally be expected (2-OH > 6-OH > 3-OH > 4-OH). However, at high alkali concentration ($\geqslant 1M$), reactivities of the 2-, 3-, and 4-hydroxyl groups are greatly lowered; no significant change occurs in that of the 6-hydroxyl. A tendency for sodium hydroxide adduct formation in the region of C-2, C-3, and C-4 could explain the decreased reactivity of the hydroxyl groups at these positions.

Ionization of the Carbon–Hydrogen Bond

Ionization of a carbohydrate hydroxyl group to form an alcoholate ion is generally favored over ionization of a C–H bond to form a carbanion (*see* Table II). Without a strongly activating substituent (such as a free carbonyl group), carbanion formation would be negligibly small. The introduction of a hydroxyl substituent at a carbon atom containing an ionizable hydrogen atom decreases the tendency for C–H ionization (*8*). In Table III the effect of hydroxyl substituents is clearly evident. Acetate and glycolate anions are included to illustrate the effect of a negative charge close to the reaction site. To explain the reducing effect of the hydroxyl substituent, Bonhoeffer and co-workers (*8*) suggested that either steric hindrance or competing ionization of a hydroxyl substituent might be responsible. Certainly these two possibilities cannot be ignored, especially the latter. However, additional explanations can be offered, and perhaps these, in addition to steric hindrance and oxyanion formation, contribute in varying degrees to the observed influence of hydroxyl substituents on carbanion formation. For example, glycolate ion may be less ionizable than acetate ion because of a possible ability to form an intramolecular hydrogen bond as shown below.

Table II. Comparison of O—H Ionization with C—H Ionization[a]

Compound	*Type of Ionization*	K	*p*K	°*C*	*Reference*
Diphenylmethane	C-H	10^{-35}	35	~25	*30*
Acetonitrile	C-H	10^{-25b}	25[b]	25	*8*
Acetone	C-H	1×10^{-20}	20.0	25	*5*
Acetaldehyde	C-H	2×10^{-20}	19.7	25	*5*
Methanol	O-H	2.9×10^{-16}	15.54	25	*4*
Glycerol	O-H	4.0×10^{-15}	14.40	25	*4*
Acetaldehyde hydrate	O-H	3.3×10^{-14}	13.48	25	*7*
D-Glucitol	O-H	2.7×10^{-14}	13.57	18	*48*
D-Glucopyranose	O-H	4.5×10^{-13}	12.35	25	*15*
Malononitrile	C-H	6.5×10^{-12}	11.19	25	*35*
2-Acetylcylohexanone	C-H	8.1×10^{-11}	10.09	25	*44*
2,4-Pentanedione	C-H	1.0×10^{-9}	9.0	25	*6, 17*
2-Acetylcyclopentanone	C-H	1.5×10^{-8}	7.82	25	*44*
1,3-Butanedione	C-H	1.2×10^{-6}	5.92	25	*35*
Malonaldehyde	C-H	$\sim 10^{-5}$	~5	25	*44*

[a] Water is solvent for all compounds except diphenylmethane, in which case ethyl ether is used. Ionization constants for carbonyl compounds and nitriles are gross acid constants, uncorrected for tautomerism.
[b] Estimated.

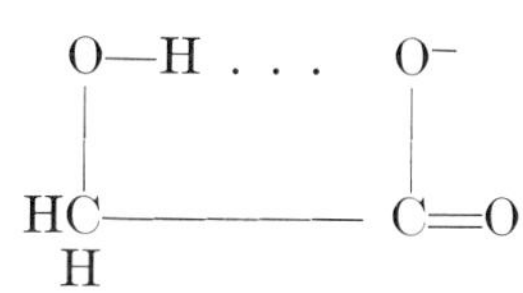

The result of this bonding would be a greater negative charge on the hydroxyl oxygen, a condition that should reduce the ionizability of the C—H bond. The low ionizabilities of glyceraldehyde and dihydroxyacetone, compared with those of acetaldehyde and acetone, respectively, could reflect a tendency for the hydroxy compounds to exist as cyclic dimers. Only the monomeric form would be expected to give a carbanion.

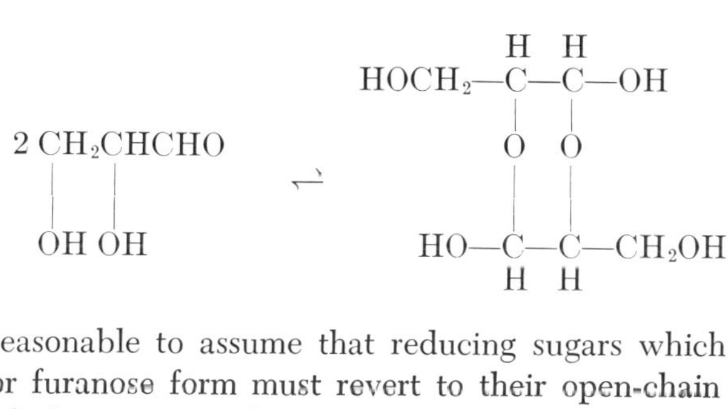

It is reasonable to assume that reducing sugars which exist in the pyranose or furanose form must revert to their open-chain aldehydo or keto forms before any significant C—H ionization occurs. This reversion is brought about by ionization of the acidic hemiacetal hydroxyl group, followed by ring opening. Subsequent carbanion formation at the carbon

Table III. Rate Constants[a] of C—H Ionization in Aqueous Alkaline Solution (*8*)

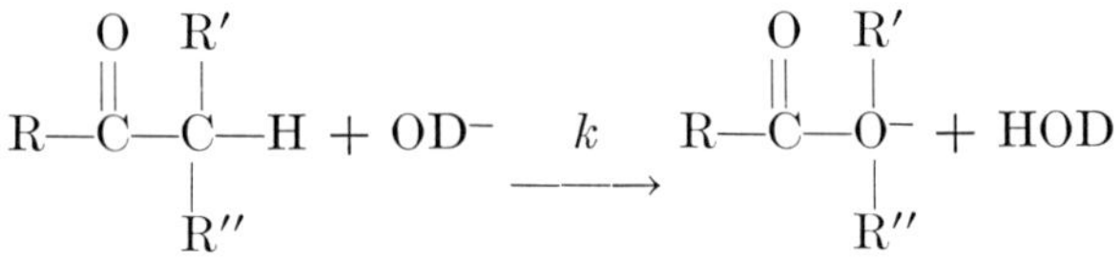

Compound	*Structure*	k (*min*$^{-1}$) *25°C*
Acetone	$CH_3\overset{\underset{\|}{O}}{C}CH_3$	14.5
Acetaldehyde	CH_3CHO	11
Dihydroxyacetone	$HOCH_2\overset{\underset{\|}{O}}{C}CH_2OH$	~5
Glyceraldehyde	$\underset{\overset{\|}{OH}}{CH_2}\underset{\overset{\|}{OH}}{CH}CHO$	1–2
Acetate ion	$CH_3\overset{\underset{\|}{O}}{C}—O^-$	10^{-6}
Glycolate ion	$HOCH_2\overset{\underset{\|}{O}}{C}—O^-$	10^{-7}

[a] These are velocity constants for deuterium exchange between aliphatic compounds and solvent D_2O at 1*N* OD^- concentration.

atom adjacent to the carbonyl group then leads to enediol formation, isomerizations, β-eliminations, and rearrangements—all of which have been given intensive, but not complete, study and have been reviewed several times (*23, 27, 46, 47, 52*). I shall treat only the earliest stages of sugar transformation in alkaline media because all others generally lie far beyond the scope of this paper.

Isbell, Frush, Wade, and Hunter (*24*) have proposed in a mechanistic scheme for alkali-induced anomerization that ring opening is a fast, momentary event which does not produce a completely free acyclic intermediate but instead a pseudoacyclic intermediate which has a conformation similar to that of the parent sugar. Isbell and associates suggested that the reaction of the pseudoacyclic intermediate with base to form a carbanion would lead to a *cis*-enediol or to a *trans*-enediol, depending upon the orientation of the aldehyde group with respect to the neighboring hydroxyl group. In Figure 9, D-glucose and D-mannose illustrate the basic features of this transformation. α-D-Glucose and β-D-mannose would

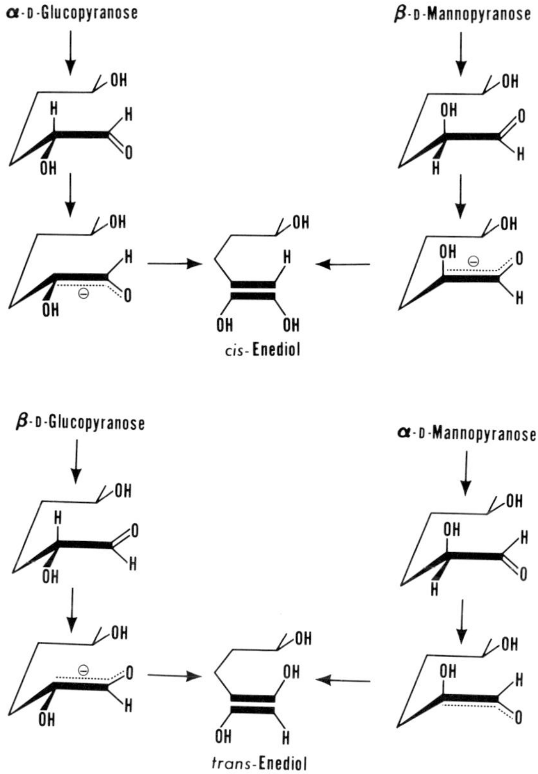

Figure 9. Formation of postulated (24) cis- *and* trans-*enediols from pseudoacyclic intermediates of two epimeric aldohexopyranoses (the hydroxyl groups have been omitted from C-3, -4, and -6 for clarity)*

give a *cis*-enediol, and β-D-glucose and α-D-mannose would give a *trans*-enediol. These two isomeric enediols, being stereochemically different, could conceivably follow different paths in subsequent transformations. Molecular orbital calculations (53) on an enediol similar to these indicate a high, positive charge on the hydrogen of the terminal hydroxyl group of the carbon–carbon double bond. If this condition does actually exist then in alkaline solution, ionization of the terminal hydroxyl group on the carbon–carbon double bond would be expected to occur preferentially to ionization of the other hydroxyl group. This occurrence would explain why in an isomerization of D-glucose to D-fructose equilibrium between sugar forms is in favor of aldose. D-Fructose would be formed only by ionization of the nonterminal hydroxyl group on the carbon–carbon double bond.

Table IV. Relative Rates of C—H Ionization (Enolization) in Water at pH 11.0–11.2 (*24*)

Sugar	*Range of Relative Reactivity*[a]
Aldo-hexoses	0.5– 4.8
Aldo-heptoses	0.5– 10.7
Aldo-pentoses	4.1– 28
2-Ketoses	2.6– 25
6-Deoxy-*aldo*-hexoses	0.3– 0.8
2-Deoxyaldoses	90 –155
Disaccharides	2.0–240

[a] Rates are relative to that of D-glucose (1.0) and are based upon rates of tritium uptake from the solvent water-*t* at 25°C.

A reasonably successful attempt to determine relative ionizabilities of the C—H bond in various reducing sugars was also made by Isbell and co-workers (*24*). By measuring rates of tritium uptake in alkaline, tritiated water, they found that the aldohexoses were among the lowest in reactivity. Table IV shows the reactivities (or rates of tritium uptake) for different types of sugars, based upon D-glucose as unity. Sugars of like configuration in the *aldo*-hexose and *aldo*-heptose series react at somewhat similar rates. The *aldo*-pentoses, as a group, react more rapidly than the *aldo*-hexoses and *aldo*-heptoses. 6-Deoxy-*aldo*-hexoses seem to have about the same reactivity as *aldo*-hexoses. The generally higher reactivity of ketoses, compared with that of aldoses, can be explained by the greater number of ionizable C—H bonds in the ketose. The 2-deoxy sugars react at high rates. Possibly these much higher rates for the 2-deoxy sugars reflect a higher proportion of open-chain aldehydo intermediate in the

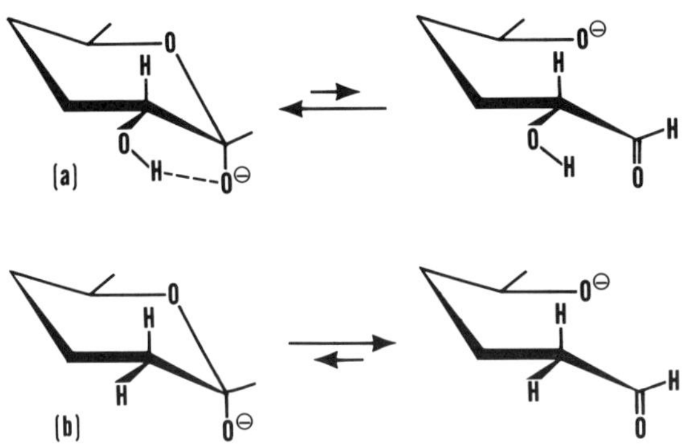

Figure 10. Hypothetical equilibria between the oxyanionic pyranose form and the oxyanionic aldehydo form of an aldo-*hexose and of a 2-deoxy-*aldo-*hexose (the hydroxyl groups have been omitted from C-3, -4, and -6 for clarity)*

reaction medium. Furthermore, differences in reactivity between members of a particular sugar group (such as aldo- or ketohexoses) might likewise be caused, at least partly, by variations in the proportion of aldehydo or keto form. It is the carbonyl form that is the precursor of the carbanion. Any factor, steric or otherwise, that would influence the concentration of carbonyl intermediates would therefore affect the observed degree of C—H ionization. The concentration of aldehydo intermediate of a 2-deoxyaldopyranose might be high compared with that of the corresponding aldopyranose because of a possibly lower stability of its precursor, the pyranose oxyanion. In Figure 10, the absence of intramolecular hydrogen bonding in the 2-deoxy sugar oxyanion (b) should favor a greater proportion of aldehydo intermediate compared with the proportion formed from aldopyranose oxyanion (a). In the aldopyranose oxyanion, intramolecular hydrogen bonding would stabilize the oxyanion structure so much that chain opening is difficult.

There are no published p*K* values (C—H ionization) for the compounds listed in Tables III and IV, except for acetone (p*K* 20) and acetaldehyde (p*K* 19.7). Consequently, in the preceding discussion relative rates of C—H ionization for these compounds were assumed to be measures of relative acidity. This assumption was not strictly valid since acidity of weak acids is accurately defined only in terms of ionization equilibria. However, in the absence of equilibrium data, one is reasonably well justified in using rate constants to obtain estimates of relative acidity. Pearson and Dillon (35) examined rate and equilibrium data for various weak carbon acids in aqueous solution and found a definite, although rough, correlation between rate constant and equilibrium (ionization) constant *K* for nearly all of the compounds. In particular, for those compounds having only one strongly activating substituent (*e.g.*, carbonyl group), there is an almost linear relationship between p*K* and the logarithm of the rate constant.

There is much to be learned about carbohydrate ionization. Only now are chemists breaking the surface to gain this understanding. Simple explanations for chemical behavior will not always suffice because with carbohydrates the systems are much too complex. Some of the concepts and interpretations of data presented here may be incorrect. A more accurate and detailed description of carbohydrate ionization in alkaline solution, however, will be possible only after much additional research.

Literature Cited

1. Allen, C. R., Wright, P. G., *J. Chem. Ed.* (1964) **41**, 251.
2. Baker, A. W., Shulgin, A. T., *J. Amer. Chem. Soc.* (1959) **81**, 4524.
3. Baker, F. W., Parish, R. C., Stock, L. M., *J. Amer. Chem. Soc.* (1967) **89**, 5677.

4. Ballinger, P., Long, F. A., *J. Amer. Chem. Soc.* (1960) **82,** 795.
5. Bell, R. P., *Trans. Faraday Soc.* (1943) **39,** 253.
6. Bell, R. P., Gelles, E., Moller, E., *Proc. Roy. Soc. (London)* (1949) **A198,** 300.
7. Bell, R. P., McTigue, P. T., *J. Chem. Soc. (London),* **1960,** 2983.
8. Bonhoeffer, K. F., Geib, K. H., Reitz, O., *J. Chem. Phys.* (1939) **7,** 664.
9. Brown, H. C., McDaniel, D. H., Häfliger, O., "Determination of Organic Structures by Physical Methods," pp. 567–662, E. A. Braude and F. C. Nachod, Eds., Academic, New York, 1955.
10. Capon, B., Overend, W. G., *Advan. Carbohyd. Chem.* (1960) **15,** 32.
11. Casu, B., Reggiani, M., Gallo, G. G., Vigevani, A., *Tetrahedron* (1966) **22,** 3061.
12. Charley, P. J., Sarkar, B., Stitt, C. F., Saltman, P., *Biochim. Biophys. Acta* (1963) **69,** 313.
13. Christensen, J. J., Rytting, J. H., Izatt, R. M., *J. Chem. Soc. (B),* **1970,** 1646.
14. Croon, I., Lindberg, B., *Svensk Papperstid.* (1956) **59,** 794; (1957) **60,** 843; Croon, I., *ibid.* (1958) **61,** 919; Croon, I., Flamm, E., *ibid.* (1958) **61,** 963; Croon, I., *Acta Chem. Scand.* (1959) **13,** 1235.
15. Degani, Ch., *Carbohyd. Res.* (1971) **18,** 329.
16. Derevitskaya, V. A., Smirnova, G. S., Rogovin, Z. A., *Proc. Acad. Sci. USSR, Chem. Sect. (English Transl.)* (1961) **136,** 1254.
17. Eidinoff, M. L., *J. Amer. Chem. Soc.* (1945) **67,** 2072.
18. Everett, D. H., Landsman, D. A., Pinsent, B. R. W., *Proc. Roy. Soc. (London)* (1952) **A215,** 403.
19. Frahn, J. L., Mills, J. A., *Aust. J. Chem.* (1959) **12,** 65.
20. Golden, R., Stock, L. M., *J. Amer. Chem. Soc.* (1966) **88,** 5928.
21. Gould, E. S., "Mechanism and Structure in Organic Chemistry," p. 201, Holt, Rinehart, and Winston, New York, 1959.
22. Hammett, L. P., "Physical Organic Chemistry," p. 256, McGraw-Hill, New York, 1940.
23. Isbell, H. S., *J. Res. Nat. Bur. Stand.* (1944) **32,** 45.
24. Isbell, H. S., Frush, H. L., Wade, C. W. R., Hunter, C. E., *Carbohyd. Res.* (1969) **9,** 163.
25. Izatt, R. M., Rytting, J. H., Hansen, L. D., Christensen, J. J., *J. Amer. Chem. Soc.* (1966) **88,** 2641.
26. Kuhn, L. P., *J. Amer. Chem. Soc.* (1952) **74,** 2492.
27. Lemieux, R. U., "Molecular Rearrangements, Part 2," pp. 743–753, P. DeMayo, Ed., Interscience, New York, 1964.
28. Lenz, R. W., *J. Amer. Chem. Soc.* (1960) **82,** 182.
29. Los, J. M., Simpson, L. B., *Rec. Trav. Chim.* (1957) **76,** 267.
30. McEwen, W. K., *J. Amer. Chem. Soc.* (1936) **58,** 1124; Conant, J. B., Wheland, G. W., *J. Amer. Chem. Soc.* (1932) **54,** 1212.
31. Musso, H., Matthies, H. G., *Ber.* (1961) **94,** 356.
32. Neale, S. M., *J. Text. Inst.* (1929) **20,** 373T.
33. Neuberger, A., Wilson, B. M., *Carbohyd. Res.* (1971) **17,** 89.
34. Norrman, B., *Acta Chem. Scand.* (1968) **22,** 1623.
35. Pearson, R. G., Dillon, R. L., *J. Amer. Chem. Soc.* (1953) **75,** 2439.
36. Peligot, E., *Ann.* (1839) **30,** 69.
37. Peligot, E., *Compt. Rend.* (1838) **7,** 106.
38. Rendleman, J. A., *Advan. Carbohyd. Chem.* (1966) **21,** 209.
39. Roberts, E. J., Wade, C. P., Rowland, S. P., *Carbohyd. Res.* (1972) **21,** 357.
40. Rowland, S. P., *Carbohyd. Res.* (1971) **17,** 393.
41. Rowland, S. P., Roberts, E. J., Bose, J. L., *J. Polym. Sci. (A-1)* (1971) **9,** 1431.
42. Salinas, A., Sproviero, J. F., *Carbohyd. Res.* (1971) **16,** 243.
43. Saric, S. P., Schofield, R. K., *Proc. Roy. Soc. (London)* (1946) **A185,** 431.

44. Schwarzenbach, G., Felder, E., *Helv. Chim. Acta* (1944) **27**, 1701.
45. Smolenski, K., Kozlowski, W., *Roczniki Chem.* (1936) **16**, 270; Smolenski, K., Porejko, S., *ibid.* (1936) **16**, 281.
46. Sowden, J. C., *Advan. Carbohyd. Chem.* (1957) **12**, 35.
47. Speck, Jr., J. C., *Advan. Carbohyd. Chem.* (1958) **12**, 63.
48. Thamsen, J., *Acta Chem. Scand.* (1952) **6**, 270.
49. Vogel, H., Georg, A., "Tabellen der Zucker und Ihrer Derivate," pp. 378–397, Julius Springer, Berlin, 1931.
50. Von Lippmann, E. O., "Die Chemie der Zuckerarten," p. 1350, Friedrich Vieweg und Sohn, Braunschweig, 1904.
51. Wells, P. R., Adcock, W., *Aust. J. Chem.* (1965) **18**, 1365.
52. Whistler, R. L., BeMiller, J. N., *Advan. Carbohyd. Chem.* (1958) **13**, 289.
53. Zhdanov, Yu. A., Minkin, V. I., Ostroumov, Yu. A., Dorofeenko, G. N., *Carbohyd. Res.* (1968) **7**, 156.

RECEIVED December 2, 1971.

5

Enolization and Oxidation Reactions of Reducing Sugars

HORACE S. ISBELL

American University, Washington, D. C. 20016

Recent progress is reported in investigations of enolization and ketonization of reducing sugars, oxidation of sugars with oxygen, and oxidation of sugars with sodium peroxide. The enolate mechanism for isomerization of aldoses and ketoses was confirmed by production of tritium-free fructose from D*-glucose-2-*t*. The proportions of* D*-glucose,* D*-fructose, and* D*-mannose obtained by alkaline rearrangements of these sugars suggest that rearrangements of the three sugars occur through slightly different enolic intermediates. Mechanisms are suggested for degradation of the sugar to* D*-arabinonic and formic acids with and without hydride transfer. Alkaline peroxides are shown to convert aldoses almost quantitatively into formic acid. Ketoses yield formic and glycolic acids, and 6-deoxy-aldoses yield formic and acetic acids. Mechanisms are presented for oxidative degradation of* α*-hydroxy carbonyl compounds by alkaline hydroperoxide.*

For many years the complex and manifold reactions of reducing sugars in alkaline solutions have fascinated chemists and engaged their attention. Notwithstanding painstaking and extensive investigations (*1, 2, 3, 4, 5*), knowledge of the enolization and oxidation reactions of reducing sugars is still fragmentary. Presumably, the reactions begin with an acyclic or carbonyl modification formed by the mutarotation reaction (*6, 7*). Spectrometric and polarographic measurements show that the equilibrium proportion of this modification is low (*8, 9, 10, 11*). The acyclic modification, which exists in ionic equilibrium with acid and base catalysts, is also hydrated. Horton and co-workers (*12*) found that acyclic sugar *derivatives* in aqueous solution exist almost entirely as hydrates. Hydration is reversible, and the acyclic forms of the sugar are available in low concentration for a variety of reactions.

Transformation Reactions of Sugars in Alkaline Solution

Enolization Reactions. In 1900 Wohl and Neuberg (*2*) advanced the concept that the interconversion of D-glucose, D-mannose, and D-fructose occurs through reversible enolization. Subsequently, this concept has been largely confirmed, but there is still doubt as to whether the enolate mechanism is the only pathway for the interconversion (*13, 14, 15*). According to the conventional reaction scheme (*3, 5, 16*), D-glucose (**1a**), D-mannose (**2a**), and D-fructose (**3a**) should give the same enediol (**4a**) and would be expected to yield the same proportions of the three sugars in the rearrangement products (Scheme I). Isbell and co-workers (*17, 18*) tried to test this hypothesis by using sugars position-labeled with carbon-14 and tritium. For each of the three sugars, they found wide differences in the proportions of the resultant sugars, previously assumed to be formed from a common enolic intermediate. Recently MacLaurin and Green (*19*) also investigated the interconversion reactions of the three sugars. They postulated a triangular reaction system, consisting of direct, reversible interconversions, and obtained satisfactory rate constants. However, they noted that the reactions are so complex that it is difficult to draw definite conclusions from kinetic data.

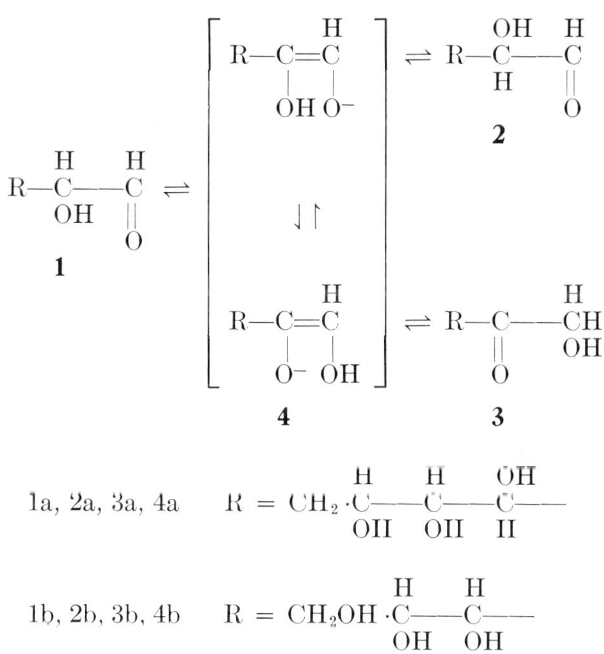

Scheme I. Conventional enolate interconversion reaction

To account for the differences in the proportions of the sugars in the rearrangement products, Isbell and co-workers (*18*) suggested that the nascent enediol derived from each sugar may retain during the reaction period stereomeric features characteristic of the original sugar. Although this type of isomerization accounts for the variation in the proportions of the rearrangement products, other reaction mechanisms need consideration.

Hydride-Transfer Reactions. The hydride-transfer mechanism for rearrangement of sugars (*20*) yields a ketose anion (**5**) by direct transfer of the C-2 hydrogen atom with its electrons to C-1. Reversal of the process leads to epimerization at C-2 (Scheme II).

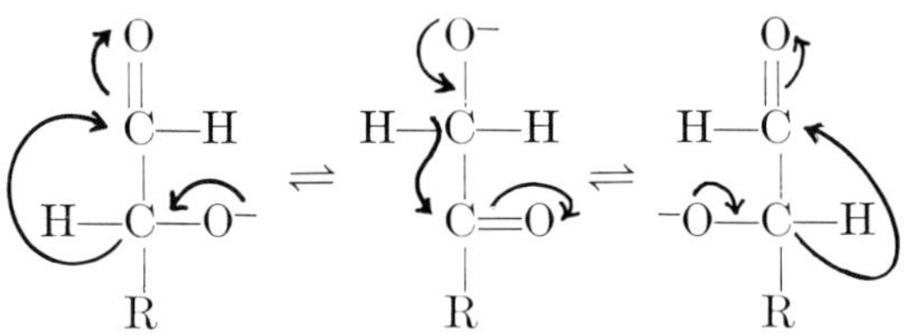

5

Scheme II. Isomerization by hydride transfer
Fieser and Fieser Mechanism (20)

Recently, Gleason and Barker studied the behavior of D-ribose in aqueous potassium hydroxide under aerobic and anaerobic conditions (*15*). They found that the D-arabinose formed from D-ribose-2-*t* contains a substantial amount of radioactivity at C-1 and concluded that rearrangement of D-ribose occurs principally by the hydride-transfer mechanism.

Rearrangement of sugars by hydride transfer rather than by enolization is contrary to much published work on the interconversions of D-glucose, D-mannose, and D-fructose in alkaline D_2O solutions (*21, 22, 23, 24*). Because of the conflicting results in the literature, a critical test was made in our laboratory for the presence or absence of the hydride mechanism in the rearrangement of D-glucose. A sample of D-glucose-2-*t* was rearranged by treatment with aqueous calcium hydroxide at 35°C for 18 hours. Aldoses were removed by oxidation with bromine; the D-fructose was separated and purified by crystallization. It was shown to be pure by paper chromatography and by its optical rotation ($[\alpha]_D$ − 93.3°). The original D-glucose-2-*t* gave 3,400 counts per minute per mg, but the resulting D-fructose gave only 0.3 count per minute per mg. Thus, hydride transfer is not significant in the rearrangement of D-glucose to D-fructose under the conditions used, and no general conclusion can be drawn at this time concerning the rearrangement pathways for all sugars.

Rates of Enolization Reactions. For a better understanding of the transformation and oxidation reactions of reducing sugars, methods have been developed to measure the primary rates of enolization (*18*). One of these methods depends on the rate at which tritium ions are released from aldoses-2-*t* to the solvent. This is measured by separation of the water-*t*, sublimation, and radiochemical assay of the water as the reaction proceeds. The rate constant is calculated from the first-order equation:

$$k_T = 1/t\,[\ln \overset{*}{A}/(\overset{*}{A} - \overset{*}{X})]$$

where A^* is the tritium in the sample at the beginning of the measurement, and X^* is the amount of tritium released to the solvent in time t. The method requires tritium-labeled sugars which are not readily available and involves a primary isotope effect.

Isbell and co-workers have now developed a more convenient method for measuring the primary rate of enolization that does not involve a primary isotope effect (*25*). The method (Scheme III) uses infrared absorption at 2.95 microns for measuring the amount of DOH formed by enolization of the deuterated sugar in alkaline D_2O. The rate constant, k_H, is calculated from the usual first-order equation:

$$k_H = 1/t \ln\,[A/(A - X)]$$

where A is the milliequivalents of enolizable hydrogen in the sample, and X is the milliequivalents of DOH formed in time t. The method gives satisfactory rate constants for short reaction periods. However, as the reaction proceeds, the rates for aldoses increase, presumably from accumulation of the more reactive ketose in the reaction mixture.

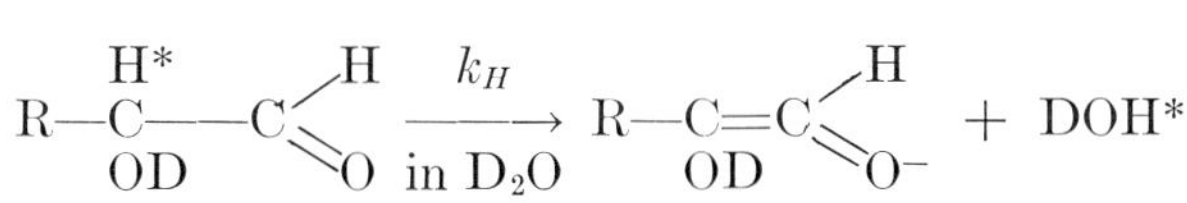

Scheme III. Measurement of primary rate of enolization

The two preceding methods have been combined to determine the tritium isotope effect. In a tritium-labeled substrate in D_2O, the change in infrared radiation absorption arises almost entirely from release of protons because the concentration of the tritium species in the reactant is small. Thus, the rate constant k_H (determined by the change in the DOH absorption) represents release of protons. The constant k_T for release of tritium to the solvent is determined from radioactivity measurements of water from the same reaction mixture. In the enolization of

D-glucose-2-*t* in 1*M* K_2CO_3 in D_2O at 35°C, values of 0.0018 and 0.00023 were found for k_H and k_T, respectively. Thus, the isotope effect, expressed as k_H/k_T is 7.8, or expressed as k_T/k_H is 0.13.

Griffiths and Gutsche (*23*) recently studied the interconversion of deuterated mandelaldehyde dimer and 2-hydroxyacetophenone in pyridine to obtain information concerning the glyceraldehyde–dihydroxyacetone rearrangement. Their results support an enolization mechanism requiring a base and an acid catalyst. They found a deuterium isotope effect of *ca.* 1.3 for the transformation of the aldehyde to the ketone. When they corrected this for the apparently differing amounts of the aldehyde form in equilibrium with the proteo dimer and the deuterio dimer, they obtained a value of 3.9. By the Swain–Schaad equation (*26*):

$$\ln (k_H/k_T) = 1.442 + \ln (k_H/k_D)$$

and values of 1.3 and 3.9, respectively, for the uncorrected and corrected deuterium isotope effects, one may estimate corresponding values for the tritium isotope effects of 5.5 and 16.5. The value of 7.8 found by Isbell and co-workers for the primary isotope effect in the enolization of D-glucose-2-*t* falls in this range.

A method developed earlier for measuring rates of enolization (*18*) depends upon the amount of tritium incorporated in the sugar by reversible enolization in water-*t*. The method, which measures the overall process of enolization and deenolization, is applicable to any available sugar. However, the process is complicated by an isotope effect that has not been evaluated. Rates of tritium uptake for monosaccharides in 0.5*M* sodium carbonate at 25°C ranged from 0.00014 for L-rhamnose to 0.072 for 2-deoxy-D-glucose (expressed as equivalents of tritium-labeled hydrogen per mole per hour). A summary of the relative rates for classes of sugars is given in Table IV of Chapter 4 of this volume. There is an approximate correlation between the rate of tritium uptake and the concentration of the acyclic sugar in solution. Table I presents a comparison

Table I. Relative Rates of Tritium Uptake and Enolization

	Tritium Uptake in H_2O-t [a]		*Enolization in D_2O* [b]	
Sugar	*Equiv/Mole of Sugar/Hr*	*Relative Rate*	k_H	*Relative Rate*
D-glucose	0.00046	1.0	0.018	1.0
D-mannose	0.00024	0.5	0.011	0.6
D-fructose	0.0049	10.7	0.292 [c]	16

[a] Conditions: 36 mg of sugar in 1 ml of 0.5*M* Na_2CO_3 at 25° C.
[b] Conditions: 90 mg of sugar in 0.6 ml of 1*M* K_2CO_3 at 30° C.
[c] The value of k_H based on the equivalent weight of the sugar with respect to two enolizable hydrogens is 0.146; for comparison with the aldoses, this constant is doubled.

of the rates of tritium uptake and the primary rates of enolization determined by the deuterium oxide method for D-glucose, D-mannose, and D-fructose. The rates of tritium uptake approximately parallel the rates of enolization. The fact that there is some variation is not surprising because ions formed by enolization may undergo reactions besides the reverse reaction leading to tritium uptake.

Oxidation of Sugars in Alkaline Solutions with Oxygen

Life as we know it on this planet depends upon reactions of compounds of carbohydrate origin with oxygen from air. Photosynthesis, respiration, and combustion have been thoroughly investigated, but there are many less spectacular processes that are not clearly understood. One of these is the reactions of sugars and polysaccharides with oxygen under alkaline conditions. The reactions are important not only in biological systems, but also in many technological and industrial operations—*e.g.*, in the treatment of wood pulp and the aging of cellulose in the rayon industry (*27, 28*). A basis for understanding these complex systems may be derived from the study of the reactions of monosaccharides with oxygen.

Ionic Enolate Mechanism. The first systematic investigation of the reactions of sugars in alkaline solutions was conducted by Nef (*4*). He found that D-glucose in alkaline solution in the presence of air yields principally formic and D-arabinonic acids, also small amounts of CO_2 and saccharinic, D-ribonic, D-erythronic, D,L-glyceric, glycolic, and oxalic acids. He suggested that these products, except for the saccharinic acids, are formed by oxidative cleavage of 1,2-, 2,3-, and 3,4-enediols. Later workers have followed this concept with modification of the mechanism for oxidative cleavage of the double bond (*29, 30, 31, 32, 33*).

Bamford and Collins (*30*) found that the auto-oxidation of D-glucose and D-fructose under high oxygen pressure and hydroxyl ion concentration results predominantly in the formation of formic and D-arabinonic acids. They represented the overall reactions by pseudo first-order equations, based on the parallel formation of ionic intermediates. They assumed that the intermediates, presumably enolate ions, react with oxygen, forming peroxides which decompose into D-arabinonate and formate ions. Other workers have not obtained as high yields for formic and arabinonic acids. They found with lower oxygen pressures and hydroxyl ion concentrations substantial amounts of the lower acids, presumably formed by oxidative cleavage of the 2,3- and 3,4-enediols and from products of a reverse aldol reaction.

Several mechanisms have been proposed for the cleavage reactions. One involves formation of an osone from the 1,2-enediol, followed by oxidation to a 2-keto-aldonic acid, which in turn gives the next lower aldonic acid by decarboxylation (*33*). This mechanism has little importance in the reaction of oxygen because the osone from D-glucose, for example, would yield D-gluconic acid by a benzylic acid rearrangement whereas little or no D-gluconic acid is formed in the reaction. 2-Keto-D-gluconic acid would yield carbon dioxide rather than formic acid by decarboxylation. A small amount of carbon dioxide is produced, but it is not a major product. In the absence of free-radical initiators the reaction may take place by the ionic mechanism of Doubourg and Naffa (*32*), as depicted in Scheme IV. In the presence of free-radical initiators, however, a free-radical chain mechanism seems more reasonable.

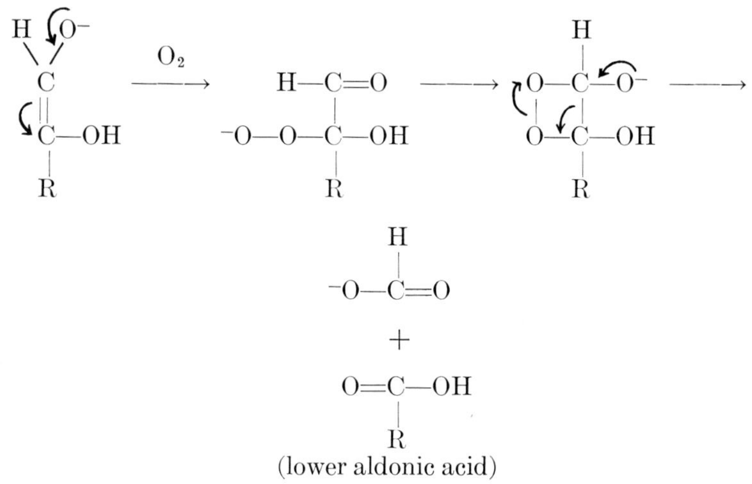

Scheme IV. Enolate mechanism for reaction of sugars with oxygen Dubourg and Naffa Mechanism (32)

Free-Radical Reactions. Much literature covering free-radical reactions is available (*34, 35, 36, 37, 38, 39, 40, 41*), but there is no comprehensive review covering the free-radical chemistry of sugars and polysaccharides. Hydrogen peroxide and alkaline peroxides have been widely used to modify starch (*42*) and to bleach textiles and cellulose products. Presumably, the peroxide acts as a source of free radicals. Control of the production and type of free radicals in solutions is important with respect to natural processes such as aging (*40*), hydroxylation of collagen, connective tissue disorders, and alteration of structural carbohydrates (*43*). In these processes carbohydrates and their derivatives play important roles. One of these is the regeneration of free-radical catalysts such as ferrous iron in systems containing oxygen and hydrogen peroxide.

Hydrogen peroxide in solution decomposes slowly at room temperature with liberation of oxygen, but it decomposes rapidly at higher temperatures. The decomposition is accelerated by trace amounts of transition metal salts. With iron salts the following reactions occur (*44*):

$$Fe^{2+} + H_2O_2 \rightarrow Fe^{3+} + OH^- + \cdot OH$$

$$\cdot OH + H_2O_2 \rightarrow H_2O + \cdot OOH$$

$$\cdot OOH + H_2O_2 \rightarrow O_2 + H_2O + \cdot OH$$

$$\cdot OH + Fe^{2+} \rightarrow OH^- + Fe^{3+} \text{ (chain-breaking process)}$$

Under suitable conditions ferrous ions may be regenerated by reduction of ferric ions, giving a continuous supply of either hydroxyl or hydroperoxide radicals. The reductant for converting ferric ions to ferrous ions may be a sugar as, for example, in mixtures used to effect polymerization of butadiene in the production of synthetic rubber (*44*). In biological systems the reductant may be a more reactive compound such as ascorbic acid. Chain reactions involving the regeneration of radicals in single-electron steps provide a way to effect structural changes and oxidations with low energy barriers. Some reaction systems are complex and yield numerous products, but the fundamental reactions are simple.

Regenerative free-radical reactions are highly important in many industrial processes. For example, in the treatment of wood pulp traces of iron salts cause loss by initiating Ruff–Meller type depolymerizations (*45*). Many biologically important polysaccharides undergo oxidative–reductive depolymerizations by free-radical reactions. Thus, Pigman and co-workers (*43*) have shown that depolymerization of hyaluronic acid and of alginic acid occurs by a series of reactions involving ferrous iron, oxygen, and ascorbic acid. Regenerative free-radical processes are common and warrant more consideration by carbohydrate chemists.

Although the reactions of reducing sugars with oxygen have been widely investigated, little attention has been given to free-radical mechanisms. An induction period at the start of an autoxidation is evidence of a possible free-radical reaction. Recently Gleason and Barker (*14*) reported a lag in the consumption of oxygen at the beginning of the reactions of numerous sugars with oxygen in alkaline solutions. De Wilt and Kuster (*46*) also observed this lag at the beginning of the reaction of D-glucose but not of D-fructose. They explained the lag by a relatively slow attainment of the enolate equilibrium of D-glucose and concluded that the enolate equilibrium of D-fructose is reached almost instantaneously. This conclusion does not seem valid because the measured rates of enolization of D-glucose and D-fructose are relatively slow (*25*). The rate of enolization of D-glucose (Table I) is about one-sixteenth that of D-fructose whereas the De Wilt and Kuster rate of oxidation of D-glucose is only one-fourth that of D-fructose.

Gleason and Barker considered the possibility that the induction period might result from a free-radical reaction but did not pursue the matter, probably because Bamford and Collins had been unsuccessful in accelerating the reaction by free-radical initiators. In a subsequent paper (*15*) Gleason and Barker found that the formic acid, produced from D-ribose-2-*t* by the action of oxygen, contains a substantial proportion of tritium. To account for this important discovery, they suggested two concurrent mechanisms: (1) enolization and oxidation and (2) hydride transfer of the C-2 hydrogen atom to C-1, followed by enolization of the resulting ketone, and oxidation of the enediol. Concurrently, Isbell and Hepner (*47*) studied the oxidation of D-glucose-2-*t* and found that it yields a small amount of tritium-labeled formic acid, as well as the expected non-labeled acid. These results, like those of Gleason and Barker, may be explained by the existence of more than one reaction path.

Two free-radical chain reactions, in addition to the ionic enolate mechanism, seem reasonable for the oxidation of the sugars by oxygen. With an aldose-2-*t* one of the free-radical mechanisms would yield non-labeled formic acid and the next lower aldonic acid; the other would yield labeled formic acid and the same aldonic acid.

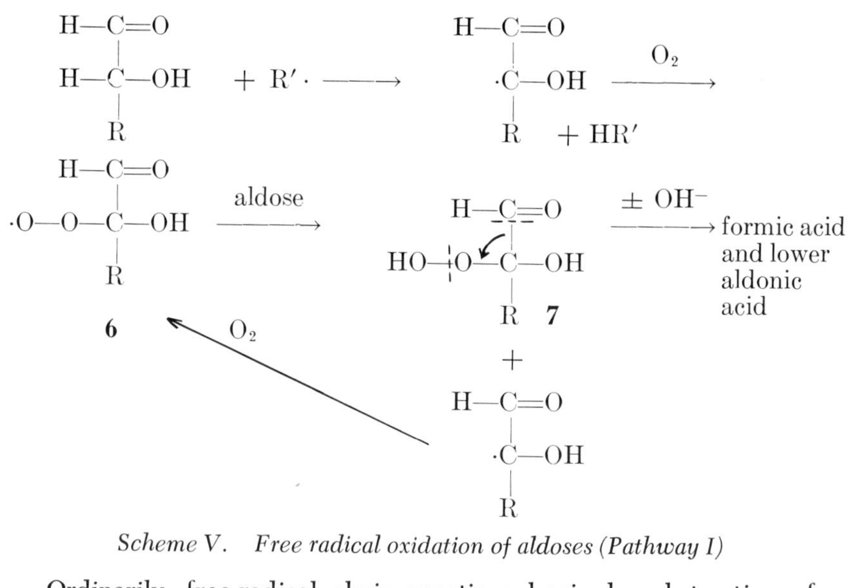

Scheme V. Free radical oxidation of aldoses (Pathway I)

Ordinarily, free-radical chain reactions begin by abstraction of a weakly bonded hydrogen atom (*34*) from the substrate by a radical which is regenerated in a subsequent step in the process. The resultant substrate radical may then rearrange or undergo decomposition reactions similar to the elimination of ions by consecutive electron displacements (*48, 49*). The path to the non-labeled formic acid begins with abstraction

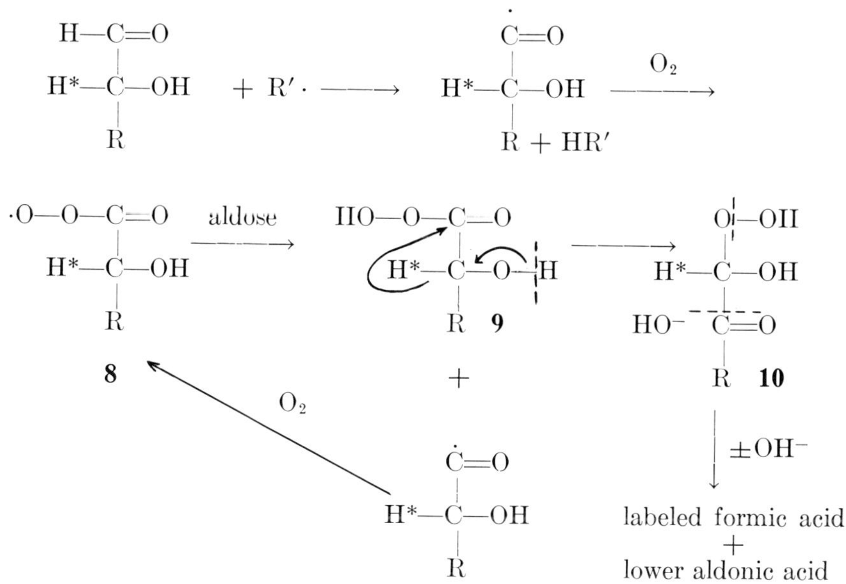

Scheme VI. Free radical oxidation of aldoses (Pathway II)

of the C-2 hydrogen atom (Scheme V). The resulting radical is continuously regenerated by reaction of the oxygen adduct (**6**) with a new supply of the aldose. By addition and elimination of either a hydroxyl ion or a hydroxyl radical, the glycosulose peroxide (**7**) yields formic acid and the next lower aldonic acid. The second pathway (Scheme VI) begins with the abstraction of the C-1 hydrogen atom, forming a glyconyl radical, which is continuously regenerated by reaction of the oxygen adduct (**8**) with a new supply of the aldose. An intermediate peraldonic acid (**9**) is formed, which rearranges to a glycosulose peroxide (**10**) by hydride transfer from C-2 to C-1. Addition of a hydroxyl ion to **10** with rupture of the C-1–C-2 bond and elimination of a hydroxyl ion yields labeled formic acid and the next lower aldonic acid.

Ketoses may be oxidized by a similar free-radical chain mechanism (Scheme VII). The reaction begins with the abstraction of a hydrogen atom from C-1. The resulting radical is continuously regenerated by reaction of the oxygen adduct (**11**) with a new supply of the ketose. The resulting peroxide (**10**) is the same as that derived from the aldose and gives by fragmentation formic acid and the next lower aldonic acid by either a free radical or an ionic mechanism.

Oxidation of Sugars in Alkaline Solutions with Sodium Peroxide

Early workers reported that treatment of D-glucose, D-mannose, and D-fructose in alkaline solutions with hydrogen peroxide yields products

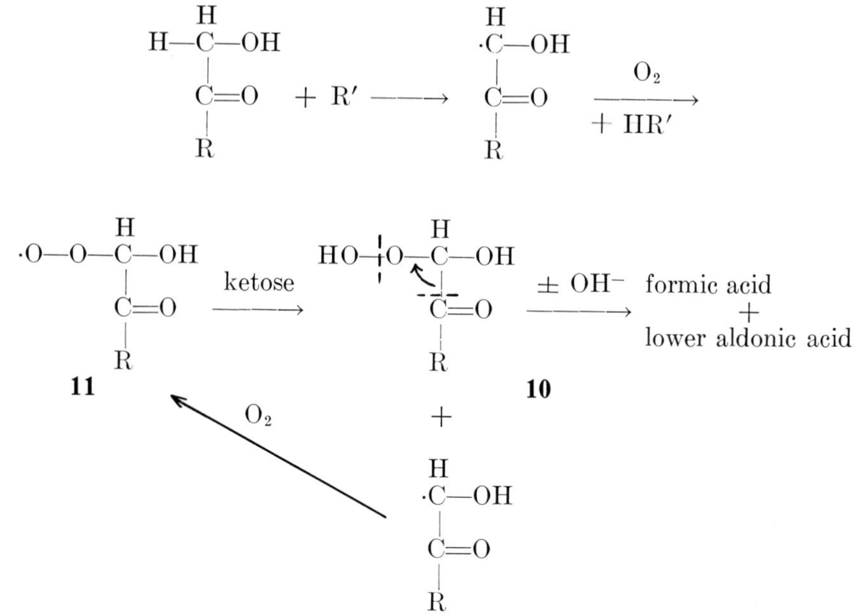

Scheme VII. Free radical oxidation of ketoses

Table II. Oxidation of Monosaccharides by Sodium Peroxide[a]

Sugar	*Reaction Time, hours*	*Total Acid*[b] *Found*	*Total Acid*[b] *Theory*	*Formic Acid*[b] *Found*	*Formic Acid*[b] *Theory*
		aldo-Hexoses			
Allose	312	5.72	6	5.61	6
Altrose	143	6.05	6	5.67	6
Galactose	120	5.63	6	5.55	6
Glucose	480	5.30	6	5.28	6
Mannose	310	5.50	6	5.84	6
Talose	46	5.86	6	6.06	6
		aldo-Pentoses			
Arabinose	261	5.19	5	4.97	5
Lyxose	93	5.04	5	5.02	5
Ribose	93	4.97	5	4.65	5
Xylose	93	4.74	5	4.89	5
		Ketoses And 6-Deoxyhexoses			
Fructose	143	4.97	5	4.38	4
Sorbose	120	4.96	5	4.05	4
Rhamnose	490	4.89	5	3.44	4
Fucose	190	4.47	5	4.09	4

[a] 0.075 gram of sugar + 0.01 gram of EDTA in 5 ml of 1*M* Na_2O_2 at 0°C.
[b] Mmoles of acid per mmole of sugar.

similar to those obtained by oxidation of the sugars by air or oxygen (*4, 50*). None of the prior workers carefully distinguished between the reaction of oxygen and those of the peroxide anion. Thus, it seems probable that the products reported in the literature arise from reactions of the sugar with oxygen and peroxide.

Isbell and co-workers (*51*) have tried to minimize the oxygen reaction and to maximize the peroxide reaction. When a large excess of peroxide and a low temperature were maintained, they found that the monosaccharides are converted almost quantitatively to formic and two-carbon acids. Table II presents results for the peroxide oxidation of 14 sugars. The total acid produced from *aldo*-hexoses under favorable conditions is about six moles, consisting almost entirely of formic acid. Aldopentoses react more rapidly than *aldo*-hexoses and yield about five moles of formic acid per mole of pentose. Fructose and sorbose yield approximately five moles of total acid of which four moles are formic acid. Glycolic acid was identified qualitatively but not determined quantitatively. L-Rhamnose and L-fucose yield about five moles of acid, including four moles of formic acid. Acetic acid was identified only qualitatively.

Consideration of reasonable mechanisms for producing formic acid from an aldose led to the hypothesis that the sugar forms an addition product with the hydroperoxide anion, comparable with an aldehyde sulfite or the addition product of aldoses with chlorous acid (*52*). The intermediate product (**12**) could decompose by a free-radical or an ionic mechanism. In the absence of a free-radical catalyst, the ionic mechanism of Scheme VIII seems probable. By either mechanism the products are formic acid and the next lower sugar. The lower sugar then repeats the process, with the result that the aldose is degraded stepwise to formic acid. Addition of the hydroperoxide anion to the carbonyl carbon is in accord with its strong nucleophilic character (*53*) and with certain reaction mechanisms suggested in the literature (*54*) for related substances.

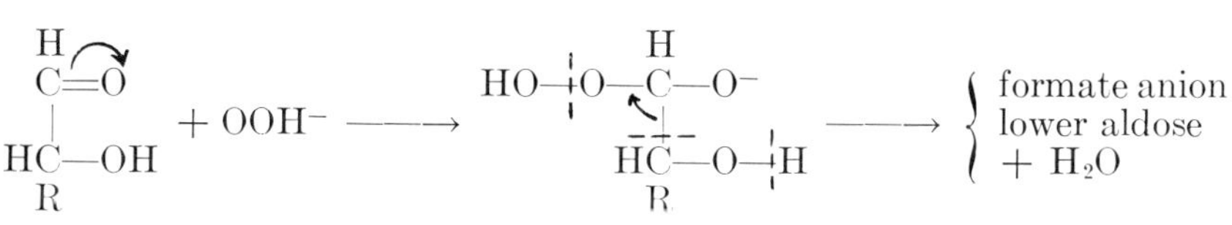

12

Scheme VIII. Peroxide oxidation of aldoses

Oxidative degradation of a ketose may proceed by two pathways (Scheme IX). Hexuloses by addition of a peroxide anion form a ketose peroxide (**13**). Cleavage of this intermediate at the C-2–C-3 bond gives glycolic acid and an *aldo*-tetrose. The *aldo*-tetrose is smoothly converted to four moles of formic acid. Alternatively, the ketose peroxide (**13**) may be cleaved between C-1 and C-2 to yield formaldehyde and an aldopentonic acid. Under the described reaction conditions, formaldehyde is oxidized to formic acid, whereas the aldopentonic acid is inert. Table II shows that the oxidation of a ketose yields largely formic and glycolic acids.

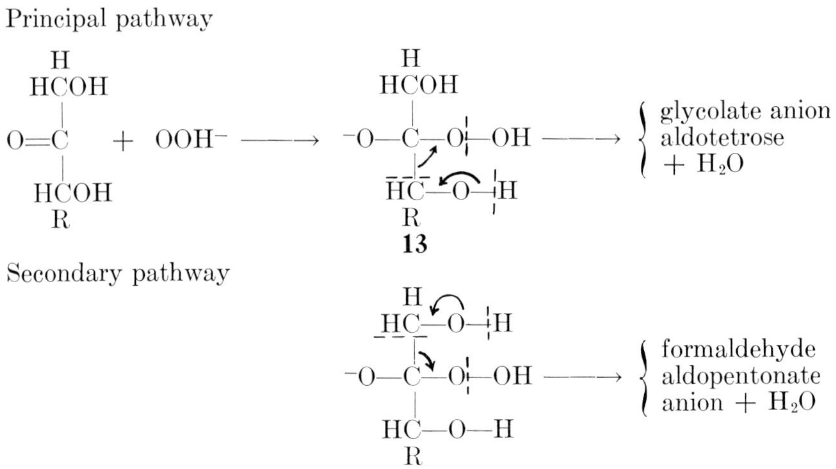

Scheme IX. Peroxide oxidation of ketoses

Concurrent Rearrangement and Oxidation Reactions

In many systems rearrangement and oxidation reactions occur concurrently and produce various products, the proportions of which are determined by the experimental conditions. An understanding of the separate reactions enables one to predict the products of the reactions and the effect of changes in the experimental conditions. Under alkaline conditions rearrangements of sugars result in glycosuloses and diuloses, which undergo benzylic acid rearrangements with formation of saccharinic acids (*48, 49, 55, 56*). Some of the intermediates are readily oxidized and give characteristic products whose formation may be explained by mechanisms similar to those proposed here for the oxidation of aldoses and ketoses.

In alkaline solutions D-glucose forms 3-deoxy-D-*erythro*-hexosulose and 4-deoxy-D-*glycero*-2,3-hexodiulose which yield saccharinic acids. Machell and Richards (*57*) have shown that 3-deoxy-D-*erythro*-hexosulose (**14**) is oxidized by 30% hydrogen peroxide to formic acid and 2-deoxy-D-*erythro*-pentonic acid (**15**). Recently Rowell and Green (*58*) found that **14** in the presence of oxygen also forms **15** in addition to the saccharinic acids. They inferred that the reactions with oxygen and hydrogen peroxide are very similar, but they did not present reaction mechanisms.

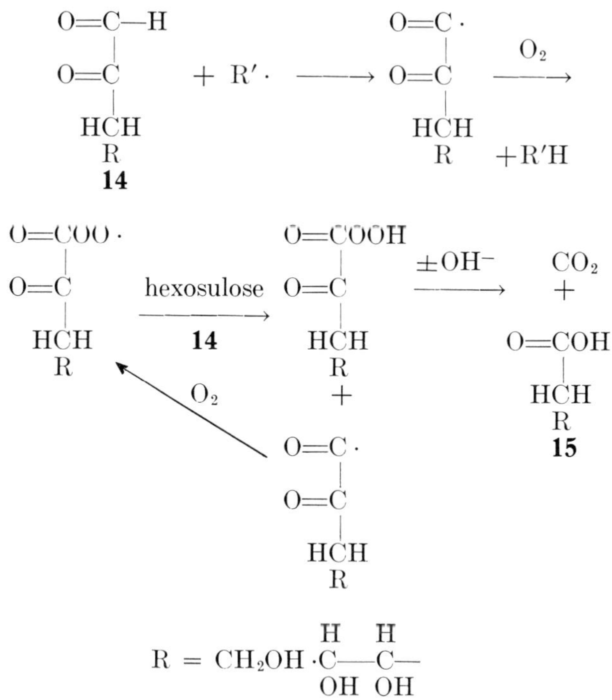

*Scheme X. Reaction of 3-deoxy-*D-erythro-*hexosulose with oxygen*

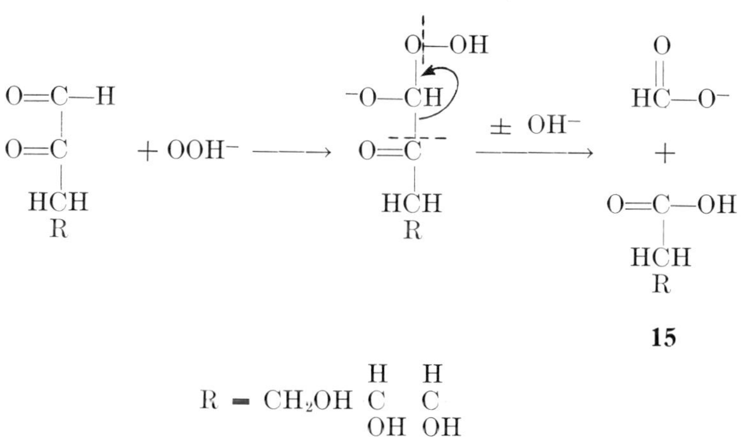

*Scheme XI. Reaction of 3-deoxy-*D-erythro-*hexosulose with alkaline hydrogen peroxide*

Referring to reaction Schemes V and VIII, one would expect the reactions with oxygen and hydrogen peroxide to be quite different. As outlined in Schemes X and XI, the reaction with oxygen should give **15** and carbon dioxide, whereas the reaction with hydrogen peroxide should give

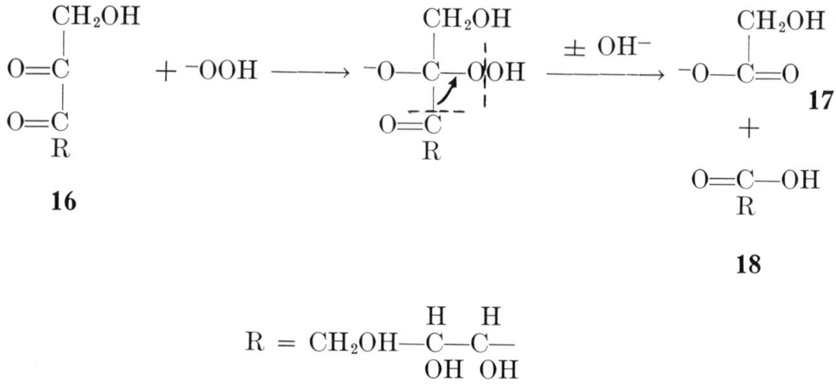

*Scheme XII. Reaction of 4-deoxy-*D-glycero-*2,3-hexodiulose with alkaline hydrogen peroxide*

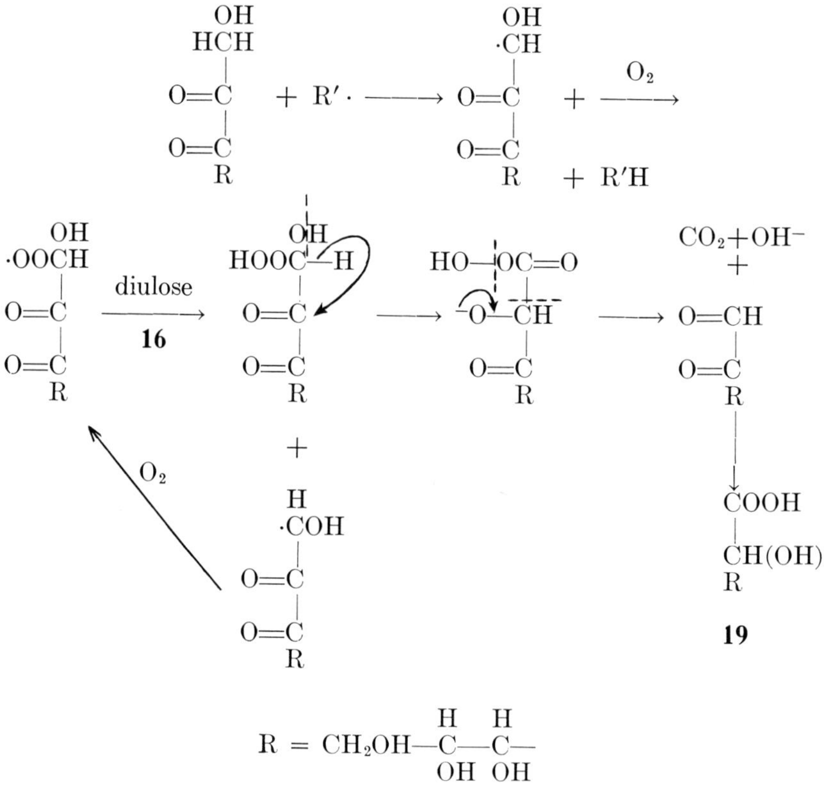

*Scheme XIII. Reaction of 4-deoxy-*D-glycero-*2,3-hexodiulose with oxygen*

15 and formic acid. Apparently, neither formic acid nor carbon dioxide has been established as a product of the oxygen reaction.

The reactions of 4-deoxy-D-*glycero*-2,3-hexodiulose (**16**) under alkaline conditions are interesting. In an inert atmosphere diulose **16** forms "D-*gluco*-isosaccharinic acid," whereas with hydrogen peroxide it is reported to give a mixture of glycolic acid (**17**) and 3,4-dihydroxybutyric acid (**18**) (*59*). The peroxide reaction may now be rationalized by Scheme XII. The reaction of diulose **16** with oxygen has not been studied, but the free-radical reaction of Scheme XIII is predicted. The expected products are carbon dioxide and the two epimeric 3-deoxy-D-*glycero*-pentonic acids (**19**) formed by benzylic acid rearrangement of the intermediate pentosulose.

Certain reactions previously considered to be ionic (*49*) may also take place by free-radical mechanisms, especially in nonaqueous solvents. Thus, the aromatization of *myo*-inosose-2 in acetic anhydride and pyridine to 1,2,3,5-tetraacetoxybenzene may take place by an ionic and a free-radical mechanism. Other free-radical mechanisms are discussed by Fatiadi in another chapter of this volume.

The author hopes that the concepts presented here will stimulate further research and lead to a better understanding of the reactions of carbohydrates in alkaline solution.

Acknowledgment

The author gratefully acknowledges Grant GP 9457 from the National Science Foundation. He is indebted to Frederick Rice, Kazimir Linek, and Harriet Frush for helpful discussions and suggestions. Thanks are also given to Karl E. Hepner, Jr., and other students at American University for assistance in the experimental work.

Literature Cited

1. de Bruyn, Lobry, van Ekenstein, Alberda, *Rec. Trav. Chim.* (1895) **14,** 203.
2. Wohl, A., Neuberg, C., *Chem. Ber.* (1900) **33,** 3900.
3. Michaelis, L., Rona, P., *Biochem. Z.* (1912) **47,** 447.
4. Nef, J. U., *Liebig's Ann.* (1907) **357,** 300; (1910) **376,** 1; (1914) **403,** 204.
5. Speck, J. C., Jr., *Advan. Carbohydrate Chem.* (1958) **13,** 63.
6. Pigman, W., Isbell, H. S., *Advan. Carbohydrate Chem.* (1968) **23,** 11.
7. Isbell, H. S., Pigman, W., *Advan. Carbohydrate Chem.* (1969) **24,** 14.
8. Tipson, R. S., Isbell, H. S., *J. Res. Nat. Bur. Stand.* (1962) **66A,** 31.
9. Swenson, C. A., Barker, R., *Biochemistry* (1971) **10,** 3151.
10. Cantor, S. M., Penniston, Q. P., *J. Amer. Chem. Soc.* (1940) **62,** 2113.
11. Los, J. M., Simpson, L. B., Wiesner, K., *J. Amer. Chem. Soc.* (1940) **78,** 1564.
12. Horton, D., Wander, J. D., *Carbohydrate Res.* (1971) **16,** 477.
13. Fredenhagen, H., Bonhoeffer, K. T., *Z. Physik. Chem.* (1938) **A181,** 392.

14. Gleason, W. B., Barker, R., *Can. J. Chem.* (1971) **49,** 1425.
15. Gleason, W. B., Barker, R., *Can. J. Chem.* (1971) **49,** 1433.
16. Pigman, W., "The Carbohydrates," p. 60, Academic, New York, 1957.
17. Isbell, H. S., Peterson, R. A., Frush, H. L., Holt, N. B., *148th Nat. Meet., Amer. Chem. Soc.*, Chicago (September 1964), 10D.
18. Isbell, H. S., Frush, H. L., Wade, C. W. R., Hunter, C. E., *Carbohydrate Res.* (1969) **9,** 163.
19. MacLaurin, D. J., Green, J. W., *Can. J. Chem.* (1969) **47,** 3947.
20. Fieser, L. F., Fieser, M., "Organic Chemistry," 3rd ed., p. 361, Reinhold, New York, 1956.
21. Sowden, J. C., Schaffer, R., *J. Amer. Chem. Soc.* (1952) **74,** 499; **74,** 505.
22. Topper, Y. J., Stetten, D., *J. Biol. Chem.* (1951) **189,** 191.
23. Griffiths, D. W., Gutsche, C. D., *J. Amer. Chem. Soc.* (1971) **93,** 4788.
24. Feather, M. S., *Carbohydrate Res.* (1968) **7,** 86.
25. Isbell, H. S., Linek, K., Hepner, K. E., Jr., *Carbohydrate Res.* (1971) **19,** 319.
26. Swain, C. G., Stivers, E. C., Reuwer, J. F., Jr., Schaad, L. J., *J. Amer. Chem. Soc.* (1958) **80,** 5885.
27. Meller, A., "The Chemistry of the Reactions with Polysaccharides in the Alkaline Extraction Stage of Bleaching of Wood Pulps," TAPPI Monograph Series No. 27, New York, 1964.
28. Meller, A., *Holzforschung* (1960) **14,** 78; **14,** 129.
29. Samuelson, O., Thede, L., *Acta Chem. Scand.* (1968) **22,** 1913.
30. Bamford, C. H., Collins, J. R., *Proc. Roy. Soc. London* (1950) **204,** 62.
31. Spengler, O., Pfannenstiel, A., *Z. Wirtschaftsgruppe Zuckerind.* (1935) **85,** 547.
32. Dubourg, J., Naffa, P., *Bull. Soc. Chim. France* (1959) 1353.
33. Staněk, J., Černý, M., Kocourek, J., Pacák, J., "The Monosaccharides," p. 139, Academic, New York, 1963.
34. Walling, C., "Free Radicals in Solutions," Wiley, New York, 1957.
35. Walling, C., "Molecular Rearrangements," p. 407, P. de Mayo, Ed., Wiley, New York, 1964.
36. Thomas, J. R., Ingold, K. U., ADVAN. CHEM. SER. (1968) **75,** 258.
37. Ingold, K. U., *Accounts Chem. Res.* (1969) **2,** 1.
38. Norman, O. C., Gilbert, B. C., *Advan. Phys. Org. Chem.* (1967) **5,** 53.
39. McNesby, J. R., Heller, C. A., Jr., *Chem. Rev.* (1954) 325.
40. Pryor, W. A., *Chem. Eng. News* (1971) **49,** 34.
41. Pryor, W. A., "Introduction to Free Radical Chemistry," Prentice–Hall, Englewood Cliffs, 1966.
42. Newton, J. M., Peckham, G. T., Jr., "Chemistry and Industry of Starch," p. 224, R. W. Kerr, Ed., Academic, New York, 1944.
43. Pigman, W., Rizvi, S., *Biochem. Biophys. Res. Commun.* (1959) **1,** 39; Matsumura, G., Pigman, W., *Arch. Biochem. Biophys.* (1965) **110,** 526; Harris, M. J., Herp, A., Pigman, W., *ibid.* (1971) **142,** 615.
44. Walling, C., "Free Radicals in Solutions," p. 565, Wiley, New York, 1957.
45. Meller, A., *Chem. Ind.* (1960) 18.
46. De Wilt, H. G. J., Kuster, B. F. M., *Carbohydrate Res.* (1971) **19,** 5.
47. Isbell, H. S., Hepner, K. E., Jr., unpublished work.
48. Isbell, H. S., *J. Res. Nat. Bur. Stand.* (1944) **32,** 45.
49. Isbell, H. S., *Ann. Rev. Biochem.* (1943) **12,** 205.
50. Spoehr, H., *Amer. Chem. J.* (1910) **43,** 238.
51. Isbell, H. S., Frush, H. L., Martin, E. T., *Carbohydrate Res.* (1973) in press.
52. Isbell, H. S., Sniegoski, L. T., *J. Res. Nat. Bur. Stand.* (1964) **68A,** 301.
53. McIsaac, J. E., Jr., Mulhausen, H. A., Behrman, E. J., *156th Nat. Meet. Amer. Chem. Soc.,* Atlantic City (September 1968) ORGN 70.
54. Steven, J. R., Ward, J. C., *Chem. Commun.* (1965) 273.

55. Feather, M. S., Harris, J. R., *Carbohydrate Res.* (1970) **15**, 304.
56. Anet, E. F. L. J., *J. Amer. Chem. Soc.* (1960) **82**, 1502.
57. Machell, G., Richards, G. N., *J. Chem. Soc.* (1960) 1938.
58. Rowell, R. M., Green, J., *Carbohydrate Res.* (1970) **15**, 197.
59. Machell, G., Richards, G. N., *J. Chem. Soc.* (1960) 1938.

RECEIVED December 2, 1971.

6

Electron Spin Resonance Studies of Chemical Changes of Phenylhydrazones and Osazones in Alkaline Solution

ALEXANDER J. FATIADI

Analytical Chemistry Division, Institute for Materials Research, National Bureau of Standards, Washington, D. C. 20234

Treatment of a solution of a sugar phenylhydrazone or osazone in methyl sulfoxide with potassium tert-*butoxide and a trace of oxygen at room temperature gives products that have a three-line electron spin resonance (ESR) spectrum characteristic of a nitroxide radical. The apparent fragmentation of the phenylhydrazine moiety under reaction conditions used does not show evidence of any paramagnetic species derived from glyoxal bis(phenylhydrazone). The latter had been reported to be the product of degradation of the sugar phenylhydrazones under more vigorous alkaline treatment. Some inosose phenylhydrazones in alkaline methyl sulfoxide solution produce stable radical-anions in which the phenylhydrazine moiety remains intact. A free radical mechanism is advanced to account for the β-elimination reactions of the sugar phenylhydrazones under mildly basic conditions.*

Phenylhydrazones of aromatic aldehydes are converted into nitriles and aniline by treatment with potassium *tert*-butoxide in boiling toluene (*1*). The significance of atmospheric oxygen in this reaction was noted, and it was suggested (*1*) that the reaction preceeds *via* a homolytic decomposition of the arylaldehyde phenylhydrazone, leading to a free-radical intermediate. The facile solubility of phenylhydrazones of aromatic ketones and aldehydes in basic solvents and in alkaline aqueous or alcohol solutions (a property observed also for sugar phenylhydrazones and osazones) suggests the possible formation of salts (*1, 2*), and this agrees with the acidic character of the imino proton in the phenylhydrazone moiety (*3*).

Literature surveys (*4, 5*) show little reported work on the action of a strong base on a sugar phenylhydrazone or osazone. Early studies in the sugar series revealed glyoxal bis(benzoylhydrazone) as a major product when, for example, D-glucose was treated with benzoylhydrazine in aqueous alkali (*6*). Glyoxal bis(phenylhydrazone) was isolated when D-*arabino*-hexulose phenylosazone was refluxed in 1% alcoholic potassium hydroxide (*7*). Wolfrom and co-workers (*8*) obtained glyoxal bis(phenylhydrazone) on heating D-galactose or D-arabinose phenylhydrazone in methanolic or ethanolic pyridine at 120°–130°C, and they suggested that homolytic fission occurs, followed by the union of two C-1 fragments (two free-radicals, 2H—C=N—NHPh) to yield the osazone. Simon and Moldenhauer (*9*) found that hot 100 m*M* alcoholic potassium hydroxide converts D-*arabino*-hexulose phenylosazone into glyoxal bis-(phenylhydrazone) (85% yield), 2,4-dihydroxybutyric acid (45% yield, in a separate experiment), and 3,6-anhydroallosazone (5% yield). They found also that similar alkaline treatment (refluxing for 4 hours under nitrogen) of glucose or mannose phenylhydrazone gives the cyclic product, 1-phenylpyrazole, in 27% yield (*via* a suggested olefinic azo-sugar intermediate) and some aniline (*via* an irreversible fission of enol-hydrazine intermediate). In a recent study by Russell and Lyons (*10*) D-*arabino*-hexulose benzoylhydrazone and D-glucose-2-^{14}C benzoylhydrazone were refluxed in 1% potassium hydroxide in ethanol for 1 hour to give glyoxal bis(benzoylhydrazone) in low yield. They concluded that in forming glyoxal bis(benzoylhydrazone) a labeled entity derived from C-1 and C-2 competes with glycoaldehyde fragments (derived from the non-labeled atoms C-3 to C-6) for reaction with the benzoylhydrazone groups. Thus, the apparent fragmentation could result from a reversed aldol mechanism (*10, 11, 12*).

In our work, treatment of a solution of glyoxal bis(phenylhydrazone) in methyl sulfoxide with a base (either aqueous potassium hydroxide or potassium *tert*-butoxide) and a trace of oxygen produced a deep-purple color, and the solution became paramagnetic (*13*). The free-radical thus observed was anionic. (It is the first anionic radical in the phenylhydrazine series that has been reported. The neutral diphenylpicrylhydrazyl radical (DPPH) and the stable tetraphenylhydrazine cation radical are known.) The radical anion is apparently formed by an electron transfer mechanism, and its relative stability in solution can partially be explained as a result of resonance. Since neither the *N*-methyl nor the *N*-benzoyl derivative of glyoxal bis(phenylhydrazone) showed either a color change or the presence of a paramagnetic species following similar treatment with base and oxygen, it was concluded that the purple radical-anion must arise following the abstraction of acidic imino protons by base. The dianionic character of this radical was shown by its conversion into the *N*-dimethyl and *N*-dibenzoyl derivatives in good yield. This discovery of the stable free-radical dianions derived from glyoxal bis(phenylhydrazone) prompted study of some sugar and inosose phenylhydrazones and osazones. The sugar phenylhydrazones and osazones under analogous reaction conditions (methyl sulfoxide–potassium *tert*-butoxide at room temperature) were expected to show the ESR spectrum characteristic of the radical-anion derived from glyoxal bis-(phenylhydrazone) which is shown in Figure 1 (A symmetrical radical-anion derived from glyoxal bis[(2,5-dichlorophenyl)hydrazone] gave the ESR features shown in Figure 1; the hyperfine structure is explained by assuming the interaction of the unpaired electron with two pairs of equivalent nitrogen atoms $a^N = 6.35$ and 6.35 G; $a^H = 6.35$, 6.35, 1.65, 1.65, 0.82, and 0.82 G (*13*). The symmetry of this radical is comparable with that of a hydrazine cation radical, *see* Reference *14*.) The ESR pattern shown is characteristic for the radical-anions derived from glyoxal bis(phenylhydrazone) or those in which the para position on the benzene ring is unsubstituted.

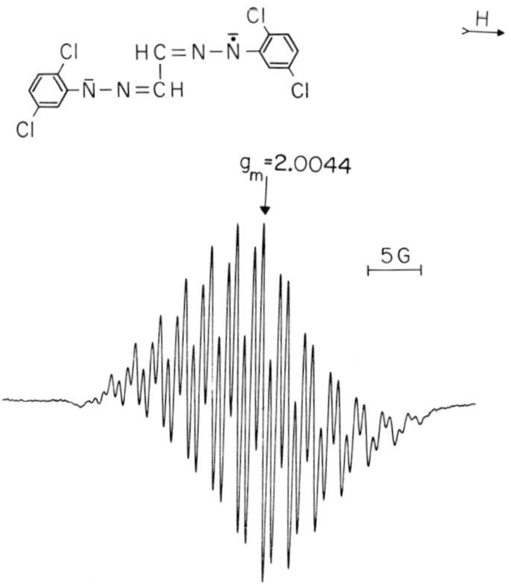

Figure 1. Radical-anion from glyoxal bis(2,5-dichlorophenyl)hydrazone (3 mM, *in Me_2SO-K-*tert-*BuO)*

Results and Discussion

Treatment of methyl sulfoxide solutions of D-mannose or other sugar phenylhydrazones or melibiose phenylosazone with potassium *tert*-butoxide and a trace of oxygen at room temperature showed no evidence of the paramagnetic species characteristic of glyoxal bis(phenylhydrazone). Instead new absorption bands appeared in the visible spectrum (as discussed below), and the solutions showed a characteristic three-line ESR spectrum (spectra R to X, Figure 2).

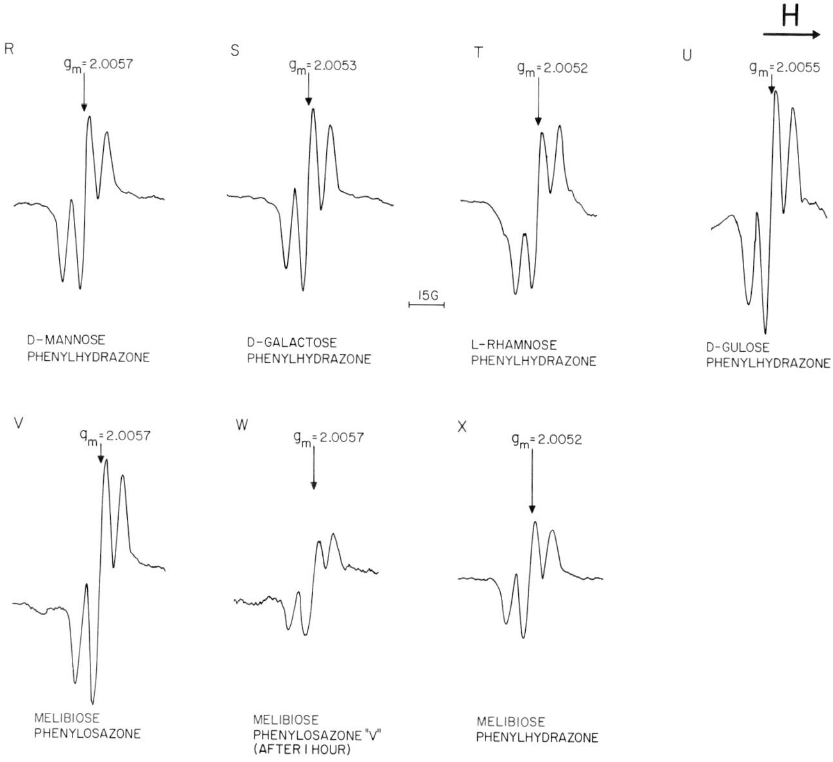

*Figure 2. Electron spin resonance spectra of sugar phenylhydrazones and osazones (2.5 m*M *to 3.5 m*M, *in Me_2SO-1% K*-tert-*BuO)*

This ESR pattern is exclusively characteristic for the sugar phenylhydrazones or osazones, and the technique may differentiate quickly between sugar and non-sugar phenylhydrazones or osazones. The direct isolation of these paramagnetic species proved to be a difficult task because of their low concentration (usually 2–5%, *see* Table I) and their limited half-life (about 60 minutes for radicals derived from osazones; compare spectra V and W, Figure 2). However, analysis of the ESR spectra, particularly the shape, intensity, stability, and the coupling constants (observed range a^N 6 to 8 gauss for compounds R to X, Figure 2 and Table I), pointed to a nitroxide radical. The reported (*16*) range for acyclic and cyclic nitroxide radicals is a^N 6–16 gauss.

Table I. Free Radicals from Sugar Phenylhydrazones and Osazones in Methyl Sulfoxide–1% Potassium *tert*-butoxide at 25 ± 3°C

Compound	*Spectrum*	*Radical Concentration*[a] *(per cent)*	*Hyperfine Structure* a^N *(gauss)*[b]
D-Mannose phenylhydrazone	R	4	7.5 ± 0.1[c, d]
D-Galactose phenylhydrazone	S	4	7.6 ± 0.3
L-Rhamnose phenylhydrazone	T	4	7.5 ± 0.2
D-Gulose phenylhydrazone	U	5	7.6 ± 0.2
Melibiose phenylosazone	V	5	7.7 ± 0.2
Melibiose phenylosazone (after 1 hr)	W	1.8	7.6 ± 0.1
Melibiose phenylhydrazone	X	2.2	7.4 ± 0.2

[a] Radical concentration, based on starting phenylhydrazone or osazone, was determined by comparison with the Fremy's salt standard solution (*15*), measured in methyl sulfoxide—1% K-*tert*-BuO.

[b] Indicated errors of all values in this table are standard deviations of the average of three determinations.

[c] A lower amplitude measurement gave a value for a^N 6.5±0.2 gauss (spectrum A).

[d] Concentrated solutions or those overexposed to air usually show a poorly resolved ESR spectrum.

The three-line patterns shown in Figure 2 have a relative intensity of 1:2:1 and not the 1:1:1 pattern characteristic of a nitroxide radical with no α-hydrogen atom. The relative intensities and somewhat lower values of the coupling constants observed (a^N 7 to 8 G, instead of $a^N > 10$ G) were affected by factors summarized below. Generally, the phenoxy radicals (*17, 18*) show a pattern with 1:2:1 intensity, but in this case they are too unstable to be considered (*19, 20*); the same can be said about possible anilino radicals (*see*, for example, *21, 22*). The factors that may affect an ESR pattern and an a^N value in an alkaline media are: (1) a solvation of the radical that can cause perturbations in a spin density distribution, (2) a close association of the radicals in solution or its low tumbling rate that can case a line broadening, and (3) an ion–dipole interaction or a delocalization of some of the unpaired spins (for example, in our case, into a radical solvated methyl sulfoxide moiety) can cause decrease in the value of the nitrogen coupling constant (a^N) (*see* Ref. *23* for a discussion of various nitroxide radicals and their a^N values).

This radical could arise by fragmentation of a phenylhydrazine moiety in the alkaline solution—*e.g.*, by splitting off a molecule of aniline. The resulting imino sugar could be converted by oxidation into an iminoxy free-radical, but these usually have larger nitrogen hyperfine coupling constants ranging from a^N 30–32 gauss (*24, 25*), and consequently, this possibility was eliminated. Also, the formation of a stable semidone type of radical (*26*), or of anilino or phenoxy radicals (*17, 18*) was unlikely under the conditions described. (The lifetime of an unsubstituted phenoxy radical in solution is in a range of milliseconds (*19, 20*).) Instead, the resulting imino sugar is presumably converted by consecutive addition and oxidation reactions into a nitroxide radical. A new approach to this problem will be discussed below.

The action of strong base on the methyl sulfoxide solutions of the sugar and inosose phenylhydrazones and osazones can conveniently be monitored in the visible spectrum. The characteristic deepening in color of the above solutions immediately after addition of base can be partially attributed to the formation of the paramagnetic species, and this was confirmed by direct ESR measurements. Some typical examples of the spectral changes shown in methyl sulfoxide containing 1% potassium *tert*-butoxide are: the initially colorless solution of D-rhamnose phenylhydrazone [0.1 m*M*, $\lambda_{max}^{Me_2SO}$ 280 and 310 (sh) nm] on mixing with base immediately acquired a cherry-red color resulting from two new absorption bands at 405 and 522 nm; the latter band rapidly changed in intensity and, as indicated by a concomitant decrease in intensity of the resulted ESR signal, must be associated with the paramagnetic species. The spectrum of a solution of D-galactose phenylhydrazone (0.1 m*M*) changed from $\lambda_{max}^{Me_2SO}$ 280 and 309nm to $\lambda_{max}^{1\%\ K\text{-}t\text{-}BuO\text{-}Me_2SO}$ 407 and 517 nm; here the two latter bands were replaced by a singlet at 362 nm in about 20 minutes.

Changes also occurred in the visible spectra of sugar osazones in alkaline solution; for example, D-glucose osazone (0.2 *mM*) with $\lambda_{max}^{Me_2SO}$ 312 and 396 nm showed on treatment with base a new band at 485 nm that rapidly changed in intensity; this band was also associated with the presence of a paramagnetic species. The initial intensity of the osazone band at 396 nm diminished about 80% in 12 minutes; similar behavior was observed with the melibiose and lactose osazone solutions. Although the color change in alkaline solutions of the phenylhydrazones of D-mannose, D-glucose, or L-arabinose continued for 24 hours, the observed changes (in the visible spectra) may be ascribed to secondary ionic reactions rather than to that of the free radical species.

Similar spectral changes were also observed for alkaline solutions of inosose phenylhydrazones and phenylosazones. For example, 2,4,6/3,5-pentahydroxycyclohexanone phenylhydrazone (*myo*-inosose-2-phenylhydrazone) [0.1 m*M*, $\lambda_{max}^{Me_2SO}$ 282 and 304 (sh) nm] in alkaline methyl sulfoxide solution gave a band at 445 nm (with rapidly changing intensity), also associated with the diminution of the paramagnetic species; this band was replaced by a new band at 382 nm. Similar behavior was observed with other inososes.

Although the ESR studies on inosose phenylosazones in alkaline solution are not included in this report, their chemical changes were convenient to follow in the visible spectrum. A methyl sulfoxide solution of 3D-3,5/4,6-tetrahydroxy-1,2-cyclohexanedione bis(phenylhydrazone) [L-inosose-1-phenylosazone 0.2 m*M*)] showed a band at 398 nm; this band in alkaline solution shifted to 482 nm and showed only a slow decrease in its intensity (about 1% per minute). This is compared with a decrease in intensity of 4.5% per minute for a band at 482 nm observed for an alkaline solution of 3D-5-*O*-methyl-3,5/4,6-tetrahydroxy-1,2-cyclohexanedione bis(phenylhydrazone) [5-*O*-methyl-*dextro*-inosose-1 phenylosazone] (0.2 m*M*, $\lambda_{max}^{Me_2SO}$ 398 nm); the alkaline solution became almost colorless in about 30 minutes; a rapid mutarotation is observed for the above phenylosazone in ethanol–dioxane or ethanol–pyridine solutions (*27*).

Since neither the *N*-methyl nor the *N*-benzyl derivative of D-mannose phenylhydrazone showed a color change or the presence of paramagnetic species following treatment with base, it was concluded that the colored radical anion from D-mannose phenylhydrazone must arise following the abstraction of the acidic imino proton by base. Thus, this must be a first reaction step before fission of the phenylhydrazine moiety can occur.

The chemical stabilities of sugar phenylhydrazones and inosose phenylhydrazones in the presence of a strong base are not equal. Wolfrom and co-workers (*28*) demonstrated with a series of sugar phenylhydrazones that these compounds can be converted into olefinic azo-sugars on treatment with acetic anhydride and pyridine. Studies in this laboratory showed that other tertiary amine bases, such as quinoline or triethylamine, are as effective as pyridine. However, in the presence of diethylamine or ethylenediamine, the yields of azo-sugars are much lower and are almost zero when bases such as sodium acetate or potassium *tert*-butoxide are substituted. In contrast, inosose phenylhydrazones treated with acetic anhydride–potassium *tert*-butoxide at room temperature gave good yields of the acetylated inosose phenylhydrazones or the acetylated olefinic azo-cyclitols, thus revealing the greater thermodynamic stability of the inosose phenylhydrazones as compared with the sugar phenylhydrazones. The chemical stability of inosose phenylhydrazones is reflected also in the shape of their ESR spectra (Figure 3), obtained by treating a solution of the compound in methyl sulfoxide with potassium *tert*-butoxide and a trace of oxygen (quick acidification (2.5*N* HCl) of the fresh alkaline solution of inosose phenylhydrazone regenerates the original phenylhydrazone in over 95% yield (by spectrophotometry).)

Since the inosose phenylhydrazones could be acylated even in the presence of the strong base and isolated, it was concluded that the radical anions must arise following the abstraction of an acidic imino proton of the phenylhydrazone group by base. On interaction with oxygen the resulting anions (by an electron-transfer mechanism (*29*)) give

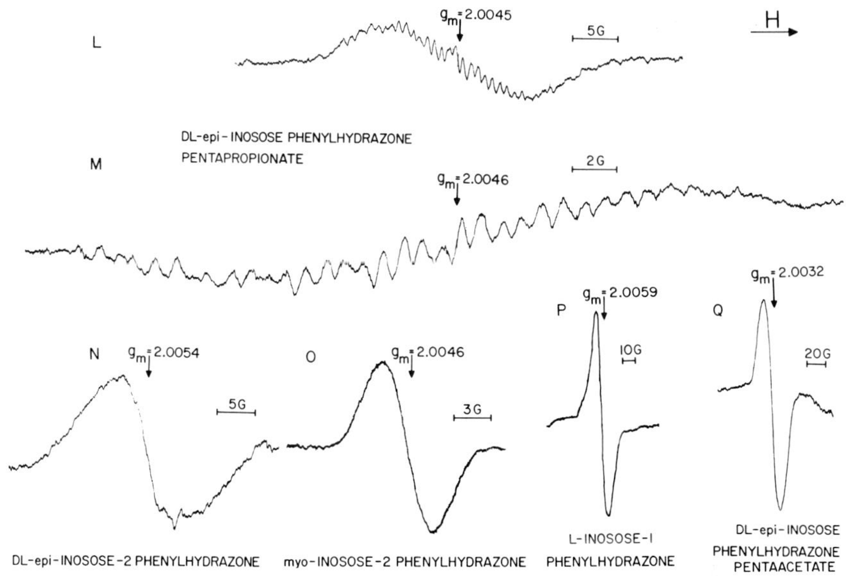

Figure 3. Electron spin resonance spectra of free radicals from inosose phenylhydrazones (2.5 mM to 3.5 mM, in Me_2SO-1% K-tert-BuO)

rise to the radical anions. Some of these ESR spectra are shown in Figure 3.

These spectra are clearly different from the three-line ESR pattern observed with the sugar phenylhydrazones after comparable treatment with base. The spectra shown in Figure 3 are now tentatively interpreted as resulting from the interaction of an unpaired electron with two magnetically equivalent nitrogen atoms of the phenylhydrazine moiety. The observed hyperfine structure in spectrum L (30 lines, with spacing 1.05 G) apparently results from additional splittings by the methyl protons of the propionate group, and in spectrum M further splittings of each component of this spectrum (linewidth 0.2 G) are evident although not satisfactorily resolved. The other inosose phenylhydrazones (spectra N–Q, Figure 3) showed a one-line pattern without a well-defined hyperfine structure.

Conversion of sugar phenylhydrazones into olefinic azo-sugars on treatment with acetic anhydride and pyridine was shown by Wolfrom and co-workers (*28*) (The acetylated forms of the acyclic phenylhydrazones of D-glucose, D-mannose, and D-galactose readily lose the elements of acetic acid to yield 1-phenylazo-*trans*-1-hexenetetrol tetraacetate when treated with warm aqueous ethanol (*28, 30*). It is assumed that atmospheric oxygen partakes in this elimination reaction.) This is a special case of base catalyzed β-elimination reactions of the type proposed by Isbell in 1943 (*31*), involving consecutive electron displacement (which actually is an electron pair displacement). This mechanism (*31, 32, 33*) gave (and is still giving) immeasurable service for interpreting various β-eliminations and aromatization reactions in sugar chemistry and in general organic chemistry. However, this mechanism can not predict the shift of an unpaired electron. To comply with present knowledge, a new free-radical mechanism is proposed that justifies the β-elimination reactions of the sugar phenylhydrazones under mildly basic conditions.

As shown, the major steps in this proposed mechanism are:

(1) formation of an anion (Deprotonation of hydrazobenzene by molecular oxygen in alkaline solution has been shown to proceed *via* a reactive dianion which rapidly loses electrons to oxygen (*34*).)

$$(\mathrm{PhNHNHPh} \xrightarrow{2B^-} \mathrm{Ph\bar{N}\bar{N}Ph} \xrightarrow{2O_2} \mathrm{PhN=NPh} + 2\dot{O}_2^- + 2\mathrm{BH})$$

(2) its interaction with oxygen to form a radical

(3) shift of a radical from nitrogen (The presence of an NH group is required for the formation of the radical ($\mathrm{PhCH{=}N{-}\dot{N}Ph} \longleftrightarrow \mathrm{Ph\dot{C}H{-}N{=}NPh}$) (*35*).) to carbon-1

(4) interaction with oxygen to form a six-membered complex *via* a peroxide ion (autooxidation of phenylhydrazones in non-polar solvents to form hydroperoxides is well known (*see* References *3, 36,* and *37;* however, autooxidation of phenylhydrazones in polar, aprotic, or nucleophilic solvents may produce peroxide ions.)

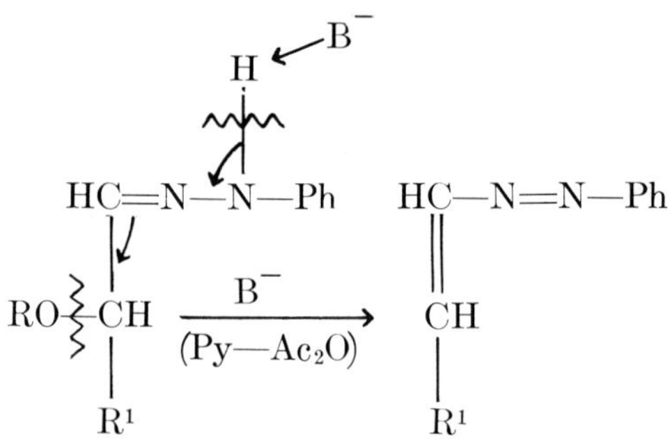

$$R = Ac \quad \text{or} \quad H$$

$$R^1 = (CHOR)_3\,CH_2OR$$

(5) intramolecular homolytic fission of this cyclic intermediate to give an olefinic product. This mechanism accounts for a base-catalyzed β-elimination reaction of sugar phenylhydrazones when (a) oxygen is present and (b) R=Ac, Pr, Bz (or any other ester group). On the con-

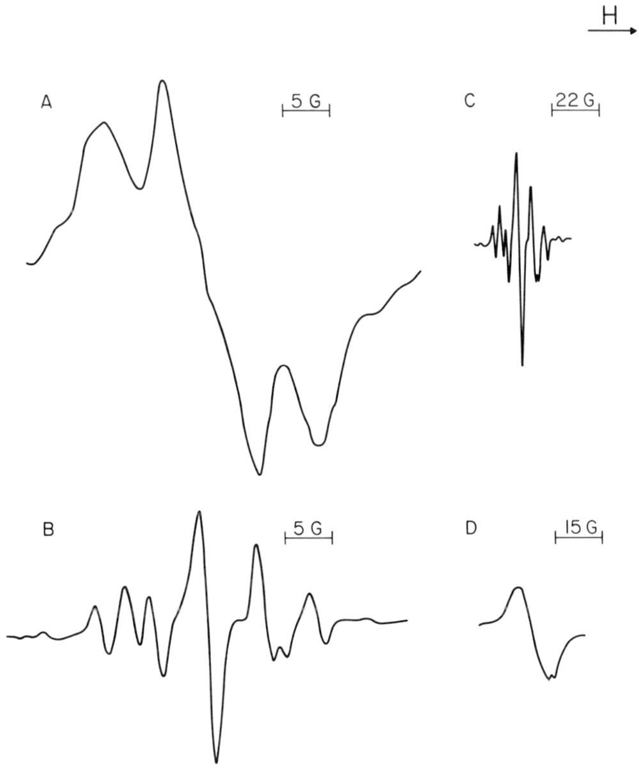

Figure 4. Electron spin resonance spectra of free radicals: (A) from D-mannose phenylhydrazone; (B and C) a rearranged radical from (A); (D) from D-mannose phenylhydrazone [$(CH_3CO)_2O$-C_5H_5–1% K-tert-BuO] (2.5 mM to 3.5 mM, in Me_2SO–1% K-tert-BuO)

trary, little or no olefinic product formation is observed when R=H or CH_3 (or OH group is substituted with the Cl- or NO_2 groups), or the reaction is conducted in the absence of oxygen, and this agrees with the proposed mechanism. Moreover, as is mentioned above, this mechanism does not predict olefinic azo-sugar formation when a sugar is treated, for example, with 1-methylphenylhydrazine; the product of this reaction can be a hydrazone, an osazone, or a poly-methylphenylhydrazone, an alkazone (*38, 39, 40*).

However, a base-catalyzed β-elimination in the cyclohexane series, for example, of the inosose phenylhydrazone, may either stop on carbon 2 to give an arylazocyclohexene derivative (*41*) or, as for the inosose 1,3-bis(phenylhydrazone), proceed further leading to a complete aromatization (*42*). (An alternative free-radical mechanism that may explain the β-elimination reactions of the sugar phenylhydrazones under mildly basic conditions is depicted below.) The important step *3a* can be regarded as a carbon-1 neighboring radical participation to render an acetoxy radical elimination on carbon-2.

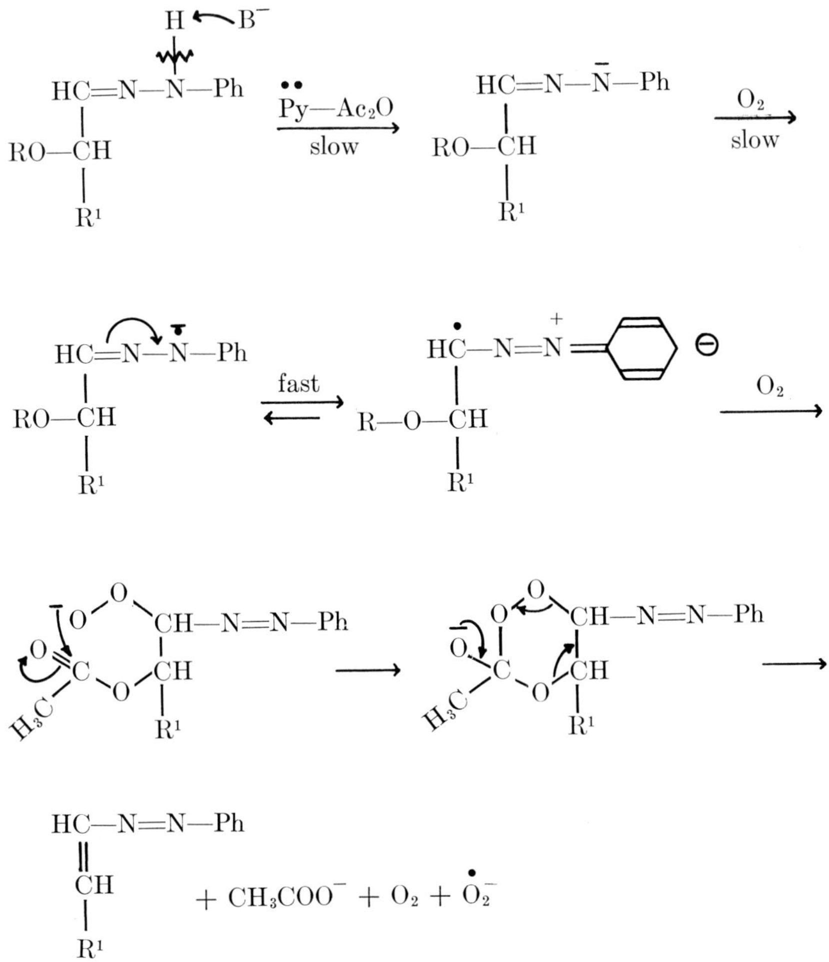

R = Ac, Pr or Bz

$R^1 = (CHOR)_3CH_2OR$

Although it is difficult to obtain an ESR spectrum from a sugar phenylhydrazone under mildly basic conditions, it was possible to record a weak ESR signal (spectrum D, Figure 4) from D-mannose phenylhydrazone in the presence of pyridine, acetic anhydride, and only a trace of potassium *tert*-butoxide; from this reaction mixture an olefinic azo-sugar was isolated.

Figure 4 (A to C) shows changes in the ESR spectra obtained with D-mannose phenylhydrazone treated with 1% potassium *tert*-butoxide in methyl sulfoxide. The original three-line pattern (spectrum A), ascribed to the nitroxide radical, rearranges to a new pattern in about 15 minutes (spectrum B); this pattern remains steady for hours if the solution is kept in the absence of oxygen (spectrum C).

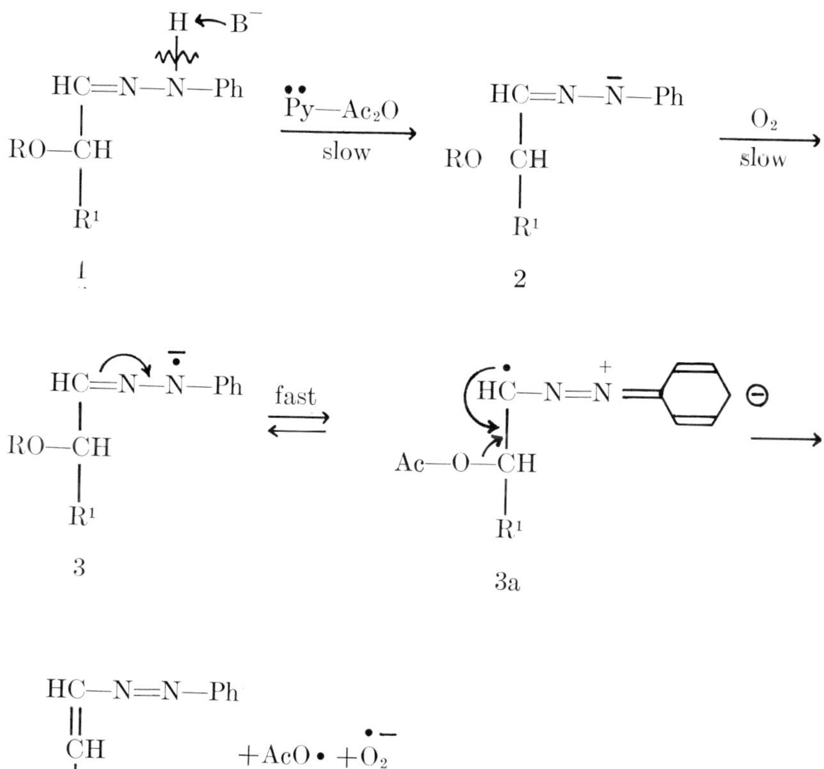

4

$$AcO\bullet + (1) \rightarrow (3) + AcOH$$

$$AcOH + B^- \rightarrow AcO^- + BH$$

$$\text{or } AcO\bullet + \dot{O}_2^- \rightleftarrows AcO^- + O_2$$

$$R = Ac, Pr \text{ or } Bz$$

$$R^1 = (CHOR)_3\, CH_2OR$$

The base-catalyzed, β-elimination reaction of D-mannose phenylhydrazone is consistent with the acyclic structure for the phenylhydrazone in solution. However, the small proportion of a nitroxide radical observed on treatment of the phenylhydrazone with a strong base may indicate the existence also of a fractional proportion in a cyclic structure in equilibrium with the open-ring structure, as was suggested by Blair and Roberts (*43*). The hydrazino moiety required for nitroxide-radical formation could be derived from the cyclic form of D-mannose phenylhydrazone in solution.

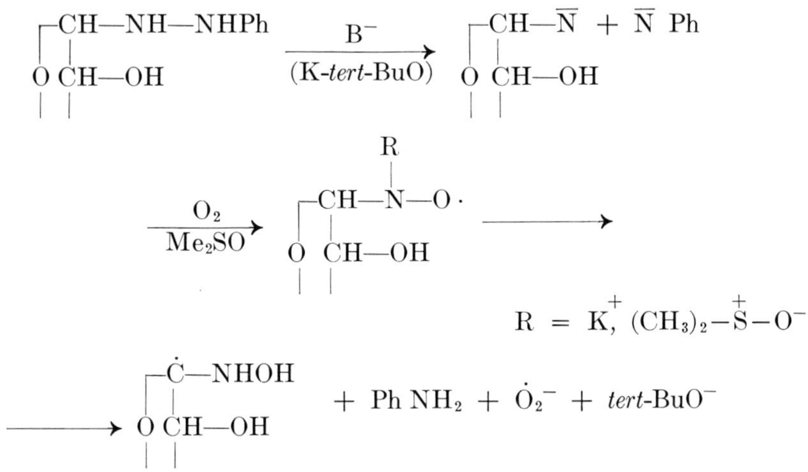

The formulas depicted may reflect radicals observed in the ESR spectra of the D-mannose phenylhydrazone in an alkaline solution (A and B, Figure 4); they show a possible path for the formation of the nitroxide radical and its rearrangement. The major steps in this pathway are:

(1) formation of the phenylhydrazine dianion

(2) homolytic split of the phenylhydrazine moiety followed by oxidation to give a nitroxide radical (spectrum A, Figure 4)

(3) rearrangement by a proton shift of this radical into a new radical (spectra B and C, Figure 4) with a relatively stable hydroxylamine structure. A somewhat overlapped nine-line pattern observed for the rearranged radical (Figure 4 B, $g = 2.0054$) may be interpreted by assuming an interaction of the unpaired electron with nitrogen (a^N 10.75 G) and three magnetically non-equivalent hydrogen atoms (a^H 3.0 G, a^H 2.6 G and a^H 1.65 G).

The observed rearrangement of the paramagnetic species derived from D-mannose phenylhydrazone in an alkaline solution (spectra A, B, and C, Figure 4) may be characteristic for the sugar phenylhydrazones having a proportion of cyclic structure in solution (*e.g.*, those from D-glucose or D-galactose (*43*)). (The change in the ESR spectrum (Figure

4) must result from a radical rearrangement derived from a sugar moiety and not from an aromatic fragment. This can be supported by the observed rise of the complex ESR pattern instead of a dimerization of the paramagnetic species.) However, there was no change in the ESR spectrum of the radical derived from melibiose osazone; its spectrum (V, Figure 2) showed only a progressive decrease in intensity (about 60% in 1 hour, W, Figure 2). The radical species observed from melibiose osazone in alkaline solution may indicate the possible existence of a small proportion of a cyclic structure in equilibrium with the open-ring structure. NMR studies of the sugar osazones in methyl sulfoxide indicate that they have mainly an open-chain chelate structure (*5, 44*). The lack of rearrangement of the radical obtained from melibiose osazone as compared with the rearrangement of the radical from D-mannose phenylhydrazone can be reconciled with the difference between the cyclic structure of the latter compound with that of the osazone. Strong evidence for the existence of the cyclic form of D-*arabino*-hexulose phenylosazone in solution was given by the work of the Percivals (*45*); this form is shown in the structure below. The structure for this osazone

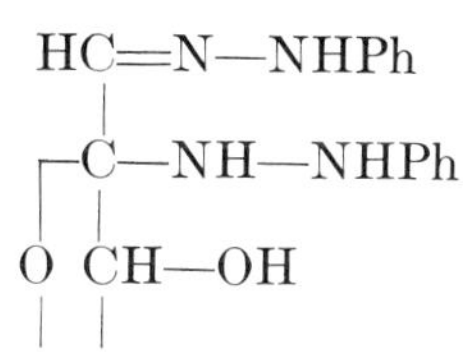

does not give a labile methine proton at C-2; whereas, in the cyclic phenylhydrazone structure the proton on C-1 is free for a radical shift. Thus, the ESR spectrum gives evidence of the presence of a small proportion of a cyclic form in equilibrium with an acyclic form in solutions of D-mannose phenylhydrazone and melibiose osazone.

Experimental

Ultraviolet and visible spectra were recorded with a Beckman DK-2 or Cary 14 spectrophotometer. ESR spectra were measured with a Varian model 4500 ESR spectrometer with 100-kHz field-modulation and detection. The klystron frequency was measured with a transfer oscillator and a frequency counter. The magnetic field was measured by a proton gauss meter monitored by the same frequency counter.

The solutions were examined in a Varian model V-4548 aqueous solution sample cell. All experiments in solution required a trace of oxygen. Freshly prepared solutions of phenylhydrazones or osazones (0.1 to 3.5 m*M*) in methyl sulfoxide were mixed with base [1 to 3% (90 m*M* to 270 m*M*) potassium *tert*-butoxide in methyl sulfoxide] and exposed to air for about 30 sec. Initial ESR spectra were recorded in about 30 sec after mixing.

Methyl sulfoxide (*46*) alone, or in the presence of electrophilic or nucleophilic activating species (acetic anhydride, dicyclohexylcarbodiimide, phosphorus pentoxide, or tertiary amines) is an efficient oxidizing agent that converts alcohols (*47, 48, 49*) or carbohydrates (*50, 51*) into keto compounds and inositols into aromatic derivatives (*52, 53*). Reviews (*46, 54, 55*) list little reported information on use of strong bases (*e.g.*, potassium-*tert*-butoxide, sodium acetate) in oxidations with methyl sulfoxide; however, sodium bicarbonate was used as an acid acceptor in the oxidation of aromatic halides or tosylates (*46*). The procedure below describes the use of a mixture of methyl sulfoxide, acetic anhydride and potassium *tert*-butoxide for acetylation of a sugar or inosose phenylhydrazone to yield a β-elimination product. A procedure for the conversion of *myo*-inosital into its hexaacetate by use of the same mixture is also described. Treatment of *myo*-inositol with methyl sulfoxide–acetic anhydride–pyridine gives pentaacetoxybenzene (*52*) (Acetylation of *myo*-inositol in methyl sulfoxide–acetic anhydride–potassium-*tert*-butoxide, instead of aromatization (as was observed for a system containing pyridine) is evidence for a general stepwise oxidation process (*46, 48*). This also indicates that, in the presence of a strong base, the rate of esterification of *myo*-inositol (a kinetically controlled process) is much faster than that of oxidation.).

DL-6-Phenylazo-(1,3/2,4)–Tetrahydroxy 5-cyclohexene Tetraacetate. To an ice-cold suspension of acetic anhydride (50 ml) and potassium *tert*-butoxide (K-*t*-BuO, 1 gram) was added a solution of 2,4.6–3,5-pentahydroxycyclohexanone phenylhydrazone (*myo*-inosose-2 phenylhydrazone) (5 grams) in methyl sulfoxide (50 ml.). The pale solution was stirred in an ice bath for 30 min and then kept in a refrigerator for 16–20 hours. Decomposition of the reaction mixture with water containing 15–20% of ethyl alcohol gave bright yellow crystals of the title compound, yield 3.8–4.1 grams. Recrystallization (90–95% aqueous ethanol) gave golden needles, mp 131°–133°C (shrinking at 112°–114°C) (literature: mp 112°–114°C, 135°–136°C (*41*)), dimorphism of this compound has been reported (*41*), λ_{max}^{MeOH} 22.5 (sh) (ϵ~12,500); 228 (ϵ~31,900); 233 (ϵ~13,000); 302 (ϵ~23,200), and 438 nm (ϵ~450); NMR ($CDCl_3$), τ 3.05 (4 H, q, $J_{4,5}$ ~1.0 Hz) τ 7.83, 7.85, and 8.05 (acetyl methyl groups). IR and other spectral properties of this compound were identical with those of the authentic compound.

Acetylation of *myo*-Inositol. A mixture of *myo*-inositol (5 grams), acetic anhydride (40 ml), methyl sulfoxide (50 ml), and K-*t*-BuO (2 grams) was stirred at 70°C (water bath, hood) for 45 min and kept overnight at room temperature. Decomposition of the brown reaction mixture with water gave *myo*-inositol hexaacetate (mp 212°–214°C) in 85% yield. Thin layer chromatography (TLC) and NMR analysis of the crude product showed only a trace proportion of an aromatic component.

Attempted Acetylation of D-Mannose Phenylhydrazone. Five grams of D-mannose phenylhydrazone was acetylated as described for *myo*-inosose-2-phenylhydrazone. Work-up of the reaction mixture (after 24 hours) gave a yellow syrup (3.6 grams) which failed to crystallize. TLC analysis of this syrup showed only a small proportion of an olefinic azo-sugar (*28*); the structure of the main portion of the syrup is under further study.

D-Mannose Phenylhydrazone. This was prepared by the method of Blair and Roberts (*43*), mp 199°–200°C (from aqueous pyridine).

D-Galactose Phenylhydrazone. This was prepared by the method of Mester (*56*), mp 159°–160°C.

L-Rhammose Phenylhydrazone. This was prepared by the method of Fischer and Tafel (*57*), mp 159°C.

D-Glucose Phenylhydrazone. This was prepared by the procedure of Fischer and Stahel (*58*), mp 143°C.

L-Arabinose Phenylhydrazone. This was prepared by the procedure of Chavanne (*59*) mp 152°–153°C.

Melibiose Phenylhydrazone and Melibiose Phenylosazone. These were prepared by the method of Scheibler and Mittelmeier (*60*), mp 145°C and 178°C, respectively.

D-Mannose 2-Methylphenylhydrazone. This was prepared by the method of Hilger (*61*), mp 181°C.

D-Mannose 2-Benzylphenylhydrazone. This was prepared by the method of Votocek, Valentin, and Leninger (*62*), mp 171°–172°C.

D-Glucose Osazone. This was prepared as reported by Fischer (*63*), mp 207°–208°C.

Lactose Osazone. This was prepared as reported by Fischer (*64*), mp 211°–212°C.

2,4,6/3,5-Pentahydroxycyclohexanone Phenylhydrazone(*myo*-Inosose-2 Phenylhydrazone); 2D-2,3,4,6/5-Pentahydroxycyclohexanone Phenylhydrazone, and 2L-2,4,5,6/3-Pentahydroxycyclohexanone Phenylhydrazone (DL-epi-Inosose Phenylhydrazone). These were prepared by the procedure of Posternak (*65*), mp 181° and 189°–191°C, respectively.

2D-2,3,5/4,6-Pentahydroxycyclohexanone Phenylhydrazone (D-*myo*-Inosose-1 Phenylhydrazone). This was prepared by the procedure of Magasanik and Chargaff (*66*), mp 196°–197°C.

3D-3,3/4,6-Tetrahydroxy-1,2-Cyclohexanedione Bis(phenylhydrazone) (D-*myo*-Inosose-1 Phenylosazone). This was prepared by the published procedure (*27*), mp 207°–209°C.

3D-5-O-Methyl-3,5/4,6-Tetrahydroxy-1,2-Cyclohexanedione Bis-(phenylhydrazone) (5-O-Methyl-*dextro*-inose-1 Phenylosazone). This was prepared by the published procedure (*27*), mp 206°–207°C.

2D-2,3,4,6/5-Pentahydroxycyclohexanone and 2L-2,4,5,6/3-Pentahydroxycyclohexanone Phenylhydrazone Pentaacetate and Pentapropionate (DL-epi-Inosose-2 Phenylhydrazone Pentaacetate and Pentapropionate). These were prepared by the published procedure (*41*), mp 134°–135°C (pentaacetate) and mp 90°–91°C (pentapropionate).

Acknowledgments

The author thanks J. F. Liebman for a useful discussion, N. Adams for the ESR measurements, and Robert Schaffer for his interest.

Literature Cited

1. Grundon, M. F., Scott, M. D., *J. Chem. Soc.* **1964**, 5674.
2. Ragno, F., *Gazetta* (1945) **75**, 175.

3. Buckingham, J., *Quart. Rev.* (1969) **23,** 37.
4. Stanek, J. M., Kočourek, J., Pačak, J., "The Monosaccharides," p. 537–565, Academic, New York, 1963.
5. Khadem, H. El, *Advan. Carbohyd. Chem.* (1965) **20,** 139.
6. Pinkus, G., *Ber.* (1898) **31,** 31.
7. Diels, O., Meyer, R., Onnen, O., *Liebigs Ann.* (1936) **525,** 94.
8. Khadem, H. El, Wolfrom, M. L., Shafei, Z. M. El, Ashry, S. H. El, *Carbohyd. Res.* (1967) **4,** 225.
9. Simon, H., Moldenhauer, W., *Chem. Ber.* (1967) **100,** 3121.
10. Russell, C. S., Lyons, R., *Carbohydrate Res.* (1969) **9,** 347.
11. Evans, W. L., *Chemical Rev.* (1942) **31,** 537.
12. Lemieux, R. U., "Molecular Rearrangements," p. 746–52, P. de Mayo, Ed., Vol. 2, Interscience, New York, 1963.
13. Fatiadi, A. J., *J. Res. Nat. Bur. Std.* (1971) **75A,** 573.
14. Nelsen, S. F., Hintz, P. J., *J. Amer. Chem. Soc.* (1971) **83,** 7105.
15. Pake, G. E., Townsend, Jr., Weisman, S. I., *Phys. Rev.* (1952) **85,** 682.
16. Forrester, A. B., Hay, J. M., Thomson, R. H., "Organic Chemistry of Stable Free Radicals" pp. 195–217, Academic, New York, 1968.
17. Ayscough, P. B., "Electron Spin Resonance in Chemistry," p. 287, Methuen, London, 1967.
18. Forrester, A. B., Hay, J. M., Thomson, R. H., "Organic Chemistry of Stable Free Radicals," pp. 281–341, Academic, New York, 1968.
19. Buchachenko, A. L., "Stable Radicals," p. 89, Consultants Bureau, New York, 1965.
20. Forrester, A. B., Hay, J. M., Thomson, R. H., "Organic Chemistry of Stable Free Radicals," p. 287, Academic, New York, 1968.
21. Terabe, S., Konaka, R., *J. Amer. Chem. Soc.* (1971) **93,** 4306.
22. Wajer, T., Mackor, A. J. W., Deboer, T. J., van Vorst, J. D.W., *Tetrahedron Lettr.* (1967) 1941.
23. Caragheogheopol, A., Caldararu, H., Constantinescu, T., Em Shini, P. V., *J. Amer. Chem. Soc.* (1971) **92,** 6766.
24. Ref. *16,* pp. 218–219.
25. Gilbert, B. C., Norman, R. O. C., *J. Chem. Soc. (B),* **1966,** 772.
26. Russell, G. A., Strom, E. T., Talaty, E. R., Weiner, S. A., *J. Amer. Chem. Soc.* (1966) **88,** 1998.
27. Fatiadi, A. J., *Carbohydrate Res.* (1968) **8,** 135.
28. Wolfrom, M. L., Thompson, A., Lineback, D. R., *J. Org. Chem.* (1962) **27,** 2563.
29. Russell, G. A., "Radicals Ions," p. 87, Kaiser, E. T., and Kevan, L., Eds., Interscience, New York, 1968.
30. Wolfrom, M. L., Fraenkel, G., Lineback, D. R., Komitzky, F., Jr., *J. Org. Chem.* (1964) **29,** 457.
31. Isbell, H. S., *Ann. Rev. Biochem.* (1943) **12,** 213.
32. Isbell, H. S., *J. Res. Nat. Bur. Std.* (1944) **32,** 45.
33. Fatiadi, A. J., Isbell, H. S., *J. Nat. Bur. Std.* (1964) **68A,** 287.
34. Blackadder, D. A., Hinshelwood, D., *J. Chem. Soc.,* **1957,** 2898.
35. Walling, C., "Free Radicals in Solution," p. 415, Wiley, New York, 1957.
36. Hawkins, E. G. E., *J. Chem. Soc.* **1971,** 1474.
37. *Idem.,* "Organic Peroxides," p. 118, Spon, London, 1961.
38. Chapman, K. J., Welstead, Jr., W. J., Murphy, T. J., King, R. W., *J. Amer. Chem. Soc.* (1964) **86,** 732.
39. *Idem., J. Amer. Chem. Soc.* (1967) **89,** 7005.
40. Simon, H., Moldenhauer, W., Kraus, A., *Chem. Ber.,* (1969) **102,** 2777.
41. Fatiadi, A. J., *Carbohyd. Res.* (1968) **7,** 89.
42. Fatiadi, A. J., Isbell, H. S., *Proc. Amer. Chem Soc. Nat. Meet, 145th,* (September 1963); NBS Technical Note 274, pp. 69–70, 1965.
43. Blair, H. S., Roberts, G. A. F., *J. Chem. Soc. (C),* **1967,** 2425.

44. Mester, L., *Angew. Chem. Int. Ed. Engl.* (1965) **4**, 574.
45. Percival, E. E., Percival, E. G. V., *J. Chem. Soc.*, **1935**, 1398.
46. Epstein, W. W., Sweat, F. W., *Chem. Rev.* (1967) **67**, 247.
47. Albright, J. D., Goldman, L., *J. Amer. Chem. Soc.* (1965) **87**, 4114.
48. *Ibid.* (1967) **89**, 2416.
49. Pfitzner, K. E., Moffatt, J. G., *J. Amer. Chem. Soc.* (1963) **85**, 3027.
50. Lindberg, B., Slessor, K. N., *Carbohyd. Res.* (1966) **1**, 492.
51. Horton, D., Jewell, J. S., *Carbohyd. Res.* (1966) **2**, 251.
52. Fatiadi, A. J., *Chem. Commun.*, **1967**, 441.
53. Ibid., *J. Chem. Eng. Data* (1969) **14**, 118.
54. Brimacombe, J. S., *Angew. Chem. Int. Ed. Engl.* (1969) **8**, 401.
55. Moffatt, J. G., "Oxidation," R. L. Augustine and D. J. Trecker, Eds., Vol. 2, Chapter 1, Marcel Dekker, New York, 1971.
56. Mester, L., "Methods in Carbohydrate Chemistry," p. 117, R. L. Whistler and M. L. Wolfrom, Eds., Vol. 2, Academic, New York, 1963.
57. Fischer, E., Tafel, J., *Ber.* (1887) **20**, 2566.
58. Fischer, E., Stahel, R., *Ber.* (1891) **24**, 528.
59. Chavanne, G., *Compt. Rend.* (1902) **134**, 661.
60. Schribler, C., Mittelmeier, H., *Ber.* (1890) **68**, 1438.
61. Hilger, A., *Ber.* (1903) **36**, 3197.
62. Votocek, E., Valentin, F., Leminger, O., *Collect. Czech. Chem. Commun.* (1931) **3**, 250.
63. Fischer, E., *Ber.* (1887) **20**, 821.
64. Fischer, E., *Ber.* (1908) **41**, 76.
65. Posternak, T., *Biochem. Prep.* (1952) **2**, 57.
66. Magasanik, B., Chargaff, E., *J. Biol. Chem.* (1948) **175**, 929.

RECEIVED January 7, 1972. Mention of tradenames in this paper does not imply recommendation or endorsement by the National Bureau of Standards.

7

Complexes of Sugars with Cations

S. J. ANGYAL

School of Chemistry, The University of New South Wales, Kensington, N.S.W. 2033, Australia

Sugars and cyclitols containing an axial-equatorial-axial (ax–eq–ax) sequence of three hydroxy groups in a six-membered ring or a cis–cis sequence in a five-membered ring form complexes with metal ions in aqueous solution. The alkaline earth metals form strong complexes; sodium and potassium form weak complexes. Complex formation will change the anomeric equilibrium of sugars like allose and gulose and the conformational equilibrium of compounds like methyl β-D-ribopyranoside and β-D-lyxopyranose. The equilibrium composition of methyl glycosides in methanol is also affected by the presence of a salt—e.g., *calcium chloride.*

Many complexes of sugars and sugar derivatives with inorganic salts and bases have been isolated in solid and often crystalline form (*1*). These compounds have constant stoichiometric composition; their isolation does not however provide evidence that such complexes exist in solution. The only compound of this type which has been examined by x-ray crystallographic analysis, sucrose sodium bromide dihydrate (*2*), has both inorganic ions attached to more than one molecule of sucrose—an arrangement unlikely to be stable in dilute solution. The structures of other sugar–metal complexes are not yet known. [The structures of several sugar–metal complexes have recently been determined by x-ray crystallography: β-D-mannofuranose, $CaCl_2$, $4H_2O$ (*28*), lactose, $CaBr_2$, $7H_2O$, α-D-galactopyranose, $CaBr_2$, $3H_2O$, and *myo*-inositol, $CaBr_2$, $5H_2O$ (*29*).]

There is considerable evidence, based on various physical measurements, that sugar–cation complexes exist in solution. Among these methods, reviewed extensively by Rendleman (*1*), are solubility, optical rotation, vapor pressure determination, and most importantly, electrophoresis. Whereas the sum total of this evidence establishes that at least some metals form complexes with some sugars and polyols in solution, the structures of these complexes have never been determined, nor have stability constants been measured. We now recognize configurational and conformational requirements for complex formation and report the measurement of stability constants.

The literature on metal complexes of carbohydrates through 1965 has been fully reviewed by Rendleman (*1*), and we shall therefore only discuss recent work. We shall not discuss the complexes formed with strong bases, such as calcium and barium oxide; these are salts in which the sugar acts as a weak acid, losing one or several protons. Nor shall we discuss the complexes formed with anions of oxyacids—*e.g.*, borate, stannate, periodate, etc. ions; all these are compounds formed by covalent bonds in alkaline solution. We are concerned only with complexes formed with cations in neutral aqueous solution; there is no evidence for the formation of complexes between sugars and simple anions in neutral aqueous solution. (For an example of complex formation between a sugar derivative and chloride ion in chloroform solution, *see* Reference *3*.)

Although nuclear magnetic resonance (NMR) spectroscopy has solved structural and conformational problems in carbohydrate chemistry (*4*), it has not been used to investigate metal complex formation until recently. McGavin, Natusch, and Young (*5*) have reported that complexing with metal ions in aqueous (D_2O) solution causes a downfield shift of the signals of some protons in the NMR spectra of several glycosides—*e.g.*, methyl α-L-arabino-pyranoside and -furanoside. Assuming the formation of 1:1 complexes, they calculated stability constants which they found too small to be biologically important (K *ca.* $0.1M^{-1}$). They concluded that two adjacent oxygen atoms (eq–eq or eq–ax) are involved in complex formation.

Mills's experiments on the electrophoresis of polyols started recent developments. Mills found (*6*) that in slightly acid solutions of many

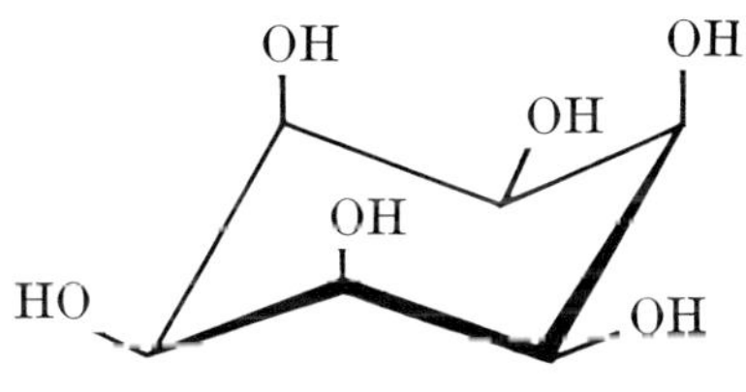

1

salts *cis*-inositol (**1**) migrated rapidly towards the cathode and epi-ino sitol (**2**), 1,2,3,4,5/0-cyclohexanepentol, and D-talose had good, though

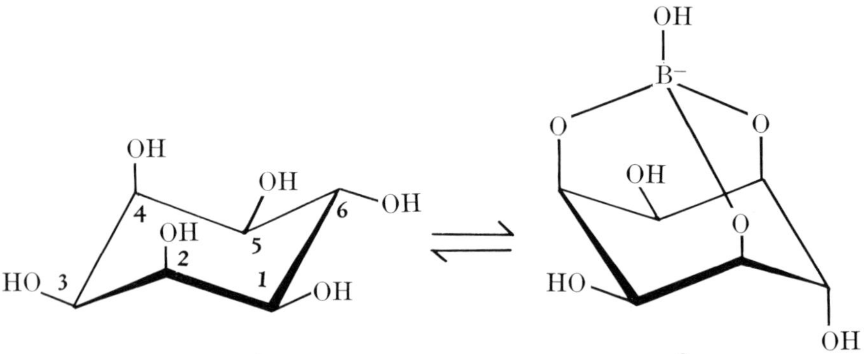

2 3

lesser, mobility (Table I). The behavior of the cyclitols recalls their electrophoresis in sodium borate solution in which they show strong mobility towards the anode (7). It has been shown that tridentate borate

Table I. Relative Mobilities[a] of Polyols on Paper Electrophoresis in Aqueous Solutions of Cations (6)

Compound	Ba^{2+}	Mg^{2+}	Na^{+}	K^{+}
cis-Inositol	82	20	10	6
epi-Inositol	25	2	3	3
L-Iditol	13	1	1	1
Allitol	5	1	1	1
D-Talose	18	0	2	2

[a] Cationic movements are given as percentages of the anionic movement (about 10 cm) of *p*-nitrobenzenesulfonic acid on the same strip with 2,3,6-tri-*O*-methyl-D-glucose as non-migrating marker. The electrolyte solution was a 0.1 *M* solution of the metal acetate in 0.2 *M* acetic acid.

anions (*e.g.*, **3**) are formed by participation of three axial oxygen atoms; *cis*-inositol, which has three axial hydroxyl groups on the same side of the molecule in either of its two chair conformations, moves much more rapidly than epi-inositol and 1,2,3,4,5/0-cyclohexanepentol, which first have to flip to their less stable chair form. Mills assumed (*6*) that participation of the three axial oxygen atoms is also responsible for complexing with cations. This hypothesis can be tested by NMR spectroscopy. On complex formation with borate anions, the NMR spectrum of epi-inositol (**2**) changes (*8, 9*), and the new values of the coupling constants show that this change results from the chair–chair interconversion. Should the same change be observed on complexing with metal cations, the extent of complexing and hence the stability constants of the complexes could be determined.

When calcium chloride was added to a solution of epi-inositol (**2**) in deuterium oxide, a significant change occurred in the NMR spectrum

(Figure 1). There was no change, however, in the splitting of the signals, indicating that the conformation of the cyclitol is not altered. The chemical shifts changed, however, with all signals moving to lower field; one signal—identified by its coupling constants as that of H-3—moved more than the others, its position depending upon the concentration of calcium ions. Similar downfield shifts have been observed in related compounds (1,2,3,4,5/0- and 1,2,3,4/5-cyclohexanepentols) on complex formation.

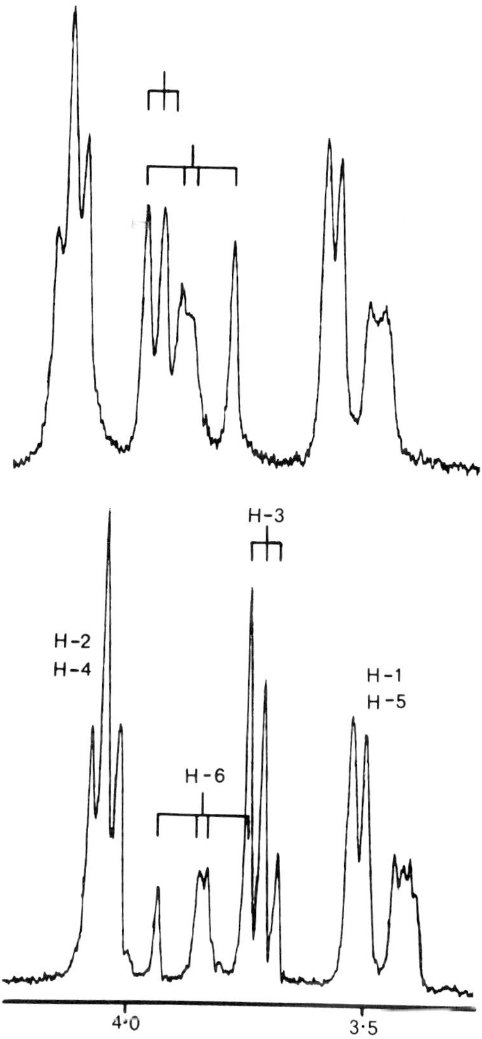

Figure 1. 100 MHz NMR spectrum of epi-inositol in D_2O (lower curve) and in 1.12 M $CaCl_2$ solution in D_2O (upper curve)

These changes in the chemical shifts recall those caused by complex formation with paramagnetic cations (shift reagents) and are being studied.

It was assumed (*10*) that the strong effect of the cation on the signal of H-3 indicates that O-3 is involved in complex formation. It was then postulated that an axial, an equatorial, and an axial oxygen atom on three consecutive carbon atoms of a six-membered ring form a suitable arrangement (**4**) for complexing with cations. Although this assumption

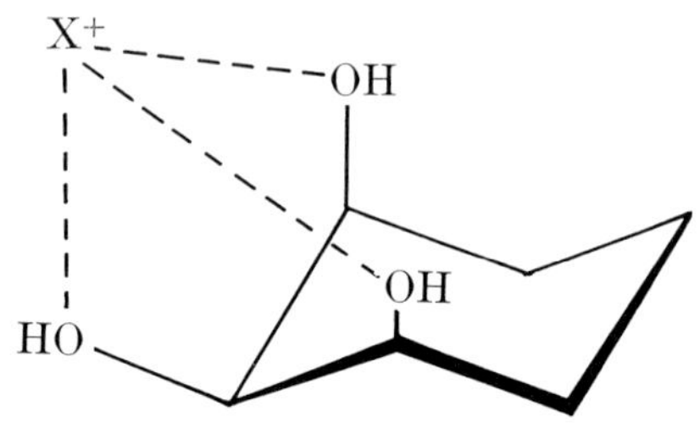

4

has not yet been directly verified, it explains all the facts known so far about complex formation of cyclic polyols; compounds which have this arrangement (such as epi- and allo-inositols, 1,2,3,4,5/0- and 1,2,3,4/5-cyclohexanepentols, D-talose and D-ribose) show good mobility on electrophoresis in salt solutions; whereas those which do not, move slowly or not at all.

Jeffrey and Kim (*11*) have shown that in the crystals of epi-inositol the two axial oxygen atoms are separated by 2.95 A, owing to their mutual repulsion which causes some flattening of the cyclohexane ring. The distance between an axial and a vicinal equatorial oxygen atom is about the same; the three oxygen atoms in an ax–eq–ax sequence therefore form an approximately equilateral triangle. This seems to be a favorable steric arrangement for complex formation.

cis-Inositol is a unique compound; it contains three ax–eq–ax sequences and also a similar triangle formed by the three axial oxygen atoms, offering four sites for complexing (**5**). Therefore, that its com-

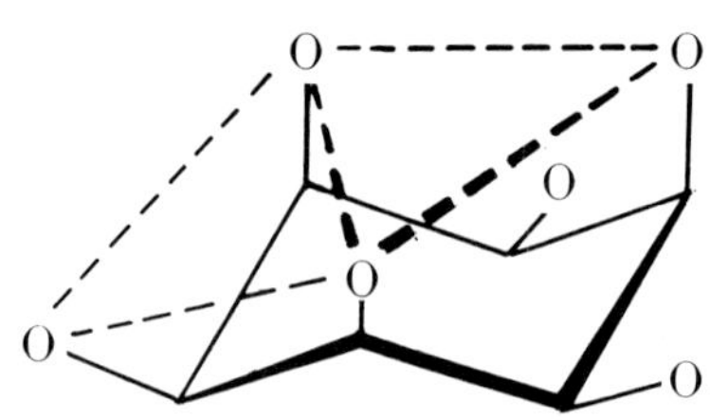

5

plexes are much more stable than those of any other polyol studied so far is not unexpected. Recent x-ray crystallographic analysis has shown (*12*) that each axial oxygen atom is equidistant from four other (two axial, two equatorial) oxygen atoms. It is not yet known whether the triaxial arrangement forms stronger or weaker complexes than the ax–eq–ax sequence. By potentiometric measurement of the calcium ion concentration, the stability constant of the *cis*-inositol–calcium ion complex was 21 M^{-1} and that of the epi-inositol complex, 3.2 M^{-1} (*13*).

Speculation is made about why epi-inositol flips into the triaxial form to form a tridentate complex with borate, but complexes without flipping with metal ions. The distances between the three axial oxygen atoms are similar to those between the three oxygen atoms in the ax–eq–ax sequence. However, in the triaxial borate complexes (**3**) the bonds are tetrahedral; if one were formed at the ax–eq–ax oxygen atoms, the bonds would be considerably distorted. However, the angles have much less effect on complex formation with cations which does not involve covalent bonds.

Alditols which have three consecutive *threo* hydroxyl groups also show good mobility, and this becomes greater when there are four or five such groups—*e.g.*, D-iditol, *meso-glycero-ido* heptitol (*15*). In the planar zigzag conformation these compounds have three (or more) oxygen atoms in the same geometrical arrangement (**6**) as in the ax–eq–ax sequence on a six-membered ring in the chair form.

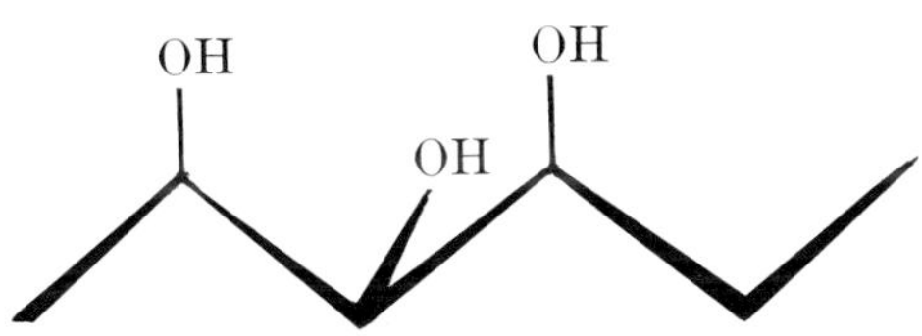

6

By extrapolation of the changes in the chemical shifts in the NMR spectrum of epi-inositol, the stability constant of the calcium complex has been calculated to be *ca.* 3M^{-1}—*i.e.*, about 70% of the cyclitol is present as a cationic complex in 1M calcium chloride solution. However, the extrapolation cannot be accurately carried out because there are secondary changes in chemical shifts caused by weaker complexing at two oxygen atoms.

To confirm the hypothesis of ax–eq–ax complexing and to obtain better values of the stability constants, a chemically mobile system was sought in which the equilibrium position can be altered by complex formation. Such a system was found in the anomeric equilibrium of D-allose.

α-D-Allopyranose in its more stable C1 conformation (*7*) contains an ax–eq–ax sequence of hydroxyl groups (at C-1–C-3), but the β-anomer does not. Addition of a complexing cation should increase the proportion

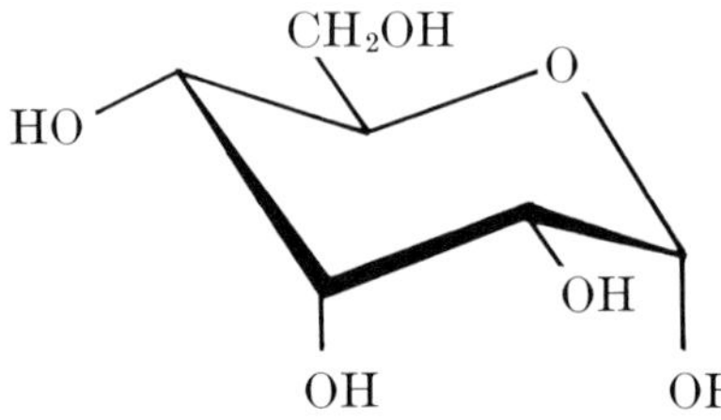

7

of the α-anomer in equilibrium. The proportion of the pyranose and furanose anomers in aqueous solution can be determined from the NMR spectrum; the signals of the anomeric protons are well separated. At 30°C the equilibrium composition is 13.8% α-pyranose, 77.5% β-pyranose, 3.4% α-furanose, and 5.3% β-furanose (*14*). In a 0.85 *M* solution of calcium chloride the composition was 37.2 : 54.5 : 4.5 : 3.8 (Figure 2). The proportion of the α-pyranose form has increased, and at higher concentrations of calcium ions it becomes greater than that of the β-form. On the assumption that a 1:1 complex is formed and that the proportion of the uncomplexed α- and β-pyranose forms is not affected by the presence of salts, the stability constant was calculated and was found to be *ca.* 6 M^{-1}. On addition of calcium chloride the anomeric proton signal of α-D-allopyranose shifted downfield by 0.1 ppm (Figure 2).

By using salts of other metals (except those which are paramagnetic), the complexing ability of various cations towards an ax–eq–ax sequence of hydroxyl groups can be compared by this method. The complexing ability depends on the charge and on the size of the cation, but these two properties do not fully determine it. Lanthanum (*K ca.* 10 M^{-1}), calcium, and strontium form the strongest complexes; barium forms somewhat less stable complexes (*K ca.* 2.9 M^{-1}), and magnesium and cadmium form much less stable (*K ca.* 0.1 M^{-1}) complexes. Other dipositive cations (zinc, lead) form complexes of intermediate strength. The alkali metals complex weakly; sodium shows the highest stability constant (*K ca.* 0.1 M^{-1}), potassium a somewhat smaller one; no complex formation was observed with lithium, rubidium, cesium, and silver. Lanthanum ion causes a greater downfield shift (0.22 ppm in 0.9 *M* solution) of the H-1 signal in the NMR spectrum (Figure 2) of the α-pyranose than any of the other metals tried, probably because of its greater positive charge.

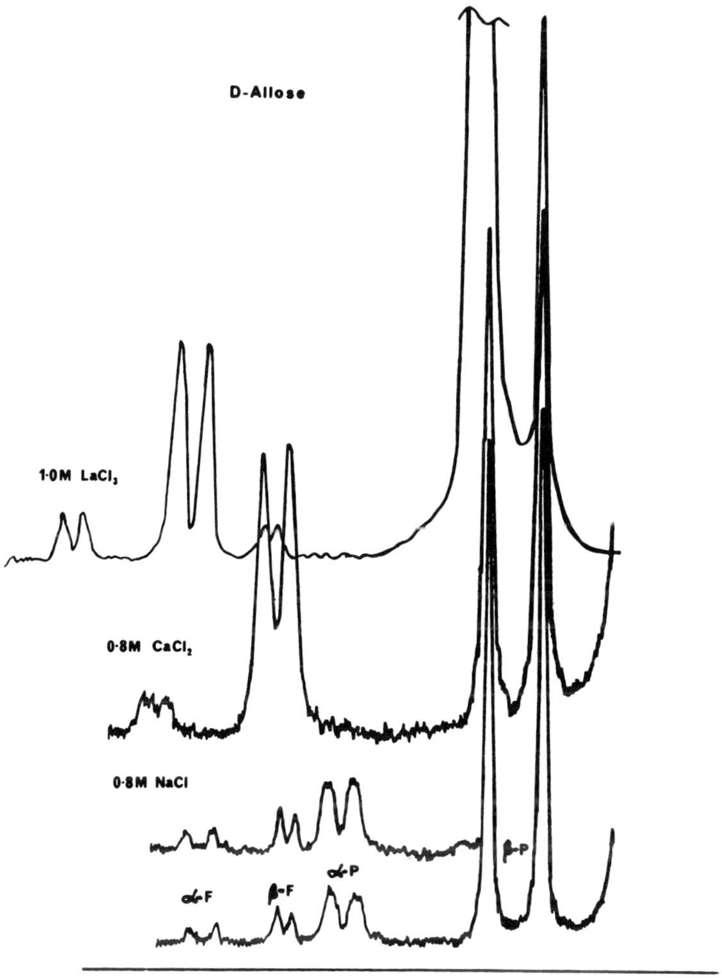

Figure 2. The anomeric region of the 100 MHz NMR spectrum of D-*allose in D_2O (lower curve), and in 0.8* M *NaCl, 0.8* M *$CaCl_2$, and 1.0* M *$LaCl_3$ solution in D_2O (successive curves) at equilibrium*

The NMR spectra of D-allose show that the signal of the anomeric proton of the α-furanose form also increases and moves downfield on the addition of metal ions. The spectrum of the β-furanose form, like that of the β-pyranose form, is not affected (Figure 2). It appears that a cis–cis sequence of three hydroxyl groups on a five-membered ring also constitutes an arrangement suitable for the formation of complexes with metal ions. Since the proportion of furanoses in the equilibrium solution of D-allose is small, the changes in their proportion cannot be accurately measured. Hence the equilibrium of 5-*O*-methyl-D-ribose was studied;

the NMR spectrum showed that there is 33% of α-furanose (**8**) and 67% of β-furanose at equilibrium in aqueous solution; on addition of barium chloride to make the solution 1.6 *M*, the composition changes to 70% α and 30% β. The stability constant therefore is 4.3 M^{-1}. The α-furanose can easily take up a conformation (**9**) in which O-1, O-2, and

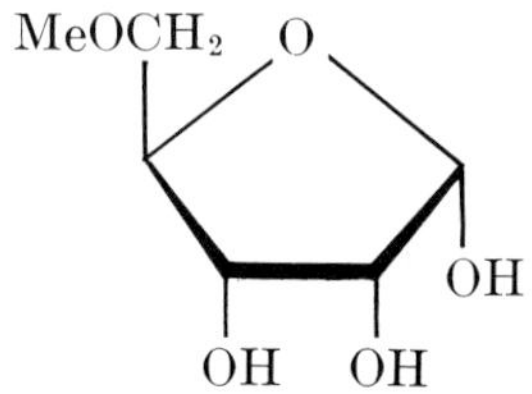

8

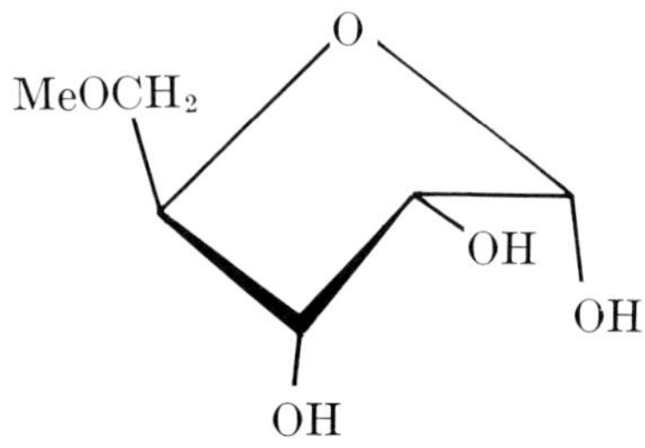

9

O-3 are quasi–axial, quasi–equatorial, and quasi–axial, respectively; this arrangement does not differ greatly from the favored ax–eq–ax arrangement in six-membered rings, but the two outer oxygen atoms are somewhat more distant from each other. Hence it may be significant that the large barium cation (ionic radius 1.29 A) seems to complex better with furanoses than with pyranoses; more examples, however, are required to confirm this hypothesis.

No substantial changes in the equilibrium composition or in the NMR spectra of D-glucose, D-mannose, and D-arabinose were observed on the addition of calcium chloride. These sugars lack the required ax–eq–ax sequence of hydroxy groups.

Some 40 years ago Isbell (*16*) observed complex formation between calcium ions and a sugar which possesses the ax–eq–ax sequence—*i.e.*, α-D-gulopyranose. He observed that an equilibrated solution of D-gulose, $CaCl_2$, underwent further mutarotation on dilution with water; and he correctly interpreted the phenomenon by postulating that α-D-gulopyranose (**10**), but not β-D-gulopyranose, forms a complex with calcium

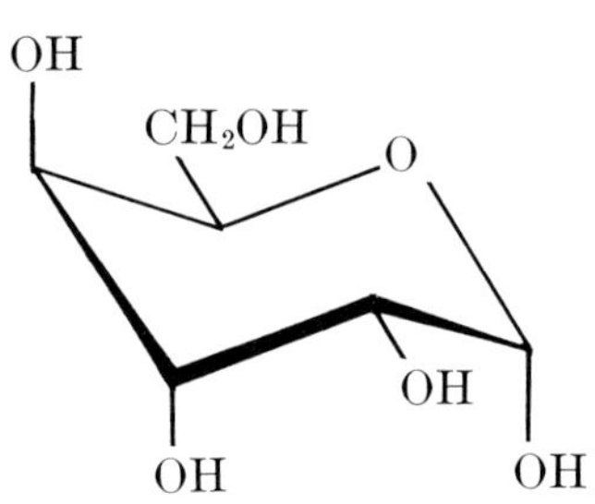

10

ions and that this complex dissociates on dilution. Isbell had no way, however, to detect the structural features required for complex formation.

In suitably selected cases metal complexing can change the conformational rather than the chemical equilibrium. Methyl β-D-ribopyranoside in aqueous solution consists of an equilibrium between the C1(D) (**11**) and 1C(D) (**12**) forms, the former predominating (*17*). Only the 1C (D) form has an ax–eq–ax sequence of hydroxyl groups. On addition of calcium chloride to the solution, the equilibrium shifts in favor of the IC(D) form, as seen from the value of $J_{1,2}$ which decreases from 5.4 Hz in D_2O to 2.5 Hz in 2.1 *M* $CaCl_2$. This corresponds to a change of the proportion of the C1(D) form from 57 to 12%.

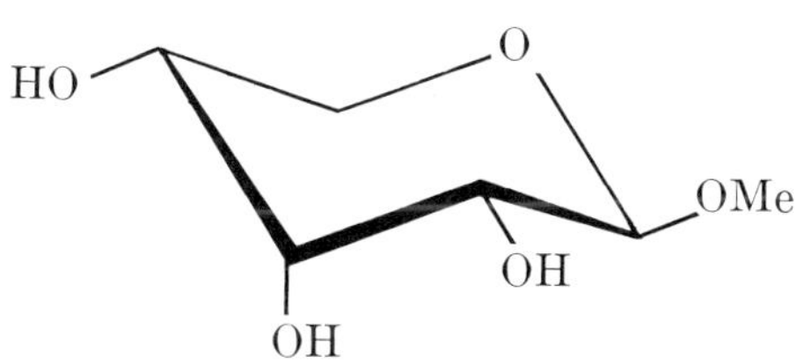

11

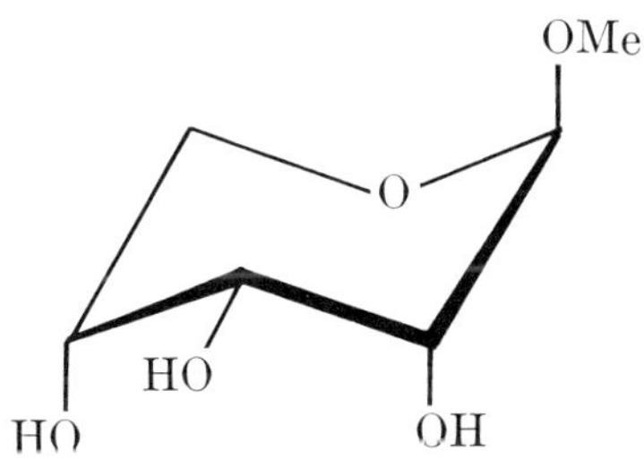

12

Chemical and conformational changes occur on complex formation of D-lyxose. The α-pyranose form lacks the ax–eq–ax sequence, but the β-pyranose form contains it in its 1C(D) conformation. This conformation, which has three axial hydroxyl groups, is not normally observed

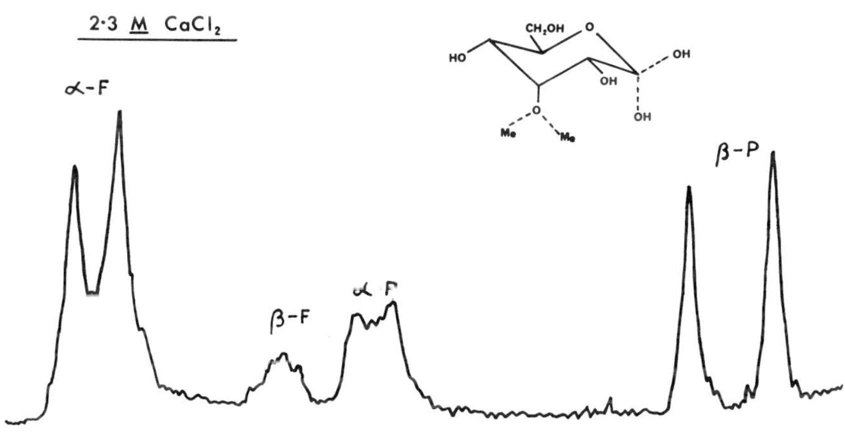

Figure 3. The anomeric region of the 100 MHz NMR spectrum of D-lyxose in D_2O (lower curve) and in 2.2 M $CaCl_2$ solution in D_2O (upper curve) at equilibrium

(*17*). On addition of calcium chloride to an equilibrium solution of D-lyxose (Figure 3), the anomeric signal of the β-pyranose form increases, moves to lower field, and acquires a larger coupling constant. All these changes are caused by the increase in the concentration of the 1C(D) form of the β-pyranose. The large change in the chemical shift of the β-anomeric proton, which causes it to cross over onto the low-field side of the α-signal, is caused not only by complexing but mainly by the change from an axial to an equatorial position. The spectrum at lower field also shows the anomeric signals of the two furanose forms; the α- and β-signals coincide in aqueous solution but not in the presence of calcium ions which cause the signal of the β-furanose to move downfield.

The hydrogen atoms of the hydroxyl groups are not involved in the formation of metal complexes. At least one of them can be replaced by a methyl group. Methyl α-D-allopyranoside shows downfield shifts in its NMR spectrum similar to those of α-D-allopyranose on addition of calcium chloride; it seems to complex to a similar extent. One feature of this NMR spectrum is that the H-2 triplet, hidden among the other protons in the uncomplexed state, appears at calcium concentrations over 1.5*M*. This is one of several cases where complex formation reveals details of NMR spectra or improves resolution by separating the signals of protons from each other.

To study the effect of a methyl substituent on complex formation, 3-*O*-methyl-D-allose was synthesized. Its NMR spectrum showed a considerably higher proportion (21%) of furanose forms than that of D-allose. The probable reason for this increase is steric hindrance of the methyl group in the pyranose forms (as shown in the formula in Figure 4). Whichever staggered conformation the methyl group takes up around the C—O bond, it will have a 1,3-parallel interaction with O-2 or O-4; hence, the pyranose forms are destabilized. This steric hindrance seems to reduce the complexing ability, and the increase in the proportion of the α-pyranose form is small (*K ca.* 0.6 M^{-1}) on addition of calcium ions. The methyl group, however, does not interfere with complex formation in the α-furanose form (Figure 4); the stability constant of the complex is *ca.* 3 M^{-1}, and at high calcium concentration the α-furanose becomes the predominant species in equilibrium.

All of these examples illustrate the complexing ability of an ax–eq–ax sequence of oxygen atoms on a six-membered or of a cis–cis sequence on a five-membered ring. Attempts have been made to find other complexing sites on sugar molecules but with little success. The O-3, O-5, and O-6 arrangement of hexofuranoses, which allows the ready formation of tridentate periodate (*18*), and orthoformate esters (*19*), does not form stable complexes with metal ions. In solution the side-chain of 1,2-*O*-isopropylidene-α-D-glucofuranose (**13**) is in the extended zig-zag form

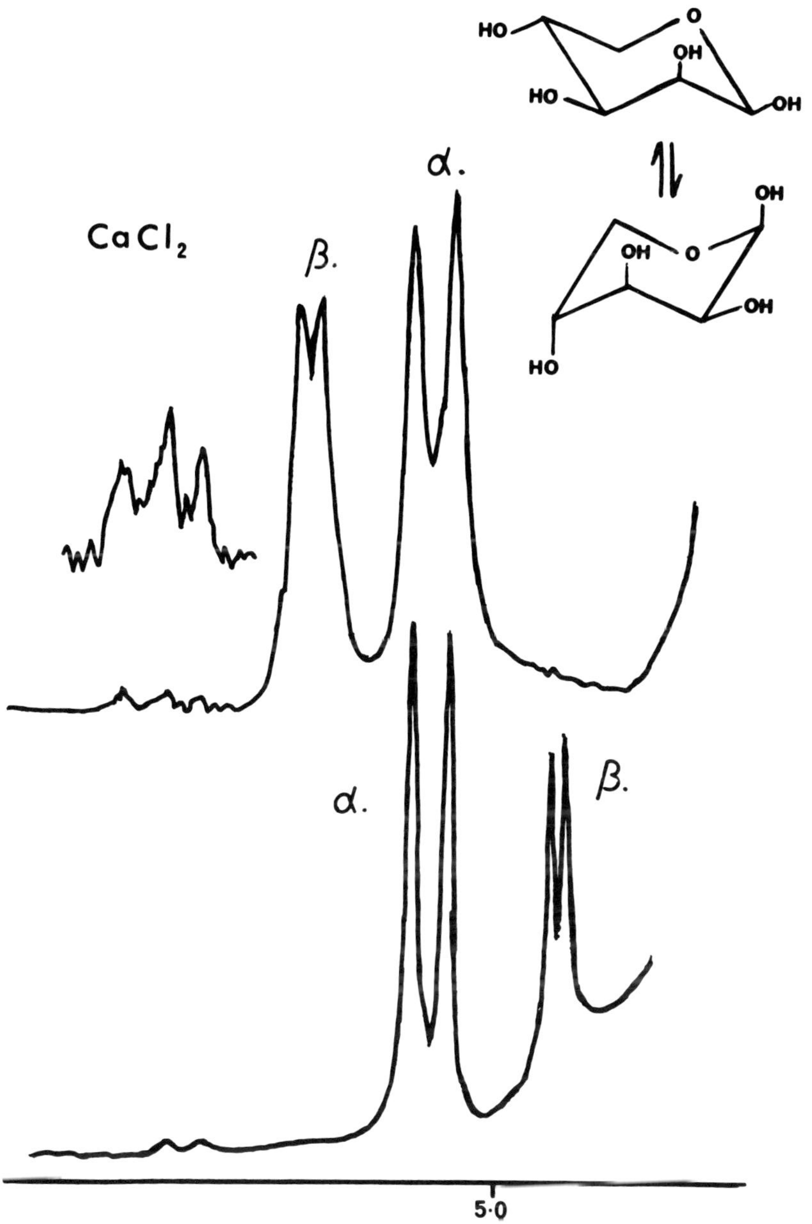

Figure 4. The anomeric region of the 100 MHz NMR spectrum of 3-O-methyl-D-allose in 2.3 M $CaCl_2$ solution in D_2O at equilibrium

in which O 3 and O-6 are not close to each other; addition of calcium chloride does not change this conformation ($J_{4,5}$ remains unchanged at 8.5 Hz).

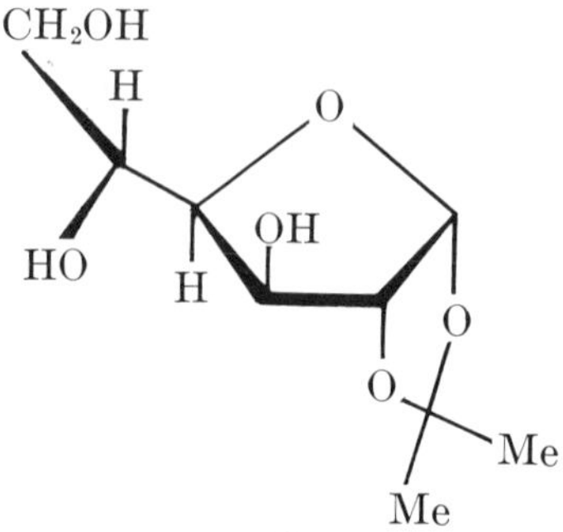

13

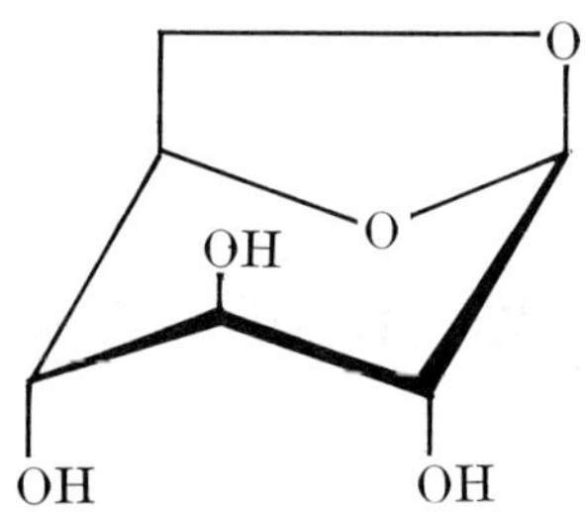

14

The two axial oxygen atoms on C-2 and C-4 and the ring oxygen atom in 1,6-anhydro-β-D-glucopyranose (**14**) are in a similar geometrical arrangement to those in an ax–eq–ax sequence. Nevertheless only weak complexes are formed, possibly because the electron density on the ring-oxygen is lower than on the others (*20*) or because the distance between O-2 and O-4 is larger (3.3 A) than in monocyclic pyranoses (*21*).

The sugar–metal ion complexes recall the crown compounds (*22*) and cryptates (*23*), tailor-made molecules with great complexing ability. However, in those compounds six (or more) oxygen atoms are arranged in a conformation suitable for complexing; the sugars with only three oxygen atoms taking part form much less stable complexes. One sugar has been found in which four oxygen atoms could be taking part in complex formation; D-*ribo*-hex-3-ulose (*24*) migrates much more rapidly on electrophoresis in a solution of calcium acetate than the other hex-3-uloses. A conformation with four oxygen atoms coordinated to the ion (**15**) is possible but has not yet been proved responsible for complex formation.

The metal complexes of sugars have many potential uses. Sometimes sugars which have not yet been obtained in the crystalline form can be

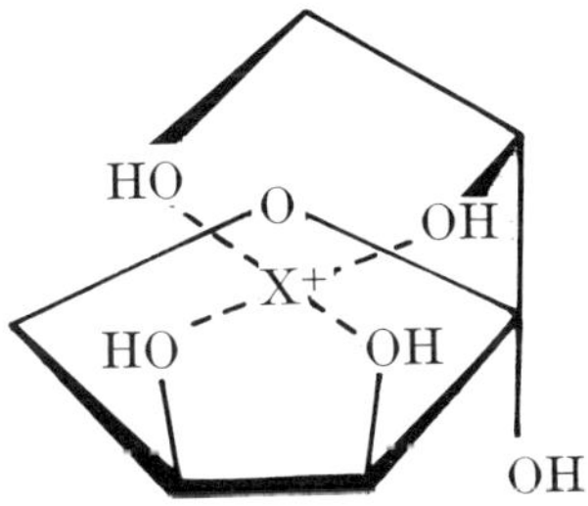

15

crystallized as complexes with a salt—*e.g.*, α-D-gulopyranose, α-D-mannofuranose. Polyols with a suitable configuration can sometimes be purified through their metal complexes. For example, epi-inositol crystallizes readily as the complex with calcium chloride and can thereby be separated from the isomeric *myo*-inositol, which remains in the mother liquors. Qualitative separation of many polyols can be done by electrophoresis in salt solutions. Preparative separation of some sugars can be achieved by chromatography on an anion-exchange resin containing a metal ion. For example, Jones and Wall (*25*) separated the products formed by the action of alkali on D-galactose—*i.e.*, D-galactose, D-tagatose, and D-talose—by chromatography on a column of Dowex 50 W in the barium form. The order of emergence of these compounds from the column is now understood and can be predicted; D-talose contains an ax–eq–ax sequence in either of its pyranose forms, D-tagatose only in the (less stable) β-pyranose form, and D-galactose not at all. Sometimes analysis of the NMR spectrum of a polyol is facilitated by the down-field shift of signals owing to complex formation with a cation. Methyl α-D-allopyranoside has already been mentioned. In the spectrum of epi-inositol (Figure 1), when lanthanum chloride is added to make the solution about 1 *M*, the triplet of H-3 moves downfield so much that it no longer overlaps that of H-6; every proton then appears well separated.

Finally, the outcome of chemical reactions which lead to equilibria can be altered by adding a cation which forms a complex with one of the products. The potential of this method is greatly increased by the fact that complexing with metal ions is much stronger in alcoholic than in aqueous solutions. When D-allose is heated with methanolic hydrogen chloride, the predominant product is methyl β-D-allopyranoside (*26*). When the reaction is carried out in the presence of calcium chloride, the α-pyranoside becomes the main product (*27*). The biological implications of the complexing of sugars with cations have yet to be studied in detail.

Acknowledgment

This work was supported by a grant from the Australian Research Grant Committee. The author thanks J. A. Mills for many helpful discussions and is grateful to K. P. Davies, Linda Ball, and Hildegarde Hufnagel for patiently running numerous NMR spectra.

Literature Cited

1. Rendleman, Jr., J. A., *Advan. Carbohyd. Chem.* (1966) **21**, 209.
2. Beevers, C. A., Cochran, W., *Proc. Roy. Soc. London Ser. A.* (1947) **190**, 257.
3. Martin, J. S., Hayami, J.-I., Lemieux, R. U., *Can. J. Chem.* (1968) **46**, 3263.
4. Hall, L. D., *Advan. Carbohyd. Chem.* (1964) **19**, 51.
5. McGavin, D. G., Natusch, D. F. S., Young, J. F., *Proc. Int. Conf. Coord. Chem., 11th,* 134 (1969).
6. Mills, J. A., *Biochem. Biophys. Res. Commun.* (1961) **6**, 418.
7. Angyal, S. J., McHugh, D. J., *J. Chem. Soc.* (1957) 1423.
8. Posternak, T., Janjic, D., Lucken, E. A. C., Szente, A., *Helv. Chim. Acta* (1967) **50**, 1027.
9. Garegg, P. J., Lindström, K., *Acta Chem. Scand.* (1971) **25**, 1559.
10. Angyal, S. J., Davies, K. P., *Chem. Commun.* (1971) 500.
11. Jeffrey, G. A., Kim, H. S., *Carbohyd. Res.* (1970) **15**, 310.
12. Freeman, H. C., Langs, D. A., Nockolds, C. E., Oh, Y. L., unpublished results.
13. Angyal, S. J., Hickman, R. J., unpublished results.
14. Angyal, S. J., Pickles, V. A., *Austral. J. Chem.* (1972) **25**, 1695.
15. Mills, J. A., personal communication.
16. Isbell, H. S., *J. Res. Nat. Bur. Stand.* (1930) **5**, 741.
17. Angyal, S. J., *Angew. Chem. Int. Ed. Engl.* (1969) **8**, 157.
18. Perlin, A. S., Rudloff, E. von, *Can. J. Chem.* (1965) **43**, 2071.
19. Freudenberg, K., Jacob, W., *Chem. Ber.* (1947) **80**, 325.
20. Park, Y. J., Kim, H. S., Jeffrey, G. A., *Acta Crystallogr. Sect. B* (1971) **27**, 220.
21. Rao, V. S. R., Vijayalakshmi, K. S., Sundararajan, P. R., *Carbohyd. Res.* (1971) **17**, 341.
22. Pedersen, C. J., *J. Amer. Chem. Soc.* (1967) **89**, 7017; (1970) **92**, 386, 391.
23. Lehn, J. M., Sauvage, J. P., *Chem. Commun.* (1971) 440.
24. Angyal, S. J., Evans, M. E., *Austral. J. Chem.* (1972) **25**, 1495.
25. Jones, J. K. N., Wall, R. A., *Can. J. Chem.* (1960) **38**, 2290.
26. Angyal, S. J., James, K., *Austral. J. Chem.* (1970) **23**, 1209.
27. Evans, M. E., Angyal, S. J., *Carbohyd. Res.* (1972) **23**, in press.
28. Craig, D. C., Stephenson, N. C., Stevens, J. D., *Carbohyd. Res.* (1972) **22**, 494.
29. Bugg, C. E., Cook, W. J., *Chem. Commun.* **1972**, 727.

RECEIVED December 2, 1971.

8

Conformational Preferences for Solvated Hydroxymethyl Groups in Hexopyranose Structures

R. U. LEMIEUX and J. T. BREWER

University of Alberta, Edmonton, Alberta, Canada

*The effect of the 4-hydroxyl group on the conformational preferences of the hydroxymethyl group in water, dimethyl sulfoxide, 1,2-dichloroethane, and mixtures of the latter two solvents was assessed by comparing 1,2-anhydro-2,3,4-trideoxy-*D-glycero-*hexitol with 1,5-anhydro-2,3-dideoxy-*D-erythro *hexitol and its mono- and dimethyl ethers and with 1,5-anhydro-2,3-dideoxy-*D-threo-*hexitol. NMR, empirical rules for optical rotation at the D-line of sodium and when possible infrared absorption were used. The effect of the ring-oxygen atom was compared with that of a methylene group by examining the diastereoisomeric 2-hydroxymethyl-cyclohexanols. Conclusions are drawn regarding solvation effects on non-bonded interactions and on intramolecular hydrogen bonding.*

The establishment of the rotamer population arising from rotation about the CH_2OH—C bond of hexopyranoid structures represents an interesting challenge to conformational analysis and a subject which may be considerably important for a proper understanding of the chemical and biological properties of these structures. It is not a simple problem. The important rotamers will strongly prefer the staggered orientations; however, as previously pointed out (*1*), these conformations, especially for an acyclic structure, are not likely to assume the idealized staggered arrangement where the torsion angles are all 60°. Instead, the minimum energy structures for these rotamers are probably a compromise between the non-bonded interactions of the neighboring groups and how these are modified through bulk and specific solvation effects, including demands for intramolecular hydrogen bonding. Thus, for example, the non-bonded interaction of the *syn*-axial groups *X* of **1** is expected to be best compromise involving ring distortion. However, for the *syn*-axial

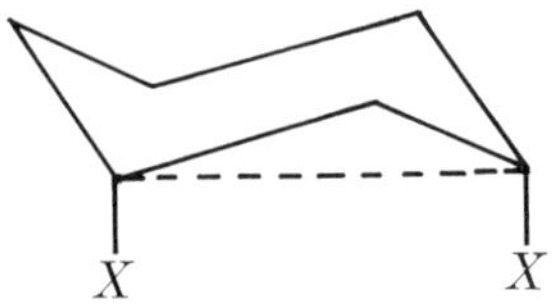

1

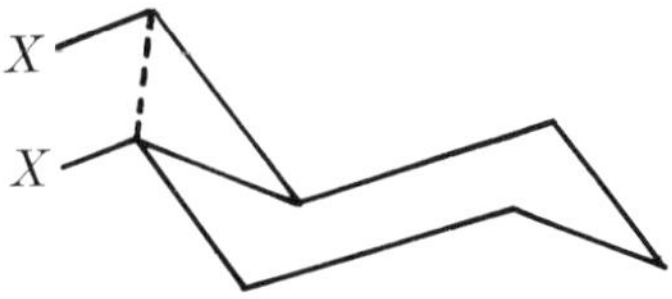

2

like situation in **2** the energetically best geometry for this rotamer may result from rotation about the single C—C bond to afford a torsion angle for the *X* substituents about the "virtual" bond (designated by the dotted line in **1** and **2**) substantially greater than the virtually zero torsion angle for these substituents in **1**. These factors complicate greatly an assessment of the rotamer population of **2** arising from rotation about the CH_2X—C by quantitative analysis of nuclear magnetic resonance (NMR) parameters. An attempt was therefore made to assess the rotamer preferences by combining NMR, infrared spectroscopy, and optical rotation. Only a semi-quantitative treatment of the data could be made, and the conclusions reached are presented to describe the problem better as much as to offer probable solutions.

To simplify the discussion of the results obtained for compounds **3** to **10** (*see* Table II), the rotamers arising from rotation about the exo-

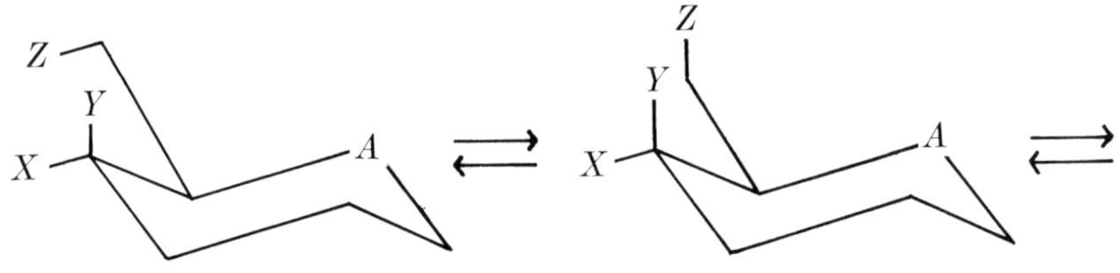

a **b**

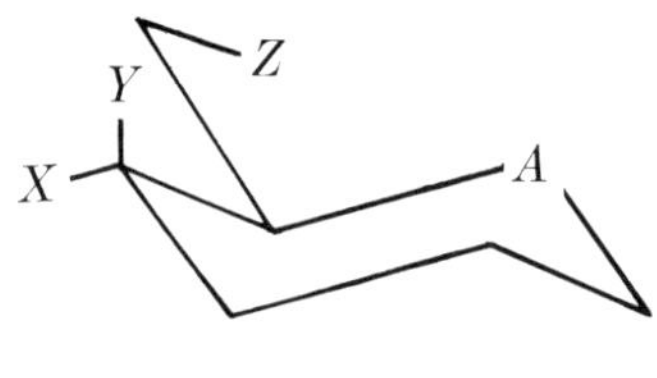

c

cyclic C—C bond will be referred to as conformers **a**, **b** and **c**. For example, for compound **3** ($A = O, X = Y = H, Z = OH$), the designation **3c** will refer to rotamer shown below.

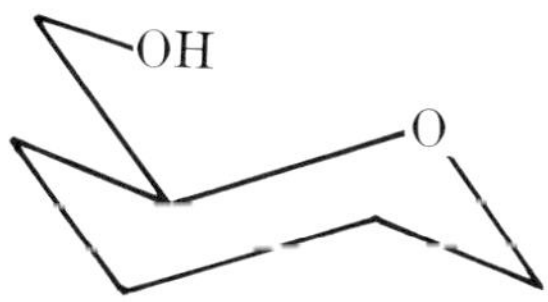

3c

It is necessary first to reappraise and extend the approach toward empirical rules for optical rotation reported by Lemieux and Martin (*1*). This approach uses parameters derived from simple model compounds rather than through those which provide the best fit for the rotations of numerous complex structures, as was done by Whiffen (*2*). The revision of the simple rules proposed by Lemieux and Martin is discussed in detail in a forthcoming publication (*3*). Table I contains the data pertinent to this research.

The increase in the OH/OH contribution from $\pm 45°$ (*2, 4*) to $\pm 55°$ seems to be demanded by the data in Table I. Brewster has used $\pm 50°$ for the OH/C contribution. This value is lowered to $\pm 45°$ as a good comprise for the rotations found for the model compounds. The introduction of the relatively high OH/O_r contribution returns to a distinction of oxygen atoms as presented by Whiffen (*2*) and assigned as a permolecular effect by Brewster (*4*). This procedure is justified by the agreement between the calculated and found rotations presented in Table I and as further elaborated by Lemieux, Daniel, Brewer, and Nagabhushan (*3*). Although the O/C contribution given by Lemieux and Martin is now known to be erroneously low, the correction has little effect on the result obtained in using the simplified rules for complex structures since the change primarily involves the uses of 45° instead of 35° for OH/O_r (endo) $-$ OH/C.

The plain optical rotatory dispersion curves for the compounds considered in this paper are such that rotations at any wavelength in the range 365–589 mμ could have been used to establish simple empirical rules for rotation (*3*). This may not be the case for glycosidic structures because of absorption to higher wavelength by the acetal grouping. Also, with regard to observed rotations for the simple model compounds listed in Table I, these rotations may be expected to be somewhat numerically smaller than the actual rotations of the compounds in the conformations shown. Table VI shows that the rotation of a compound generally decreases with increasing temperature. Thus, the development

Table I. Molecular Rotations in Water at 25°C

Compound	*Analysis*[a]	*Calc.*	*Found*
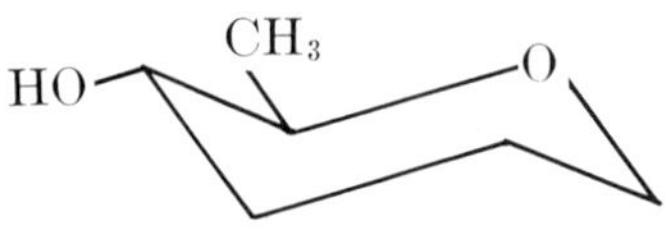	OH/C	45°	39°
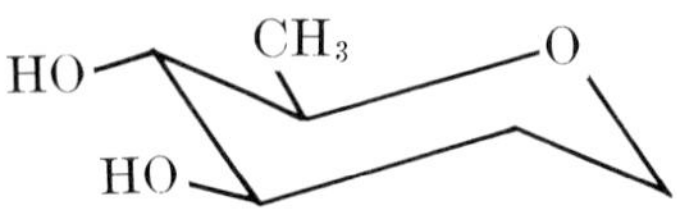	OH/C—OH/OH	−10°	−14.5°
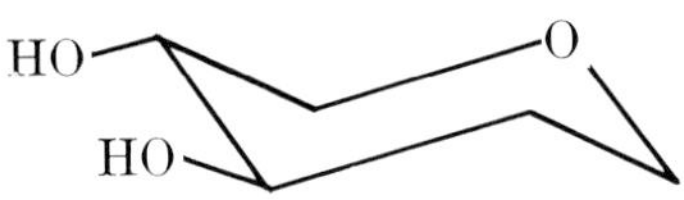	OH/OH	−55°	−46° (3) −53° (5)
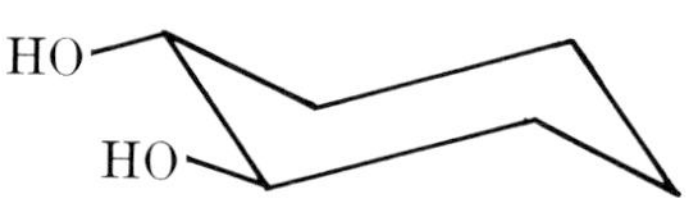	OH/OH	−55°	−54° (6)
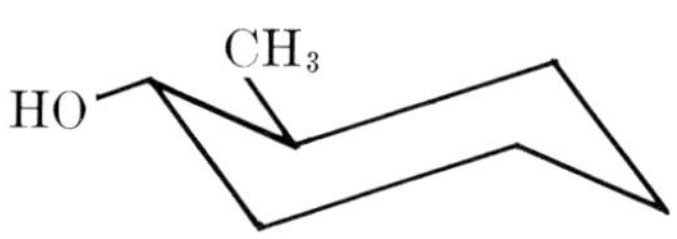	OH/C	45°	49° (7) 44° (8)
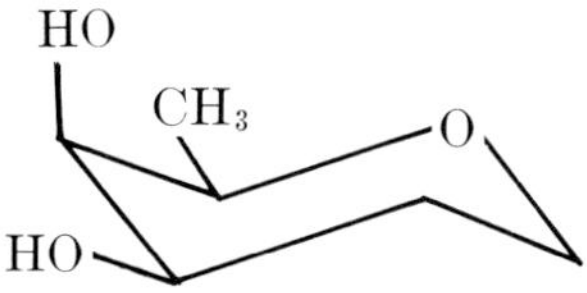	OH/OH + OH/O_r − 2 OH/C	55°	55°

[a] OH/C = ± 45°, OH/OH = ±55°, OH/O_r = ±90°.

and use of these empirical rules for rotation are subject to considerable arbitrariness. The main justification for their use is that, regardless of the many assumptions (*1*), the rules clearly show promise as an additional basis for assessing conformational equilbria, and it is considered that it is only through continuing experimental assessment that a proper understanding will eventually be achieved.

The problem which results clearly from the above revised rotation rules is whether or not the OH/O_r parameter should be applied to conformations involving a gauche interaction between the ring oxygen and the hydroxyl group of the hydroxymethyl group of compounds such as **3**, **5** and **7**. This would apply if the exaltation of the OH/O parameter from 55° to 90° when O is O_r instead of OH results from the different chemical nature and relatively fixed orientation of O_r with respect to the OH group as compared with two gauche hydroxyl groups. However, permolecular effects in the sense of higher order units of conformational asymmetry may be involved in the high apparent value for OH/O_r (endo). A purpose of this research was to gain information on this matter, but no definitive insight was achieved.

Table III presents the rotational analysis for compounds **3–7**, in the three staggered orientations for the hydroxymethyl side chain to give conformers of types **a**, **b**, and **c**. The ranges in rotation expected for certain conformations results from the above-mentioned uncertainty in the value for OH/O_r (exo).

The purpose of this research was to compare the effect on the conformational equilibrium for the hydroxylated compounds listed in Table II of changing the solvent from dimethyl sulfoxide, which is expected to minimize intramolecular hydrogen bonding, to 1,2-dichloroethane which should promote such bonds. These solvation effects on conformational equilibria were then to be compared with those of water which can serve as a hydrogen donor and hydrogen acceptor in hydrogen bond formation. As will be seen, the conformational equilibria generally appear similar for water and dimethyl sulfoxide but often different from those in 1,2-dichloroethane.

First the intramolecular hydrogen bonding possibilities, as shown by infrared absorption of solutions in 1,2-dichloroethane are examined. von R. Schleyer (9) predicted that although the absorptions for the hydroxyl group are to lower wave numbers than for solutions in carbon tetrachloride, the frequency differences for hydrogen bonded hydroxyls should be about the same. From the infrared data in Table IV it is apparent that the frequency shift between hydroxyl groups that are free and those engaged in the 1,2-type hydrogen bridge is too small with 1,2-dichloroethane as solvent for these absorptions to be adequately resolved. The absorptions at 3900 cm^{-1} for compounds **4**, **5**, and **6** must arise from the free hydroxyl groups. The near equal absorbances at 3900 and 3505 cm^{-1} for compound **4** and at 3900 and 3500 cm^{-1} for compound **6** indicate that these compounds have, as expected, nearly one-half the hydroxyl groups free and the other half in the 1,3-type intramolecular hydrogen bond (*11, 12, 13*). Thus, the 3580 cm^{-1} absorption for compound **3** may be assigned to free hydroxyl. However, since the compound is extensively involved in

Table II. Solvation Effects on Molecular Rotation (c,0.4)

Compound		Molecular Rotation at 25°C		
		Water	*Dimethyl Sulfoxide*	*1,2-Dichloroethane*
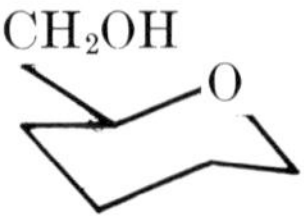	**3**	26	−6.5	12[a]
	**4**	72	67	13
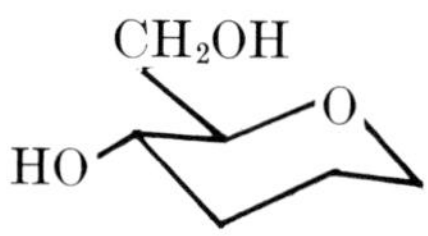	**5**	71	64	42
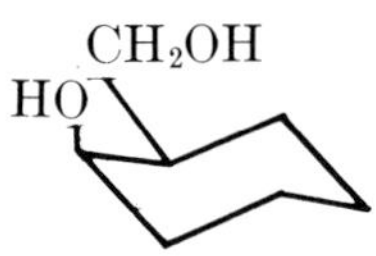	**6**	−47	−50	−47
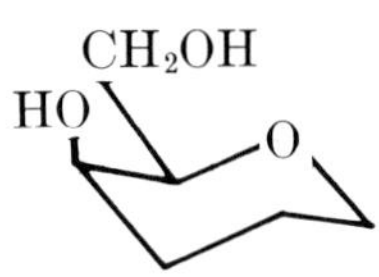	**7**	7	−10	5
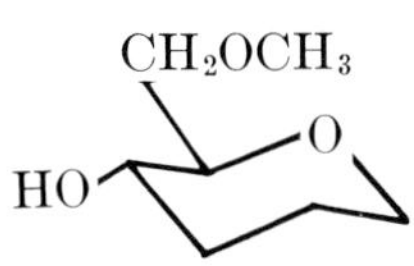	**8**	84	70	0
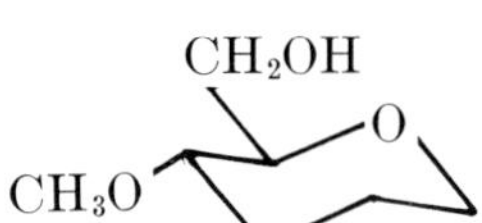	**9**	136	138	138
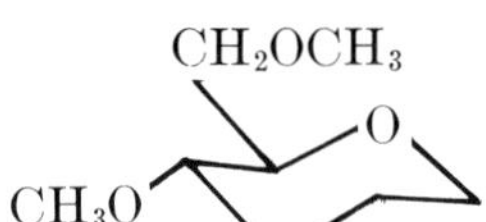	**10**	147	137	134

[a] + 23° in CCl_4.

the 1,2-type bonding in carbon tetrachloride, this is probably also the case for the compound dissolved in 1,2-dichloroethane. Therefore, this strong absorption is probably related to an intramolecular hydrogen bond. This conclusion seems to be supported by the inequality in the absorbances at 3590 cm^{-1} and 3510 cm^{-1} for compound **5**. About two-thirds the absorption at 3590 cm^{-1} is probably related to 1,2-type intramolecularly hydrogen-bonded conformer since, as will be seen later, this is the expectation based on molecular rotation. The relatively high content of free hydroxyl, as indicated by the infrared spectrum for the methyl ether **8**, is unexpected from a consideration of its molecular rotation, assuming little contribution to rotation by the CH_3O—C6—C5 unit of conformational asymmetry (*see* below).

As pointed out by Lemieux and Martin (*1*), the levorotation of **3** in dimethyl sulfoxide (Table II) probably arises from a lower population for **3c** than for **3a** (Table III). The conformer **3b** is expected on the basis of non-bonded interactions to be least stable. It is possible that the interpretation of the rotations for this compound is seriously complicated by the conformers in the alternate chair form. Nevertheless, using Lemieux's symbols for non-bonded interactions (*16*), it follows that in dimethyl sulfoxide OH/O is less favorable than OH/H, a situation which would not be surprising in view of the dipole–dipole repulsion between the gauche C—O bonds in this solvent. The low molecular rotation of 12° was unexpected for **3** in 1,2-dichloroethane since the rotation of **3c** is expected (*see* Table III) to be between 55°–90°, depending on the magnitude of the OH/O_r (exo) contribution to rotation, and in this solvent intramolecular hydrogen bonding is favorable. The molecular rotation in carbon tetrachloride is 23° which agrees with a relatively high population of **3c**. The rotation of 26° for **3** in water, as compared with −6.5° in dimethyl sulfoxide, suggests clearly that the non-bonded interaction OH/O_r (DMSO) is greater than $OH/O_r(H_2O)$.

The molecular rotations estimated for conformers **4a**, **4b**, and **4c** are 0°, 45°, and 90°, respectively (*see* Table III). The high rotations of

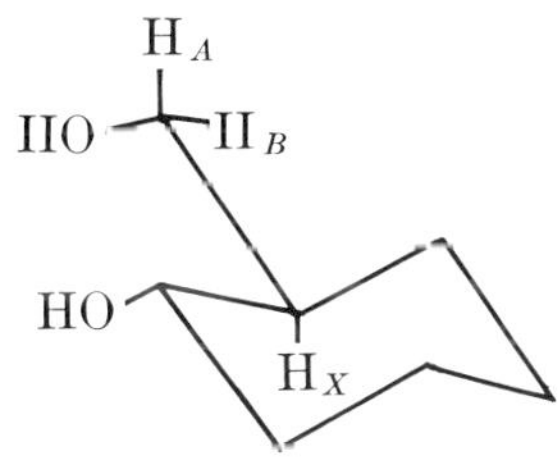

4a

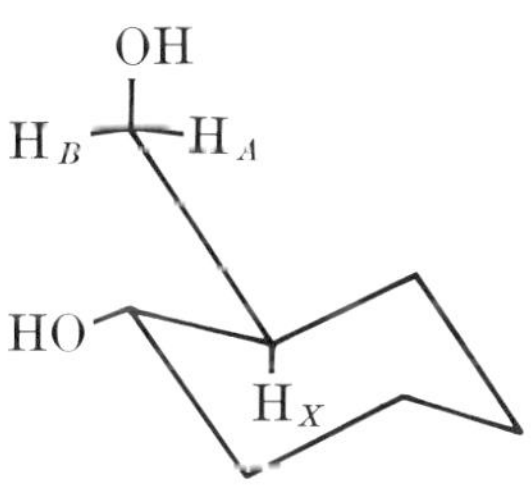

4b

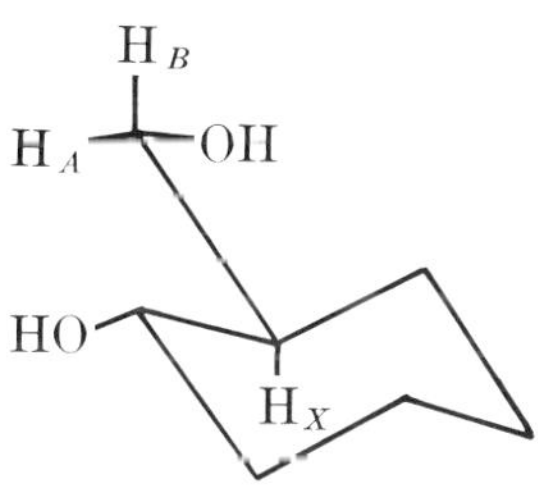

4c

Table III.

Compound		*a*
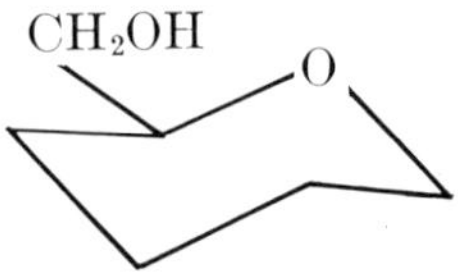	*3*	$-OH/C = -45°$
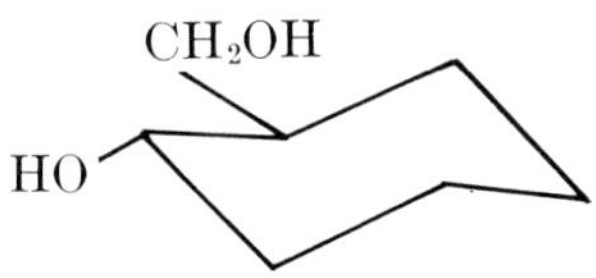	*4*[b]	$OH/C - OH/C = 0°$
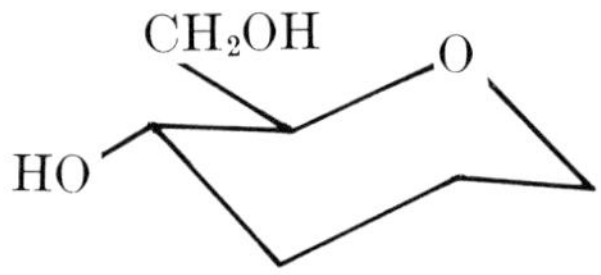	*5*	$OH/C - OH/C = 0°$
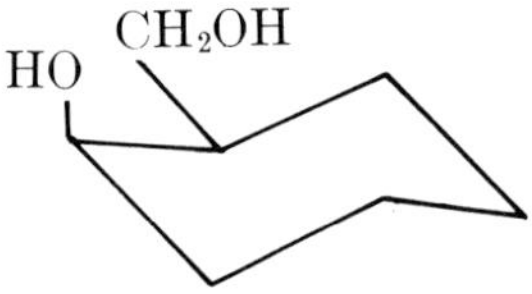	*6*	$-2OH/C = -90°$
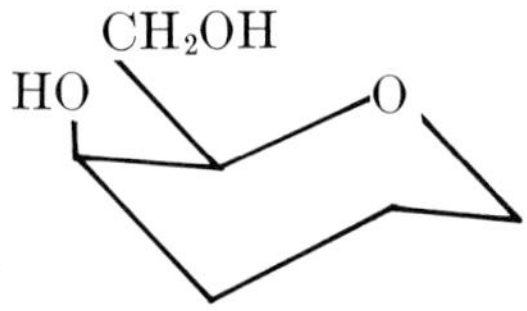	*7*	$-3OH/C + OH/O_r$ (endo) $= -45°$

[a] $OH/C = \pm 45°$, OH/O_r (endo)[b] $= \pm 90°$, OH/O_r (exo) $= \pm 55°$ or $\pm 90°$.

the compound in water and dimethyl sulfoxide (*see* Table II), therefore, indicate low populations of **4a** and relatively high populations of **4c**. The *syn*-axial-like relationship of the hydroxyl groups in **4a** is expected (*17*) to be destabilized strongly in these solvents. Also, conformational analysis requires **4c** to be thermodynamically more favorable than **4b**. The implication of **4c** as the preponderant conformer in water was confirmed by NMR. Using the labelling for the relevant hydrogens provided in the formulas on p. 127 we expected that in **4a**, $J_{AX} > J_{BX}$, in **4b**, $J_{AX} \cong J_{BX}$,

Rotational Analyses[a]

Rotamer Type	
b	***c***
$OH/C - OH/O_r$ (exo) $= -10°$ to $-45°$	OH/O_r (exo) $= 55°$ to $90°$
$2OH/C - OH/C = 45°$	$2OH/C = 90°$
$2OH/C - OH/O_r$ (exo) $= 0°$ to $35°$	$OH/C + OH/O_r$ (exo) $= 100°$ to $135°$
$-2OH/C + OH/C = -45°$	$OH/C - OH/C = 0°$
OH/O_r (endo) $- OH/C - OH/O_r$ (exo) $= -10°$ to $-45°$	OH/O_r (endo) $+ OH/O_r$ (exo) $- 2OH/C = 55°$ to $90°$

[b] The enantiomer was the form prepared and examined in this research.

and in **4c**, $J_{AX} < J_{BX}$ (*18*). The fact that one of these vicinal coupling constants is substantially larger than the other (*see* Table V) implicates a preponderance of either **4a** or **4c**. From the chemical shifts, the major rotamer is evidently **4c**. H_A and H_B are expected to have about the same chemical shift in **4a**, but H_B must be strongly deshielded by the opposing hydroxyl in **4b** and give its signal to lower field than H_A (*19*). The opposite is true for **4c**—*i.e.*, H_A should provide the lower field signal. Therefore, since the strong averaged coupling is associated with the higher field signal, the preponderant rotamer is **4c**, as inferred by the rotational data. The rotational data with water and dimethylsulfoxide as solvent are compatible with a conformational equilibria of about

$$0.1\ \mathbf{4a} \rightleftarrows 0.3\ \mathbf{4b} \rightleftarrows 0.6\ \mathbf{4c}$$

and generally agree with expectations based on conformational analysis.

As seen above, infrared absorption showed intramolecular hydrogen bonding to be extensive for **4** in 1,2-dichloroethane. Thus, the low molecular rotation of 13° (Table II) in this solvent is consistent with a high population of **4a**. To keep the relative amounts of **4b** and **4c** constant, a plausible equilibrium for **4** in 1,2-dichloroethane is,

$$0.85\ \mathbf{4a} \rightleftarrows 0.05\ \mathbf{4b} \rightleftarrows 0.1\ \mathbf{4c}$$

In view of the hydrogen bond conjugation effect reported by Lemieux and Pavia (*17*), the gradual addition of dimethyl sulfoxide to a solution of **4** in 1,2-dichloroethane should initially enrich the population

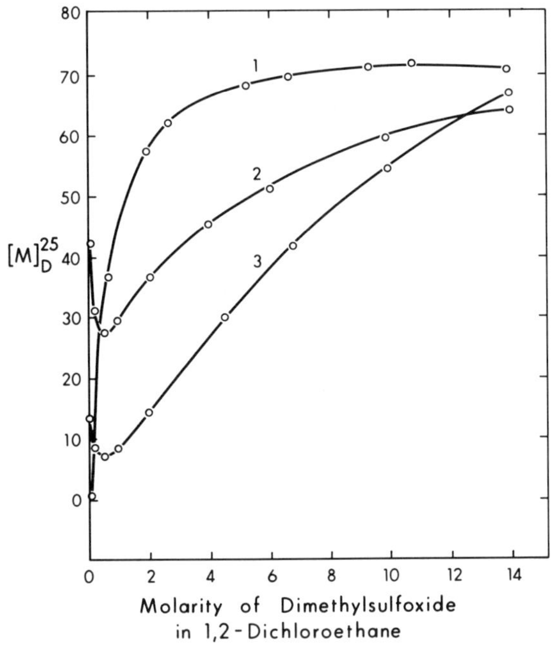

*Figure 1. Variation in molecular rotation with increasing concentrations of dimethyl sulfoxide in 1,2-dichloroethane: (1) 1,5-anhydro-2,3-dideoxy-6-O-methyl-*D-erythro-*hexitol* (**8**); *(2) 1,5-anhydro-2,3-dideoxy-*D-erythro-*hexitol* (**5**); *(3) 1*R,2S*-hydroxymethylcyclohexanol* (**4**).

of conformer **4a**. Therefore, the rotation of the solution should first decrease to a minimum before rising again toward the rotation in pure dimethyl sulfoxide. This was observed as seen in Figure 1. The plot seems to confirm the zero rotation expected for **4a** on the basis of the empirical rules for rotation.

In view of the above discussion, it is not surprising that compounds **4** and **5** have similar rotations in water and in dimethyl sulfoxide. The NMR data for **5** presented in Table V implicate **5c** as the preponderant rotamer in water for the same kind of reasons mentioned above regarding compound **4**. That the rotation of **5** is somewhat less in dimethyl sulfoxide than in water seems to agree with the observations made above concerning compound **3**. In contrast to **4** where only one mode (1,3-type) of intramolecular hydrogen bonding is possible, two modes exist for **5**. As mentioned earlier, the infrared absorption data for **5** in 1,2-dichloroethane showed less 1,3-type intramolecular hydrogen bonding than for compound **4**. It was expected that **5** would adopt the *c*-type conformation more extensively than **4** with 1,2-dichloroethane as solvent. The higher rotation of **5** in this solvent (*see* Table II) agrees with this expectation. The uncertainty in the value for OH/O_r (exo) handicaps attempts to analyze the rotation for **5** in water. Nevertheless, a mole fraction for **5c** between 0.5–0.6 is indicated. The low value of about 2.5 Hz for J_{AX} (*see* Table V) requires a very low abundance of **5a**. The magnitude of J_{BX} requires a substantial population of a conformer other than **5c**. Therefore, J_{AX} is apparently very small, $\sim$ 1 Hz in conformer **5c**. The torsion angles defined by H_X with H_A and H_B are expected to be near 60° because of the buttressing effects the H-4 and O-5 atoms must have on the orientation of O-6. Consequently, J_{AX} and J_{BX} for conformer **5b** are expected to be about 3.5 Hz (*20*).

The NMR and rotation data (using OH/O_r (exo) $= \pm 55°$) for **5** in water seem reasonably well accommodated by the following equilibrium,

$$0.1\ \mathbf{5a} \rightleftarrows 0.3\ \mathbf{5b} \rightleftarrows 0.6\ \mathbf{5c}$$

The decrease in rotation observed on changing the solvent to dimethyl sulfoxide is probably reflecting a decreased population of **5c**. The molecular rotation of 42° for **5** in 1,2-dichloroethane suggests a change in the equilibrium toward a situation such as,

$$0.5\ \mathbf{5a} \rightleftarrows 0.15\ \mathbf{5b} \rightleftarrows 0.35\ \mathbf{5c}$$

The gradual addition of dimethyl sulfoxide to a solution of **5** in 1,2-dichloroethane is expected initially to populate **5a** further and therefore lead to a decrease in rotation. As seen in Figure 1 and as previously noted by Lemieux and Martin (*1*), this was the case.

These interpretations of the rotational data for **5** seem confirmed by the behavior of its 6-mono-*O*-methyl and di-*O*-methyl derivatives, compounds **8** and **9**. The monoethyl ether (*8*) was found by infrared absorption to be extensively intramolecular hydrogen bonded in the 1,3-type of bond when dissolved in 1,2-dichloroethane. As a consequence, **8** should show a low rotation in this solvent. As Figure 1 shows, a low rotation was observed, indicating a relatively high preference for the **8a** conformer. However the contributions to rotation arising from asymmetric units of conformation involving the *O*-methyl group cannot be expected with certainty. The infrared spectrum (Table IV) requires a considerable population of free hydroxyl group and conformations **8b** and **8c**. Disruption of the intramolecular hydrogen bond by gradual addition of dimethyl sulfoxide should disfavor the **8a** conformer and lead, as was found in comparable situations by Lemieux and Pavia (*17*), to increasing optical rotation reflecting an increase in the relative populations of the dextrorotatory conformers **8b** and **8c**. As seen in Figure 1, this was observed. The conformational equilibrium for the di-*O*-methyl derivative **10** should not be influenced by change of solvent except through effects arising from changes in dielectric constant. Therefore, the rotation of the compound should remain nearly constant on changing the solvent as was observed (Table II).

The infrared absorption spectrum for the 4-*O*-methyl compound **9** in ethylene chloride showed much less absorption for the 1,3-type of intramolecular hydrogen bond than did compound **8**. Evidently, the orientation of the methoxy group required to enable hydrogen bonding with the C-6 hydroxyl is strongly demanding in energy, and the 1,2-type hydrogen bond with ring oxygen is preferred. The near constant high dextrorotation with change in solvent probably reflects a high total population of the **b** and **c** type conformations in all solvents. The high rotations of compounds **9** and **10** in water as compared with compounds **4** and **8** probably (*1*) result from the 4-*O*-methyl group defining on the average a right-handed screw pattern of asymmetry with the C-3 grouping. The relatively weak C-6OH $\leftarrow$ OC-4 hydrogen bond of compound **9** is important with regard to the conformational properties of 1 $\rightarrow$ 4 linked disaccharide structures.

The model compounds **6** and **7** (*see* Table II) gave rotations substantially unaffected by change in solvent. This is understandable on the basis of the empirical rules for rotation. Since the molecular rotation of **6b** is expected to be equal to the average of the rotations for **6a** and **6c** and the latter are expected to be of about equal population, no change in rotation is expected on changing the solvent from water to

1,2-dichloroethane. The infrared spectrum of **6** requires a high abundance of the hydrogen-bonded conformer **6b**. As seen in Figure 2, the addition of dimethyl sulfoxide to a solution of **6** in 1,2-dichloroethane caused an initial decrease in rotation. This phenomenon seems to occur primarily because of an enrichment of **6b** through hydrogen-bond conjugation (*17*); however, the subsequent rise in rotation was unexpected. The possibility exists that solvation of the 1,3-hydrogen bonded form by dimethyl sulfoxide increases the abundance of the conformer in the alternate chair form (**6b′**) for which an about 0° molecular rotation is expected. Inspection of **6b** and **6b′** shows that their conformational free energies may not differ greatly, and this difference may be decreased by dimethyl sulfoxide since the *A*-value for cyclohexanol is reported (*21*) to increase substantially on changing the solvent from a solvent inert to hydrogen bonding to dimethyl sulfoxide.

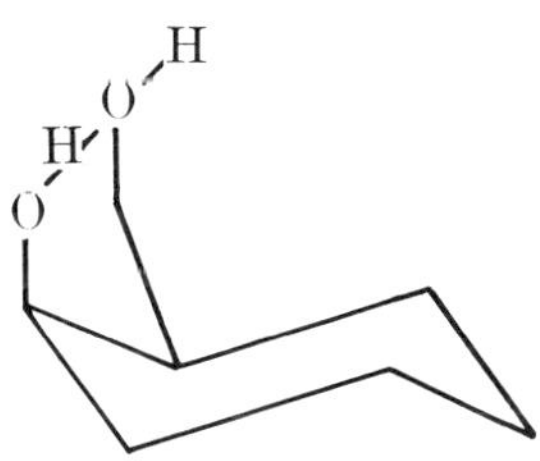

6b

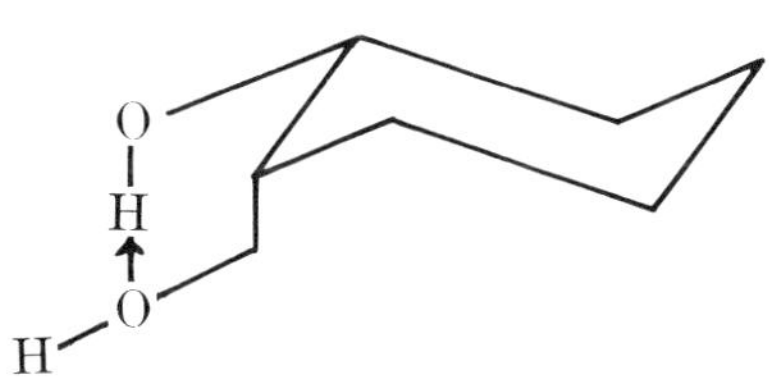

6b′

As seen above with reference to Table IV, the hydroxyl groups of compound **7** in 1,2-dichloroethane seem to be hydrogen bonded extensively. This precludes a substantial amount of conformer **7a**. Thus, the compound in this solvent likely exists primarily in an equilibrium between the hydrogen-bonded conformers **7b** and **7c**. On this basis and the calculated rotations for these conformers (*see* Table III) using OH/O_r (exo) $= \pm 55°$, a ratio of about 4:1 for the population of **7b** and **7c** is indicated and represents a ratio agreeing with the infrared spectrum. On addition of dimethyl sulfoxide, the rotation of the solution slowly decreased, as seen in Figure 2, to the levorotatory value of $-10°$ in pure dimethyl sulfoxide. This phenomenon agrees with expectation since the addition of dimethyl sulfoxide should depopulate the dextrorotatory conformer **7c**. An initial sudden increase in the population of **7b** was not apparent from the change in rotation on adding dimethyl sulfoxide and may not occur since the existing intramolecular bond would have to be sacrified in favor of an intermolecular hydrogen bond. How-

Table IV. Assignment of Infrared Absorption

Compound		*Fundamental Hydroxyl Stretching Frequencies,*[a]		
			Intramolecular H-Bond	
		Free	*1.2-Type*	*1.3-Type*
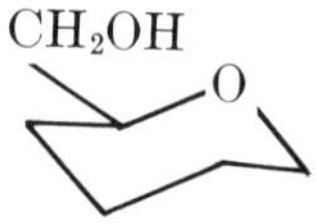	***3***	—	3580 (0.41)	—
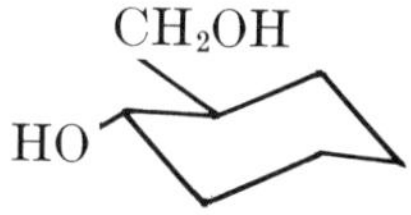	***4***	3590 (0.24)	—	3505 (0.22)
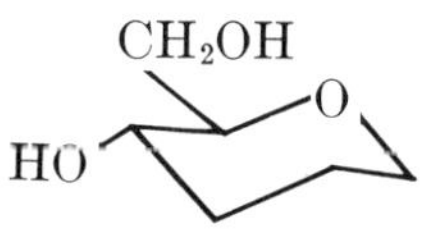	***5***	3590 (0.39)	—	3510 (0.10)[b]
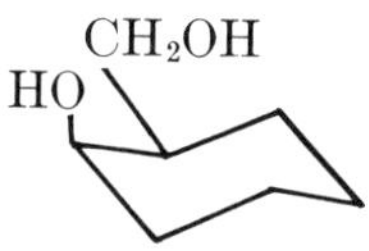	***6***	3590 (0.13)	—	3500 (0.11)
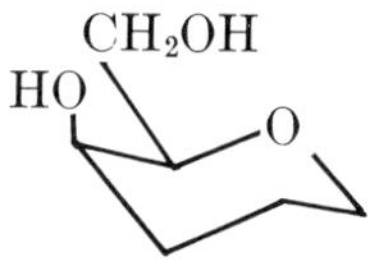	***7***	—	3565 (0.22)	3500 (0.18)
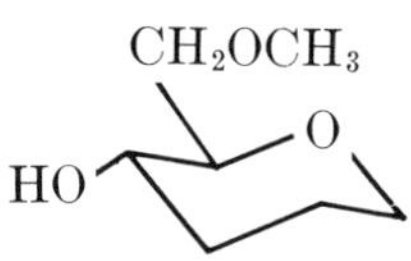	***8***	3585 (0.12)	—	3500 (0.14)
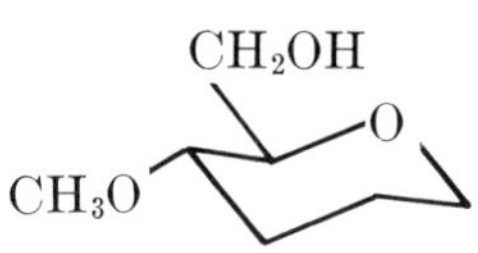	***9***	—	3565 (0.14)	3510 (0.07)[c]

[a] For 0.01 *M* solutions in carbon tetrachloride; ***3,*** 3637 (0.04), 3595 (0.26); ***4,*** 3618 (0.17) with shoulder at 3630, 3535 (0.22), 3375 (0.06)[b]; ***5,*** 3630 (0.19), 3600 (0.21), 3525 (0.11)[b] and 3450 (0.09)[b].

Bands for Compounds in 1,2-Dichloroethane (0.03*M*)[a]

(cm⁻¹), and Absorbances
Intermolecular H-Bond
3460 (0.06)[b]
3400 (0.04)[b]
—
—
—
–
—

[b] Broad.
[c] Shoulder.

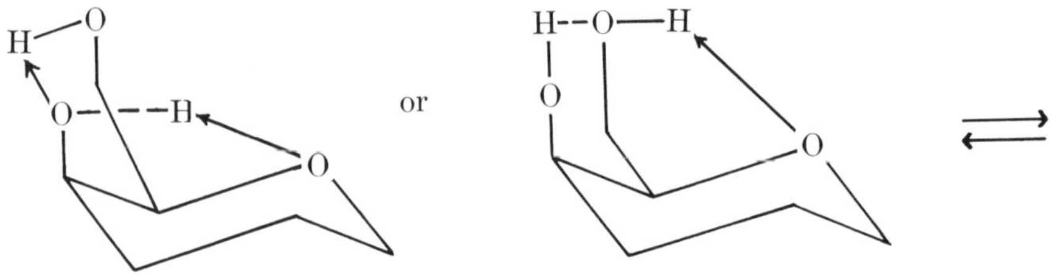

7b

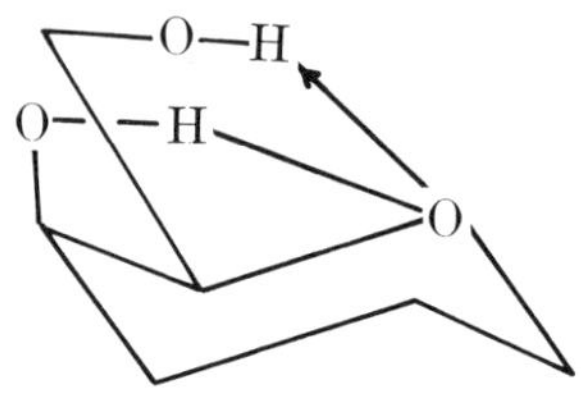

7c

ever, as the concentration of dimethyl sulfoxide increases, the equilibrium should tend toward increasing amounts of the equilibrium mixture of di-solvated forms of **7a** and **7c** with the levorotatory **7a** conformer being favored. That is, the cause of the change in rotation for **7** from 5° in 1,2-dichloroethane to −10° in dimethyl sulfoxide is probably similar to that stated above to account for change in rotation observed for compound **3**, +12° in 1,2-dichloroethane to −6.5° in dimethyl sulfoxide.

It was impossible to measure the coupling constants J_{AX} and J_{BX} for **7** in D_2O; however, for **6** (*see* Table V) these constants were almost equal at 7.0 Hz. The magnitudes of these constants require almost equal populations for **6a** and **6c** and relatively low abundance of **6b**. The difference in chemical shifts for H_A and H_B, as the chemical shifts, are temperature independent. The following conformational equilibrium is expected to approximate that which actually exists since it agrees well with the NMR and the rotational data and general expectations based on conformational analysis,

$$0.45\ \mathbf{6a} \rightleftarrows 0.1\ \mathbf{6b} \rightleftarrows 0.45\ \mathbf{6c}$$

Assuming that for **7**, the ratio of the populations for **7a** and **7c** is about the same as that suggested above for **3a** and **3c**, (*i.e.*, **7c** is much more abundant than **7a**), then there seems to be no acceptable basis for rationalizing the low molecular rotation of 7° observed for **7** in water. Either the mole fraction of **7a** or that of **7b** is unexpectedly high. Since

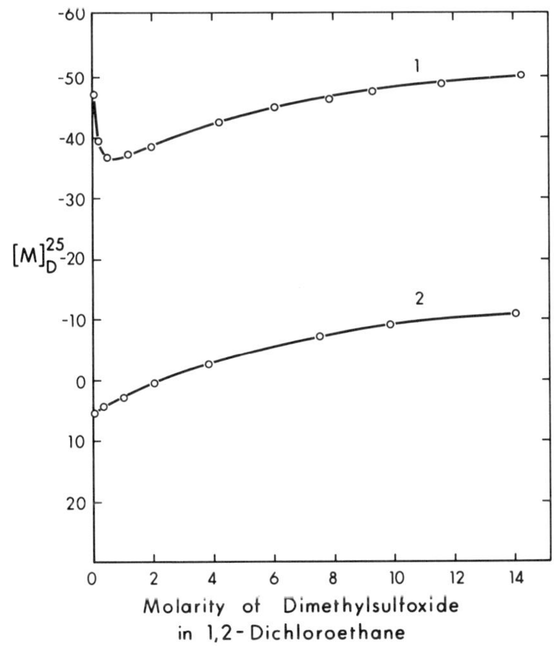

*Figure 2. Variation in molecular rotation with increasing concentrations of dimethyl sulfoxide in 1,2-dichloroethane: (1) 1*R*,2*R*-hydroxymethylcyclohexanol* (**6**)*; (2) 1,5-anhydro-2,3-dideoxy-*D-threo-*hexitol* (**7**).

no unexpected specific solvation effect seems to influence the relative populations of **6a** and **6c**, it seems unlikely that such an influence would appear for **7a** and **7c**. Therefore, the possibility seems to exist that the mole fraction of **7b** is unexpectedly high in water. This possibility warrants serious attention since of the model compounds studied only structure **7b** can form a completely chelated structure. Thus, the stabilization to the system gained by **7b** donating a hydrogen to a hydrogen bond

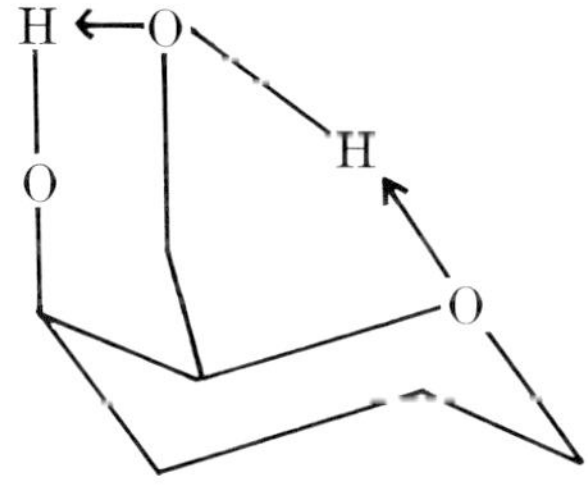

7b

Table V. Nuclear Magnetic Resonance

Compound	*Temperature, °C*	*Chemical Shifts*[a] H_A	H_B
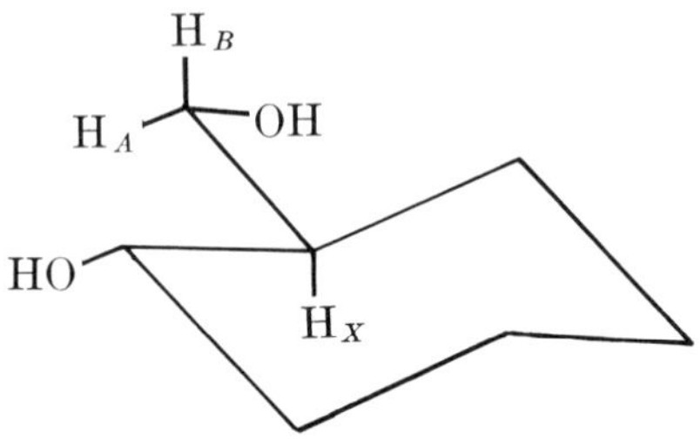 **4**	5	6.04	6.22
	40	5.99	6.16
	80	5.94	6.09
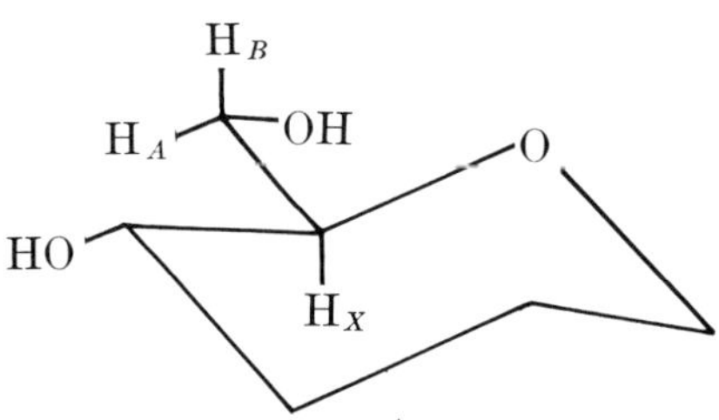 **5**	5	5.94	6.16
	40	5.89	6.09
	80	5.81	5.99
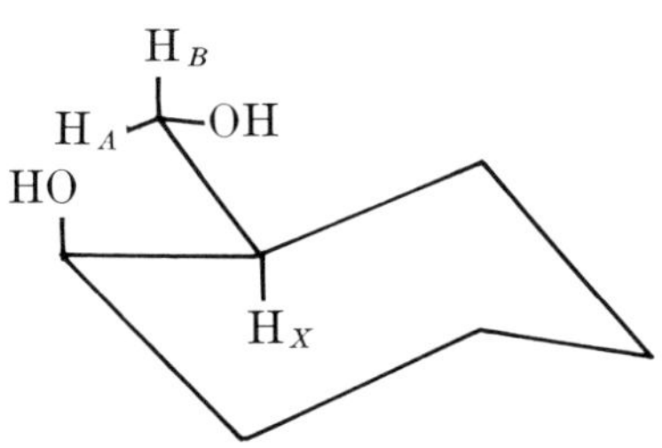 **6**	5	6.28,6.41	
	40	6.23,6.36	
	80	6.16,6.29	

[a] In τ values with tetramethylsilane as external standard.

with water is minimal. In other words, in the conformer **7b** the hydrogen accepting power of the ring oxygen can be satisfied intramolecularly while strengthening the O-4 to O-6 hydrogen bond through hydrogen bond conjugation (*17*) as shown below. If this is true, this phenomenon would have an important implication on the conformational properties of galactopyranosides.

Parameters Measured at 220 MHz in Deuterium Oxide

Coupling Constants, Hz		
J_{AX}	J_{BX}	J_{AB}
4.3	6.5	10.5
4.5	6.2	10.5
5.0	6.0	10.5
2.4	6.6	12
2.6	6.4	12
2.8	6.2	12
7.0	7.0	11
7.0	7.0	11
7.0	7.0	11

Although the above rationalizations of the rotations for compounds 3, 5, and 7 are inconclusive, most of the evidence seems to indicate that OH/O_r (exocyclic) has a smaller numerical value than OH/O_r (endocyclic). If so, then as suggested by Brewster (*4*), there is a permolecular effect to rotation arising from the presence of axial substituents at the 1, 2, 4, and 5 positions of the pyranose ring. This problem may be solved through a knowledge of the coupling constants J_{AX} and J_{BX} for compounds 3 and 7 in aqueous solution.

The effect of temperature on the optical rotations of several compounds is presented in Table VI. The temperature coefficients were

Table VI. Variation of Molecular Rotation with Temperature

Compound	$\Delta[M]_D°$ *per 10°C × 10*
1,5-Anhydrohexitols	
2,3,6-Trideoxy-D-*glycero*	~0
2,6-Dideoxy-D-*threo*	0.1
2,6-Dideoxy-D-*arabino*	0.3
2-Deoxy-D-*arabino*	0.4
2,3-Dideoxy-D-*erythro* (**5**)	0.9
2,3-Dideoxy-D-*threo* (**7**)	1.3
2,3,4-Trideoxy-D-*glycero* (**3**)	1.5
2-Deoxy-D-*lyxo*	1.9
Cyclohexanols	
cis-2-Hydroxymethyl (**6**)	1.0
trans-2-Hydroxymethyl (**4**)	1.7

determined by drawing the best line through a plot of the rotations at 5°, 25°, 40°, 60°, and 80°C. In all cases the rotations tended to lower values numerically as the temperature was increased. Table VI shows that the conformationally more rigid structures underwent little change. However, temperature increase should tend to populate the less stable conformers for the conformationally labile hydroxmethyl compounds, and the rotations of these compounds were relatively much more sensitive to temperature. The temperature effects on the NMR data presented in Table V show that the increase in temperature caused a change in conformational equilibria about the CH_2OH bond of compounds 4 and 5. The coupling constants J_{AX} and J_{BX} approach the same value with increasing temperature, and this agrees with the lowering of the magnitude of the rotations for these compounds with increasing temperature.

Experimental

All optical rotations were measured at the D-line of sodium with a Perkin-Elmer polarimeter (model 141) using a thermostatted 10-cm polarimeter tube dried by a stream of filtered, dried air. Precaution was necessary to exclude water from the solutions in 1,2-dichloroethane and dimethyl sulfoxide. In this connection all glassware was dried at 120°C and stoppered while hot. All transfers of solvents or solutions were made with syringes using serum caps to exclude moist atmosphere. The compounds which could not be prepared in crystalline condition and recrystallized to purity were obtained from a pure crystalline derivative and purified by distillation *in vacuo* under conditions expected to exclude water. All solutions expected to be anhydrous when possible were checked in this regard by infrared absorption spectroscopy.

The silicic acid used for column chromatography and the Amberlite resin were Silicar CC-7 100-200 mesh purchased from Mallinckrodt Chemical Works, St. Louis, Mo. The 5% palladium on charcoal catalyst was purchased from Engelhard Industries Ltd., Newark, N. J. The tri-*O*-acetyl-D-glucal and the tri-*O*-acetyl-D-galactal (as a 70% solution in benzene) were purchased from Raylo Chemicals Limited, Edmonton, Alberta.

All thin layer chromatograms (TLC) were made on silica gel G purchased from E. Merck A.G., Darmstadt, W. Germany and developed with the solvent system reported for the column chromatogram. The gas liquid partition chromatograms (GLPC) were performed on a 6-foot, 1/4 inch column packed with 10% Carbowax M on 60-80 mesh Ultraport.

1,5-Anhydro-2,3,4-trideoxy-D-*glycero*-hexitol (*3*). 3,3,6-Tri-*O*-acetyl-1,5-anhydro-2-deoxy-D-*lyxo*-hex-1-enitol (tri-*O*-acetyl-D-galactal) (40 grams) in 200 ml of ethyl acetate was hydrogenated in the presence of diethylamine (12 grams) over 4.4 grams of 5% palladium on charcoal at room temperature and atmospheric pressure. The catalyst was changed after 5, 16, and 24 hours. After a total hydrogenation of 40 hours, the catalyst was removed by filtration and washed with ethyl acetate, and the combined filtrates were concentrated at reduced pressure to about 100 ml. Chloroform (500 ml) was added, and the solution washed successively with *N* hydrochloric acid, saturated aqueous bicarbonate, and water before drying over sodium sulfate. Solvent removal left 24 grams of reddish brown syrup. GLPC analysis at 175°C revealed three volatile components in the ratios about 4:1:5 in order of their appearance at the detector.

The low boiling component was isolated by fractional distillation at 80°–85°C (13 mm pressure). The material was impure *s*-2 acetoxymethyltetrahydropyran (*1*). The isolation and characterization of the two other components are described in relation to the preparation of 7 and 1,5-anhydro-2-deoxy-D-*lyxo*-hexitol.

The above low boiling compound was deacetylated in methanol using a catalytic amount of sodium methoxide. The distilled product (2.75 grams) was dissolved in 50 ml of pyridine, and 6.7 grams of triphenylmethyl chloride was added. After 48 hours at room temperature, the solution was poured onto an ice and water mixture. The crystalline solid which separated was first recrystallized from 20 ml of methanol and then recrystallized twice from 98% ethanol to give 4.9 grams of a product, mp 107°–108°C, $[\alpha]_D^{25}$ 20.7° (*c*, 0.56 in chloroform), with physical constants that remained unchanged on further recrystallization. *Anal.* Calcd. for $C_{25}H_{26}O_2$: C, 83.76; H, 7.31%. Found: C, 83.56; H, 7.25.

The above compound, 1,5-anhydro-2,3,4-trideoxy-6-*O*-triphenylmethyl-D-*glycero*-hexitol (4.0 grams) was dissolved in 75 ml of chloroform. Dry hydrogen bromide was bubbled through the solution kept at 0°C until TLC examination showed all the starting material had disappeared. The solution was concentrated at 30°C and 13 mm pressure to a semi-solid mass. The pressure was then reduced to 0.08 mm, and the material which distilled at 30°C was condensed in a receiving flask cooled in liquid nitrogen. The distillate was made slightly basic with trimethylamine and redistilled at 0.1 mm. The material (620 mg) appeared to be pure **3** on examination by GLPC and NMR. Its rotation in water, $[\alpha]_D^{25}$ + 22.5° (*c*, 0.5 in water) was higher than that (+19.2°) previously reported (*1*). *Anal.* Calcd. for $C_6H_{12}O_2$: C, 62.04; H, 10.41%. Found: C. 61.52; H, 10.57%.

1,5-Anhydro-2,3-dideoxy-D-*threo*-hexitol (7). Residues from several preparations of crude s-2-acetoxymethyltetrahydropyran (*see* the above experiment) were combined and subjected to fractional distillation using a spinning band column. About 75% of the fraction which distilled in the range 90°–100°C (0.6 to 0.1 mm pressure) was the di-*O*-acetyl derivative of **7**. The material, 5.7 grams, was deacetylated in methanol using a catalytic amount of sodium methoxide. After neutralization with Amberlite IR 120 (H^+) resin, the solvent was removed *in vacuo* to leave a syrup which was triturated with ether. Crystalline 1,5-anhydro-2-deoxy-D-*threo*-hexitol formed and was removed by filtration. The ether was removed from the filtrate, and the residue, 2.4 grams, was dissolved in 30 ml of pyridine with 7.6 grams of *p*-nitrobenzoyl chloride. The mixture was warmed with stirring until it became homogeneous and was left on the steam bath for a further hour. The light brown crystalline product, 6.85 grams, was isolated by the usual method. After two recrystallizations first by dissolving the compound in dichloromethane, then adding ethanol, and removing the dichloromethane by distillation, an almost colorless material was obtained, mp 150°–151°C, $[\alpha]_D^{25}$ − 89.3° (*c*, 0.8 in chloroform). *Anal.* Calcd. for $C_{20}H_{18}O_9N_2$: C, 55.81; H, 4.22; N, 6.51%. Found: C, 55.73; H, 4.32; N, 6.48%.

The above compound (4.8 grams) was dissolved in 60 ml of dichloromethane, and a solution of 10 mg of sodium methoxide in 100 ml of anhydrous methanol was added. After 2 hours the solvents were removed *in vacuo,* and the residue twice was extracted with 40 ml of water. Sodium hydroxide (3.0 grams) was added, and the solution was left for 24 hours. Continuous extraction of this solution with ether gave after solvent removal 1.39 grams of a colorless oil. Distillation at 0.2 mm pressure and bath temperature of 95°C gave a fraction, 1.19 grams, which solidified. The compound was recrystallized from anhydrous ether to give 0.87 gram of pure **7**, mp 51°–53.5°C, $[\alpha]_D^{25}$ + 5.2° (*c*, 0.5 in water). These constants were unaffected by a further recrystallization. *Anal.* Calcd. for $C_6H_{12}O_3$: C, 54.53; H, 9.15%. Found: C, 54.84; H, 9.20%.

1,5-Anhydro-2,3-dideoxy-D-*erythro*-hexitol (5). Hydrogenation of 3,4,6-tri-*O*-acetyl-1,5-anhydro-2-deoxy-D-arabino-hex-1-enitol (tri-*O*-acetyl-D-glucal) in the presence of diethylamine as previously reported (*22, 23*) gave a product which was subjected to fractional distillation at 0.07 mm pressure using a spinning band column. The fraction which distilled in the range 70°–73°C represented a 70% yield and solidified on standing at room temperature. After two recrystallizations from toluene at −20°C, the compound appeared pure, mp 35°–36°C, $[\alpha]_D^{25}$ + 42° (*c*, 0.6 in chloroform). The specific rotation in ethanol was +39.9° as compared with +35° (*24*) for the compound previously reported (*23, 24*) as a syrup. The NMR spectrum was the same as that previously reported (*23*) for the di-*O*-acetyl derivative of **5**.

The above compound was deacetylated using a solution of sodium methoxide in methanol. The sodium ion was removed using carefully washed Amberlite IR 120 (H^+) resin before solvent removal. The methanol was removed *in vacuo*, and the residue was dissolved in 1,2-dichloroethane. The solvent was then removed by distillation at atmospheric pressure to ensure the absence of water and methanol. The residue was then distilled at a bath temperature of 90°–100°C at 0.2 mm pressure. The oily product (**5**), $[\alpha]_D^{25}$ + 53.6° (*c*, 0.4 in water), appeared pure by TLC and NMR. The NMR parameters are presented in Table V.

1,5-Anhydro-2,3-dideoxy-6-O-methyl-D-*erythro*-hexitol (8). Triphenylmethyl chloride (11.5 grams) was added to 5.0 grams of **5** in 100 ml of pyridine. After the solution had stood at room temperature for 48 hours, 18.0 grams of *p*-toluenesulfonyl chloride were added. After standing another 2 days, the mixture was poured into 500 ml of ice water. After vigorous stirring for 30 min., chloroform was added, and the chloroform layer was washed first with water and then with 5% aqueous sodium bicarbonate solution. After drying over sodium sulfate, most of the solvent was removed by evaporation *in vacuo*. Toluene was then added and removed by distillation *in vacuo* until the residue was free of pyridine. The syrupy residue was dissolved in 40 ml of ether, and 60 ml of *n*-hexane were added. On cooling to 0°C, 10.0 grams of a precipitate were formed and collected by filtration. An additional 5 grams of this product were obtained by chromatography of the material in the mother liquor using 1:19 ethyl acetate–benzene mixture as eluting agent and silicic acid as adsorbent. The combined material, 75% crude yield, was recrystallized to purity from ethyl acetate, mp 145°–147.5°C, $[\alpha]_D^{25}$ + 37.6° (*c*, 0.7 in chloroform). The NMR spectrum agreed with the formation of a mono-*O*-*p*-toluenesulfonyl-mono-*O*-triphenylmethyl derivative of **4**.

The above compound was de-triphenylmethylated using hydrogen bromide in chloroform at 0°C for 3 hours. The product resisted crystallization and was purified by chromatography on silicic acid using a 1:4 mixture of ethyl acetate-benzene as eluting agent. The slowest moving material was the desired product which was obtained in 75% yield (6.2 grams). The material (4.4 grams) was dissolved in 50 ml of dry dimethylformamide, and 5.0 grams of barium oxide and 10 ml of methyl iodide were added. No reaction occurred even after refluxing for 1 hour. The mixture was cooled to 0°C, and 0.6 gram of sodium hydride was added. After stirring for 4 hours at room temperature, methylation appeared complete. The excess sodium hydride was destroyed by the addition of methanol, and then the mixture was poured into 200 ml of water for extraction with chloroform. The combined chloroform extracts were washed with water, dried, and then concentrated *in vacuo* to a syrupy product (4.4 grams) which was chromatographed on a 1.75 × 30 inch column of silicic acid, using a one to one mixture of ethyl acetate and benzene as eluting agent. The main fraction (3.8 grams) crystallized and was purified by recrystallization from methanol (mp 48°–50°C, $[\alpha]_D^{25}$ + 49.3° (*c*, 0.5 in chloroform)). The NMR spectrum agreed with the expected structure. *Anal.* Calcd. for $C_{14}H_{20}O_5S$: C, 55.98; H, 6.71; S, 10.67%. Found. C, 56.06; H, 6.44; S, 10.54%.

The above prepared 1,5-anhydro-2,3-dideoxy-6-*O*-methyl-4-*O*-*p*-toluenesulfonyl-D-*erythro*-hexitol (2.25 grams) was dissolved in 75 ml of methanol. Portions of 2% sodium amalgam were then added with vigorous stirring until TLC analysis showed the absence of starting material. The mercury was removed, and the solution concentrated to an oil which was dissolved in chloroform. The chloroform solution was dried over sodium sulfate before concentration to 1.3 grams of an oil, which was applied to a column of silicic acid for chromatography using a 1:4 mixture of acetone-chloroform as eluting agent. The main fraction was isolated and dissolved in 3 ml of ether, and 3 ml of *n*-pentane were added. After a day at −25°C, 0.6 gram of crystals was deposited. This low melting solid (below 10°C) appeared to be a pure sample of the desired compound **8**. It was distilled at about 60°C (0.5 mm) for optical rotation studies and analysis. The compound, $[\alpha]_D^{25}$ + 57.7° (*c*, 0.5 in water) gave an NMR spectrum which agreed completely with the expected structure. *Anal.* Calcd. for $C_7H_{14}O_3$: C, 57.51; H, 9.65%. Found: C, 57.40; H, 9.39%.

1,5-Anhydro-2,3-dideoxy-4-O-methyl-D-*erythro*-hexitol (*10*). Triphenylmethylation of **4** (18.0 grams) as described above for 48 hours followed by isolation of the product by the usual method gave 20 grams of a crystalline product that was a 1:1 complex of the desired compound and pyridine (mp 60°–100°C, $[\alpha]_D^{25}$ − 33° (*c*, 0.5 in chloroform)).

The complex (17.7 grams) was added to a suspension of 1.5 grams of sodium hydride in 50 ml of dry dimethylformamide kept at 0°C. Methyl iodide (10 ml) was then added. The resulting mixture was allowed to warm to room temperature, and the stirring continued for 6 hours. The product was isolated in the usual manner and after five recrystallizations from methanol-ether appeared pure (mp 76°–77°C, $[\alpha]_D^{25}$ + 41.0° (*c*, 0.7 in chloroform)). *Anal.* Calcd. for $C_{26}H_{28}O_3$: C, 80.38; H, 7.24%. Found: C, 80.19; H, 7.15%.

The above compound (6.0 grams) was detriphenylmethylated using hydrogen bromide in dichloromethane at 0°. The product was isolated by a method effectively the same as that described above for the regeneration of **3** from its triphenylmethyl ether. The yield after the second distillation (about 65°C at 1 mm pressure) was 1.5 grams (75%) of an oil which appeared pure on examination by TLC and GLPC, $[\alpha]_D^{25}$ + 93.5° (*c*, 0.57 in water). The NMR spectrum confirmed the presence of one methoxyl group at τ 6.64. *Anal.* Calcd. for $C_7H_{14}O_3$: C, 57.51; H, 9.65%. Found: C, 57.74; H, 9.63%.

1,5-Anhydro-2,3-dideoxy-4,6-di-O-methyl-D-*erythro*-hexitol (*10*). Compound **4** was O-methylated in the usual manner using sodium hydride and methyl iodide but using tetrahydrofuran as solvent. The product was purified by distillation at about 70°C (3 mm pressure). The oily material appeared pure by TLC and GLPC; the infrared spectrum was free of absorption for hydroxyl group, and the NMR showed signals for two O-methyl groups. *Anal.* Calcd. for $C_8H_{16}O_3$: C, 59.98; H, 10.07%. Found: C, 60.06; 10.04%.

1R,2R-Hydroxymethylcyclohexanol (*6*). 1S,2R-Hydroxycyclohexanecarboxylic acid was obtained by resolution of the racemic mixture using the procedure reported by Torne (*25*). The rotation agreed with that previously reported (*25, 26*). The acid was converted to the methyl ester using diazomethane, and the ester was reduced by the usual method using lithium aluminum hydride in ether. The product crystallized readily and was purified by three recrystallizations from ether (mp 49°–50°C, $[\alpha]_D^{25}$ − 36.0° (*c*, 0.42 in water)). *Anal.* Calcd. for $C_7H_{14}O_2$: C, 64.58; H, 10.84%. Found: C, 64.44; H, 10.83%.

NMR parameters are listed in Table V.

1R,2S-Hydroxymethylcyclohexanol (Enantiomer of 4). 1R,2R-Hydroxycyclohexanecarboxylic acid was prepared by base-catalyzed isomerization of the 1S,2R-isomer following the procedure described by Torne (*25*). The literature values (*25*, *27*) for the melting point and specific rotation were reproduced. Lithium aluminum hydride reduction of the methyl ester gave a syrupy product which crystallized after chromatographic purification on silica gel using a 3:7 acetone-chloroform mixture as eluting agent. The compound appeared pure after three recrystallizations from ether (mp 23°–24°C, $[\alpha]_D^{25}$ − 55.0° (*c*, 0.34 in water)). *Anal.* Calcd. for $C_7H_{14}O_2$: C, 64.58; H, 10.84%. Found: C, 64.30; H, 10.70%.

Acknowledgments

The authors are indebted to the National Research Council of Canada for grants-in-aid of this research to R. U. Lemieux (A-179) and postgraduate scholarships to John Brewer. The NMR, infrared, and micro analyses were provided by the service laboratories of this Department. The 220 MHz spectra were purchased.

Literature Cited

1. Lemieux, R. U., Martin, J. C., *Carbohyd. Res.* (1970) **13**, 139.
2. Whiffen, D. H., *Chem. Ind.* (1956) 964.
3. Lemieux, R. U., Brewer, J. T., Daniel, A., Nagabhushan, T. L., in preparation.
4. Brewster, J. H., *J. Amer. Chem. Soc.* (1959) **81**, 5483.
5. Gehrke, M., Obst, F., *Ber.* (1931) **64**, 1724.
6. Wilson, N., Read, J., *J. Chem. Soc.* (1935) 1269.
7. Backstrom, R., Sjoberg, B., *Arkiv. Chim. fur Kemi* (1967) **B26**, 549.
8. Gough, G. A. C., Hunter, H., Kenyon, J., *J. Chem. Soc.* (1926) 2052.
9. Allerhand, A., Schleyer, P. von R., *J. Amer. Chem. Soc.* (1963) **85**, 371.
10. Kuhn, L. P., *J. Amer. Chem. Soc.* (1952) **74**, 2492.
11. Kuhn, L. P., *J. Amer. Chem. Soc.* (1954) **76**; 4323.
12. Cole, M. R. H., Jeffries, P. R., *J. Chem. Soc.* (1956) 4391.
13. Mori, N., Tsuzuki, Y., *Bull. Chem. Soc. Japan* (1966) **39**, 2454.
14. Barker, S. A., Foster, A. B., Haines, A. H., Lehman, J., Webber, J. M., Zweifel, G., *J. Chem. Soc.* (1963) 4161.
15. Foster, A. B., Harrison, R., Lehmann, J., Webber, J. M., *J. Chem. Soc.* (1963) 4471.
16. Lemieux, R. U., "Molecular Rearrangements, Part 2," P. de Mayo, Ed., Interscience, New York, 1964.
17. Lemieux, R. U., Pavia, A. A., *Can. J. Chem.* (1969) **47**, 4441.
18. Lemieux, R. U., Kulling, R. K., Bernstein, H. J., Schneider, W. G., *J. Amer. Chem Soc.* (1958) **80**, 6098.
19. Lemieux, R. U., Stevens, J. D., *Can. J. Chem.* (1966) **49**, 249.
20. Lemieux, R. U., Lown, J. W., *Can. J. Chem.* (1964) **42**, 893.

21. Hirsch, J. A., "Topics in Stereochemistry," Vol. I, N. L. Allinger and E. L. Eliel, Eds., Interscience, New York, 1967.
22. Gray, G. R., Baker, R., *J. Org. Chem.* (1967) **32**, 2764.
23. Lemieux, R. U., Pavia, A. A., Martin, J. C., Watanabe, K. A., *Can. J. Chem.* (1969) **47**, 4427.
24. Maki, T., Tejima, S., *Chem. Pharm. Bull.* (1968) **16**, 2242.
25. Torne, P. G., *Chem. Abs.* (1967) **66**, 55082W.
26. Ferrer, M. A., Gomis, P., Pascual, J., *An. Fis. y Quim.* (1967) **63b**, 449.
27. Pascual, J., *An. Fis. y Quim.* (1967) **63b**, 727.

RECEIVED December 2, 1972.

9

Conformational Equilibria of Acylated Aldopentopyranose Derivatives and Favored Conformations of Acyclic Sugar Derivatives

P. L. DURETTE, D. HORTON, and J. D. WANDER

The Ohio State University, Columbus, Ohio 43210

By using low-temperature NMR spectroscopy and averaging of spin couplings, the relative proportions of the two chair conformers in solution were determined for various configurational series of acylated aldopentopyranosyl halides, glycosides, esters, and thioesters. The effects on these equilibria of solvent polarity and the nature of the aglycon in the glycosides have been examined and are discussed in terms of steric and electronic interactions between substituent groups in their influence on conformational equilibria and rate of conformational inversion. In acyclic systems, NMR-spectral studies show that the extended, planar, zigzag arrangement of the carbon atoms in the sugar chain is the favored conformation in solution except when such an arrangement would generate a parallel, 1,3-interaction between substituents. In the latter situation, the interaction is alleviated by rotation about a C—C bond to give a bent (sickle) form as the favored conformation.

The conformational studies (*1*) on acyclic sugar derivatives and on aldopentopyranose derivatives that have been conducted in our laboratories during the last few years are surveyed, and some of our more recent results in each of these areas are introduced. For each aspect the sugar derivatives were examined in solution by proton magnetic resonance (PMR) spectroscopy, and the data obtained were used to provide conformational information. Acyclic systems will be treated first.

Acyclic Systems

It has long been supposed that a hydrocarbon chain favors an extended conformation in which the carbon atoms lie in an approximate plane in a zigzag arrangement that results from maximum separation of the largest groups along each carbon—carbon bond. This rationale for unsubstituted chains has been widely used to express conformational formulas for acyclic sugar derivatives by simple extension of the principle of assigning favored conformations that agree with the maximum staggering of small–medium–large sets of groups along each carbon—carbon bond in the chain (2). However, direct chemical or physical evidence for such assignments until recently has been largely lacking, except for a few experiments on reactivity of acyclic sugar derivatives that are not necessarily valid as a direct measure of the conformation in the ground state. A point that is particularly questionable is the conformational behavior of extended-chain systems in which there would be two substituent groups, separated by one intervening carbon atom, on the same side of an extended chain—namely, a parallel 1,3-relationship of substituents. Such an arrangement bears a formal resemblance to the syn-diaxial disposition of these substituents on a six-membered ring, a situation that is conformationally unstable.

To test experimentally the validity of the planar, zigzag rationalization of conformation in acyclic-sugar systems, various types of derivatives having acyclic sugar-chains present were examined by PMR spectroscopy. To obtain satisfactory dispersion of the signals of the methine and methylene protons on the chain, acyclic sugar derivatives having dissimilar end-groups were utilized. In each instance these spectra were recorded at a field strength sufficient to allow reliable spin-couplings to be determined, either by first-order approximation or by appropriate calculation, and the spin couplings are accurate to within ±0.3 Hz or better.

The first example illustrated (Figure 1) is a quinoxaline derivative, having a four-carbon, acetylated carbohydrate chain attached to it. The

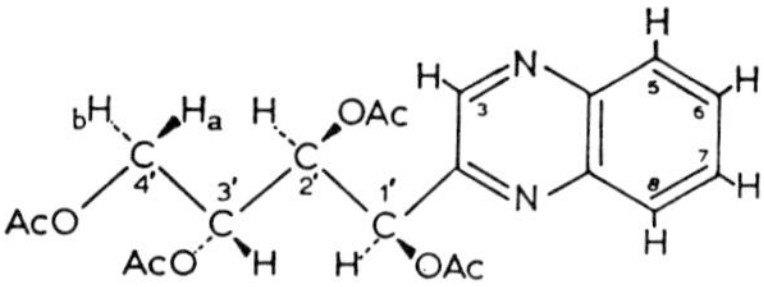

*Figure 1. 2-(*D*-arabino-Tetrahydroxybutyl)quinoxaline*

$J_{1',2'}$	3.0 Hz	$J_{3',4b'}$	5.5 Hz
$J_{2',3'}$	8.5 Hz	$J_{4a',4b'}$	12.0 Hz
$J_{3',4a'}$	3.0 Hz		

chain has the *arabino* stereochemistry, and from the small coupling of the protons at positions 1′ and 2′ of the side chain, it can be inferred that these protons are in gauche disposition, whereas protons 2′ and 3′, which show a large spin coupling of 8.5 Hz, are evidently in antiparallel disposition. Consideration of these and the other couplings for the compound in chloroform solution leads to the conclusion that the favored conformation of the chain is as shown in Figure 1 and corresponds to a planar, zigzag arrangement of the carbon atoms in the chain, resulting from maximum staggering of large–medium–small sets of groups along each carbon—carbon bond (*3*).

Before this rationalization obtained with one example is extended to other systems, it should be remembered that this compound has the *arabino* stereochemistry, and in the extended conformation it has no parallel 1,3-interaction between acetoxyl groups. To make generalizations on the conformational behavior of such derivatives and obtain data that might be extended to other systems, it is important to study systems having other configurations. This was performed initially with a series of triazole derivatives having a sugar chain attached, by varying the stereochemistry in the side chain (*4*).

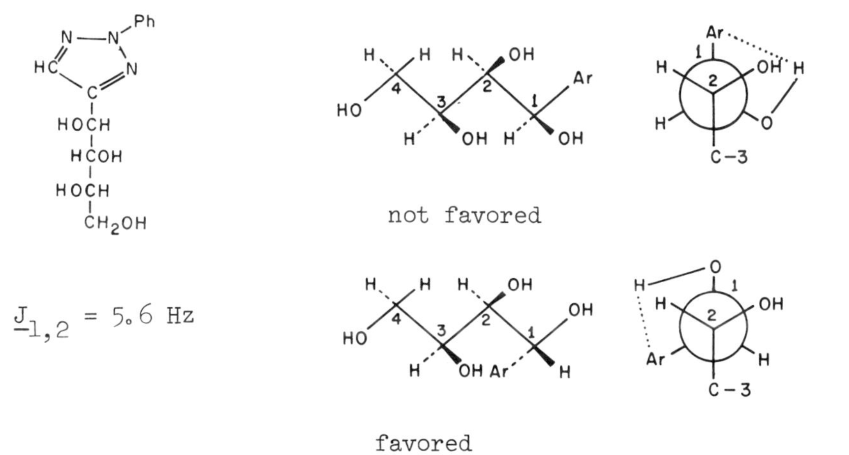

Figure 2. Destabilization by 1,3-interactions

In the triazole derivatives having a tetrahydroxybutyl side-chain, the behavior of the L-*xylo* derivative is noteworthy. Figure 2 shows that the observed spin-coupling of H-1 with H-2 on the side chain is 5.6 Hz. However, the fully extended conformation of the carbon chain in such a system would bring H-1 and H-2 into gauche disposition, and for this arrangement a small value of $J_{1,2}$ would be expected. The magnitude actually observed indicates that there is a substantial contribution from

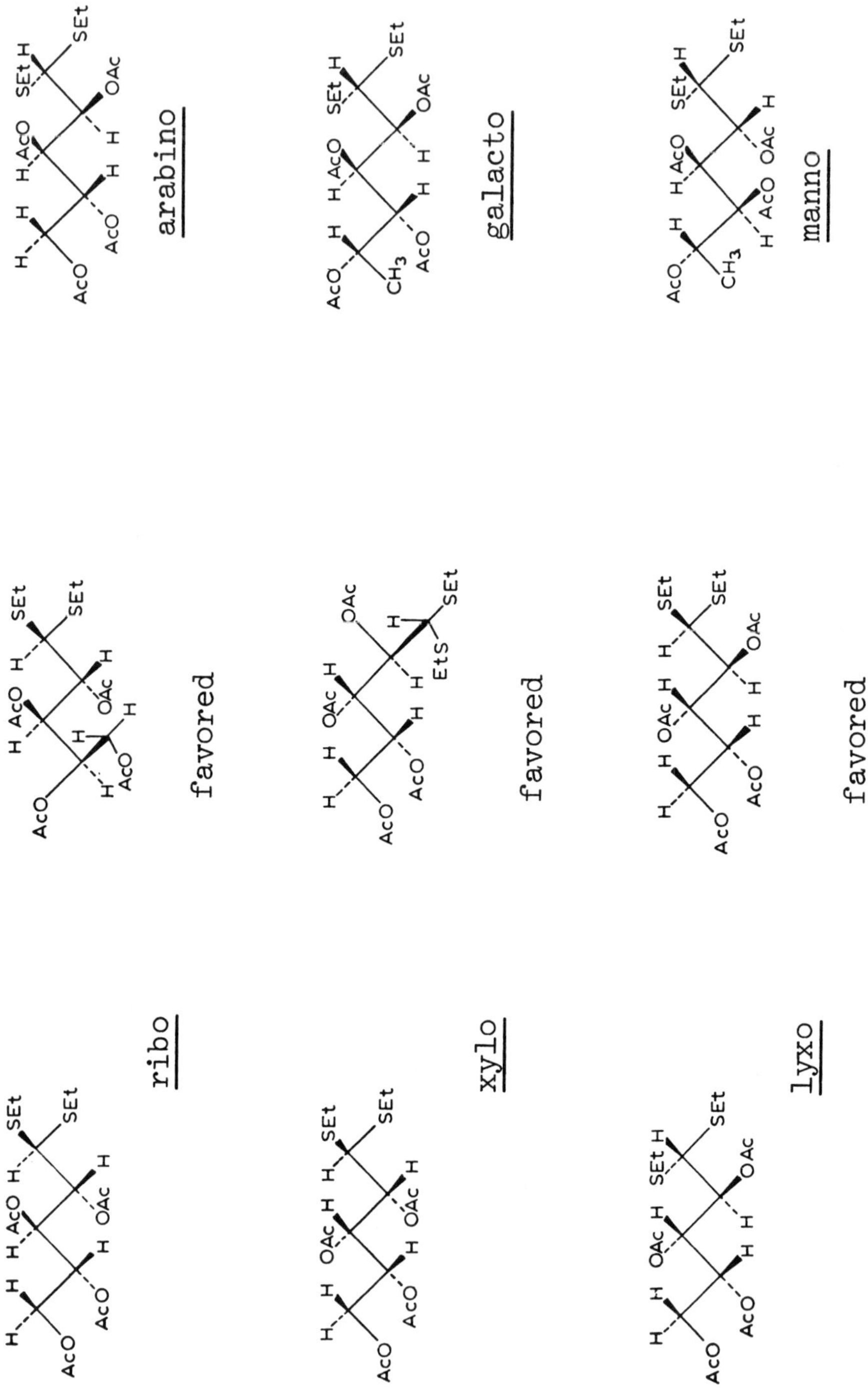

Figure 3. Conformations of dithioacetal peracetates

a conformation having H-1 and H-2 antiparallel, as shown in the lower part of the figure; such a bent conformation is called a sickle conformation since the chain of carbon atoms is bent around so that, in this example, the carbon atom of the aryl system can be considered as the point of the sickle and C-3 and C-4 of the side chain constitute the handle. Results entirely analogous to these have been observed with the corresponding acetylated derivatives of these substituted triazoles (*5*).

Extending these studies to systems not having a heterocycle at the end of the chain, a series of dithioacetal peracetates was examined (*6*). Generally, (Figure 3) these derivatives adopt favored conformations having the fully extended, planar zigzag chain only when such an arrangement would not lead to a parallel 1,3-interaction between substituent groups. Extended zigzag conformations are indicated by the spin-couplings in the *arabino*, *galacto*, and *manno* series. However, the coupling data for the *ribo*, *xylo*, and *lyxo* derivatives were not consistent with the fully extended arrangement. For example, in the *ribo* series, the $J_{3,4}$ coupling was small, corresponding to a gauche disposition of H-3 and H-4 and not to the antiparallel arrangement required in the extended conformation. The favored conformation shown is a sickle form derived from the extended form by rotating C-5 out of the plane of the other carbon atoms so that it becomes the point of the sickle. Likewise, in the *xylo* series a sickle conformation is adopted. In this instance it is C-1 that is rotated out of the plane of the other carbon atoms. This operation gives rise to the conformation shown in which the protons at C-2 and C-3 are essentially antiparallel, which agrees with the large coupling constants observed; the extended arrangement would have led to a small value for $J_{2,3}$. In the *lyxo* series it is one of the end groups that is in 1,3-interaction with the acetoxyl group at C-3 in the fully extended conformation, and the value observed for $J_{1,2}$ agrees with a favored conformation in which there is rotation about the C-1–C-2 bond to alleviate this interaction without generating another similar one.

Closely similar results were obtained with various diphenyl dithioacetal acetates (*7*), with the unsubstituted diethyl dithioacetals (*8*), and with the *aldehydo*-pentose peracetates (*9*) and the tetra-*O*-acetylaldopentose dimethyl acetals (*10*). Subsequent work in other laboratories has shown the same general principles for the methyl 5-hexulosonates (*11*) and the pentononitrile tetraacetates (*12*), two examples where a full series of stereoisomers has been studied. Other workers have investigated isolated examples or partial series (*13, 14, 15, 16, 17, 18*), and parallel work by x-ray crystallography (*19, 20, 21, 22*) on acyclic sugar derivatives in the solid state has shown excellent correlation with the general principles outlined here for the molecules in solution.

Some results of a recent collaborative study (*10*) on the tetraacetates of the aldopentose dimethyl acetals are shown in Table I. The *arabino*

Table I. The 2,3,4,5-Tetra-O-Acetylpentose Dimethyl

Compound	*Configuration*
1	*ribo* [b]
2	*arabino*
3	*xylo*
4	*lyxo*

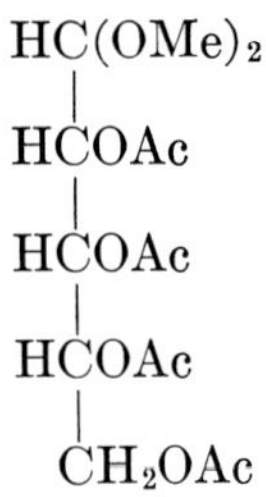

1

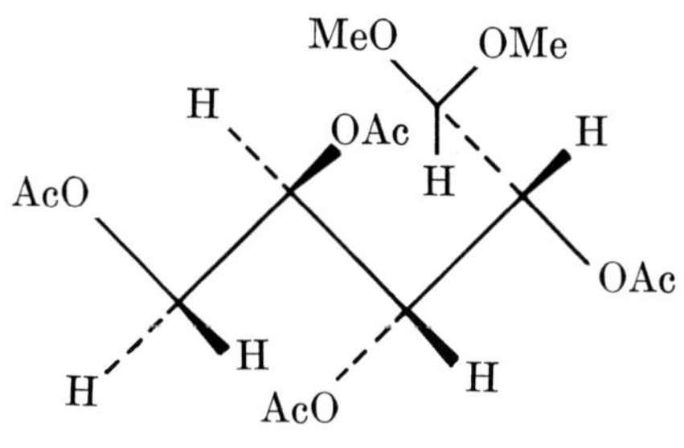

1a

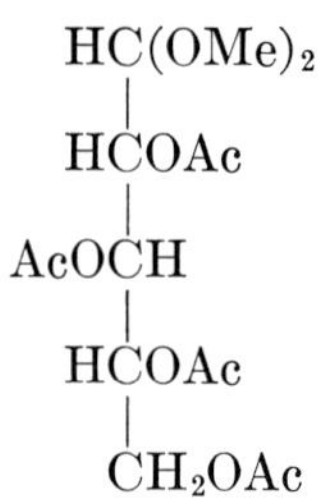

3

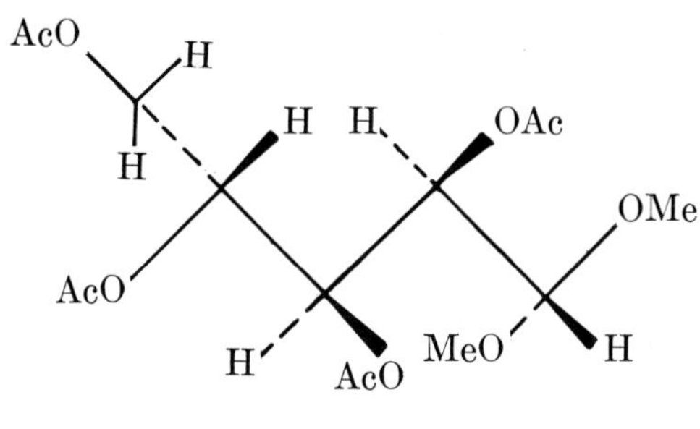

3a

[b] From 220-MHz spectrum at ambient temperature, measured in chloroform-*d*.

derivative favors the fully extended form, whereas the *ribo* and *xylo* derivatives adopt sickle conformations as the favored forms. The *lyxo* derivative adopts a favored conformation derived from the fully extended form by rotation about C-1–C-2 to a rotamer having H-1 and H-2 antiparallel as the favored, but not exclusive, form.

These conformational tendencies have also been correlated with the ease of irreversible cyclization of various acyclic derivatives (*23, 24*).

Acetals in Solution[a]

Coupling Constants, Hz					
$J_{1,2}$	$J_{2,3}$	$J_{3,4}$	$J_{4,5}$	$J_{4,5'}$	$J_{5,5'}$
6.5	3.9	5.5	2.5	6.3	12.2
6.7	2.6	8.3	2.9	5.1	12.3
5.5	4.7	5.4	4.3	5.9	12.0
5.6	6.5	3.3	4.4	6.3	11.6

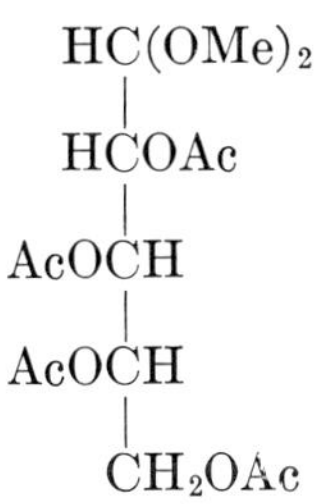

2

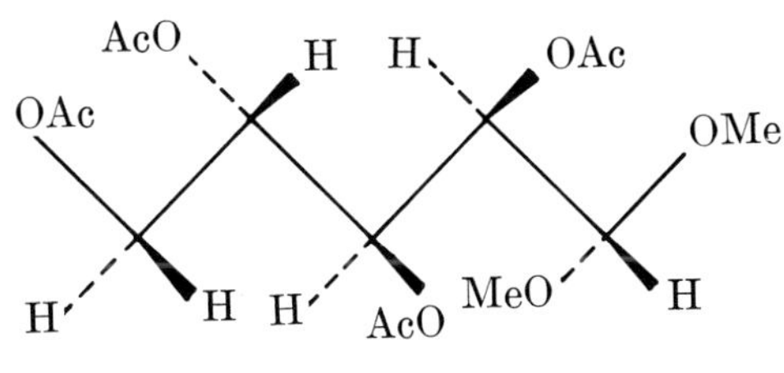

2a

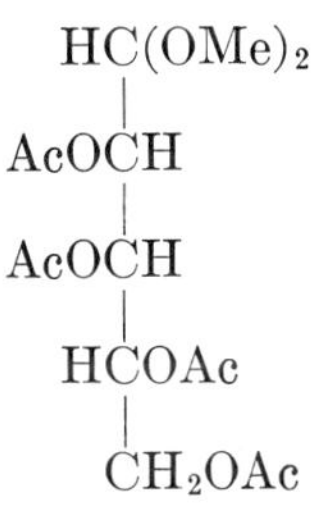

4

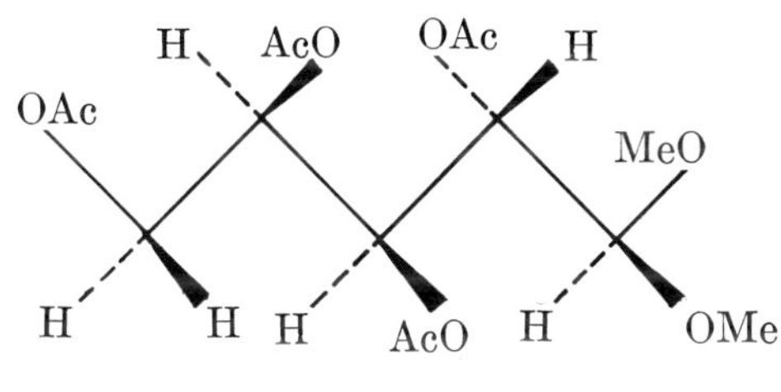

4a

[a] From 100-MHz spectra at ambient temperature, unless otherwise stated.

Thus, when the aldopentose diethyl dithioacetals are treated with 1 mole of *p*-toluenesulfonyl chloride in pyridine, cyclization to form a 2,5-anhydride is observed in each instance except in the *arabino* series, where a 5-*O*-*p*-tolylsulfonyl derivative can be isolated. It may be supposed (23) that the energy of activation for the cyclization reaction, being the difference between the ground-state and the transition state energies, is smaller in those stereoisomers where conformational factors bring O-2 into close proximity with C-5 in a low-energy conformation. In the *arabino* series, extra energy has to be expended to bring the molecule from its favored, extended conformation into an orientation in which O-2 approaches C-5 from its rear side to permit displacement of the *p*-tolylsulfonyloxy group. For the *arabino* derivative there is an additional elevation of the transition-state energy for cyclization because of the necessity to develop the all-syn arrangement of three substituents at the ring-closure step.

A related study on ring closure (*24*), this time under conditions of acid-catalyzed methanolysis, is illustrated in Figure 4, which shows a recent collaborative effort. The four stereoisomeric 1,2-*O*-isopropylidene-5-*O*-*p*-tolylsulfonylaldopentofuranoses were refluxed in methanolic hydrogen chloride. Within a few minutes each starting material had become converted into an anomeric mixture of methyl 5-*O*-*p*-tolylsulfonylaldopentofuranosides. Further refluxing of this mixture for 5 hours led, as illustrated in the *lyxo* series, to the corresponding 2,5-anhydropentose dimethyl acetals. A high yield of this product was obtained in the *lyxo* series after 5 hours, and the *xylo* series behaved likewise; in the *ribo* series a similar result was observed although the yield of product was somewhat lower. However, in the *arabino* series the product after 5 hours was exclusively the mixture of pyranosides, and none of the 2,5-anhydride could be detected. In the *arabino* series it was necessary to extend the period of reflux to 72 hours to obtain the 2,5-anhydro dimethyl acetal, even in very low yield. This observed difference in behavior as a function of configuration can again be interpreted in conformational terms. In this instance the relative transition-state energies for ring-closure are presumably the most significant. Since, in at least two of these examples, it would be impossible for the anhydride ring to form from the furanose precursor directly, it can be supposed that the small proportion of dimethyl acetal statistically present in equilibrium with the more favored furanoside undergoes irreversible ring-closure to form the 2,5-anhydride, and that the steric requirements for the ring closure are particularly unfavorable for the arabinose derivative.

Cyclic Systems

Our studies on cyclic systems set out to answer questions such as the following, with respect to multisubstituted tetrahydropyran rings:

(a) What effect does the heteroatom have on the conformational behavior?

(b) Are sugars and their derivatives in rapid conformational equilibrium in solution?

(c) Are steric effects additive?

The general type of molecule studied is shown in Figure 5. In the 2,3,4,5-tetrasubstituted, tetrahydropyran ring-system the anomeric substituents (R) were systematically varied by the groups OAc, OBz, OMe, halogen, and SAc while the substituents at the other three positions were acetoxyl or benzoyloxyl groups. Again the key tool was PMR spectroscopy, and particular emphasis was placed on comparing behavior throughout various whole series of stereoisomeric examples.

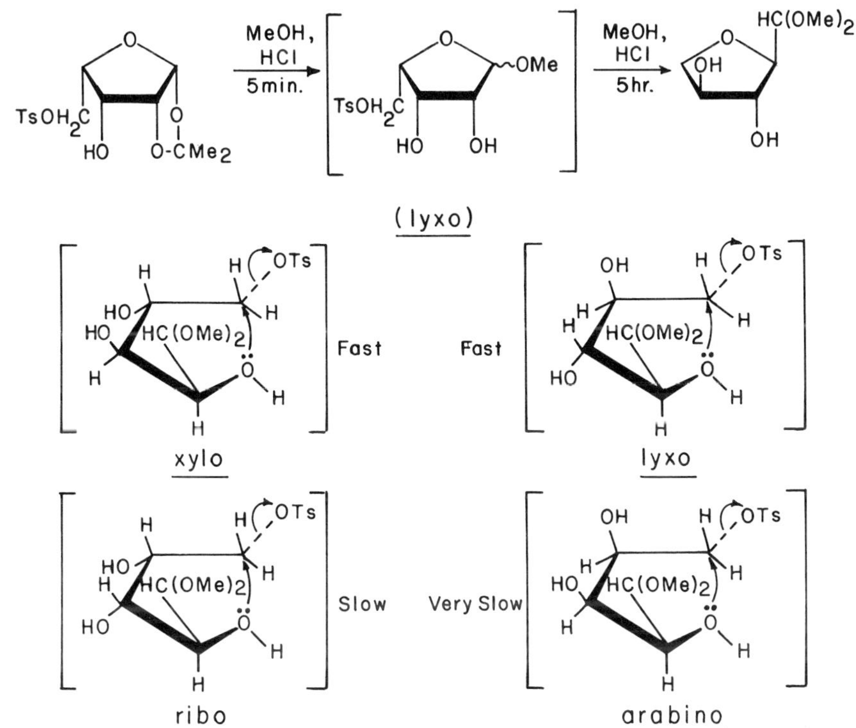

Figure 4. Ring closure of the four stereoisomeric 1,2-O-isopropylidene-5-O-p-tolylsulfonylaldopentofuranoses

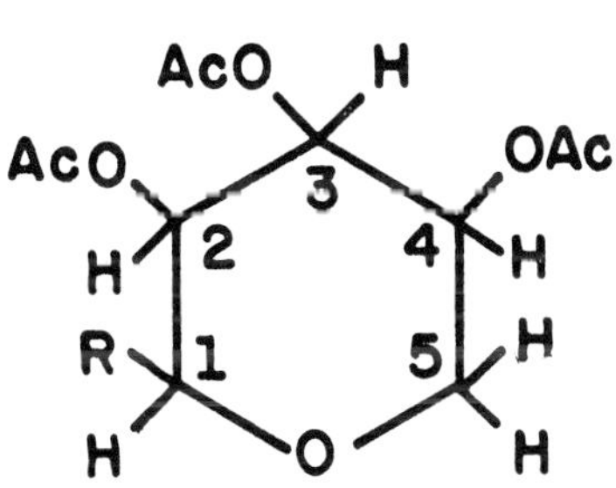

Figure 5. 2,3,4,5 Tetrasubstituted, tetrahydropyran ring-system

First, the anomeric equilibria of the aldopentopyranose tetraacetates were examined by NMR spectroscopy and by optical rotation (*25*). Each of the acetylated anomeric forms of the aldopentopyranose tetraacetates was allowed to attain equilibrium in a 1:1 mixture of acetic anhydride and acetic acid containing perchloric acid as a catalyst (Table II). The

Table II. Anomeric Equilibria of D-Aldopentopyranose Tetraacetates at 27°C in 1:1 Acetic Anhydride–Acetic Acid, 0.1*M* in Perchloric Acid

Anomeric pair	*Equilibrium constant* K = β/α	ΔG°, *kcal mole*$^{-1}$, *for* α ⇄ β *at 27°C*
Tetra-*O*-acetyl-α,β-D-ribopyranose	3.4	−0.73 ±0.03
Tetra-*O*-acetyl-α,β-D-arabinopyranose	5.4	−1.01 ±0.03
Tetra-*O*-acetyl-α,β-D-xylopyranose	0.23	+0.89 ±0.03
Tetra-*O*-acetyl-α,β-D-lyxopyranose	0.20	+0.98 ±0.05

equilibrium constants determined in the ribose, arabinose, and xylose series agreed with literature values (*26, 27*), but the value for the lyxose series was somewhat larger than that previously (*26*) reported. From the equilibrium constants it is evident that the α-anomer in the xylose series is more stable than the β-anomer, even though the α-anomer undoubtedly has one of its substituents—the C-1 acetoxyl group—in axial orientation whereas the corresponding β-anomer could adopt a conformation having all four groups equatorial. However, the question arises not only as to what is the position of the configurational equilibrium in these examples, but, for each individual compound, what might be the equilibrium between the two chair conformers of the molecule. For this reason each compound was examined to determine whether it adopts one favored conformation or whether a conformational equilibrium exists in which a substantial proportion of more than one conformer is present.

A representative spectrum from the series is illustrated in Figure 6, which shows the NMR spectrum of the tetraacetate of 1-thio-β-D-xylopyranose at 100 MHz in three different solvents (*28*). The spin-couplings observed agree fully with a favored conformation having all four substituents equatorial, and consequently H-1, H-2, H-3, and H-4, all axial, with H-4 antiparallel to the axial proton at C-5. The large magnitude of the $J_{1,2}$ spin-coupling, and the corresponding large magnitude of the $J_{4,5}$ diaxial coupling, indicate directly this conformation, which is supported by further analysis of the other couplings in the ring system. Entirely analogous results are obtained for 1-thio-β-D-ribopyranose tetraacetate, except that here the C-3 acetoxyl group is axially oriented.

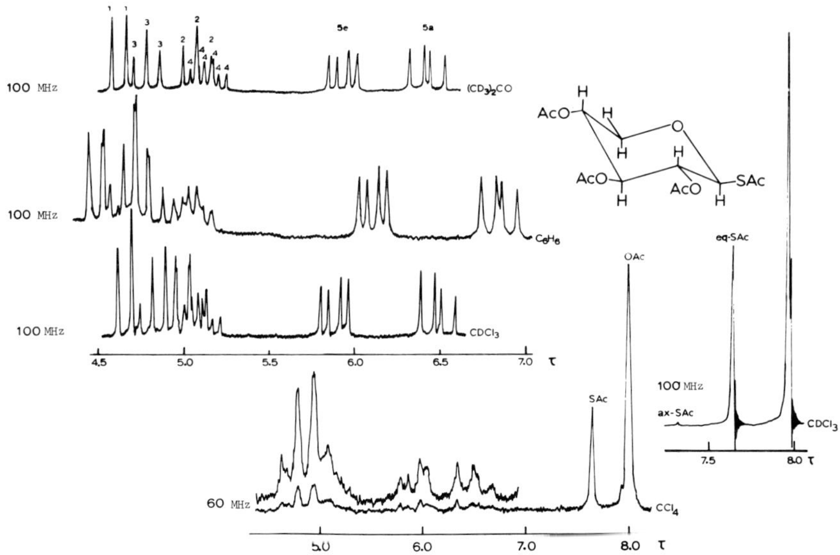

Figure 6. NMR spectrum of 1-thio-β-D-xylopyranose tetraacetate

The acylated glycosyl halides were examined in detail (*29, 30*). The spectrum of tri-*O*-acetyl-α-D-xylopyranosyl bromide (Figure 7) gives spin-couplings that agree completely with a conformation for this molecule having all of the substituents except the bromine atom equatorial (*29*). Since this anomeric form is the thermodynamically favored one, it can be concluded that the anomeric halogen atom is thermodynamically more stable in the axial than in the equatorial orientation. This is an illustration of the well-known anomeric effect, which has been interpreted (*31, 32*) as an electronic interaction between the dipolar vector of the C-1—halogen bond and a net dipole generated as a result of the ring heteroatom; this interaction would create an electronic destabilization of the halogen atom in equatorial orientation of sufficient magnitude to overcome the steric destabilization caused by the halogen atom's adopting the axial orientation. The importance of this effect is readily shown by referring to tri-*O*-acetyl-β-D-ribopyranosyl chloride (Figure 8), a halide that is again the thermodynamically stable anomeric form (*29*). The spin couplings show that H-1 and H-2 are in gauche orientation (diequatorial) and that H-4 bisects the angle of the two protons at C-5, as seen from the very narrow signals observed for H-1 and from the small $J_{4,5}$ and $J_{4,5'}$ couplings. This compound is the more-stable anomeric form, and this favored conformation has three of the four substituents axial, and a syn-diaxial interaction exists between the acetoxyl groups at C-2 and C-4. Such a conformation would be regarded as strongly destabilized were steric considerations alone invoked, and if the anomeric effect is respon-

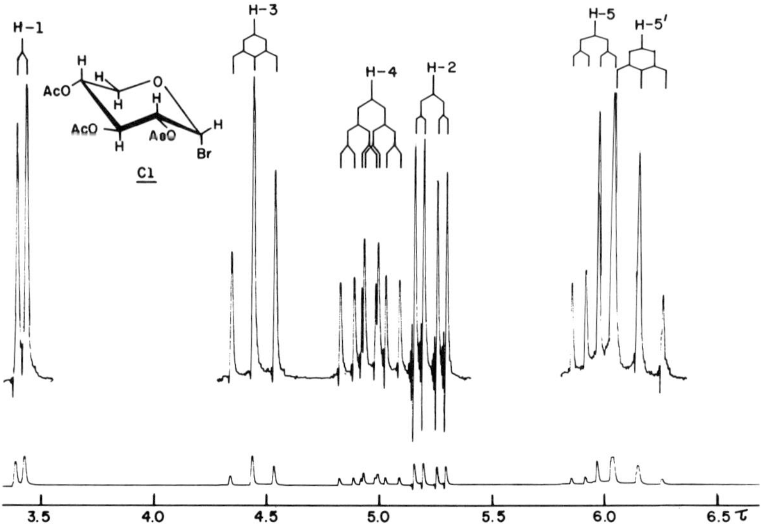

Figure 7. NMR spectrum of tri-O-acetyl-α-D-xylopyranosyl bromide in chloroform-d

sible for the adoption of the ring conformation observed, its magnitude must considerably outweigh the combined steric destabilization of three axial groups and one syn-diaxial interaction.

In an extensive, comparative study of a range of glycosyl halide derivatives in their thermodynamically more-stable forms, it was uniformly found that the conformation adopted in solution is the one having the halogen group axial, suggesting that the anomeric effect must be of considerable magnitude (*29, 30*).

When some of the thermodynamically unstable glycosyl halides, obtained by kinetic reactions in which equilibration to the stable form did not take place, were examined, some still-more-striking results were obtained. For example, when tri-*O*-acetyl-β-D-xylopyranosyl chloride was examined in various solvents (Table III), the $J_{1,2}$ coupling constants and the coupling of H-4 with the protons at C-5 were found to be uniformly small, indicating that H-1 and H-2 are diequatorial and that H-4 bisects the angle of the C-5 protons in the favored conformation (*29, 33*). This favored conformation has all four substituents axially disposed. The fact that such a conformation is favored, despite the presence of four axial groups and two sets of syn-diaxial interactions—an arrangement highly destabilized from conventional steric considerations—could be rationalized by the operation of an anomeric effect of surprisingly high magnitude. Predictive treatments that have been advanced (*26, 27*) to determine conformational preference, on the basis of additive interactions between various substituents in conjunction with an estimated magnitude for the anomeric effect,

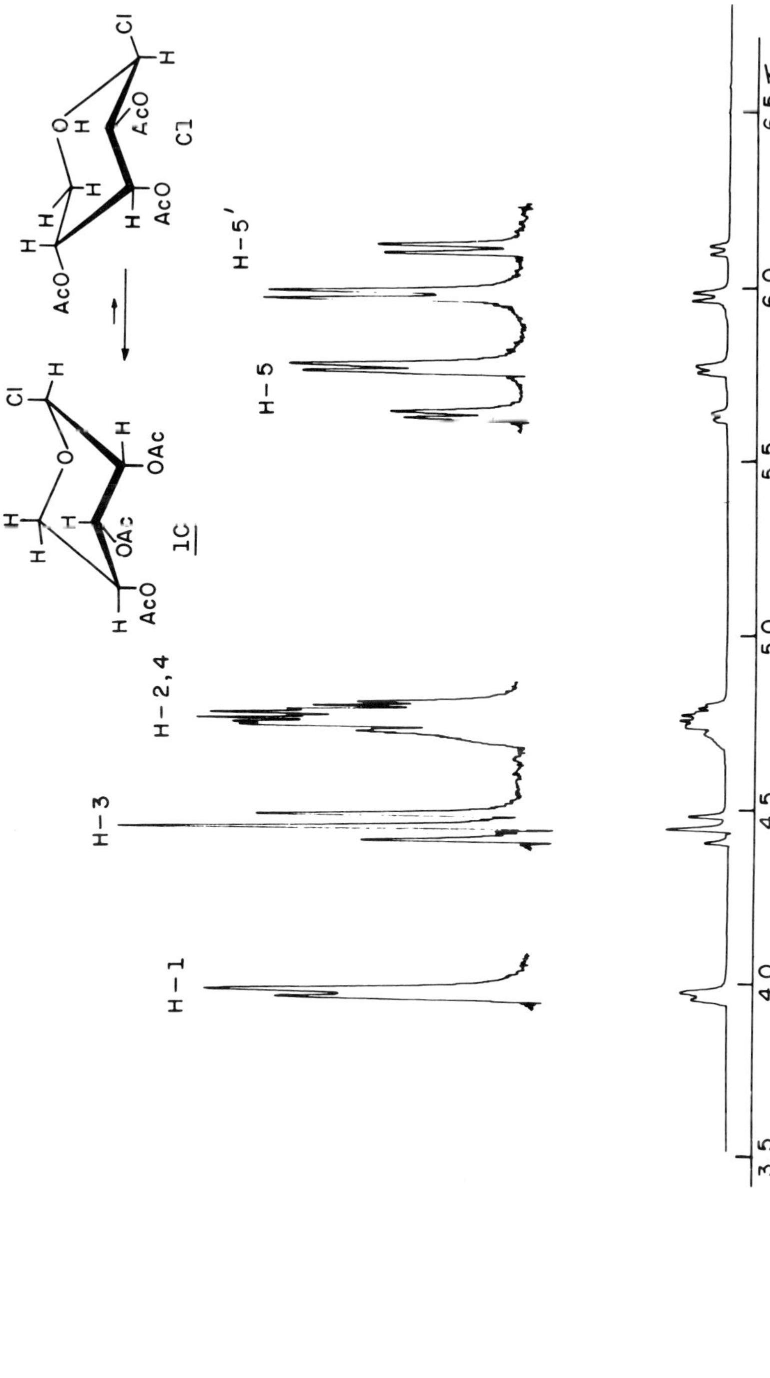

Figure 8. NMR spectrum of tri-O-acetyl-β-D-ribopyranosyl chloride in chloroform-d

do not agree with the experimental facts observed with compounds such as the one illustrated in Table III, even with adjustment of some of the values of the interactions considered.

Table III. Tri-O-Acetyl-β-D-Xylopyranosyl Chloride

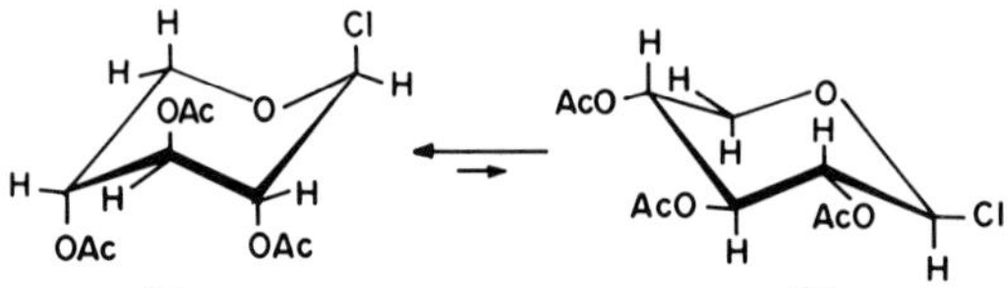

1C C1

Solvent	Coupling Constants, Hz $J_{1,2}$	$J_{4,5a}$	$J_{4,5e}$
C_6D_6	3.3	3.1	3.9
$CDCl_3$	[a]	2.8	3.7
$(CD_3)_2CO$ [b]	[a]	3.1	4.4

[a] Small couplings, but second-order effects prevent specific determination.
[b] In $(CD_3)_2CO$ the H-1 signal is at τ 4.09 (31°C), 3.94 (−65°C), 3.87 (−80°C).

Studies on conformational equilibrium, described below, establish that for tri-*O*-acetyl-β-D-xylopyranosyl chloride the all-axial *1C* (D) conformation comprises about 80% of the total conformational population, in an equilibrium wherein approximately 20% of the all-equatorial *C1* (D) form is present.

A general conclusion from the data on the halides is that the anomeric effect strongly outweighs conventional steric factors in determining conformational behavior.

Concerning the possibility of conformational equilibration between chair forms in various pyranoid sugar derivatives, it has been noted in early work that the observed NMR parameters are usually in fair accord with those expected for pure conformational states (*34*). It has been speculated that the multiplicity of large substituents around the tetrahydropyran ring in pyranoid sugar derivatives might decrease the tendency of the ring to undergo inversion because of passing interactions between substituents, so that conformational interconversion at room temperature might be slow on the NMR time-scale, so that the spectrum of a pure conformer is observed. Another viewpoint is that such derivatives exist in a rapid conformational equilibrium, but that for various reasons most of the derivatives studied favor one conformation over the other to a sufficient extent that the coupling values observed are close to those expected for a pure conformer. Experimental data from our laboratory establish that the latter viewpoint is the correct one (*35, 36*).

A hypothetical equilibrium between two chair conformers of a tetrahydropyran derivative is illustrated in Table IV. The equilibrium between the two forms, having mole fraction N_e of the form having H-1 equatorial, in equilibrium with a mole fraction N_a of the form having H-1 axial, is related by the standard thermodynamic expression relating the equilibrium constant (K) with the free-energy difference (ΔG°) for the transformation. If such an equilibrium is rapid on the NMR time-scale at room temperature, the vicinal coupling observed between the adjacent protons will be an averaged value, and the chemical shifts of the signals will be at positions intermediate between those for protons in exclusively equatorial or axial orientation. At temperatures sufficiently low that interconversion becomes slow on the NMR time-scale, separate signals will be observed for protons in the individual conformers. The conformer having diequatorial protons will show the H*e* resonance as a narrow doublet (small value of $J_{vicinal}$), and the conformer having diaxial protons will show the H*a* resonance at somewhat higher field as a wide doublet (large value of $J_{vicinal}$). The integrated intensities of the signals for H*e* and H*a* are directly proportional to the relative mole fractions, N_e and N_a, of the two conformers, and thus by integration of the spectrum below the temperature of "conformational freeze-out," it is possible to determine the equilibrium constant (K) and the free-energy difference (ΔG°) for the conformational interconversion. At a certain temperature intermediate between that required for conformational freeze-out and the high temperature at which an averaged pattern is observed, there will be visible a transitional pattern established when the separate signals of the individual conformers just coalesce into a single, broad peak. At this "coalescence temperature" (T_c) and from the separation of the signals of the individual conformers, it is possible by Gutowsky's equation (*37, 38*) to determine the rate of conformational inversion. From that value, by using the Eyring equation, the free-energy of activation of the ring interconversion can be determined (*36*). The Gutowsky equation is strictly applicable only for equal population of the two conformational states although it can be modified to take unequal distribution into account (*25, 39*).

The free-energy difference for the conformational interconversion, which can be calculated directly if the mole fractions of the two forms are determined from a spectrum recorded at low temperature, can also be obtained by the method of averaging of coupling constants, by relating the observed coupling in the time-averaged spectrum with the couplings observed at low temperature for the isolated conformers. The values for N_a and N_e thereby determined allow the calculation of the

Table IV. Measurement of Conformational Equilibrium and Rate of Ring Inversion

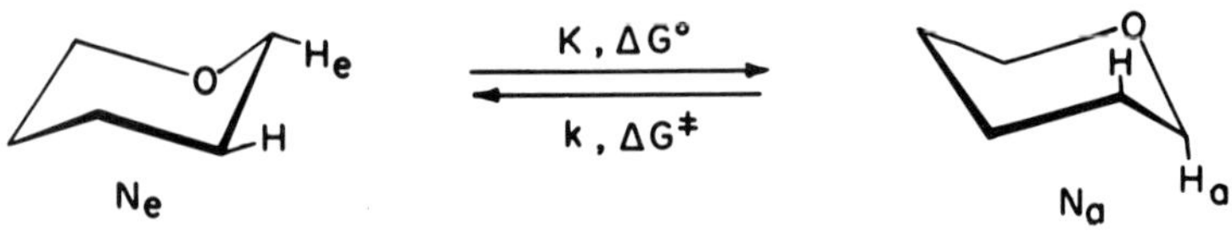

$$\Delta G^\circ = -RT\ ln\ K = -RT\ ln\ \frac{N_a}{N_e}$$

$N_e + N_a = 1$

Determination of ΔG°

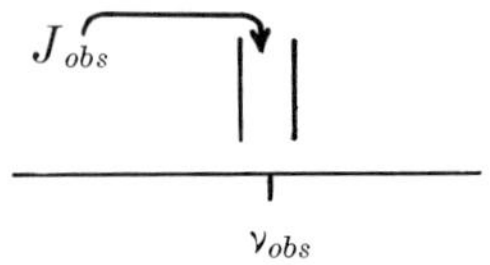

Time averaged spectrum

(a) From coupling constants $J_{obs} = N_e J_e + N_a J_a$

(b) From chemical shifts

$\nu_{obs} = N_e \nu_e + N_a \nu_a$

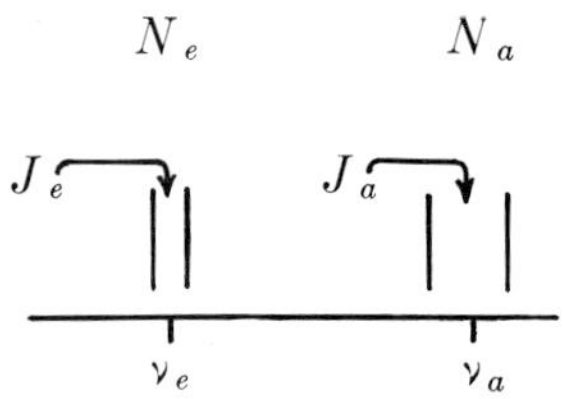

Spectrum of individual conformers

(c) From spectral integrals at low temperature, to give N_e and N_a directly

Inversion rate k

Spectrum at coalescence temperature, T_c

$$k = \frac{(\nu_a - \nu_e)\pi}{\sqrt{2}}\ \text{sec}^{-1} \text{ at } T_c$$

$$k = \frac{\kappa k_B T_c}{h}\ e^{-\Delta G^{\ddagger}/RT_c}$$

equilibrium constant at the temperature of the averaged spectrum and thus allow calculation of ΔG° for this temperature. A similar approach for determining ΔG° might, in principle, be applied with the chemical-shift values, but our results show that this particular method is unreliable because the chemical shifts of the pure conformers themselves are not necessarily independent of temperature (*36*).

It is not easy to find examples in the sugar field to illustrate directly a conformational equilibrium by the method of conformational freeze-out. To show such behavior, it is necessary to have a sugar derivative that has substantial proportions of the two forms present at low temperature, so that the signals of the minor conformer are not lost in the background "noise" of the spectrum. It is also necessary to have a suitable solvent that does not freeze above the temperature of conformational freeze-out, from which the compound does not crystallize at low temperature, which does not become too viscous at low temperature, and in which the spectrum of the compound in question is sufficiently well resolved to allow detailed interpretation. Because of these stringent limitations, it is desirable to work at the highest possible field-strength. Figure 9 illustrates the 220 MHz NMR spectrum of β-D-ribopyranose tetraacetate in acetone-d_6 at various temperatures (*35*). This compound was selected because its $J_{1,2}$ value of 4.8 Hz at room temperature suggests that it exists as an equilibrium mixture of the form having H-1 and H-2 antiparallel and that having H-1 and H-2 diequatorial. The spectrum observed at room temperature shows little change as the temperature is lowered until about −50°C when a progressive broadening of certain peaks is observed. This observation is not especially significant because such broadening may merely reflect a lower rate of tumbling of the molecules. However, as the temperature is lowered still further, a progressive sharpening of the spectrum is observed, and at −84°C the original sharp doublet for the anomeric proton has become resolved into two, discrete signals, a wide doublet at somewhat higher field than the original doublet and a narrow signal at somewhat lower field than the original H-1 doublet. The higher-field signal can be interpreted as that of H-1 of the *C1* (D) conformation, having H-1 and H-2 axial, and the lower-field signal can be ascribed to the equatorial H-1 of the compound in the *1C* (D) conformation, in which the $J_{1,2}$ spin-coupling is on the order of only 1 Hz. The relative intensities observed indicate the the triaxial *1C* (D) form preponderates to the extent of about 2:1 over the *C1* (D) conformation having three groups equatorial (*35, 36*). From these data, in comparison with the behavior observed for the corresponding 1-halo derivative (*29, 30*), and the corresponding 1-acetylthio derivative (*28, 40*), a ranking can be made of the relative axial-directing effect of the group at position 1, in the order Br ≈ Cl > OAc > SAc.

The data obtained from Figure 9 are tabulated in Table V. The 2:1 equilibrium ratio of the *1C* (D) and the *C1* (D) forms indicates a relative free-enery difference of about 0.3 kcal mole^{-1} at −84°C. From the coalescence temperature (−60°C) the rate of interconversion of the *C1* (D) to the *1C* (D) conformer can be calculated to be about 117 times per sec., corresponding to a free-energy of activation for ring inversion of about 10.3 kcal mole^{-1}. This last value (*25, 35, 36*) indicates that the tetrahydropyran ring in this derivative has approximately the same flexibility as tetrahydropyran (*41*) and is somewhat more flexible than cyclohexane, for which a free-energy of activation for ring inversion of approximately 11 kcal mole^{-1} has been determined (*42, 43*). Presumably, therefore, the ring-system in monocyclic pyranoid-sugar derivatives is not

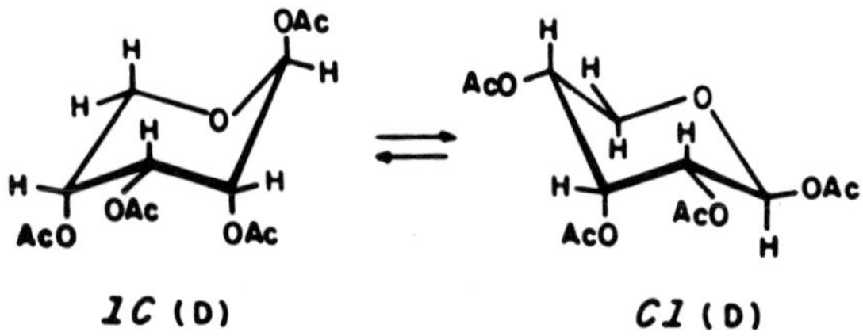

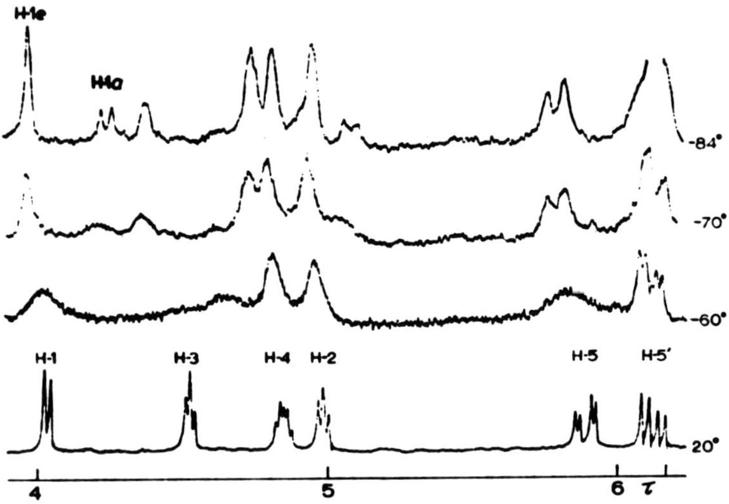

Figure 9. The low-field portion of the 220 MHz NMR spectrum of β-D-ribopyranose tetraacetate in acetone-d_6 at 20, −60, −70, −84°C

significantly less flexible than unsubstituted tetrahydropyran, and "passing interactions" of the substituents seem to have no marked effect in decreasing the rate of conformational interconversion.

For comparative purposes it is inconvenient to use equilibrium data that apply to a low temperature, such as the value of −84°C used in the experiment establishing the conformational interconversion of β-D-ribopyranose tetraacetate. However, the values determined at low temperature for $J_{1a,2a}$ (8 Hz) and $J_{1e,2e}$ (~1 Hz) that apply to the separate chair conformers, when taken in conjunction with the averaged value for $J_{1,2}$ (4.8 Hz) observed at room temperature, permit the calculation by the method of averaging of spin coupling of the equilibrium at room temperature (Table VI). At room temperature the equilibrium constant is almost unity. The method of averaging of spin coupling is applied here to the coupling between H-1 and H-2, but the coupling of H-4 with the protons at C-5 could equally well be utilized for this calculation since the relevant couplings for the pure conformers can be readily determined, as can the averaged values at room temperature. The use of the 1,2 coupling for this purpose is limited to those compounds having the substituents at C-1 and C-2 in trans orientation, whereas use of the coupling between the C-4 and C-5 protons can be utilized for all stereochemical variations, regardless of the configuration at the anomeric position.

Study of the eight stereoisomeric aldopentopyranose tetraacetates showed that significant population of both chair conformers is the rule

Table V. β-D-Ribopyranose Tetraacetate

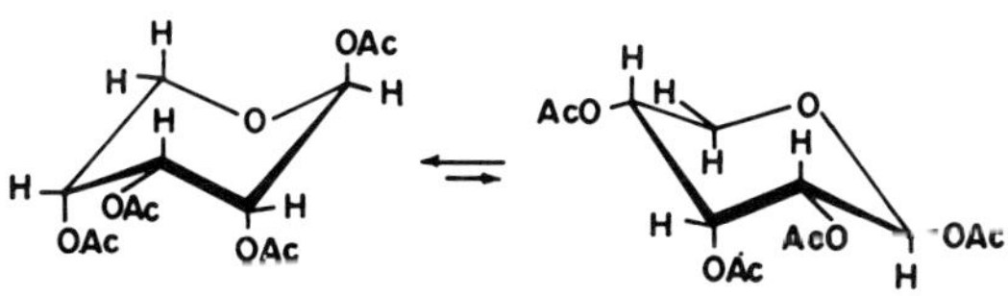

1C C1

Chemical shifts (τ) in $(CD_3)_2CO$ at 31°C

H-1	H-2	H-3	H-4	H-5	H-5′
4.04	5.00	4.54	4.86	5.90	6.16

At −84°C H-1*e* τ 3.93
H-1*a* τ 4.19

1C:*C1* = 2:1 at −84°C

ΔG° = 0.3 kcal mole^{-1}

$k_{C1 \rightarrow 1C}$ = 117 sec^{-1} at −60°C

$\Delta G^\ddagger_{C1 \rightarrow 1C}$ = 10.3 kcal mole^{-1}

Table VI. β-D-Ribopyranose Tetraacetate

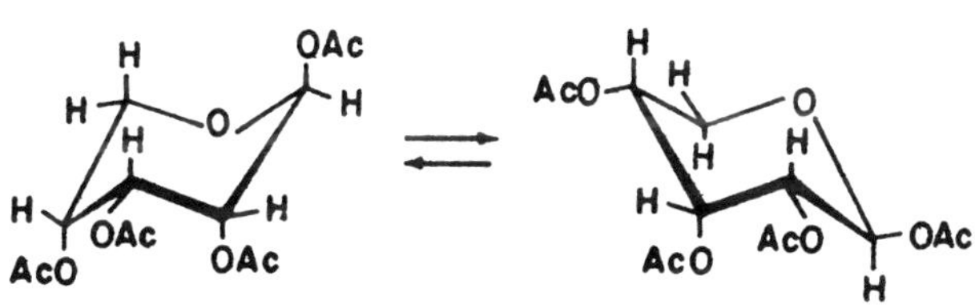

1C *C1*

$(CD_3)_2CO$

$J_{1a,2a}$ 8 Hz

$J_{1e,2e} \sim 1$ Hz

$J_{1,2}$ av 4.8 Hz

220 MHz, 20°C

$J_{4,5'}$ av 5.8 Hz

$K = \frac{C1}{1C} - 1.2$

$\Delta G^\circ = -0.11 \frac{\text{kcal}}{\text{mole}}$

rather than the exception (*25, 44*). Two limiting cases were observed in which one conformation is favored very strongly. These are illustrated in Figure 10. For α-D-xylopyranose tetraacetate, the *C1* conformation

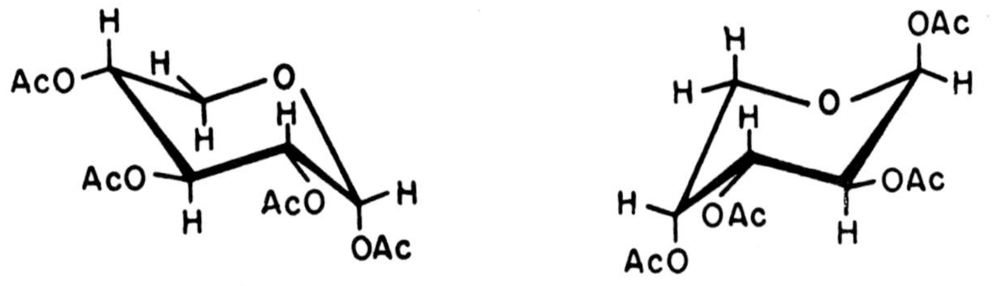

α-D-Xylopyranose tetraacetate-C1
$J_{1,2}$ = 3.5 Hz
$J_{4,5}$ (cis) = 5.5 Hz

β-D-Arabinopyranose tetraacetate-1C
$J_{4e,5a}$ = 0.8 Hz
$J_{4e,5e}$ = 1.5 Hz

Figure 10. Structures of α-D-xylopyranose tetraacetate-C1 and β-D-arabinopyranose tetraacetate-1C

was favored almost exclusively, to at least 98%. No variations in the spin couplings nor in the signal positions of the various protons on the ring system were observed over a wide range of temperatures. Accordingly, this compound can be utilized to provide model values for spin couplings for the *C1* (D) conformation in the series of aldopentopyranose tetraacetates. Thus, 11.6 Hz was used for the limiting value of the coupling between H-4 and the axial proton at C-5.

The other example in this series that showed limiting behavior was β-D-arabinopyranose tetraacetate, which adopts the *1C* (D) conformation to greater than 95% at −25°C. At this temperature the $J_{4,5'}$ value reached a lower limit of 1.5 Hz, and, accordingly, this value was used for the coupling of H-4*e* with H-5*e*. From these model values and the corresponding time-averaged values observed with the other six stereoisomers, the conformational equilibria for the whole series of eight tetraacetates were determined (Figure 11). The conformational equilibria illustrated diagrammatically (Figure 11) are quantitatively tabulated in Table VII. Only in the α-D-xylose series is the *C1* (D) conformation favored almost exclusively, and only in the β-D-arabinose series is the *1C* (D) conformation favored very strongly. The other six examples all have one conformation as the favored form, but a substantial proportion of the minor form is also present at equilibrium.

An interesting correlation is observed (Table VIII) between the magnitude of $J_{4,5}$ (*cis*) and the conformational equilibrium constants (*45*). A similar correlation is also observed (*45*) with the value of the geminal $J_{5e,5a}$. The value for $J_{4,5}$ (*cis*) extends from about 1.0 Hz when the *1C* (D) conformation is favored exclusively, to a value of 5.5 Hz when the *C1* (D) conformation is favored exclusively. Intermediate values, therefore, give a rough estimate of the conformational equilibrium. The geminal couplings for $J_{5e,5a}$ range from −13.2 Hz for the pure *1C* (D) conformation to −11.2 Hz for the pure *C1* (D) conformation; thus, the magnitude of the geminal coupling also provides a rough measure of the equilibrium constant (*45*).

The effect of temperature upon the conformational equilibrium was examined by using β-D-xylopyranose tetraacetate (Table IX), which at room temperature has about 72% of the all-equatorial conformer in equilibrium with about 28% of the all-axial form. As the temperature is lowered, the conformation tends more exclusively towards the *C1* (D) form, so that at temperatures where conformational freeze-out would be expected the single *C1* (D) conformation becomes so preponderant that any contribution from the minor conformer would be lost in the background noise of the spectrum (*36*). This result indicates that the entropy change in the conformational interconversion is not zero and explains why it is difficult to find many examples where a conformational freeze-out can be shown directly with such derivatives (*25*).

Although a ranking of relative axial-directing effects of various groups at the anomeric position has already been shown (halogen > AcO > AcS), it was interesting to retain the same atom directly attached to the anomeric position and to vary the nature of the attached substituent. Table X shows the conformational behavior of the analog of β-D-xylopyranose tetraacetate in which the 1-acetoxyl group has been replaced by a benzoyloxyl group (*46*). The data (*46*) indicate that 39% of the molecules exist in the all-axial conformation, as compared with only 28% in this conformation for β-D-xylopyranose tetraacetate. It can be inferred that the axial-directing effect of the benzoyloxyl group is higher than that of the acetoxyl group, and for the series of derivatives having a β-D-xylopyranose structure, the relative order of axial-directing effects of the 1-substituent falls in the order halogen > OBz > OAc ≈ SAc. Such a trend is not necessarily adopted for other configurations.

In the next example (Table XI) the effect of changing the substituents at the secondary positions on the ring is examined. The compound has the same overall structure as that in Tables IX and X, except that now all of the four substituents are benzoyloxyl groups. The equilibrium data indicate that the all-axial *1C* (D) conformer is present in an approximate 1:1 equilibrium with the all-equatorial *C1* form. Since the all-axial form occurs to the extent of 51%, as compared with only 39% for the corresponding tri-*O*-acetyl 1-benzoate and only 28% for the corresponding tetraacetate (*25*), this indicates that the replacement of acetate groups around the ring by benzoate groups correlates with an enhancement of

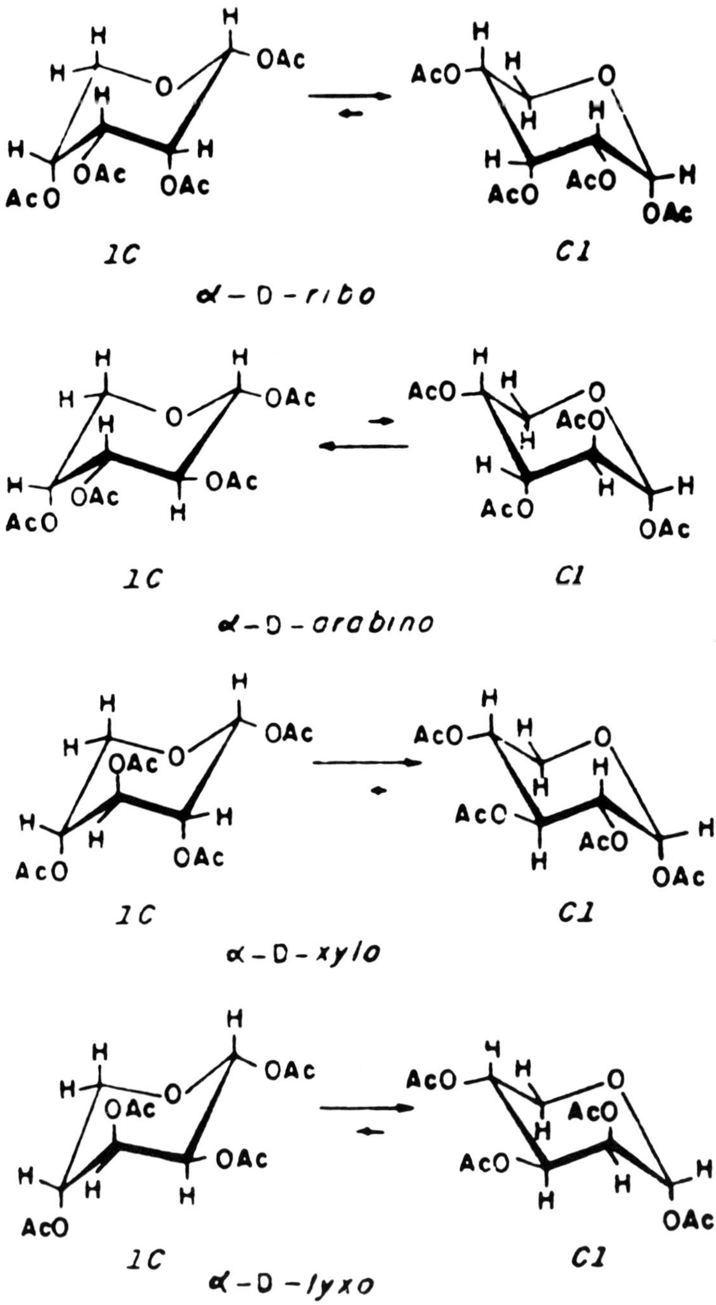

Figure 11. Conformational equilibria for the

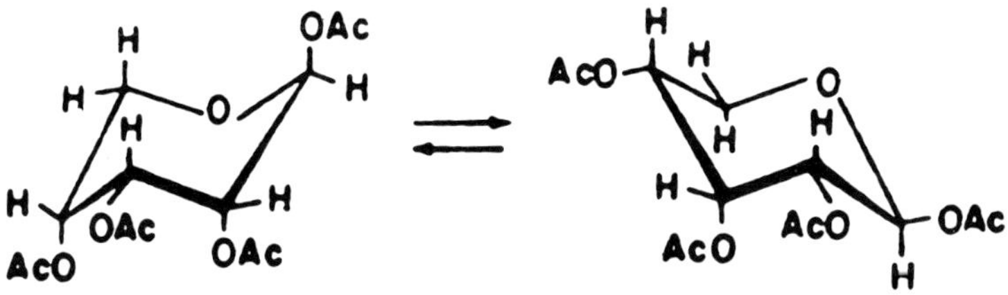

1C C1

β-D-ribo

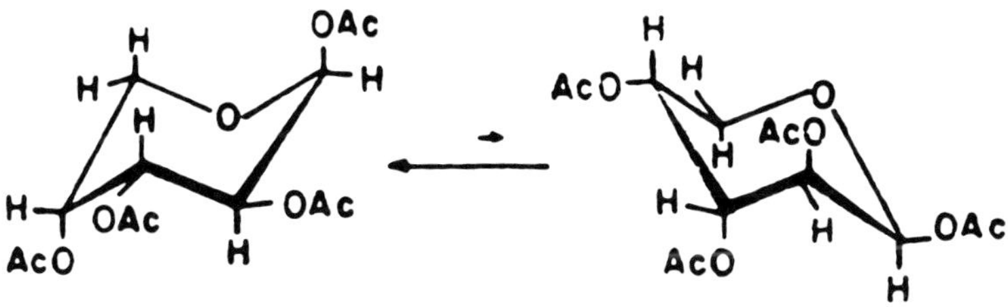

1C C1

β-D-arabino

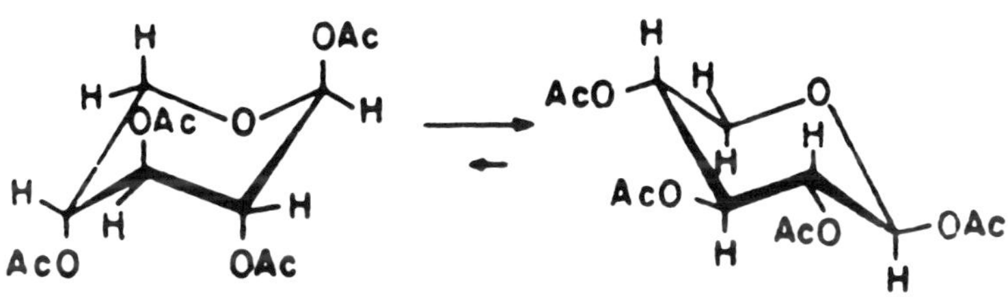

1C C1

β-D-xylo

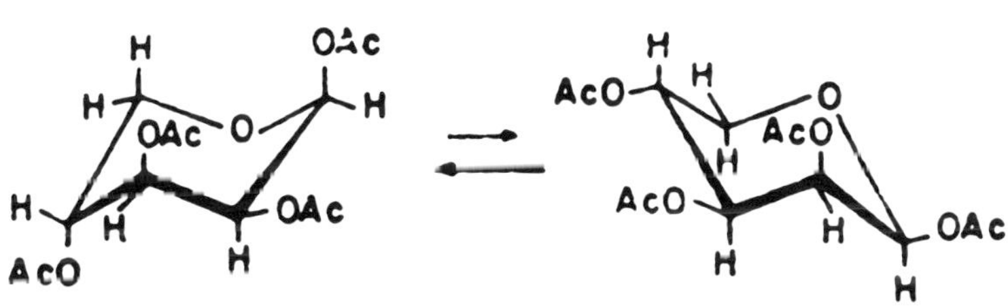

1C C1

β-D-lyxo

eight D-*aldopentopyranose tetraacetates*

Table VII. Conformational Equilibria of D-Aldopentopyranose Tetraacetates in Acetone-d_6 at 31°C

Configuration	*Coupling Constants, Hz*[a]			*Equilibrium Constant*	$\Delta G°_T$, *kcal mole*$^{-1}$, *for*
	$J_{1,2}$	$J_{4,5}$	$J_{4,5'}$	K = *C1/1C*	*1C*(D)⇄*C1*(D)
α-D-*ribo*	3.6	9.3	4.7	3.4	−0.74 ±0.33
β-D-*ribo*	4.6 (8.0)[b]	3.4	5.8	0.74	+0.18 ±0.26
α-D-*arabino*[c]	6.4 (7.5)	3.6	2.0	0.26	+0.81 ±0.34
β-D-*arabino*	2.9	1.0	1.9 (1.5)	0.04	+1.9 ±1.0
α-D-*xylo*	3.5 (3.5)	5.5 (5.5)	11.6 (11.6)	>50[d]	<−2.4
β-D-*xylo*	6.7 (8.1)	4.9	8.8 (10.5)	2.6	−0.58 ±0.30
α-D-*lyxo*	3.0	4.4	8.7	2.5	−0.55 ±0.30
β-D-*lyxo*	2.5	3.3	5.4	0.63	+0.28 ±0.27

[a] Values in parentheses are limiting values observed at low temperature (~−85°C). The C-5 proton resonating at lower field is designated H-5, that resonating at higher field is designated H-5′.

[b] $J_{1a,2a}$.

[c] In $CDCl_3$.

[d] Almost exclusively *C1*(D) at 31°C.

Table VIII. Experimental $J_{4,5}$ (cis) and $J_{5,5'}$ Values as a Function of the Equilibrium Constant

Series	*Configuration*	$J_{4,5}$ (cis)	$J_{5e,5a}$	K = *C1*(D)/*1C*(D)
D-Aldopentopyranose	β-D-*arabino*	1.0	−13.2	0.04
tetraacetates	α-D-*arabino*	2.0	−13.0	0.26
	β-D-*lyxo*	3.3	−12.4	0.63
	β-D-*ribo*	3.4	−12.4	0.74
	α-D-*lyxo*	4.4	−11.6	2.5
	β-D-*xylo*	4.9	−11.8	2.6
	α-D-*ribo*	4.7	−11.2	3.4
	α-D-*xylo*	5.5	−11.2	>50

Table IX. Temperature Dependence of the Conformational Equilibrium for β-D-Xylopyranose Tetraacetate in Acetone-d_6

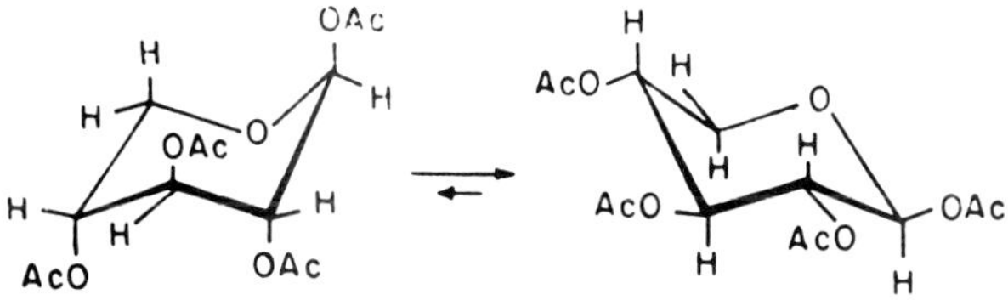

1C (D) *C1* (D)

β-D-*Xylopyranose Tetraacetate*

Temperature[a], *degrees C*	$J_{1,2}$, *Hz*[b]
−65	7.5
−49	7.3
−25	7.1
−16	7.0
+37	6.6
+62	6.4
+68	6.3

[a] Within ±2°C.
[b] Within ±0.1 Hz.

Table X. 2,3,4-Tri-O-Acetyl-1-O-Benzoyl-β-D-Xylopyranose[a]

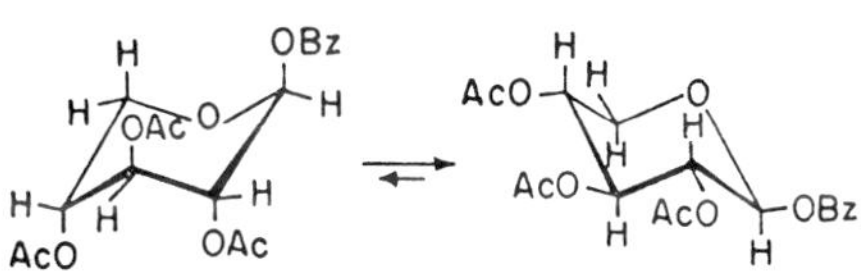

1C C1

Coupling Constants, Hz			*Equilibrium Constant*	ΔG°
$J_{1,2}$	$J_{4,5}$	$J_{4,5'}$	$K = \frac{C1}{1C}$	$\frac{(\text{kcal})}{(\text{mole})}$
5.9	4.5	7.7	1.6	−0.28

Order of size of axial-directing effect for various aglycons in the β-*xylo* series:

$$\text{Halogen} > \text{OBz} > \text{OAc} \approx \text{SAc}$$

[a] 100 MHz, $(CD_3)_2CO$, 31°C.

Table XI. β-D-Xylopyranose Tetrabenzoate[a]

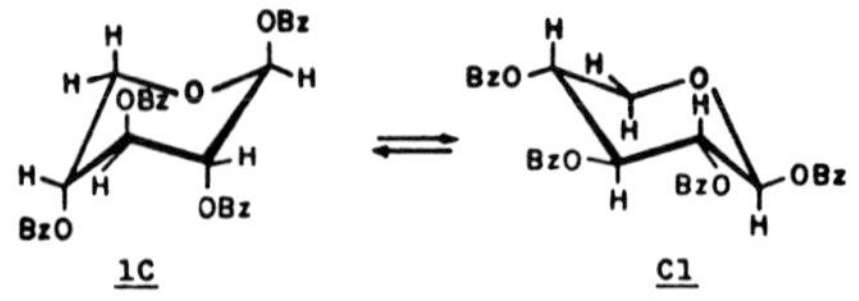

Coupling Constants, Hz				Equilibrium Constant	ΔG°
$J_{1,2}$	$J_{2,3}$	$J_{4,5}$	$J_{4,5'}$	$K = \frac{C1}{1C}$	$\frac{\text{(kcal)}}{\text{(mole)}}$
5.1	6.7	4.0	6.6	0.98	+0.01

[a] 100 MHz, $(CD_3)_2CO$, 31°C.

the axial-directing effect of the substituent at C-1. In related studies some other examples of the same general effect in which the axial-directing effect of the 1-substituent seems greater when the ring substituents are changed from acetates to benzoates have been noted. Such instances include the acylated methyl aldopentopyranosides (*47*) and various halides (*29*). For example, tri-*O*-benzoyl-β-D-xylopyranosyl chloride exists in chloroform-*d* to greater than 95% in the all-axial form, as compared with about 80% with the tri-*O*-acetyl analog (*29*).

When the equilibrium positions for the peracetylated aldopentopyranoses are compared with those for their perbenzoylated analogs (Table XII), a general overall correlation suggests that the syn-diaxial arrange-

Table XII. Comparison of Equilibrium Positions for the Peracetylated and Perbenzoylated Aldopentopyranoses

(a) β-D-*xylo*, β-D-*ribo*, and β-D-*lyxo* perbenzoates	—	favor *1C* form more strongly than do corresponding peracetates
(b) α-D-*arabino* and α-D-*lyxo* perbenzoates	—	*C1* form favored more strongly
(c) β-D-*arabino*, α-D-*xylo*, and α-D-*ribo* perbenzoates	—	equilibria approximately the same

ment of substituent groups in the ring, if this can be regarded as a destabilizing effect, is less destabilizing with the benzoates than with the acetates; alternatively, it might be supposed that a negligible repulsion or an actual attraction exists at a certain atomic distance between benzoate substituents when they are in syn-diaxial orientation (*25*).

Another question concerns the effect of solvent polarity on the conformational equilibrium. The results in Table XIII show for β-D-xylopyranose tetrabenzoate (which is approximately a 1:1 mixture of chair conformers in chloroform) that changes in the solvent polarity over a

Table XIII. β-D-Xylopyranose Tetrabenzoate

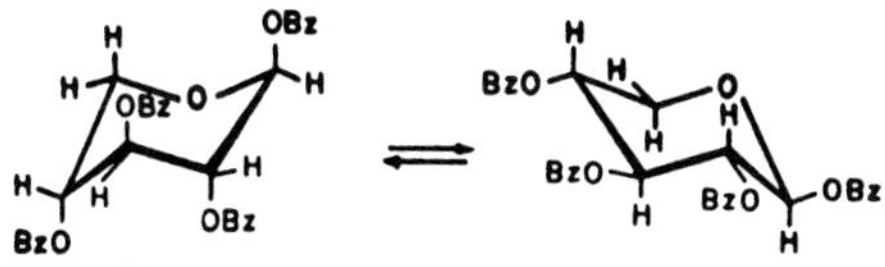

Solvent	ε	*Coupling Constants, Hz at 29°C*	
		$J_{1,2}$	$J_{2,3}$
C_6D_6	2.3	5.0	6.7
$C_6D_5CD_3$	2.4	5.2	6.9
$CDCl_3$	4.8	4.3	5.9
C_5D_5N	12.3	4.3	5.9
$(CD_3)_2CO$	20.7	5.1	6.7
$(CD_3)_2SO$	48.9	5.3	6.9

wide range, from that of benzene to that of dimethyl sulfoxide, causes a negligible change in the various coupling-constants observed for the molecule. From this result it is concluded that the polarity of the solvent has very little effect on the magnitude of the anomeric effect and that approximately the same distribution of conformers is observed over the whole range of solvent polarity. By contrast with these data established for the tetra-*O*-acyl derivatives, a regular dependence of conformational preference upon solvent polarity is observed with the corresponding methyl glycoside triacetates (*47*).

In a parallel investigation the effect of the steric bulk of substituents beyond O-1 was examined (compare Ref. *48*). The conformational behavior (*49*) of the methyl, ethyl, and isopropyl tri-*O*-acetyl-β-D-ribopyranosides was almost identical, but for the *tert*-butyl glycoside (Figure 12) there was some destabilization of the form having the alkoxyl group axial. This can be attributed to the fact that the C 1—oxygen bond loses a measure of rotational freedom in the *tert*-butoxyl derivatives (*49*).

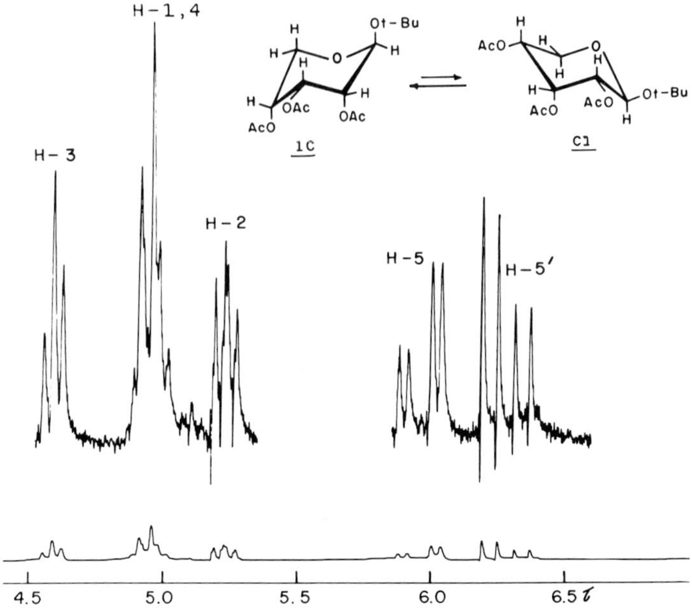

Figure 12. NMR spectrum of tert-*butyl tri-*O-*acetyl-β-*D-*ribopyranoside in acetone-*d_6

Summary

This work has shown that in acyclic sugar systems the planar zigzag arrangement is the favored conformation in solution unless such an arrangement would lead to parallel 1,3-interaction between substituent groups, in which case the molecule adopts a "sickle" conformation to alleviate this interaction.

Pyranoid-ring sugars exist in rapid conformational equilibrium, and the flexibility of the ring differs little from that of tetrahydropyran.

In the pentopyranose series, population of both chair conformers to a significant extent is the rule rather than the exception.

The anomeric effect dominates the conformational preference, but quantitative conformational predictions by presently described methods, based on additive treatments of steric effects taken in conjunction with the anomeric effect, do not accord with the observed data.

The axial-directing effect of the C-1 substituent depends on the total stereochemistry and mode of substitution and falls roughly in the order of polarity of the substituents. The axial-directing effects of the C-1 substituent is augmented when acetates at the other ring positions are replaced by benzoates; possibly there are attractive syn-diaxial interactions between benzoate groups.

Solvent polarity has a negligible effect on conformational equilibrium in the aldopentopyranose tetrabenzoates, but a regular effect is observed for methyl glycoside derivatives.

The conformational equilibrium position of alkyl glycosides is almost independent of the nature of the alkyl group, except when the latter is a *tert*-butyl group, in which case the form having the aglycon axial becomes somewhat less favored.

Acknowledgment

This work was supported by the National Science Foundation, Grant No. GP-9646.

Literature Cited

1. Durette, P. L., Horton, D., *Advan. Carbohyd. Chem. Biochem.* (1971) **26,** 49.
2. Hough, L., Richardson, A. C., "Rodd's Chemistry of Carbon Compounds," Ch. 22, 2nd ed., S. Coffey, Ed., Vol. 1, Part F, Elsevier, Amsterdam, 1967.
3. Horton, D., Miller, Martha J., *J. Org. Chem.* (1965) **30,** 2457.
4. El Khadem, H. S., Horton, D., Page, T. F., Jr., *J. Org. Chem.* (1968) **33,** 734.
5. El Khadem, H. S., Horton, D., Wander, J. D., *J. Org. Chem.* (1972) **37,** 1630.
6. Horton, D., Wander, J. D., *Carbohyd. Res.* (1969) **10,** 279.
7. Horton, D., Wander, J. D., *Carbohyd. Res.* (1970) **13,** 33.
8. Horton, D., Wander, J. D., unpublished data.
9. Horton, D., Wander, J. D., *Carbohyd. Res.* (1970) **15,** 271.
10. Defaye, J., Gagnaire, D., Horton, D., Muesser, M., *Carbohyd. Res.* (1972) **21,** 407.
11. Angyal, S. J., James, K., *Aust. J. Chem.* (1970) **23,** 1223.
12. Binkley, W. W., Diehl, D. R., Binkley, R. W., *Carbohyd. Res.* (1971) **18,** 459.
13. Angyal, S. J., Gagnaire, D., Le Fur, R., *Carbohyd. Res.* (1972) **23,** 121, 135.
14. Chilton, W. S., Krahn, R. C., *J. Amer. Chem. Soc.* (1968) **90,** 1318.
15. Horton, D., Philips, Kerstin D., *Carbohyd. Res.* (1972) **22,** 151.
16. Lee, J. B., Scanlon, B. F., *Tetrahedron* (1969) **25,** 3413.
17. Lyle, Gloria G., Piazza, M. J., *J. Org. Chem.* (1968) **33,** 2478.
18. Williams, J. M., *Carbohyd. Res.* (1969) **11,** 437.
19. Berman, Helen M., Rosenstein, R. D., *Acta Crystallogr.* (1968) **B24,** 435.
20. Hunter, F. D., Rosenstein, R. D., *Acta Crystallogr.* (1968) **B24,** 1652; Jeffrey, G. A., Kim, H. S., *Carbohyd. Res.* (1970) **14,** 207; Kim, H. S., Jeffrey, G. A., *Acta Crystallogr.* (1969) **A2553,** 197; (1969) **B25,** 2607.
21. Kim, H. S., Jeffrey, G. A., Rosenstein, R. D., *Acta Crystallogr.* (1968) **B24,** 1449.
22. Snyder, R. L., Rosenstein, R. D., Kim, H. S., Jeffrey, G. A., *Carbohyd. Res.* (1970) **12,** 153.
23. Defaye, J., Horton, D., *Carbohyd. Res.* (1970) **14,** 128.
24. Defaye, J., Horton, D., Muesser, M., *Carbohyd. Res.* (1971) **20,** 305.
25. Durette, P. L., Horton, D., *J. Org. Chem.* (1971) **30,** 2058.
26. Lemieux, R. U., "Molecular Rearrangements," pp. 735–743, Part 2, P. de Mayo, Ed., Interscience, Wiley, New York, 1964.

27. Lemieux, R. U., Chü, N. J., *Abstr. Amer. Chem. Soc. Meet.* (1958) **133,** 31N.
28. Bhacca, N. S., Holland, C. V., Horton, D., Miller, Martha J., *J. Org. Chem.* (1967) **32,** 3077.
29. Durette, P. L., Horton, D., *Carbohyd. Res.* (1971) **18,** 57.
30. Horton, D., Turner, W. N., *J. Org. Chem.* (1965) **30,** 3387.
31. Edward, J. T., *Chem. Ind. (London)* (1955) 1102.
32. Wolfe, S., Rauk, A., Tel, L. M., Csizmadia, I. G., *J. Chem. Soc. (B)* (1971) 136; *cf.* Jeffrey, G. A., Pople, J. A., Radom, L., *Carbohyd. Res.* (1972) **25,** 117.
33. Holland, C. V., Horton, D., Jewell, J. S., *J. Org. Chem.* (1967) **32,** 1818.
34. Hall, L. D., *Advan. Carbohyd. Chem.* (1964) **19,** 51.
35. Bhacca, N. S., Horton, D., *J. Amer. Chem. Soc.* (1967) **89,** 5993.
36. Durette, P. L., Horton, D., Bhacca, N. S., *Carbohyd. Res.* (1969) **10,** 565.
37. Gutowsky, H. S., Holm, C. H., *J. Chem. Phys.* (1956) **25,** 1228.
38. Pople, J. A., Schneider, W. G., Bernstein, H. J., "High-Resolution Nuclear Magnetic Resonance," p. 223, McGraw–Hill, New York, 1959.
39. Shanan-Atidi, H., Bar-Eli, K. H., *J. Phys. Chem.* (1970) **74,** 961.
40. Durette, P. L., Horton, D., *Carbohyd. Res.* (1971) **18,** 419.
41. Gatti, G., Segre, A. L., Morandi, C., *J. Chem Soc. (B)* (1967) 1203.
42. Anet, F. A. L., Ahmad, M., Hall, L. D., *Proc. Chem. Soc.* (1964) 145.
43. Bovey, F. A., Hood, III, F. P., Anderson, E. W., Kornegay, R. L., *Proc. Chem. Soc.* (1964) 146.
44. Durette, P. L., Horton, D., *Chem. Commun.* (1970) 1608.
45. Durette, P. L., Horton, D., *Org. Magn. Resonance* (1971) **3,** 417.
46. Durette, P. L., Horton, D., *Carbohyd. Res.* (1971) **18,** 389.
47. Durette, P. L., Horton, D., *Carbohyd. Res.* (1971) **18,** 403.
48. Pierson, G. O., Runquist, O. A., *J. Org. Chem.* (1968) **33,** 2572.
49. Durette, P. L., Horton, D., *Carbohyd. Res.* (1971) **18,** 289.

RECEIVED December 2, 1971.

10

Conformational Studies in the Solid State: Extrapolation to Molecules in Solution

G. A. JEFFREY

Department of Crystallography, University of Pittsburgh, Pittsburgh, Pa. 15213

Numerical data from crystal structure determinations give convenient starting points for calculating the geometry of conformations consistent with bond rotational barriers and non-bonding interactions. In carbohydrates this approach to conformational analysis must be applied reservedly because of the common occurrence of intra- and inter-molecular hydrogen bonding. In molecules where there are many free hydroxyls, the hydrogen bonding in the solid state is predominantly intermolecular, and these crystal field forces have only a secondary influence on conformation. The hydrogen bonding of such a molecule in the solid must closely resemble that in an aqueous or polar solvent, except for the difference between static and dynamic environment. Intramolecular hydrogen bonding is much more significant in determining conformation and is the most serious reservation in computer extrapolation from the conformation in the solid to the flexible molecule.

Crystal structure determination by diffraction methods is the principal and most powerful method for obtaining numerical data on bond lengths, valence angles, and the characteristic geometry of functional groups. It is the only method for the direct measurement of these quantities in carbohydrates, which are not suitable compounds for microwave spectroscopy or gas-phase electron diffraction. The present limits of accuracy for the most productive of the diffraction methods, single crystal x-ray diffraction, is usually 1 or 2 picometers in bond distances and a few degrees in valence angles. (The picometer is the most convenient SI unit for reporting crystal structural data although most crystallographers continue to use the Å unit; p = 0.01 Å.) Within these limits these data are molecular properties, which in the absence of chemical change can be extrapolated from the solid-state to other phases including solutions.

A recent compilation (*1*) of organic crystal structures completed between 1939–1969 lists over 4,000 entries covering representative types of the most important compounds. This number has increased from 24 in 1952 to 250 in 1964 and 800 in 1968, resulting mainly from the development of computers in the Fifties and the introduction of automatic diffractometers in the Sixties. It is estimated that there are now more than 1000 new organic crystal structures being completed every year. It is not surprising therefore that sufficient data are available to predict bond lengths and angles and much of the molecular geometry of all but the unusual organic compound. The crystallographer has therefore studied in recent years the configurational analysis for newly isolated natural products and the conformational analysis of molecules where the configuration is known. A protein crystal structure determination is an experiment in conformational analysis since the amino acid sequence is generally known and the precision of the observed electron density distribution rarely allows the identification of a pair of atoms or a bond, on the basis of the observed atomic peaks or interatomic distances.

All the molecular data shown below are implicit in the atomic coordinates and lattice parameters reported in a molecular crystal structure determination.

Crystal Structure Data

Molecular	Bond Distances Valence Angles Geometry of Functional Groups
Molecular or Crystal	Torsion Angles Intramolecular Non-bonding and H-bonding Distances
Crystal	Intermolecular Non-bonding and H-bonding Distances

Until recently it was not customary to report measurements such as torsion angles and intramolecular non-bonding distances because they are not necessarily characteristic molecular properties. The bond torsion angles, which are of special interest in conformational analysis, are often only reported and discussed when there is a special reason for doing so which is connected with the objectives of the structure determination. Tradition dies hard, and one still finds tables of bond distances and angles in crystal structure papers which are not significantly different from the expected values, whereas torsion angles and other conformational data which are of special interest to the chemist are not reported and have to be computed by the reader from the atomic parameters and unit cell dimensions. (The conformational chemist using such crystal structural data is cautioned because some crystal structure papers report coordinates for one molecule while others report the asymmetric contents of one particular unit cell. These two sets of coordinates are not necessarily the same, and it is seldom stated in the paper which set is given.) Even when this structural information is a characteristic of the crystal rather than of the molecule *per se,* it can, if used judiciously, give valuable clues to the conformational behavior of the molecules in the liquid state. Since most of carbohydrate chemistry takes place in solution between room temperature and 100°C, the carbohydrate crystallographer is conscious of an obligation to examine the abundance of conformational information that is implicit in the crystal structures he is capable of producing to assist his less fortunate colleagues who are concerned with the much more complex problem of the structure of flexible molecules in solution.

One attractive and convenient approach to conformational analysis in the liquid state is to start with the numerical data given by a set of atomic coordinates from a crystal structure determination and, by computer, calculate the geometry of all conformations which are consistent with the concepts of single-bond rotations and van der Waals radii, thereby generating the numerical conformational data for the free molecule (this idea has been applied mainly to nucleosides (*2, 3*)). In the absence of any experimental means of comparing these conformational possibilities with observations, this is more an exercise in computer programming than a contribution to conformational analysis. The nuclear magnetic resonance spectra, particularly through the application of the Karplus Equation (*4*), do give information relating directly to torsional angles, but these usually relate to the hydrogen atoms which are the least well-defined from the x-ray structure analysis. (Distances and angles involving hydrogens obtained from x-ray data are only reliable to about 10 p and 10°. To obtain accuracy comparable with that of the carbons and oxygens, a neutron diffraction analysis is necessary (*5*).) Nevertheless, the concept of computing all likely conformational geometries and systematically eliminating those which are inconsistent with the NMR spectra could lead to a feasible approach in favorable cases. It has the advantage that it combines the information from two independent and distinctly different techniques. An elegant example of the application of this type of approach to the mononucleotides using the lanthanide ion as a probe has been recently reported (*5*). This work shows that the average solution conformations which fitted the NMR data were generally close to the conformations observed in the crystal structures.

The computor input for such a conformational calculation consists of (1) the set of atomic coordinates, (2) the bonds about which rotation is permitted, (3) the angular intervals over which this rotation will be explored, and (4) a set of limiting non-bonding atomic radii. A more quantitative evaluation of the most likely conformations can be obtained if a provision is also made for calculating non-bonding atom-pair potential energy sums of the Lennard–Jones type (*6*).

If the molecule can be treated as several rigid units linked by two or three rotameric bonds, then the computations can be handled on a small or moderate-sized computer. As the number of rotameric bonds increases so do the demands on computer memory and paper output. There seems to be a real advantage to the use of the graphical display equipment to give rapid pictorial computer output (*7*). If the rotameric bonds have atoms other than hydrogens attached to them, it becomes reasonable to assume that the staggered bond conformations will have significantly lower conformational energies than the eclipsed. The rotations can then be restricted to $2\pi/3 \pm 10°$; this greatly reduces the computation and allows consideration of more rotameric bonds.

This approach, which may offer advantages particularly for interpreting C^{13} NMR spectra of carbohydrates, has the basic assumption that the crystal structure atomic coordinates are a valid starting point for calculating conformations in solution. (The alternative method, which is commonly used for predicting polypeptide and polysaccharide conformations, is to construct an ideal model and use its atomic coordinates as the starting point (*8, 9, 10, 11*). Since the dimensions of the ideal molecule are generally based on a personal selection of data from crystal structure determinations of related molecules, there is a more objective element about starting with a set of actual observed atomic parameters for the particular molecule under consideration.) This is a premise that requires careful scrutiny for carbohydrate molecules and is the principal theme of this chapter. At the same time we will review the conformational data from some carbohydrate crystal structures studied recently in our department and elsewhere.

There are several reasons for reservations about applying the computer extrapolation of crystal structure data for carbohydrates. One is that much of the crystal structure data refer to unsubstituted sugars which are only soluble in hydroxylic or polar solvents where the conformational analysis may be complicated by hydrolysis, isomerism (mutarotation) (*12*), or stereospecific solvent interactions which require a more sophisticated model. However, assuming that such chemical changes do not occur or can be suppressed, there still remain questions to be answered before the conformation observed in the crystal can be accepted as a close enough approximation to that of one or more of the rotomers which may predominate in the solution state. (α-L-Sorbose gives an example of the coexistence of two primary alcohol rotameric conformers in the crystal (*13*), and the same phenomenon is observed in 1-kestose (*14*)). Molecules with free hydroxyl groups cohere in the solid state by systems of intermolecular H-bonds which frequently form infinite chains (to a crystallographer an infinite chain is one which extends to the domain boundaries of a single crystal, usually about 10^{-4} mm.) which link the molecules in a much more stereospecific and directional manner than do van der Waals forces. We would expect that the intermolecular H-bonds determine the torsion angles of the O—H bonds and they may place significant conformational restraints on the shape of the rest of the molecule. The methylated or acetylated sugar derivatives that the NMR spectroscopist uses to simplify *his* problem make the crystal structure determination more difficult and less attractive to the crystallographer since the additional atoms, n, increase the number of parameters by $9n$ and the computing by $81n^2$. More importantly, it is usually the free sugar which is the molecule with biological function, and this is often the long term objective of the crystallographic research. (An example of a crystal structure determination directed solely toward testing a NMR interpretation is that of 1,2-*o*-aminoisopropylidine α-D-glycopyranoside hydroiodide (*15*).)

The second difficulty peculiar to carbohydrates is the allowance for conformational flexibility in the pyranose and furanose rings, which is not easily accounted for by the comparatively simple computer program described above.

A third question which is important for the unsubstituted sugars is whether it is necessary to provide for additional conformational stability resulting from the formation of intramolecular H-bonds in solution. (Intramolecular H-bonds are difficult to incorporate into limiting van der Waals radii because of their vectorial character and uncertainty about their potential energy function.)

These three questions will be examined in relation to the conformational data which have been given by recent carbohydrate crystal structure determinations. Because of the added reliability and accuracy arising from the use of automatic diffractometers only those numerical data will be quoted where these instruments were used.

How Important is the Influence of Intermolecular H-Bonds on Molecular Conformation in the Solid State?

The answer to this question will vary from structure to structure since it depends upon the flexibility of the molecules and the degree of intermolecular H-bonding.

In any one class of compounds where a number of related structures are known, it is possible to evaluate the molecular distortions arising

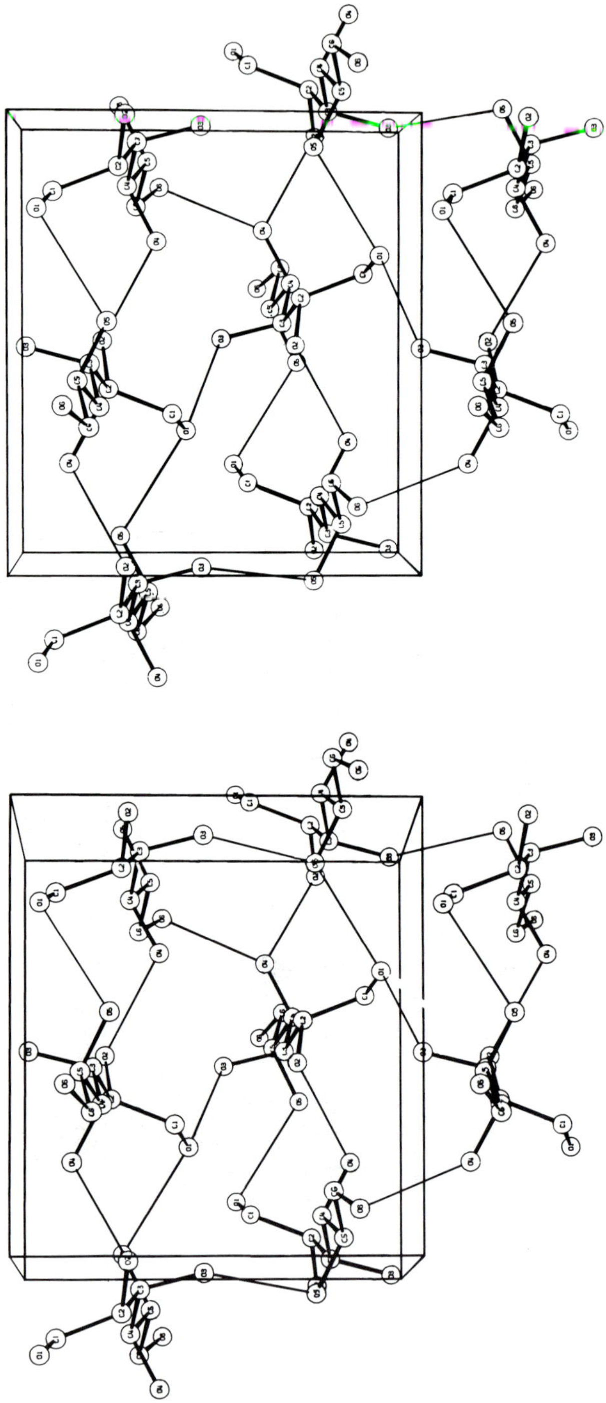

Figure 1. Stereoscopic view of the molecular packing in the crystal structure of D-*glucitol. The molecules, indicated by circles linked by heavy lines, are viewed down the carbon chain axis. The thin lines are the intermolecular H-bonds.*

from the crystal fields by comparing the observed conformations among themselves and with those expected for the free molecules. The sugar alcohols (*16*) give an excellent example for this approach since they are comparatively flexible molecules, being composed solely of single bonds. Because of the many hydroxyl groups, the molecules cohere in the solid state by an extensive system of H-bonds which form chains or spirals throughout the crystal structure. In the majority of cases, each hydroxyl acts as H-bond donor and acceptor so that each hexitol molecule, for example, coheres to its neighbours by twelve H-bonds. Figure 1 shows a stereoscopic view of the H-bonding and molecular packing in the crystal structure of D-glucitol viewed in the direction of the carbon chain. When seen in this direction, each molecule has six nearest neighbours in a typical hexagonal packing arrangement. Table I shows the variation in

Table I. Torsion Angles in Pentitols and Hexitols

	C/C[a b]	*C/O*[a b]	*O/O*[a b]	*Chain Conformation*	*Reference*
D, L-Arabinitol	180(0)	60(1)	59(1)	straight	(*17*)
Ribitol	174, 62	60(24)	60(13)	bent	(*18*)
Xylitol	184, 70	62(0)	60(16)	bent	(*19*)
D-Mannitol (B)	180(8)	60(8)	63(6)	straight	(*20*)
D-Mannitol (K)	181(11)	60(12)	59(3)	straight	(*21*)
Galactitol	187(4)	63(3)	62(6)	straight	(*22*)
D-Glucitol	179(2), 51	58(3)	61(11)	bent	(*23*)
D-Iditol	177(10), 62	58(2)	65(15)	bent	(*24*)
Allitol	70	55(4)	64(10)	bent	(*24*)
D-Gluconate	180(3)	61(15)	64(7)	straight	(*25*)
D-Glucitol-pyridine	170(19), 70	66(23)	53(18)	bent	(*26*)
Mean	180.4, 64	59.7	61.7		

[a] X/X refers to the torsion angle X-C-C-X.
[b] The figures are mean values in degrees. The figure in parenthesis is the spread in degrees between the smallest and largest torsion angles. Only in the O/O torsion angles does there seem to be a systematically greater spread for the bent conformations relative to the straight conformations.

the torsion angles in 11 crystal structures of the pentitols and hexitols. The torsion angles involving hydrogen atoms were omitted because the precision of location of the hydrogens was not comparable with that of the carbon and oxygens, except for D-glucitol where the structure determination used x-ray and neutron diffraction. The results show that the torsion angles, γ, vary from the ideal staggered value of $n\pi/3$ on the average of $\pm 5°$. Using $V_r = 1/2\, V_o(1 + \cos 3\gamma)$ with the ethane value of about 3 kcal for V_o, this amounts to about 150 cal/C–C bond, which is an order of ten less than the energy interactions which are generally considered to be important in carbohydrate conformational problems (*27*,

28). These results suggest that in the pentitols and hexitols, at least, the intermolecular H-bonds have only a secondary influence on the shape of the molecules in the solid state. The observed conformations in the solid state can be rationalized (*16*) in terms of the intramolecular interactions between hydroxyls which are believed to involve energies of the order of several kilocalories (*27*); (the interpretation of the NMR spectra of acetylated acyclic sugar derivatives in solution is based on the same type of hypothesis (*29*)).

Another approach is to compare polymorphic forms of the same compound. This allows the comparison of molecules with the same constitution and configuration in different crystal fields. Do these different potential environments lead to different conformations? There are few comparative crystal structural studies of organic polymorphs. For D-mannitol the molecules had the same conformation in all three polymorphic forms studied with differences in torsion angles for the two that were studied in detail of only a few degrees (*see* Table I).

The rarity of the phenomenon of conformational polymorphism (one of the rare examples of conformational polymorphism is D,L-methionine (*30*)) is an indication that it is unusual for the crystal field forces to exercise a dominant influence on molecular conformation. It appears that the differences in the lattice sums of the crystal field forces for different conformations are in most cases an order of magnitude too small to induce conformational changes—*i.e.*, tenths of kilocalorie per interaction rather than kilocalories.

Another comparison which can be made is between a carbohydrate molecule in homomolecular crystal structure and that of a heteromolecular H-bonded complex. Examples are D-glucitol and D-glucitol–pyridine; α-D-glucose, α-D-glucose hydrate, and α-D-glucose–urea (*31*). In these examples conformational differences are found in the hydroxy methyl groups. In D-glucitol, the terminal hydroxyls are anti to the vicinal carbon (gauche–trans) whereas in the complex they are anti to the hydrogen (gauche-gauche). In α-D-glucose the primary alcohol oxygen is anti to the ring carbon, C-4, whereas in the hydrate and urea compounds it is anti to the hydrogen, H(C-5). These are the two preferred conformations since the third staggered orientation involves a peri interaction between oxygens (*32*)—*i.e.*, O-6 to O-5. A recent report (*33*) of the conformational statistics for the orientation of the primary alcohol group in 23 carbohydrate structures showed 14 anti to carbon and nine anti to hydrogen. An equilibrium between these two preferred staggered conformations for —CH_2OH groups can be expected in solution and can be easily incorporated into a computer calculation such as described above.

The stronger influence of the crystal field in complex formation vis-a-vis polymorphism is shown in the torsion angles of the glucitol molecule in the pyridine complex which vary over $\pm 10°$—*i.e.*, more than twice the mean variations in the homo-molecular sugar alcohol structures. This difference is not apparent, however, in the comparison between glucose and glucose–urea (Table II), but it is not surprising that the cyclic hexose is more rigid than the acyclic hexitol.

Table II. Ring Torsion Angles in Simple Pyranose Monosaccharides[a]

	C—C—O—C	C—C—C—C	C—C—C—O	*Reference*
β-D-gluco	64.5 (3.5)	52.1 (2.6)	56.7 (6.1)	(*34*)
α-D-gluco	61.6 (1.3)	52.3 (2.0)	55.8 (4.2)	(*35*)
α-D-gluco/urea	61.2 (1.5)	54.8 (0.7)	56.4 (2.0)	(*36*)
methyl α-D-gluco	58.0 (0.0)	54.9 (1.4)	56.2 (4.1)	(*37*)
methyl α-D-manno	58.8 (0.7)	53.5 (0.7)	55.0 (0.9)	(*38*)
methyl α-D-galacto (H_2O)	59.7 (3.5)	53.7 (3.1)	55.2 (6.1)	(*39*)
methyl 1 thio-α-D-ribo	61.4 (2.8)	55.6 (2.1)	59.3 (7.2)	(*40*)
α-L-sorbose	58.0 (0.5)	52.5 (1.5)	54.4 (1.6)	(*13*)
α-D-tagatose	57.2 (1.9)	55.9 (1.0)	55.5 (2.5)	(*41*)
mean value	60.0 (1.7)	53.9 (1.7)	56.1 (3.9)	

[a] The torsion angle data on earlier structures completed prior to the use of automatic diffractometers are reported elsewhere (*42*).

What to Do About Ring Conformations

About 150 crystal structures of carbohydrates have been reported with pyranose or furanose rings of which about one-third are nucleosides and nucleotides containing ribofuranose (For a recent review of the stereochemistry of the nucleosides and nucleotides *see* Ref. *43*.) The results indicate a clear distinction in conformational variety with the furanose rings much the more flexible. This seems to agree generally with the experience of the carbohydrate chemist (*27*). The pyranose rings are generally observed in C1, more rarely in 1C, exactly as Reeves predicted (*44*) 20 years ago. The ring torsion angles seldom lie outside a range of 50°–65°, as compared with the strain-free range of 55.8°–61.7° (*12*). The most recent data on comparatively simple pyranose derivatives, given in Table II, agree surprisingly well in the ring torsion angles with a variation which is generally less than $\pm 2.5°$. The angles about the C—O bonds vary on average less than 1° either side of the ideal 60° angle. The torsion angles about the C—C bonds show wider variation with a well-defined distinction between C—C—C—C and O—C—C—C, the former averaging 53.9° and the latter 56.1° (angles which are $\pm 1°$ either side of the commonly accepted value of 55° for cyclohexane).

The 60° C—O torsion angle corresponds to the sharpening of the oxygen head of the chair. The flattest pyranose ring observed in the solid state, except for anhydro and other fused-ring systems, is that in sucrose where the torsion angles lie in an exceptionally narrow range of 55°–56° (*45*). This may result from the two intramolecular H-bonds between the glucose and fructose rings. This flattening is not observed in the other disaccharides which have only one intramolecular H-bond nor in the trisaccharides which have no intramolecular H-bonding. In the trisaccharides, raffinose pentahydrate (*46*), and planteose dihydrate (*47*), the range of torsion angles in the pyranose rings is 52° – 62°, and in 1-kestose (*14*) it is 50° – 65°, all of which are greater than those of the monosaccharides shown in Table II. The boat and intermediate twisted conformations that are theoretically possible (*28, 48*) have not been observed, except where there are some primary bonding constraints in the molecules, such as an anhydro ring. In 1,6-anhydro-β-D-glucopyranose (*49*), for example, the torsion angles of the pyranose ring vary from 35°–75°. As would be expected, the pyranose ring is sharpened at the ring oxygen which is also part of the anhydro ring and flattened away from the anhydro ring. In 1,6:2,3-dianhydro-β-D-gulopyranose (*50*) the additional three-membered ring introduces further strain into the pyranose ring, and the torsion angles range between 1° – 82°—*i.e.*, almost flat to almost perpendicular. Both molecules are extremely rigid, and the only conformational freedom expected in going into solution is in the orientation of the O—H bonds.

In contrast, the furanose rings occur in several variations of twist and envelope. Most of the experimental data refer to the deoxy ribofuranoside unit of the nucleotides, the conformations of which have been summarized recently elsewhere (*43*). Taking the ring oxygen and its adjacent carbon atoms as the reference plane, all varieties of twist are observed including the intermediate envelope forms. (In the pentafuranoses the displacements of C(2) and C(3) are referred to C(1)O(1)C(4) as the reference plane, whereas in the hexoses it is C(3) and C(4) with reference to C(2)O(2)C(5).) This variation of the furanose ring with environment is well illustrated in Figure 2, which shows the comparison of the fructofuranose conformations in sucrose and the three trisaccharides, raffinose, planteose, and 1-kestose, all of which contain the sucrose unit. In these four structures the fructofuranose rings of the sucrose components vary from ${}^4T_3$ to E_3 while the second fructose ring on 1-kestose is ${}^3T_4$. Irrespective of the conformation in the solid state, a comparable dynamical equilibrium of ring shapes might be expected in solution. As is shown in Figure 2, comparatively small differences in ring shape lead to greatly amplified differences in the positions of the substituents.

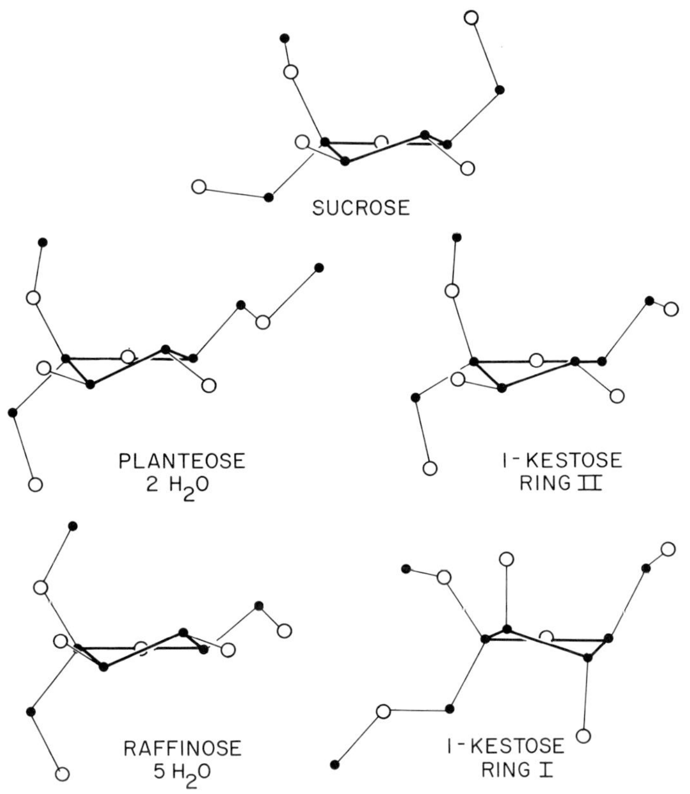

Figure 2. Conformations of the furanose rings in sucrose and the sucrose-containing trisaccharides, raffinose, planteose, and 1-kestose

How Important is the Influence of Intramolecular Hydrogen Bonding on Conformational Stability?

It is generally assumed that in hydroxylic solvents the opportunities for intermolecular H-bond formation preempt the formation of intramolecular H-bonds in unsubstituted carbohydrates. Only in special cases where the molecule is such that the possibility for intramolecular H-bonding is combined with solubility in a non-polar solvent does this factor become important (*51*).

The geometry of the hydroxyls on adjacent carbon atoms is comparatively unfavorable for intramolecular H-bond formation. Although the OH · · · O distance is appropriate (~200 p), the O-H · · · O angle of less than 120° is not. The same considerations apply to intramolecular H-bond formation between a hydroxyl and a ring oxygen in the same monosaccharide unit. Neither of these have been observed in the crystalline state although they have been mentioned in the discussion of certain cyclic glycols in non-polar solvents (*52*). Much more favorable is the stereochemistry of the syn-diaxial hydroxyl groups on alternate carbon atoms; an optimum OH · · · O distance of 160–180 p can be obtained with an acceptable O · · · H · · · O angle of about 150°. This so-called peri interaction is associated with conformational instability when it involves two methoxy groups or a hydroxyl and methoxy group. The interaction between two syn-diaxial hydroxyls was not specifically mentioned as one of Reeves' original instability factors (*44*) although it is clearly related to the Hassel–Otter effect (*53*). This was probably because of uncertainty in assigning a van der Waals radius to oxygen. Of all the first row elements oxygen in its singly-bonded state has the least spherical electronic distribution as a result of its lone pairs of electrons which are most sensitive to polar and dipole interactions, changing the effective non-bonding radius. Depending upon the direction of approach relative to the C—O—H bonds, oxygen can have a van der Waals radius varying between 120–170 p, a difference which corresponds to a distinction between attraction and repulsion of 1,3 syn-diaxial oxygens attached to a pyranose ring.

Crystal structure determinations have tended to confirm that intramolecular H-bonds are rare in carbohydrates. The notable exceptions are the disaccharides where intramolecular H-bonds between hydroxyls and oxygens on the different monosaccharide units are the rule as in sucrose (*45*), maltose monohydrate (*54*), β-cellobiose (*34*), α-lactose monohydrate (*33*), and methyl-β-D-maltoside (*55*) and methyl-β-D-cellobioside (*56*) (but not in α-α-trehalose) (*57*). There are no intramolecular H-bonds in the three trisaccharides that have been studied (*14, 46, 47*), but they have been reported in cellotetrose (*58*) and are frequently postulated in model structures for polysaccharides (*59, 60*). There are several examples of molecules with syn-axial hydroxyl groups where intermolecular hydrogen bonding clearly predominates in the crystalline state. When this occurs, the orientation of the O—H bonds is such that the lone pairs on the oxygens are directed towards each other, resulting in marked repulsion. The oxygen atoms move apart from the unstrained distance of 253 p (based on carbon–carbon distances and angles of 152.3 p and 112°) to about 290 p. This repulsion effect is shown in the structure of epiinositol (*61, 62*) where the effect of the repulsion of the syn-diaxial hydroxyls results in well-defined distortions of the torsional angles of the C—O bonds and within the cyclohexane ring. Other examples are found in the structures of 1,6-anhydro-β-D-glucose (*49*) and methyl-α-altropyranoside (*63*). In 1,6-anhydro-β-D-glucose, there are two opportunities for intramolecular H-bonds resulting from the strained 1C pyranose ring induced by the five-membered anhydro ring. Instead, the H-bonds are to adjacent molecules, and the O · · · O distances are 280 p and 290 p. In methyl-α-altropyranoside the repulsion of axial O(1) and O(3) in the Cl conformation distorts the pyranose ring so that the ring torsion angles range between 44°–62°—*i.e.*, the ring flattens as O(2) · · · O(3) opens to the observed separation of 294 p.

There are, however, three recent examples of intramolecular H-bonds which seem to be special cases. One of these is the crystal structure of potassium gluconate monohydrate (*25*). The gluconate ion, originally studied in 1953 in the structure of the anhydrous potassium salt (*64*), is the only example in a series of 11 crystal structure determinations (*16*) where the conformation of the polyhydroxyalkyl chain in the crystalline state cannot be successfullly predicted by eliminating those conformations with the peri interactions between hydroxyls—*i.e.*, those with parallel C(*n*)—OH to C(*n* + 2)—OH bonds. Examination of the structure of the gluconate monohydrate with modern methods (*25*), which revealed the positions of the hydrogen atoms, showed that there was an intramolecular H-bond between the O(4)H and the O(2), and presumably it is this bond which stabilized the observed planar zig-zag chain conformation. It was also clear that the crystal field of the gluconate ion was dominated by the oxygen coordination around the K^+ ions. This is eight-fold with two water oxygens and six hydroxyl oxygens from neighboring gluconate ions in the first cationic coordination sphere. Surprisingly, perhaps, the carboxylate oxygens are not included in the first eight nearest neighbors to the K^+ ion and are solely involved in H-bonding to the water and hydroxyls. The arrangement of the gluconate ions in the crystal structure, which is secondary to the hydration of the cations, is not favorable for intermolecular H-bond formation by O(4)H, which then settles for intramolecular bonding to O(2). This example is a solid-state analog of the enhancement of intramolecular H-bond formation by non-polar solvents and suggests that strong ionic concentrations in aqueous solutions might have a similar effect.

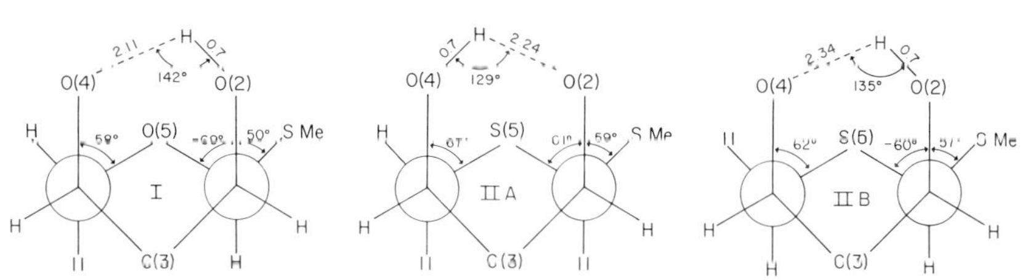

METHYL 1-THIO-α-D-RIBOPYRANOSIDE (I) METHYL 1,5-DITHIO-α-D-RIBOPYRANOSIDE (II)

Figure 3. Intramolecular H-bonding in some thio-ribopyranosides

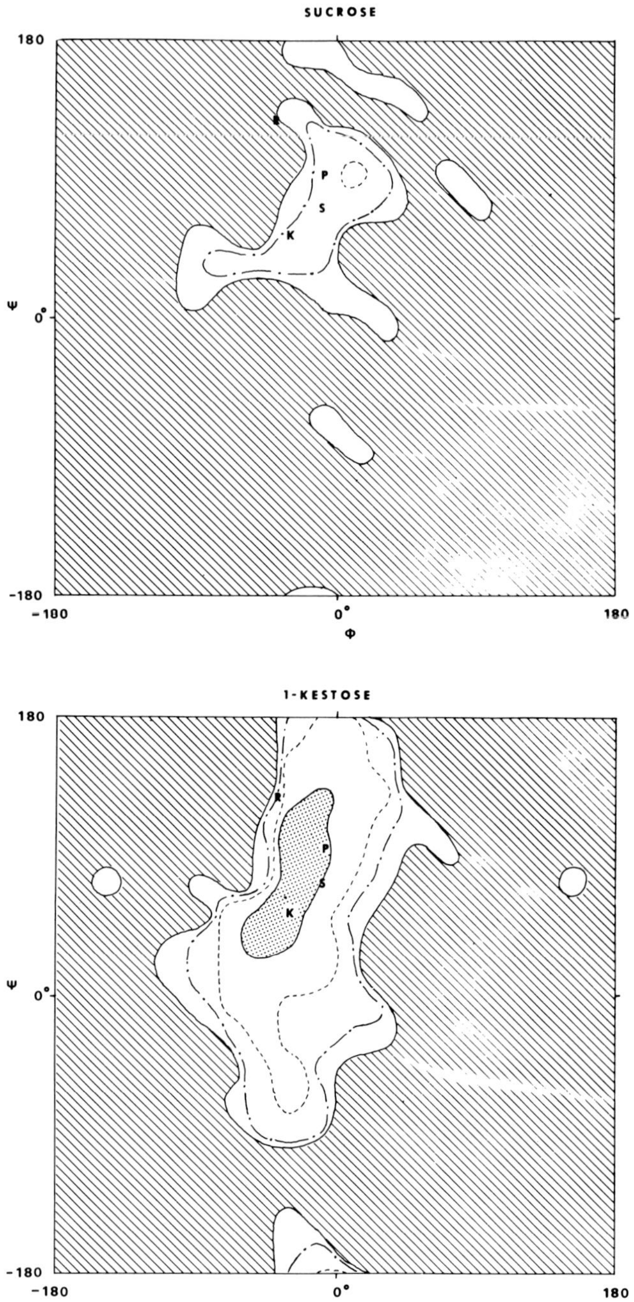

Figure 4. Conformational potential energy maps for rotation
Φ: H(C1)—C(1)—O(1)—C′(2)

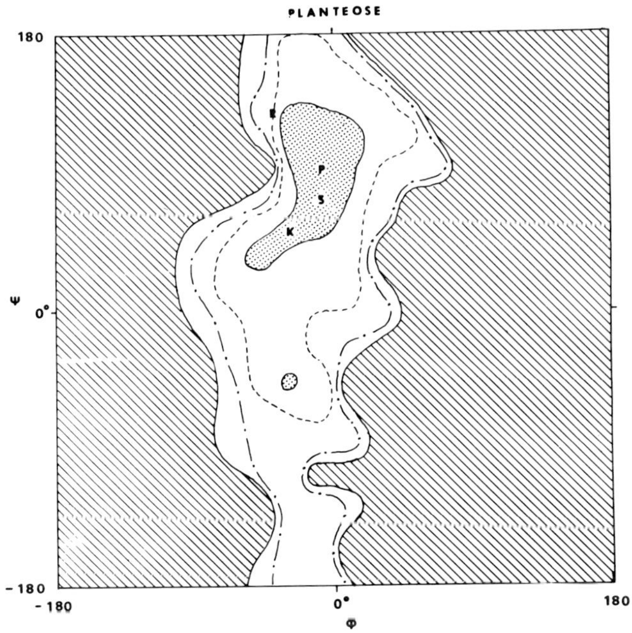

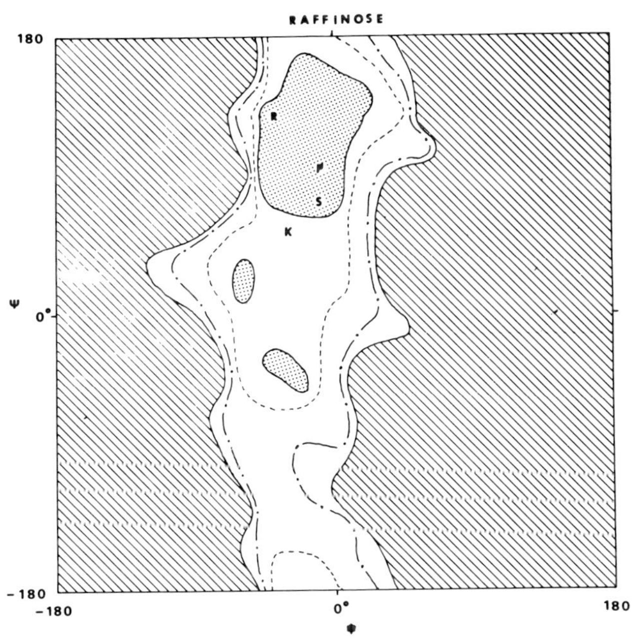

about C(1)—O(1), Φ, and O(1)—C′(2), ψ, of the sucrose molecule

ψ: C(1)—O(1)—C′(2)—C′(1)

The other two known examples of intramolecular H-bonds within a monosaccharide unit in the crystalline state occur in two thio-sugars, methyl-1-thio and methyl-1,5-dithio-α-D-ribopyranoside (*40*) (illustrated in Figure 3). As far as we know, these are the first examples of syn-diaxial O-H · · · O bonding in a cyclic carbohydrate in the solid state. It is not certain now, whether any particular significance should be associated with the fact that these are thio sugars.

Conclusion

From the above discussion we infer that apart from the complexities of hydrolysis and mutarotation the most serious reservations concerning

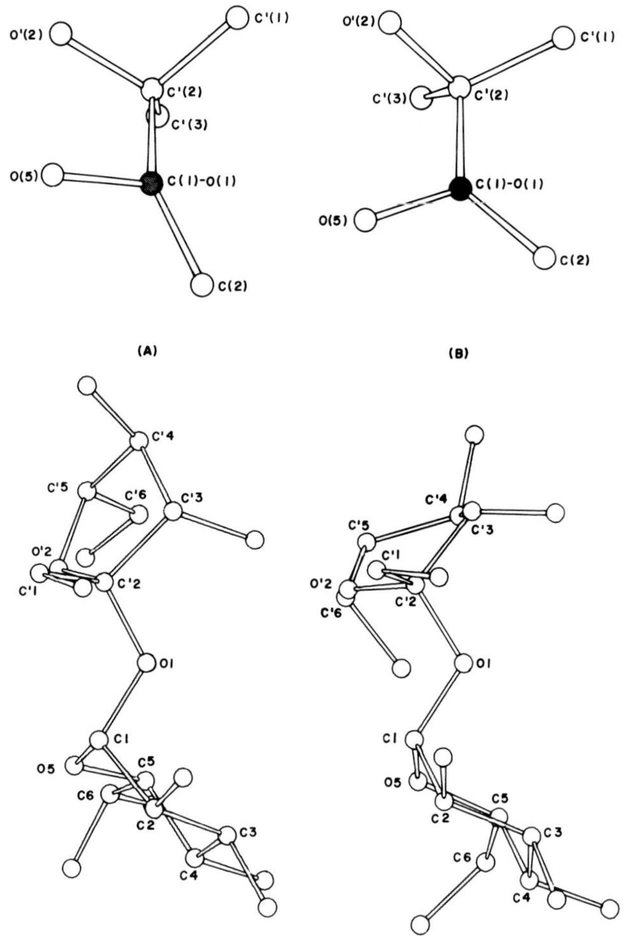

Figure 5. The conformation of the sucrose molecular unit C(1)—O(1) perpendicular to the paper (above.

the extrapolation from crystal structures to carbohydrate conformations in solution will apply to structures with furanose rings and to solutions where the solvent species cannot form intermolecular H-bonds. The latter complication is serious only when there can be intramolecular H-bond formation and arises from the difficulty of including an appropriate stereospecific potential energy function in the conformational analysis of the rotamer states.

Considering these problems, we tested a computer extrapolation (*6*) for sucrose, shown in the conformation maps in Figure 4. Four sets of crystal parameters were used, those from the structure of sucrose (S) (*45*) and those from the sucrose component of raffinose pentahydrate

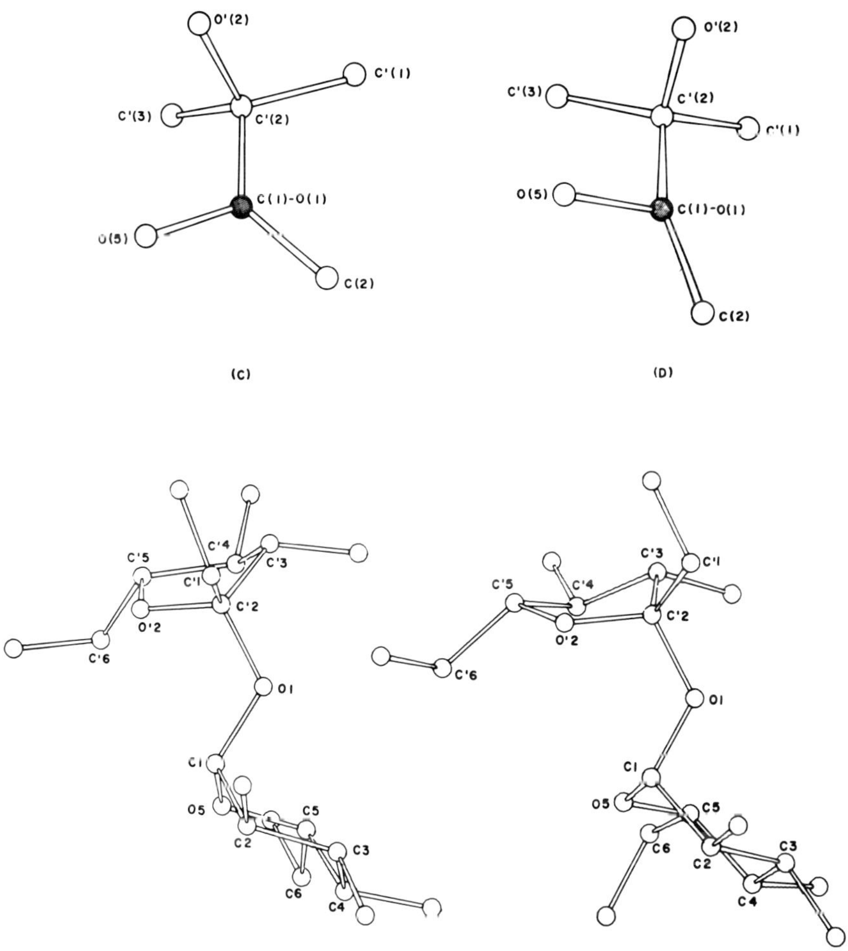

in 1-kestose (A), sucrose (B), planteose (C), and raffinose (D) viewed with with overlapping atoms shaded) and in the plane of the paper (below)

(*R*) (*46*), planteose dihydrate (*P*) (*47*), and 1-kestose (*K*) (*14*). As pointed out above, the principal difference in these four sets of starting coordinates was in the shape of the furanose rings, the intramolecular H-bonds in the sucrose structure, and the differences in torsion angles between the linkages shown in Figure 5. The rotameric bonds considered were the torsion angles Φ, H(C1)—C(1)—O(1)—C′(2) and ψ, C(1)—O(1)—C′(2)—C′(1) (*see* Figure 5). The cross-hatched areas are forbidden since they correspond to conformations with at least one non-bonding contact distance less than half the sum of the van der Waals radii (H 110 p, C 160 p, O 170 p). A Buckingham potential energy function was used as described elsewhere (*65*). The shade-and-dot area corresponds to a potential energy sum of zero or small negative (*i.e.*, weak attraction); the two intermediate contours are 10 and 50 kcal/mole, respectively. The letters S, *R*, *P*, and *K* show the Φ, ψ values for the four structures.

The abnormal case is sucrose. Apparently the geometry of the molecule as observed in the crystal structure places constraints on the rotations which are not found in the sucrose moiety of the three trisaccharides. These constraints arise from the positions of the hydrogen atoms involved in the two intramolecular H-bonds, which are present in

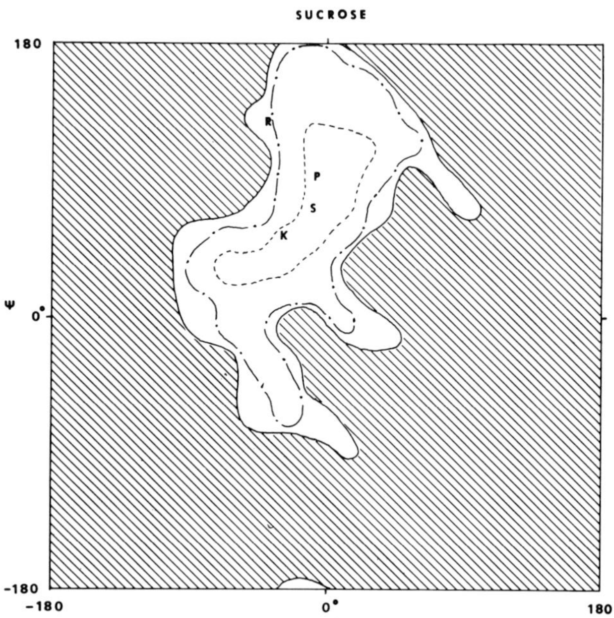

Figure 6. Conformational potential energy map from sucrose parameters with hydrogens of intramolecular H-bonds removed

sucrose but not in the trisaccharides. The sucrose conformation map with these two hydrogens removed is shown in Figure 6; it closely resembles that of the 1-kestose model. Since no provision was made to rotate the hydrogens about the O—H bonds, all the models are probably too constrained. Taking this into account, one should expect in solution almost 360° of rotameric freedom about ψ, whereas Φ is restricted to about 60° as indicated by the maps.

Literature Cited

1. Kennard, O., Watson, D. G., "Molecular Structures and Dimensions," Vol. 1, NVA Oostoek, Utrecht, 1970.
2. Donohue, J., Trueblood, K. N., *J. Mol. Biol.* (1960) **2**, 363.
3. Haschemeyer, A. E. V., Rich, A., *J. Mol. Biol.* (1967) **27**, 369.
4. Karplus, M., *J. Chem. Phys.* (1959) **30**, 11.
5. Barry, C. D., North, A. C. T., Glasel, J. A., Williams, R. J. P., Xavier, A. V., *Nature* (1971) **232**, 236.
6. Rohrer, D., Molecular Rotation Energy Map Program, Department of Crystallography, University of Pittsburgh.
7. Nievergelt, J., Secrest, D., Eds., "Emerging Concepts in Computer Graphics," Benjamin, New York, 1968.
8. Ramachandran, G. N., Ed., "Conformation of Biopolymers," Academic, London, 1967.
9. Rees, D. A., Skerrett, R. J., *Carbohyd. Res.* (1968) **7**, 334.
10. Rees, D. A., Skerrett, R. J., *J. Chem. Soc.* (1970) **B**, 189.
11. Rees, D. A., Scott, W. E., *J. Chem. Soc.* (1971) **B**, 496.
12. Pigman, W., Isbell, H. S., *Advan. Carbohyd. Chem.* (1968) **3**, 11.
13. Kim, S. H., Rosenstein, R. D., *Acta Crystallogr.* (1967) **22**, 648.
14. Jeffrey, G. A., Park, Y. J., *Acta Crystallogr.* (1972) **B28**, 257.
15. Trotter, J., Fawcett, J. K., *Acta Crystallogr.* (1966) **21**, 366.
16. Jeffrey, G. A., Kim, H. S., *Carbohyd. Res.* (1970) **14**, 207.
17. Hunter, F. D., Rosenstein, R. D., *Acta Crystallogr.* (1968) **B24**, 1652.
18. Kim, H. S., Jeffrey, G. A., Rosenstein, R. D., *Acta Crystallogr.* (1969) **B25**, 2223.
19. Kim, H. S., Jeffrey, G. A., *Acta Crystallogr.* (1969) **B25**, 2607.
20. Berman, H. M., Jeffrey, G. A., Rosenstein, R. D., *Acta Crystallogr.* (1968) **B24**, 442.
21. Kim, H. S., Jeffrey, G. A., Rosenstein, R. D., *Acta Crystallogr.* (1968) **B24**, 1449.
22. Berman, H. M., Rosenstein, R. D., *Acta Crystallogr.* (1968) **B24**, 435.
23. Park, Y. J., Jeffrey, G. A., Hamilton, W. C., *Acta Crystallogr.* (1971) **B27**, 2393.
24. Azarnia, N., Jeffrey, G. A., Shen, M. S., *Acta Crystallogr.* (1972) **B28**, 1007
25. Jeffrey, G. A., Fasiska, E. J., *Carbohyd. Res.* (1972) **21**, 187.
26. Kim, H. S., Jeffrey, G. A., Rosenstein, R. D., *Acta Crystallogr.* (1971) **B27**, 307.
27. Eliel, E. L., Allinger, N. L., Angyal, S. J., Morrison, G. A., "Conformational Analysis," Wiley, New York, 1965.
28. Stoddart, J. F., "Stereochemistry of Carbohydrates," Chapt. 3, Wiley–Interscience, New York, 1971.
29. Horton, D., Wander, J. D., *Carbohyd. Res.* (1969) **10**, 279.
30. Mathieson, A. McL., *Acta Crystallogr.* (1952) **5**, 332.
31. Snyder, R. L., Rosenstein, R. D., Kim, H. S., Jeffrey, G. A., *Carbohyd. Res.* (1970) **12**, 153.

32. Sundaralingam, M., *Biopolymers* (1968) **6,** 189.
33. Fries, D. C., Rao, S. T., Sundaralingam, M., *Acta Crystallogr.* (1971) **B27,** 994.
34. Chu, S. S. C., Jeffrey, G. A., *Acta Crystallogr.* (1968) **B24,** 830.
35. Brown, G. M., Levy, H. A., *Science* (1965) **147,** 1038, and private communication.
36. Snyder, R. L., Rosenstein, R. D., *Acta Crystallogr.* (1971) **B27,** 1969.
37. Berman, H. M., Kim, S. H., *Acta Crystallogr.* (1968) **B24,** 897.
38. Gatehouse, B. M., Poppleton, B. J., *Acta Crystallogr.* (1970) **B26,** 1761.
39. Gatehouse, B. M., Poppleton, B. J., *Acta Crystallogr.* (1971) **B27,** 654.
40. Girling, R. L., Jeffrey, G. A., *Carbohyd. Res.* (1971) **18,** 339.
41. Takagi, S., Rosenstein, R. D., *Carbohyd. Res.* (1969) **11,** 156, and private communication.
42. Kim, S. H., Jeffrey, G. A., *Acta Crystallogr.* (1967) **22,** 537.
43. Sundaralingam, M., *Biopolymers* (1969) **7,** 821.
44. Reeves, R. E., *J. Amer. Chem. Soc.* (1950) **72,** 1499.
45. Brown, G. M., Levy, H. A., *Science* (1963) **141,** 921.
46. Berman, H. M., *Acta Crystallogr.* (1970) **B26,** 290.
47. Rohrer, D. C., *Acta Crystallogr.* (1972) **B28,** 425.
48. Isbell, H. S., Tipson, R. S., *J. Res. Nat. Bur. Stand.* (1960) **64A,** 171.
49. Park, Y. J., Kim, H. S., Jeffrey, G. A., *Acta Crystallogr.* (1971) **B27,** 220.
50. Berking, B., Seeman, N. C., *Acta Crystallogr.* (1971) **B27,** 1752.
51. Lemieux, R. U., Pavia, A. A., *Can. J. Chem.* (1969) **47,** 4441, and Lemieux, preceding paper.
52. Barker, S. A., Brimacombe, J. S., Foster, A. B., Whiffin, D. H., Zweifel, G., *Tetrahedron* (1959) **7,** 10.
53. Hassel, O., Ottar, B., *Acta Chem. Scand.* (1947) **1,** 929.
54. Quigley, G. J., Sarko, A., Marchessault, R. H., *J. Amer. Chem. Soc.*, submitted for publication.
55. Chu, S., Jeffrey, G. A., *Acta Crystallogr.* (1967) **23,** 1038.
56. Ham, J. T., Williams, D. G., *Acta Crystallogr.* (1970) **B26,** 1373.
57. Brown, G. M., Rohrer, D. C., Berking, B., Beevers, C. A., Gould, R. O., Simpson, R., *Acta Crystallogr.*, submitted for publication, 1972.
58. Poppleton, B. J., Mathieson, A. McL., *Nature* (1968) **219,** 1946.
59. Hermans, P., *Kolloid. Z.* (1943) **102,** 169.
60. Carlström, D., *Biochem. Biophys. Acta* (1962) **59,** 361.
61. Jeffrey, G. A., Kim, H. S., *Carbohyd. Res.* (1970) **15,** 310.
62. Jeffrey, G. A., Kim, H. S., *Acta Crystallogr.* (1971) **B27,** 1812.
63. Gatehouse, B. M., Poppleton, B. J., *Acta Crystallogr.* (1971) **B27,** 871.
64. Littleton, C. D., *Acta Crystallogr.* (1953) **6,** 775.
65. Liquori, A. M., Damiani, A., Elefante, G., *J. Mol. Biol.* (1968) **33,** 439.

RECEIVED December 2, 1971. Carbohydrate structural determinations reported from our department are supported by Research Grant GM-11293 from the U. S. Public Health Service, National Institute of Health.

11

Stability Constants of Some Carbohydrate And Related Complexes by Potentiometric Titration

REX MONTGOMERY

Department of Biochemistry, University of Iowa, Iowa City, Iowa 52240

The hydroxy α-amino acids L-*serine and* L-*threonine, used as models for the 2-amino-2-deoxy glyconic acids, have been complexed with Ni(II) at 37°C in aqueous solutions of* 0.15M *potassium nitrate. Values for the stability constants were obtained from* iso-*pH titration data which were collected by alternate, small, incremental additions of metal ion and potassium hydroxide being made such that the pH of the solution remained nearly constant. The data were consistent with the predominance of* ML_n *species, along with additional protonated and hydrolyzed complexes. There was no evidence for the involvement of the hydroxyl group in chelation. By the same iterative computations the complexes formed between borate and mannitol have been analyzed, and the stability constants have been calculated. Complexes with mannitol:borate stoichiometries of 1:1, 1:2, 1:3, and 2:1 were proposed.*

Many commercial uses of carbohydrates depend upon reversible interactions of the hydroxyl groups with ions to form complexes. The processes are intricate, and analyses of the equilibria involved have been difficult or impossible (*1*). For example, the nature of the complexes involving polysaccharides and polyanions, such as borate, can only be hypothesized (*2*). Similarly the gelling of pectates with Ca(II) (*3*), the reactions of heptono-γ-lactones with Fe(III), or the nature of Fe(III)–dextran complexes in intramuscular injectable solutions are awaiting proper description. Although the systems involving monomeric components are more amenable to study, as for the reactions of metal salts or hydroxides with monosaccharides or their simple derivatives (*4, 5, 6, 7*) (some of which are discussed in other parts of this volume), even here the reactions are complicated either by small equilibrium constants or by extreme conditions of pH. Such systems are not yet amenable to study by potentiometry. Also the complex composition of aqueous solutions of reducing sugars resulting from equilibria of the various ring forms (*8*) presents a major analytical problem when each isomer may form several metal complexes. The studies described below were therefore directed towards simple acyclic sugar derivatives.

Relatively few determinations of stability constants have been reported in which the ligand is carbohydrate in nature (*9*). Since many of the essential trace metals in living systems form stable chelates with bidentate ligands containing amino and carboxylate groups (*10*), the present study considers the simplest of the amino sugar acid ligands, 2-amino-2-deoxy-L-glyceric acid (L-serine) and 2-amino-2,4-dideoxy-L-threonic acid (L-threonine), studying particularly the involvement of the hydroxy group, the nature of the possible complex species, and the variation of the major species with pH.

Another type of carbohydrate complex is the cyclic ester formed between a glycitol and a polyanion such as borate or molybdate (*4*). The possible species present at equilibrium have not previously been evaluated by iteration. The system mannitol–borate has been so analyzed, and it was possible to use the same approach as that for the metal chelates.

Analysis of Complex Equilibria by Iterative Methods

The possible complexes formed between a ligand L and M, which may be Ni(II) or mannitol, are given in Equation 1. These complexes include the simple forms such as ML, those with protonated ligands—*e.g.*, ML(HL)—the hydrolyzed complexes M(OH)L, and polynuclear species such as $M_2(OH)_2L_2$. Most of the published stability data for metal-complexes have been obtained, based on $\bar{n}$-functions (*11*), such systems being Fe(II)-2-amino-2-deoxy-D-gluconic acid (*12*), Fe(III)-D-gluconic acid (*5*), or Ni(II)-2-amino-2-deoxy-L-glyceric acid (*13, 14*), which are only a few of many (*9*). The many equilibria represented by Equation 1 are coupled together and coupled to those of the protonation of the free ligand. To obtain values for the equilibrium constants of the individual reactions, it is necessary to derive the concentrations of each component without disturbing the system. This is rarely possible because only one property is usually being measured experimentally, such as $[H^+]$, heat generation, solubility, or liquid/liquid partition (*11*). Spectroscopic methods offer study of the system at several wavelengths, but the spectral changes upon complexation are usually not great, and one is still left without a knowledge of the number of species producing an envelope of absorptions or the ϵ_{max} and λ_{max} for each species. All approaches leave too many possibilities for too few data so that a unique solution is frequently impossible, and computations by iterative methods are involved.

A set of equilibria such as those discussed above may be generalized:

$$m_jM + l_jL + w_j\,(OH) \overset{\beta_j}{\rightleftharpoons} M_{m_j}\,L_{l_j}\,OH_{w_j} \quad (1)$$

The concentration of the complex, $[M_{m_j}\ L_{l_j}\ OH_{w_j}]$, represented as C_j, may be expressed:

$$C_j = \beta_j[M]^{m_j}[L]^{l_j}[OH]^{w_j} \quad (2)$$

Since

$$[OH^-] = K_w \times 10^{pH} \quad (3)$$

and

$$[M]_{total} = [M]_{free} + \sum_{j=1}^{n} m_j C_j \quad (4)$$

$$[L]_{total} = [L]_{free} + \sum_{j=1}^{n} l_j C_j \quad (5)$$

then the concentration of each complex, of which there are n, can be calculated from the pH of the solution, $[M]_{free}$, $[L]_{free}$, and the equilibrium constant β_j. $[M]_{total}$ and $[L]_{total}$ are known. The functions:

$$f_M = [M]_{total} - ([M]_{free} + \sum_{j=1}^{n} m_j C_j) \quad (6)$$

$$f_L = [L]_{total} - ([L]_{free} + \sum_{j=1}^{n} l_j C_j) \quad (7)$$

will become zero when proper values are given for $[L]_{free}$, $[M]_{free}$, and β_j. The iterative process substitutes differing values in Equations 6 and 7 until f_M and f_L are zero, but a parameter for these iterations must be given whereby an experimental determination can be matched to a calculated value. The commonly used procedure involves a least squares function R (Equation 8), where P_d and P_c are the two parameters of the determined and calculated values, respectively, for k experimental points and a weight factor ω

$$R = \sum_{i=1}^{k} \omega(P_d - P_c)^2 \quad (8)$$

The degree to which these parameters can be unambigously minimized determines whether or not the model is unique. The parameter may be any property that changes during the experiment; a potentiometric procedure provides a suitable set of data for minimization.

The potentiometric procedure requires that the mass balance of hydrogen ion or metal ion be known at each condition of pH or pM (*15, 16*). Considering the analysis using pH, it is necessary that the pH of the solution containing M and L be accurately determined throughout the titration. The hydrogen ion concentration $[H^+]$ is related to pH as follows:

$$p[H^+] = pH + \log F \tag{9}$$

where the factor F allows for the activity coefficient of the hydrogen ion, f_+, the residual liquid junction potential, ΔE_j, and any other experimental errors in the pH measurements, ΔpH, as follows:

$$\log F = \log f_+ - \Delta E_j + \Delta pH \tag{10}$$

By providing the computer with starting estimates of the β_j values and the data for $[M]_{total}$, $[L]_{total}$, and pH at each titration point, a non-linear curve fitting procedure—such as the method of Newton–Raphson—results in the minimization of the least squares function R, which is the sum of the squares of the difference between the experimental amount of alkali titrated and the calculated amount for each experimental point.

The quantity F is most variable when the experimental design calls for titrations over a relatively wide range of pH and is best known at those points on the pH scale where the pH meter has been calibrated with primary standard buffers.

Experimental Procedure

Details of equipment and general procedure were those described by Perrin and co-workers (*15*). However, instead of the normal titration in which standard KOH was added to a mixture of M and L from about pH 3 to some higher pH, it was found advisable in this study to add alternate increments of M and KOH to maintain the pH around the same value. The pH-meter was then being used as a null-instrument at chosen values of pH over the range to be studied.

The titration data were analyzed by a computer program, which was a modification of the program SCOGS (*17*). The systems Ni(II) and L-serine or L-threonine, as analogs of amino-sugar acids, and boric acid and mannitol–borate were evaluated. For each of these systems it was convenient to refine first the formation constants for the simple complexes, such as ML and ML_2, adding other more complex species later until the best fit to the data had been obtained.

Complex Formation of* L*-Serine and* L*-Threonine with Ni(II)

Table I lists the refined values of the log formation constants for Ni(II) and L-serine and L-threonine. Under the conditions of the study where the ligands may be present either as the zwitterion or the anion, both forms could interact with the Ni(II). The zwitterions, HL, are similar to the aliphatic hydroxy acids, and the anions of such compounds possess a carboxylate group through which unindentate complexes may be formed—*e.g.*, Ni(II)-lactate, log *K* 2.216 (*18*). The values for the protonated Ni(II)–threonine and Ni(II)–serine complexes in this study were comparable (Table I).

Table I. Log Formation Constants of Ni(II) amino acid Complexes at 37°C in 0.15 *M* KNO_3

	Solutions[a]	
	Ni(II) L *Threonine*	*Ni(II)*-L-*Serine*
$\log K_{a_1}$	8.709 [b]	8.841 [b]
$\log K_{a_2}$	2.200 [b]	2.180 [b]
$\log K_{NiL}$	5.410 ± 0.005	5.353 ± 0.006
$\log K_{NiL_2}$	4.553 ± 0.004	4.434 ± 0.003
$\log K_{NiL_3}$	3.135 ± 0.055	3.297 ± 0.018
$\log K_{Ni(HL)}$	1.522 ± 0.018	1.184 ± 0.037
$\log K_{NiL(HL)}$	6.818 ± 0.015	
$\log K_{Ni(OH)L}$	$^{-}6.882 \pm 0.032$	

[a] The constants were computed using a value of 0.80 for factor *F* (Equation 9).
[b] The pK_a values for serine and threonine at 37° in 0.15*M* KNO_3 were taken from Perrin *et al.* (*32*) and Sharma (*19*), respectively.

The major complexes were those involving the anionic form of the ligands, giving the bidentate complexes by stepwise addition, where the positive charge on the complex is progressively reduced and the stability constants become less going from $[NiL]^+$ to $[NiL_2]^0$ and finally $[NiL_3]^-$. In agreement with an earlier finding (*19*) there was no evidence for complexation through the hydroxyl groups, and only when limiting amounts of ligand were present, was there significant formation of a hydrolyzed species, Ni(OH)Thr.

The dependence of the relative concentrations of the complexes in Ni(II)–threonine and Ni(II)–serine systems is presented in Figures 1 and 2 for solutions containing varying proportions of metal ion and ligand but not exceeding 30 *mM* total Ni(II) or total ligand concentration. The distributions show the variability of the different species at different pH values. At the higher threonine–nickel ratios there is a relatively small range of pH, 7.2–9.2, where only the simple complexes are present. In this range computations based on the $\bar{n}$ function can be made, $\bar{n}$ being defined as the average number of ligands, L, bound per

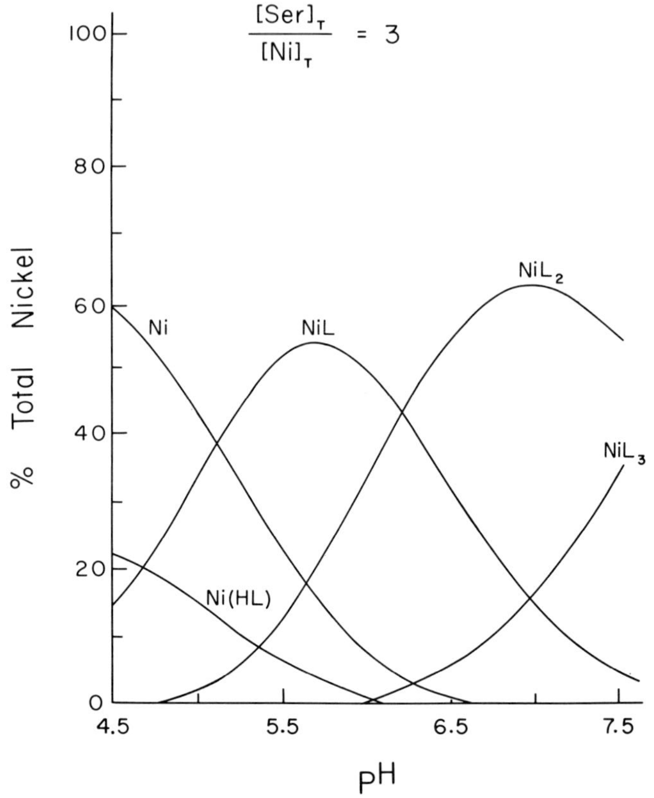

*Figure 1. pH dependence of the computed composition of a solution, which was 0.01*M *nickel(II)nitrate and 0.03*M *L-serine, at 37°C and 0.15*M *potassium nitrate, using the values in Table I*

metal ion. For a total threonine to total Ni(II) ratio of 10 (Figure 2), $\bar{n}$ changes from 1.8 at pH 7.2 to 2.8 at pH 9.2. Perhaps this accounts for the good agreement between the results of other workers (*14*), who used $\bar{n}$, and those from the present study. Out of this pH range, however, the values of $\bar{n}$ are difficult to interpret because of the presence of protonated and hydrolyzed species.

Analysis of the data for the complexation of Ni(II) by serine exclusively in terms of the simple species ML, ML_2, and ML_3 gives the values for the equilibrium formation constants log K_{NiL} 5.27_5, log K_{NiL_2} 4.53_4, and log K_{NiL_3} 3.33_1, which are near those given in Table I where all possible species had been considered. The fit of the data is not as good, however, as when the specie M(HL) is considered. Earlier data (*12*) for D-glucosaminic acid and Ni(II), analyzed graphically by $\bar{n}$ function gave log K_{NiL} 5.6 and log K_{NiL_2} 4.4 which have the same order as the constants found for L-serine and L-threonine. These values suggest that the hydroxyl groups of D-glucosaminic acid are not significant in the chelation with Ni(II).

Mannitol–Borate Complexes

In studies with amino acids and similar compounds, the ligand is nearly always monomeric. This situation changes greatly in those cases where the ligand can form dimers and higher condensation *n*-mers, such as phosphate (*20*), borate (*21, 22, 23*), and where polynuclear species are possible, such as with the copper(II)ion (*24*) and the Fe(III)ion (*25*). To extend these studies and to consider the analyses of other types of systems that reversibly form complexes, the mannitol–borate system which has previously been studied (*26, 27*), but without taking all forms of the borate ligands into account was examined.

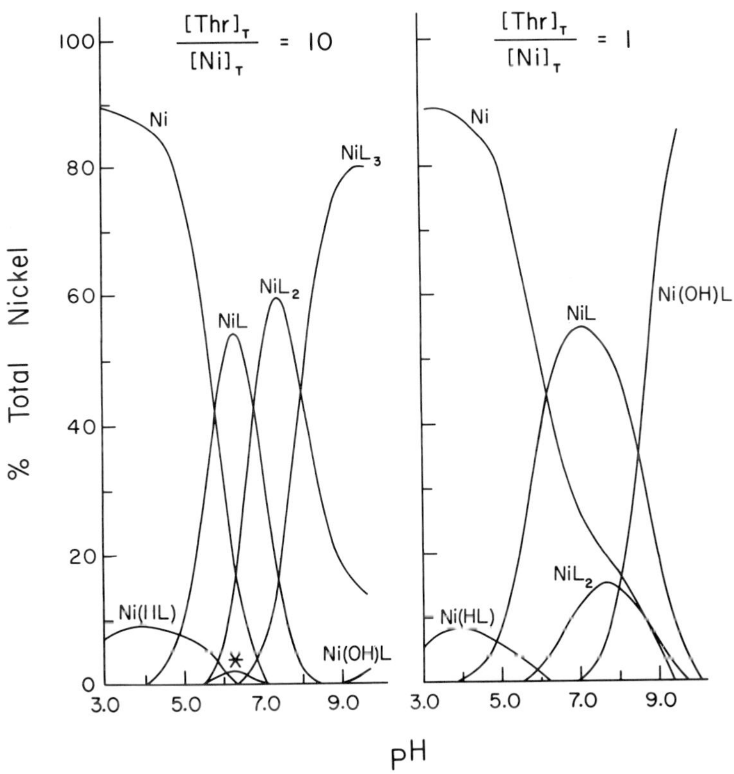

Figure 2. pH dependence of the computed composition of solutions, which were 0.003M in nickel(II)nitrate and 0.003M or 0.0003M in L-threonine, at 37°C and 0.15M in potassium nitrate, using the values of constants summarized in Table I

The composition of aqueous boric acid–borate solutions depends upon pH, the concentration of the total borate, and the nature of the background electrolyte if one occurs (*21, 22, 23, 28, 29, 30*). The simplest equilibrium may be represented by the reaction of boric acid with OH^- to form the borate anion.

$$H_3BO_3 + H_2O \rightleftharpoons B^-(OH)_4 + H^+ \tag{11}$$

At concentrations of total boron above 0.025*M*, polynuclear species are significant (*23*). The present study, using 0.05*M* total boron, gave evidence for equilibria involving $[B_3O_3(OH)_5]^{2-}$ and $[B_3O_3(OH)_4]^{1-}$. The existence of other polyborates was not consistent with the data, except perhaps for $B_4O_4(OH)_4^{2-}$ at high pH conditions (>9.5). The results are summarized in Figure 3, with the possible structures of the related

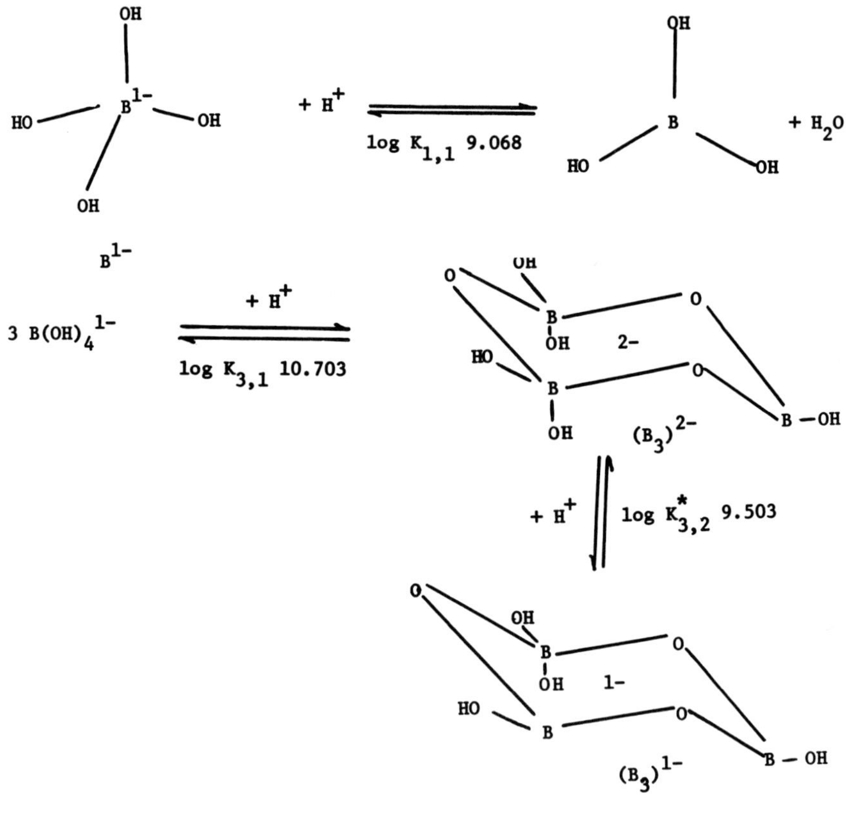

*Figure 3. Structures of the polyborates with the log formation constants for 0.05*M *boric acid at 25°C in 0.15*M *potassium nitrate*

polyborate ions. Some of these structures are supported by x-ray analysis of boron-containing minerals (*31*), but except for requiring that species with certain stoichiometric relationships be coupled, the structures cannot be proved by potentiometric studies.

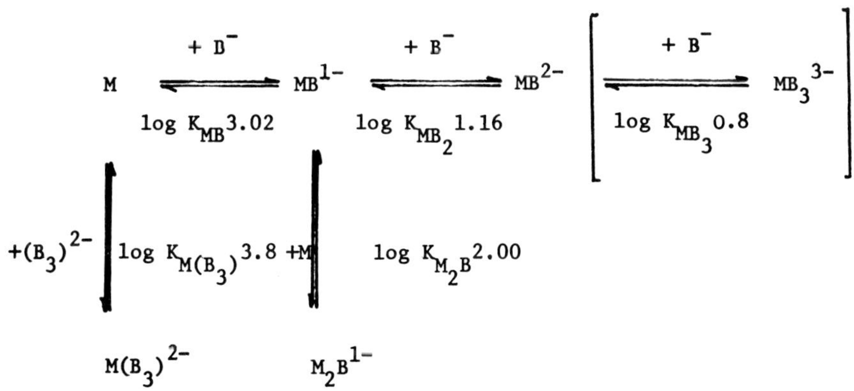

*Figure 4. Reactions of borate with mannitol with the log formation constants of the complexes at 25°C and 0.15*M *in potassium nitrate*

Considering the species proposed in Figure 3 to be the forms of borate available to complex with mannitol, the potentiometric data were best fit by the complexes summarized in Figure 4. The same modified computer program, SCOGS, was used as it had been applied to the data from the Ni(II) complexation. Before discussing these results we consider the concentrations of these species as they are found in solutions with different mole ratios of total borate and total mannitol. With excess mannitol (*see* Figure 5(c)) the principal complex is always M_2B^{1-}, MB^{1-} never exceeding 15% with reference to total mannitol; the composition is nearly constant from pH 8.8–10.0. With equimolar amounts of mannitol and borate (Figure 5(b)), there is one additional complex MB_2^{2-} above pH 8.5, but the proportions of MB^{1-} and M_2B^{1-} are quite different. Over a pH range of 7.0–10.0 the $\bar{n}$ ratios change from 0.35–0.84 reflecting the change in concentration of M_2B^{1-}. Finally in a large excess of borate (Figure 5(a)) the principal complex is always MB^{1-}, and above pH 8.0 it is present with all other proposed complexes except M_2B. Such a mixture cannot be adequately analyzed based on Equation 12

$$HB + nM \rightleftharpoons H + M_nB \tag{12}$$

except under certain conditions of pH and with excess mannitol. In such a study (*26*) the log overall formation constants for MB^{1-} and M_2B^{1-}, 2.79 and 4.98, respectively, are similar to those determined in this study, 3.02 and 5.02 (Figure 4).

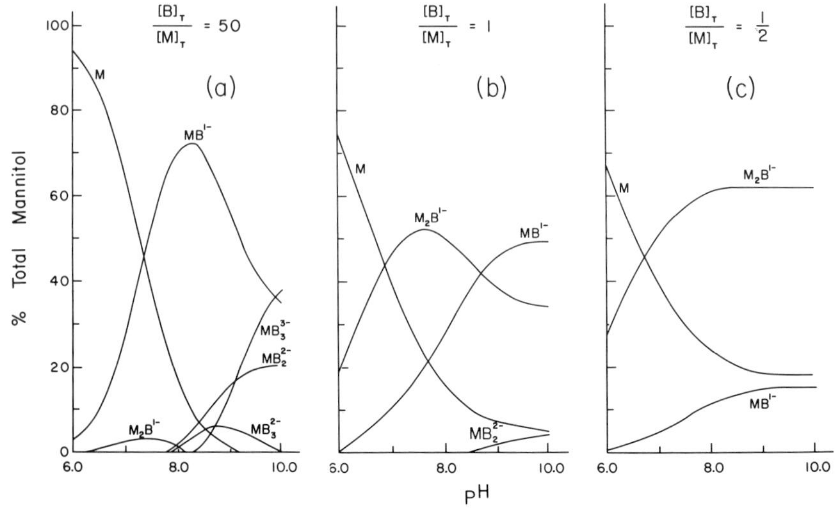

*Figure 5. pH dependence of the computed composition of solutions which were 0.05*M *in borate and (a) 0.001*M, *(b) 0.05*M, *and (c) 0.1*M *in mannitol at 25°C and 0.15*M *in potassium nitrate, using the values given in Figures 3 and 4*

A question remains in the analysis of the mannitol–borate system. A species MB_3^{3-} was used to fit the data at pH > 8.5 in a large excess of total borate. There is no way in this type of study to determine whether this species is a complex between mannitol and three borate ions or mannitol and the triborate ion, $B_3O_3(OH)_6^{3-}$ or some other structure. The possibility always exists in iterative procedures that the fit of data by a certain combination of equilibria will be fortuitous. Previous work would suggest otherwise (*20*), but it will be necessary to search for other supporting evidence for any new species, for which reason the equilibrium involving MB_3^{3-} in Figure 4 is enclosed in brackets.

Systems as complex as those between metal ions, polyhydroxy compounds, and polyanions will rarely have a parameter for following the change in concentration of each species. In the absence of enough data, either some approximations must be made to ignore minor species, or an iterative approach to a best fit of the data is necessary. The resulting solution may not describe a unique model, but with other supporting chemical or physical properties, it is possible to discriminate between reasonable and improbable equilibria.

Acknowledgment

The author acknowledges the receipt of a Special Senior Fellowship from the National Institutes of Health for study during 1969–1970 at the

John Curtin School for Medical Research, Australian National University, Canberra, Australia. I am grateful to D. D. Perrin for providing research facilities and helpful discussions, to C. W. Childs, P. S. Hallman, and W. A. E. McBryde for their assistance, and to H. Kinns for guidance in computational matters.

Literature Cited

1. BeMiller, J. N., in "Starch," R. L. Whistler and E. F. Paschall, Eds., Vol. 1, Academic, New York, 1965.
2. Deuel, H., Neukom, H., *Makromol. Chem.* (1949) **3**, 13.
3. Schweiger, R. G., *J. Org. Chem.* (1962) **27**, 1789.
4. Bourne, E. J., Nery, R., Weigel, W., *Chem. Ind. London* (1959) 998.
5. Pecsok, R. L., Sandera, J., *J. Amer. Chem. Soc.* (1955) **77**, 1489.
6. Rendleman, Jr., J. A., *Advan. Carbohyd. Chem.* (1966) **21**, 209.
7. Rendleman, Jr., J. A., ADVAN. CHEM. SER. (1973) **117**, 51.
8. Anderson, L., ADVAN. CHEM. SER. (1973) **117**, 20.
9. Sillén, L. G., Martell, A. E., "Stability Constants of Metal-Ion Complexes," *Chem. Soc. (London)* Special Publication No. 17 (1964).
10. Williams, R. J. P., *Endeavour* (1967) **26**, 96.
11. Rossotti, F. J. C., Rossotti, H., "The Determination of Stability Constants," McGraw-Hill, New York, 1961.
12. Miyazaki, M., Senshu, R., Tamura, Z., *Chem. Pharm. Bull.* (1965) **14**, 114.
13. Pelletier, S., *Compt. Rend.* (1957) **245**, 160.
14. Letter, J. E., Jr., Bauman, J. E., Jr., *J. Amer. Chem. Soc.* (1970) **92**, 437.
15. Childs, C. W., Perrin, D. D., *J. Chem Soc.* (1969) **A**, 1039.
16. McBryde, W. A. E., *Analyst (London)* (1969) **94**, 337.
17. Sayce, I. G., *Talanta* (1968) **15**, 1397.
18. Evans, W. P., Monk, C. B., *J. Chem. Soc.* (1954) 550.
19. Sharma, V. S., *Biochem. Biophys. Acta* (1967) **148**, 37.
20. Childs, C. W., *Inorg. Chem.* (1970) **9**, 2465.
21. Ingri, N., Lagerstrom, G., Frydman, M., Sillén, L. G., *Acta Chem. Scand.* (1957) **11**, 1034.
22. Ingri, N., *Acta Chem. Scand.* (1963) **17**, 573.
23. Ingri, N., *Acta Chem. Scand.* (1963) **17**, 581.
24. Perrin, D. D., Sharma, V. S., *J. Inorg. Nuclear Chem.* (1966) **28**, 1271.
25. Perrin, D. D., *J. Chem. Soc.* (1958) 3125.
26. Knoeck, J., Taylor, J. K., *Anal. Chem.* (1969) **41**, 1730.
27. Acree, T. E., ADVAN. CHEM. SER. (1973) **117**, 208.
28. Edwards, J. O., Morrison, G. C., Ross, V. F., Schulta, J. W., *J. Amer. Chem. Soc.* (1955) **77**, 266.
29. Dale, J., *J. Chem. Soc.* (1961) 922.
30. Weser, U., *Struct. Bonding (Berlin)* (1968) **2**, 160.
31. Christ, C. L., Clark, J. R., *Acta Cryst.* (1956) **9**, 830.
32. Perrin, D. D., Sayce, I. G., Sharma, V. S., *J. Chem. Soc.* (1967) **A**, 1755.

RECEIVED December 2, 1971.

12

The Chemistry of Sugars in Boric Acid Solutions

TERRY E. ACREE

New York State Agricultural Experiment Station, Cornell University, Geneva, N. Y. 14456

The reactions of boric acid solutions with diols have been used for almost a century to examine structural differences among carbohydrates. The complexity of these reactions seems to arise not only from simple structural differences but also from differences in carbohydrate configuration and conformation. The precise nature of these reactions is not clear. Recent studies of the chemistry of polyol–boric acid solutions have clarified some aspects of these reactions that have important bearing on the structure of carbohydrates in solution. Nevertheless, some of the most fundamental questions about the nature of the reaction are still unanswered.

In 1842 Biot (*1*) observed that the addition of certain sugars increased the acidity of boric acid solutions. Research during the next one hundred years revealed that this phenomenon results from the formation of sugar–boric acid esters which were stronger acids than boric acid and that there are certain configurational requirements which sugars must satisfy to react with boric acid. Reviews of the research done during this period have been presented by Roy *et al.* (*2*), Böeseken (*3*), Weser (*4*), and Nies and Campbell (*5*).

Some of the findings of this early work are:

1) Boric acid will react with diols which satisfy certain spatial requirements

2) For vicinal diols to react, the dihedral angle between the hydroxyl groups must be between 0° and 60°. (Using paramagnetic resonance (PMR) spectroscopy, Lenz and Heeschen (*6*) reported that the complexes of D-glucose and D-xylose with boric acid had a dihedral angle of 48°. This angle was derived from the coupling constant observed for the anomeric proton, and such a derivation has yet to be substantiated. Probably a more generally useful description of the configurational re-

striction of the diol is that the oxygen–oxygen distance must be between 2.2–2.5 A. Such a description also includes 1,3 diols, etc.)

3) At equilibrium the diol–boric acid ester or complex can have a diol to boron ratio of 1 to 1, 2 to 1, or both.

4) The rate of esterification and hydrolysis of these complexes is rapid, and equilibrium occurs within one second.

Most of the work on the boric acid–diol reaction during the last twenty years has been done to determine the coordination number of the diol (number of diol molecules) in the complex and to evaluate the equilibrium constant (often called a stability constant) for a number of diol–boric acid reactions. Several techniques have been used to study these questions, including polarimetry (*7*), optical rotatory dispersion (*8*), polarography (*9*), conductivity (*3*), vapor pressure osmometry (*10*), and electrochemistry (*11, 12, 13*). The most frequently studied system has been the electrochemical (pH) titration of boric acid or borax solutions with various diols.

During this recent research several conflicting reports and opposing hypotheses have appeared. Three questions about the diol–boric acid reaction which have been particularly baffling are discussed here. These are:

1) Does the diol react with trigonal boric acid or with the tetrahedral borate anion?

2) How are data from titrations with diols to be interpreted to indicate the diol coordination number or numbers and the equilibrium constants for the reaction?

3) What does the diol–boric acid reaction imply about the chemistry of sugars in solution?

Before examining these three questions, some of the properties of boric acid in solution should be discussed.

Some Aspects of the Chemistry of Boric Acid Solutions

In concentrated solutions boric acid forms a complex mixture of polyborate ions. In such solutions data obtained from titrations are complicated by the presence of these complex ions. Aqueous boric acid solutions below 0.2*M* seem to have negligible amounts of polyborate ions (*14*). Much of the early data reported in the literature was collected using more concentrated solutions (*3*). Thus, the ionization of boric acid can be described by the following equation:

$$B(OH)_3 + H_2O = B(OH)_4^- + H^+ \qquad \text{I}$$

and the equilibrium constant is

$$K_a = \frac{(H^+)\,(B(OH)_3)}{(B(OH)_4^-)} \qquad \text{II}$$

The pK value for this dissociation is about 9.14. Reports of lower values (*15*) may result from the formation of polyborate ions or from the presence of chloride ions. If a knowledge of K_a is important in a particular study, the possible effect of chloride ion should be investigated before it is introduced into the system. Dilute boric acid solutions probably contain only two boroxy species, trigonal boric acid and tetrahedral borate anion.

The Reactive Boroxy Species. Most of the previous work on the boric acid–diol reaction has included the assumption that a diol reacts with the borate anion and not with undissociated boric acid. Whether or not this assumption is correct in most cases would not have affected the general conclusions of the work. However, the assumption has been stated in most of the previous literature for one reason or another.

An example of some recent data which was interpreted to support this assumption is the work of Knoeck and Taylor (*12*). Using PMR spectra of mannitol–boric acid solutions, they observed that a decrease in pH resulted in a decrease in the mannitol–boric acid complex concentration. These results are supported by ^{11}B nuclear magnetic resonance (NMR) spectroscopy (*16*) which showed that the complex between mannitol and boric acid increased with increasing pH. However, Knoeck and Taylor (*12*) reasoned that since an increase in pH resulted in an increase in the borate anion concentration, as well as an increase in the complex concentration, the diol reacts only with the borate ion to produce the complex.

To understand how these conclusions are not justified by the data used to support them, consider the overall equilibrium expressions for the two possible reactions,

$$\text{HB} + \text{D} = \text{BD} + \text{H} \qquad \text{III}$$

$$\text{B} + \text{D} = \text{BD}, \qquad \text{IV}$$

where HB, B, H, D, and BD are boric acid, borate ion, hydrogen ion, diol, and complex, respectively. The equilibrium constants for the two reaction schemes are

$$K_1 = \frac{(\text{BD})\ (\text{H})}{(\text{B})\ (\text{D})} \qquad \text{V}$$

and

$$K_2 = \frac{(\text{BD})}{(\text{B})\ (\text{D})} \qquad \text{VI}$$

In these reactions the diol coordination number was assumed to be one. The following argument would be true no matter what the coordination number actually is.

If Equations V and VI are rearranged and the dissociation equation for boric acid is substituted into V, Equations VII and VIII are obtained.

$$\frac{(\mathrm{B})K_1}{K_a} = \frac{(\mathrm{BD})}{(\mathrm{D})} \qquad \text{VII}$$

and

$$(\mathrm{B})K_2 = \frac{(\mathrm{BD})}{(\mathrm{D})} \qquad \text{VIII}$$

Expressions VII and VIII are identical in form; they differ only in the meaning of the constants they contain. In both cases an increase in the borate ion concentration would result in an increase in the diol–boric acid complex. Therefore, an examination of the effect of pH on the equilibrium concentrations of various components of the system cannot be used to determine which of the two boroxy species actually reacts with the diol.

This question is essentially a mechanistic one in which the two candidates reaction with the diol mechanism are in rapid equilibrium with each other. Such mechanistic questions can be more properly answered by studying the kinetics of the reaction. A kinetic study of the reaction boric acid and tartaric acid has been reported (*17*). It was found that tartaric acid reacts with boric acid and not the borate ion. Also, the relaxation time for this reaction is near 20 msec.

Such rapid reactions imply a low transition state energy, and mechanisms proposed for this reaction should take this into account. A reasonable mechanism can be postulated in which the reactive species is boric acid.

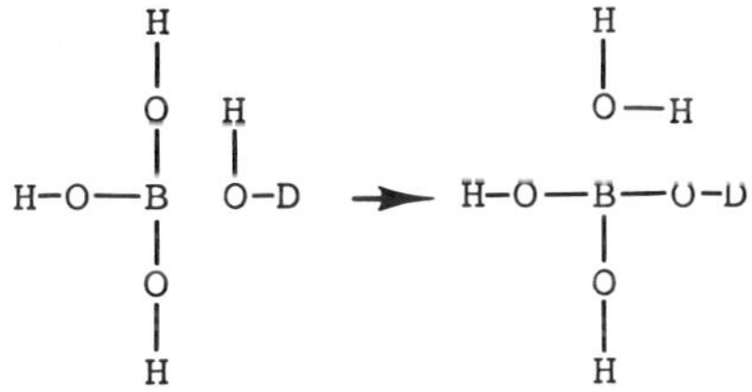

Figure 1. A mechanism for the reaction between trigonal boric acid and a diol to produce a trigonal product

Isotope labeling experiments indicate that the B—O bond is broken and not the C—O bond in the formation of the diol–boric acid complexes (*18*). This indicates that the initial step in the mechanism may be an attack on the boron atom by an oxygen of the diol, followed by the release of water. This could occur without developing any charge separation. If such a mechanism were correct, it would seem that an attack on the boron atom would be easier for trigonal boric acid than for the tetrahedral borate anion (Figure 1).

However, it remains to be proved which of the boroxy species reacts with diols other than tartaric acid.

Determination of the Coordination Number from Diol-Titration Experiments. Early in the research on the diol–boric acid reaction it was recognized that two possible complexes could form (*3*), one with a diol coordination number of one, the other with a coordination number of two. A generalized expression for the overall complex reaction can be written as follows,

$$HB + nD \;{}_{n}{=}\; BD_n^- + H^+ \qquad \text{IX}$$

where n is the coordination number for the diol. The expression for the equilibrium condition is

$$K_n = \frac{(BD_n^-)\ (H^+)}{(HB)\ (D)_n} \qquad \text{X}$$

The evaluation of this equilibrium using pH titration requires the use of certain approximations. Most of the experiments in the literature have dealt with systems which consist of pH measurements of solutions consisting of dissociable monoboric acid, a diol, and sometimes an additional cation such as sodium. Which approximation is made depends upon the examination of these three conservation equations for the titration in question.

The conservation of charge:

$$H^+ + Na^+ = BD_n^- + B^- + OH^- \qquad \text{XI}$$

The conservation of boron:

$$B_t = B^- + BD_n^- + HB \qquad \text{XII}$$

The conservation of diol:

$$D_t = BD_n^- + D \qquad \text{XIII}$$

Through the years there has been some conflict in the literature over the determination of the coordination numbers of various diol–boric acid complexes. Most of these conflicts can be resolved by making the proper approximations in these three conservation equations as determined by the experimental conditions.

Consider, for example, the titration of sodium tetraborate solution with a diol and suppose the following experimental conditions are met: $(Na^+) >> (H^+)$ and that $(D_t) >> (B_t)$, where D_t and B_t are the total diol and boron concentrations respectively (*e.g.*, $0.01M$ $Na_2B_4O_7$, $0.1–1M$ diol, and between pH 4–8). Applying these conditions to the conservation equations, XI becomes

$$Na^+ = BD_n^- + B_n^- \qquad \text{XIV}$$

Substituting XIV into XII yields

$$B_t = HB + Na^+ \qquad \text{XV}$$

Since (Na^+) and (B_t) are constants, HB is constant throughout the titration. For sodium tetraborate solutions the pH equals the pK_a. Substituting K_a for H^+ in the dissociation for boric acid yields

$$B_o^- = HB$$

where B_o^- is the initial concentration of borate ion in the absence of diol. In the presence of diol (or more properly when the pH $(pK - 2)$), B^- is about zero. Applying these last approximations to equation XIV yields

$$Na^+ = B_o^- = HB = BD_n^-$$

and Equation XIII becomes

$$D_t = D \qquad \text{XVI}$$

By applying these approximations to Equation X, taking the negative logarithm, and finally rearranging, Equation X becomes

$$pH \cong pK - n\log(D_t). \qquad \text{XVII}$$

Thus, a plot of the pH *vs.* D_t will give at high D_t a slope (n) equal to the negative coordination number and an intercept (pK_n) equal to minus the log of the overall equilibrium constant.

In many cases it has been found that complexes with coordination numbers of one and two are present at equilibrium. However, in sodium tetraborate solution when the same general approximations are used, only the resulting equations for the relationship between (D_t) and (H^+) are somewhat more complicated (*see* Reference *11*).

Many of the titration experiments, however, have dealt with boric acid solutions in which the only positive ion is H^+. In this case the conservation of charge Equation XI becomes

$$H^+ = BD_n^- + B^-,$$

and at low pH and high D_t

$$H^+ = BD_n^-$$

and

$$D_t = D,\ B_t = HB.$$

Therefore, Equation IX becomes

$$pH = -\frac{(\log K_n + \log(B_t))}{2} - \frac{n}{2}\log D_t \qquad \text{XVIII}$$

In this case the coordination number is —2 times the slope of the line for a plot of pH *vs.* log D_t. The difference between slopes in Equations XVII and XVIII resulted in much of the confusion over the interpretation of diol titrations of boric acid and borax solutions (*11, 12, 19*).

Sugar Structure and the Boric Acid-Diol Reactions

The Mutarotation of D-Glucose in pH 7 Boric Acid–Sodium Hydroxide Solution. For years the major interest in the diol–boric acid reaction has been to determine the implications this reaction holds about the structure of carbohydrates in solution (*3, 6, 10*). D-Glucose reacts with boric acid in solution (*3*); however, which tautomer is the reactive form has never been completely determined. Since only cis-vicinal diols with a dihedral angle of less than 50° appear to react with boric acid, β-D-glucopyranose should not react with boric acid no matter what conformation it is in. However, α-D-glucopyranose could react with boric acid at the 1,2-position if it could assume a boat conformation. It is also equally possible that the reactive form of D-glucose in solution is α-D-glucofuranose. The existence of the α-furanose form of D-glucose in the borate complex has been postulated more seriously (*3, 10*) than the boat or twist conformations of the α-pyranose tautomer (*6*). Knowing which of these two isomers actually does react with boric acid is extremely important to understand the behavior of D-glucose in boric acid solution. To demonstrate the possible implications, consider the mutarotation (tautomerization) of α- and β-D-glucopyranose in pH 7 boric acid–sodium hydroxide solution.

When crystalline α- or β-D-glucopyranose is dissolved in water, the following reaction takes place

$$\alpha\text{-D-glucopyranose} \rightleftarrows \beta\text{-D-glucopyranose}$$

If this reaction is observed polarimetrically, the equilibrium rate constant obtained is equal to the sum of the tautomerization rate constants ($K_1 + K_{-1}$) for the overall reaction, as described above (*20*). Therefore, no matter what the path or the mechanism of the mutarotation phenomenon, the data obtained polarimetrically are completely explained by the above overall configuration change.

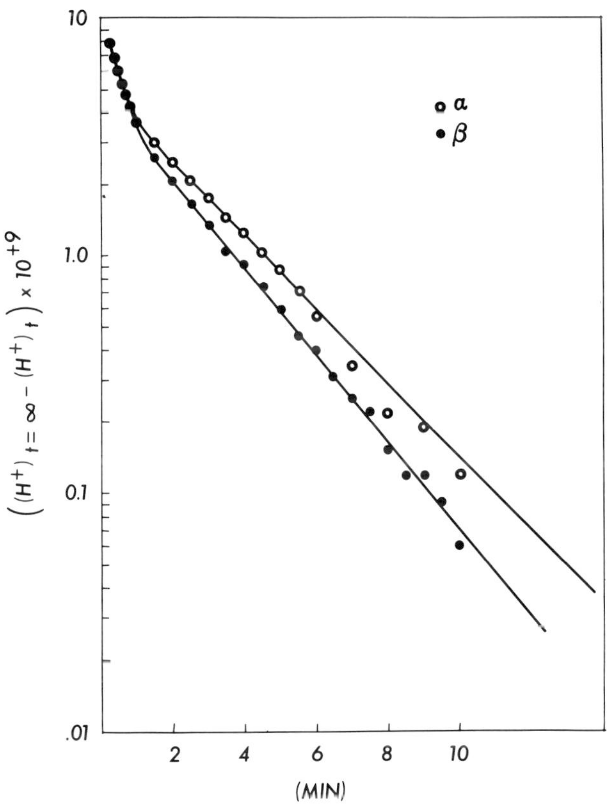

*Figure 2. Equilibrium rate plots of the hydrogen ion concentration vs. time during the mutarotation of 0.2*M *α- and β-D-glucopyranose in 0.1*M *borax, 25°C*

Examination of the mutarotation of D-glucose in boric acid solutions has certain experimental problems. First, in a sodium tetraborate solution the high initial pH is in the region in which the mutarotation reaction is extremely sensitive to changes in hydrogen ion concentration. For example, dissolving 0.2*M* D-glucopyranose in 0.1*M* sodium tetraborate solution results in the same kinetic change in the pH for the α-anomer and the β-anomer (Figure 2). A similar result is obtained in dilute boric acid solution where the low pH also disturbs the tautomerization constants.

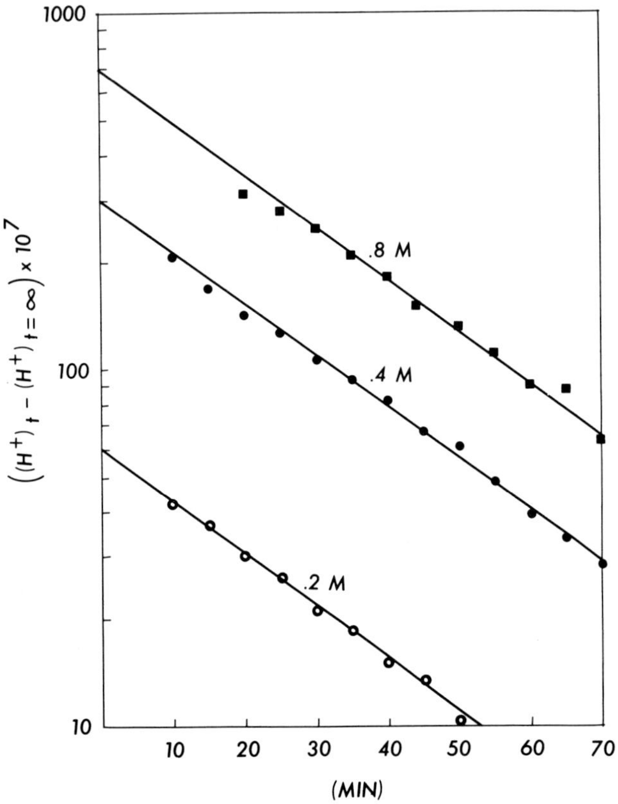

Figure 3. Equilibrium-rate plots of the hydrogen ion concentration vs. *time during the mutarotation of α-D-glucopyranose at different concentrations in 0.1*M *boric acid at pH 7 (w/NaOH) and 25°C*

In the region between pH 7 and 4 the mutarotation constants are relatively unaffected by pH. Figure 3 shows the kinetic data obtained for the change in hydrogen ion concentration of a 0.01*M* boric acid solution adjusted to pH 7 with sodium hydroxide when different amounts of

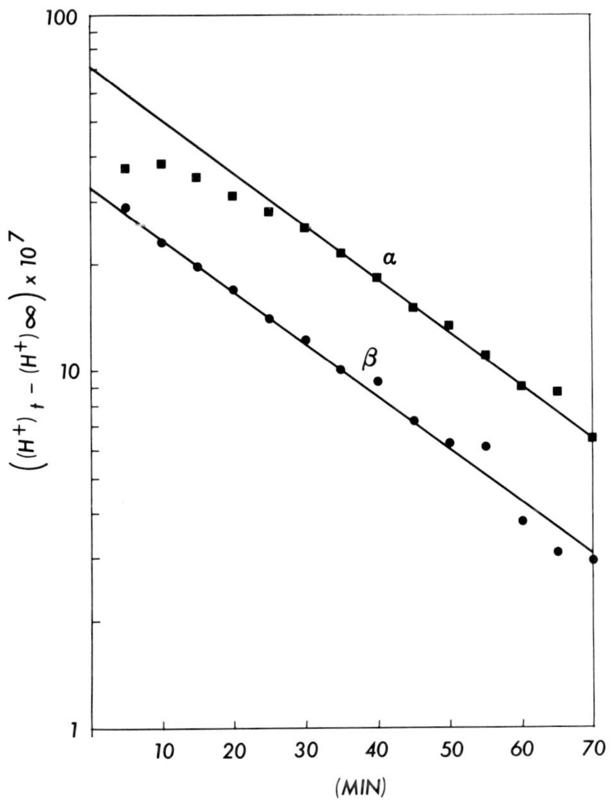

*Figure 4. Equilibrium-rate plots of the hydrogen ion concentration vs. time during the mutarotation of 0.8*M, *α- and β-D-glucopyranose in 0.01*M *boric acid adjusted to pH 7 with NaOH and 25°C*

α-D-glucopyranose are added. The equilibrium rate constant obtained in these experiments are unaffected by the D-glucose concentration and are comparable with values reported for $(K_1 + K_{-1})$ of D-glucose anomerization at 25°C (*20*). Figure 4 shows the results obtained for α- and β-D-glucopyranose. Again the same equilibrium rate constant is obtained when the starting material is the β-form as is obtained with the α-form.

If it is true that the structural form of D-glucose which reacts with boric acid is the α-D-pyranose form, then that form probably exists in a boat or twist conformation in the complex. This implies that the study of the stability constants of sugar borate ester might give information about the ability of various carbohydrates to form such boat or twist conformations (*10, 21*).

However, if the reactive form of D-glucose is not the α-pyranose structure but the α-furanose structure, then the implications of this reaction are quite different. The study of the stability of various sugar esters could give information about the ability of various carbohydrates to form furanose isomers. What is more important is that the most generally accepted view of the mechanism for the tautomerization reaction is that all the tautomers are in equilibrium with the same intermediate. This central intermediate model would be difficult to apply to the tautomerization of D-glucose in boric acid solutions in which one is observing the overall $\alpha \rightleftarrows \beta$-pyranose reaction by studying the formation and disappearance of α-furanose. Recognizing this implication, Böeseken (*3*) proposed a cyclic mechanism for D-glucose tautomerization which did not involve the existence of a central intermediate. He suggested that α- and β-furanose are intermediates in the transformation of α- to β-pyranose. (In a recent article by Isbell and Pigman (*22*), a mechanism is presented which has more than one intermediate. Collectively, however, these intermediates are central to the four major ring isomers.)

The study of the structure of the sugars in solution can be approached chemically if the chemistry used is as sensitive to sugar structure as sugars are diverse in their structure. The boric acid–sugar reaction could serve this purpose if some of the fundamental aspects of this reaction were understood. Priority should probably be given to: (1) the determination of the structure of the borate–sugar complex, (2) the evaluation of the stability constants for these complexes, and (3) the determination of the mechanism of the reaction.

Literature Cited

1. Biot, M., *Compte Rend., Acad. Sci., Paris* (1842) **14,** 49.
2. Roy, G. L., LaFerriere, A. L., Edwards, J. O., *J. Inorg. Nucl. Chem.* (1957) **4,** 106.
3. Böeseken, J., *Advan. Carbohyd. Chem.* (1949) **4,** 189–219.
4. Weser, U., *Struct. Bonding Berlin* (1967) **2,** 160–80.
5. Nies, N. P., Campbell, G. W., "Boron, Metallo-Boron Compounds, and Boranes," R. M. Adams, Ed., Wiley, New York, 1964.
6. Lenz, R. W., Heeschen, J. P., *J. Polymer Sci.* (1961) **51,** 247–61.
7. Isbell, H. S., Brewster, J. F., Holt, N. B., Frush, H. J., *J. Res. Natl. Bur. Std.* (1948) **40,** 129.
8. Katzin, L. I., Gulyas, E., *J. Amer. Chem. Soc.* (1966) **88,** 5209.
9. Swann, W. B., McNabb, W. M., Hazel, J. F., *Anal. Chim. Acta* (1963) **28,** 441–9.
10. Mazurek, M., Perlin, A. S., *Can. J. Chem.* (1963) **41** (10), 2403–11.
11. Conner, J. M., Bulgrin, V. C., *J. Inorgan. Nucl. Chem.* (1967) **29** (8), 1953–61.
12. Knoeck, John, Taylor, J. K., *Anal. Chem.* (1969) **41** (13), 1730–4.
13. Antikainer, P. J., *Suom. Kemistielehti* (1958) **31B,** 255.
14. Kemp, P. H., "The Chemistry of Borates, Part I.," Borax Consolidated, London, 1956.
15. How, M. J., Kennedy, G. R., Mooney, E. F., *Chem. Commun.* (1969) **6,** 267.

16. Onak, T. P., Landesman, H., Williams, R. E., Shapiro, I., *J. Phys. Chem.* (1959) **63**, 1533.
17. Kustin, Kenneth, Pizer, Richard, *J. Amer. Chem. Soc.* (1969) **91** (2), 317–22.
18. Gardner, H. K., Lucas, H. J., *J. Amer. Chem. Soc.* (1950) **72**, 5497.
19. Nickerson, R. F., *J. Inorgan. Nucl. Chem.* (1968) **30** (6), 1447–55.
20. Lee, C. Y., Acree, T. E., Shallenberger, R. S., *Carbohyd. Res.* (1969) **9**, 356.
21. Foster, A. B., *Advan. Carbohyd. Chem.* (1957) **12**, 81–115.
22. Isbell, H. S., Pigman, W. W., *Advan. Carbohyd. Chem.* (1969) **24**, 36.

RECEIVED December 2, 1971. Approved by the Director of the New York State Agricultural Experiment Station for publication.

13

Uses and Misuses of Bases in Studies of Glycoproteins

WARD PIGMAN and JOHN MOSCHERA

Biochemistry Department, New York Medical College, New York, N. Y. 10029 and Institute of Dental Research and Departments of Physiology and Oral Biology, University of Alabama, Birmingham, Ala. 35233

The mucous glycoproteins are the components of the mucous secretions of the internal passages of higher animals that cause the high viscosity of the secretions, and they may be responsible for many immunological phenomena. Although the number of highly purified mucus glycoproteins is still limited, evidence suggests that they are tissue and species specific in their composition and reactions. They seem to be related closely to the glycoproteins and glycolipids of cell membranes. In the development of their structures, the use of alkalies has provided valuable information but improperly used has caused several misinterpretations. The glycoproteins are composed of a protein core to which are attached numerous oligosaccharides. Studies of the primary structure of bovine submaxillary glycoprotein show that it consists of many covalently bound repetitive sequences of short peptides of similar composition. Other mucus glycoproteins seem to have similar primary structures for the protein core.

Reducing sugars are extremely labile to the action of bases in the general sense, and profound isomerizations, rearrangements, and chain cleavages are possible. These effects have been reviewed by Pigman and Anet (*1*). The initial effects are first, anomerizations and ring changes, second, enolizations, and, third, β-eliminations at a position next to a hydrogen atom activated by a carbonyl or other electron withdrawing group. The first two types of effects are catalyzed by acids and bases, bases usually being the best catalysts. As a result, reducing sugars show their maximum stability normally in the range of pH 3–4. β-Elimination, however, is primarily catalyzed only by bases.

Most glycosides are stable to the actions of alkalies, but when the aglycon contains groups that tend to withdraw electrons from the glycosidic linkage, the linkage becomes labile to alkali (2). Two types of reactions are common: (1) formation of anhydro rings, usually under conditions of strong alkalinity and temperature; and (2) the equivalent of a β-elimination which involves the removal of the free glycon (sugar) and the creation of a double bond in the aglycon (which also may be a sugar residue).

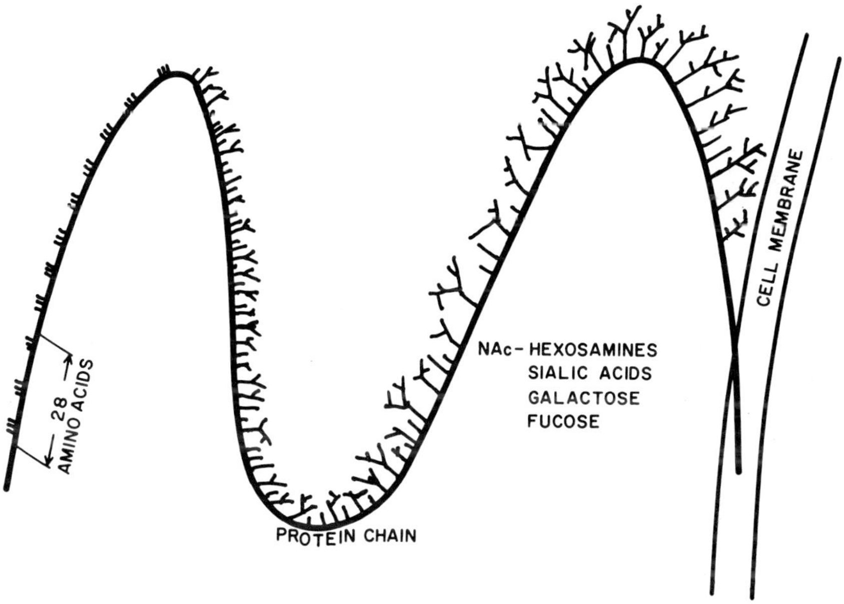

Figure 1. Schematic structure of mucus glycoproteins.

The continuous heavy line represents the protein core (MW, ca. 300,000). The short lines represent oligosaccharides or small polysaccharides as side chains. These side chains may be as small as one or as large as five sugar residues for the types found in BSM, OSM, and PSM. The longer side chains are more of the type formed in the human blood-group substances, any one may be as long as 20 units.

For glycoproteins, the formation of anhydro rings from hexosamine residues is important in chromogen formation, especially in relation to the Ehrlich reaction and its modifications (3). β-Elimination is involved in the removal of oligosaccharide or polysaccharide chains from the protein core of glycoproteins when they are attached by O-glycosidic linkages serine or threonine residues. A subsequent β-elimination may also occur if the free oligosaccharide side chains contain 1 → 3 glycosidic linkages between the first and second sugar units of the side chain.

This paper discusses mainly the glycoproteins of the mucus secretions of higher animals, the so-called mucins and blood-group substances which are chemically closely related if not identical. Several of these compounds have been highly purified and extensively characterized (*4, 5*). They have high molecular weight (*ca.* 1–5 $\times$ 10^6) and are rich in carbohydrates, often comprising about 65–75% of the total weight. These glycoproteins consist of a long peptide core to which numerous oligosaccharide chains are attached. A generalized diagram is shown in Figure 1. The continuous line represents the peptide core. The side chains are oligosaccharides of different compositions and sizes. Even individual glycoprotein molecules may contain numerous types of oligosaccharides (*6, 7*). The proteoglycans will also be briefly considered. These materials are the well-known mucopolysaccharides bound to a protein core.

In the elucidation of the structures of the glycoproteins and proteoglycans, their behavior with respect to alkalies has played an important part. Although misinterpretations were made, the proper use of alkalies has given important information. The purpose of this paper is to show these applications and the problems involved.

Table I. Original Sources of Sialic Acids

Sialic Acid (crystalline):	$C_{14}H_{24}O_{11}N$	Blix (8, 9)
By mild acid hydrolysis of BSM	$C_{13}H_{23}O_{11}N$	Blix et al. (10)
Methoxyneuraminic Acid (crystalline): By acid methanolysis of brain glycolipids (and later BSM)	$C_{11}H_{21}O_9N$	Klenk (11)
	$C_{12}H_{21}O_{10}N$	Klenk and Lauenstein (12)
"Split Product": From glycoproteins treated with *Vibrio cholerae* enzyme or influenza virus		Gottschalk and Lind (13)
		Gottschalk (14)
Lactaminic Acid (crystalline methyl ester): From cow colostrum	$C_{11}H_{19}NO_9$	Kuhn and Brossmer (15)
	$C_{12}H_{21}NO_9$	Heimer and Meyer (16)

The Sialic Acids

The Ehrlich reagent, developed originally for the colorimetric assay of pyrrole derivatives, was shown by Ehrlich to give a color with certain glycoproteins, with and without prior alkali treatment (*3*). The reagent consists of dimethylaminobenzaldehyde in strong hydrochloric acid. This assay later became the basis for the Morgan–Elson reaction for the determination of *N*-acetylhexosamines (after a prior treatment with alkali) and of the Elson–Morgan reaction for hexosamines (after treatment with acetylacetone and alkali). The reaction without pretreatment with alkali became known as the direct Ehrlich reaction and that involving alkali as the indirect Ehrlich reaction.

Blix (*8, 9, 10*) after mild acid hydrolysis of bovine submaxillary glycoprotein isolated a crystalline material called sialic acid which gave the direct and indirect Ehrlich reactions. Table I shows that similar substances were isolated by other workers from glycolipids (*11, 12*), other glycoproteins (*13, 14*), and cow colostrum (*15, 16*). All gave direct and indirect reactions, whereas *N*-acetylhexosamines did not give the direct reaction. The compounds were finally recognized as structurally related and, by agreement, were called sialic acids, whereas the individual acids were named as substituted neuraminic acids—*e.g.*, *N*-acetylneuraminic acid (*17*). The known sialic acids, all of which occur in bovine submaxillary glycoprotein, are shown in Figure 2 (*18*). They differ in the number of *O*-acetyl groups and in having either *N*-acetyl and/or *N*-glycolyl groups.

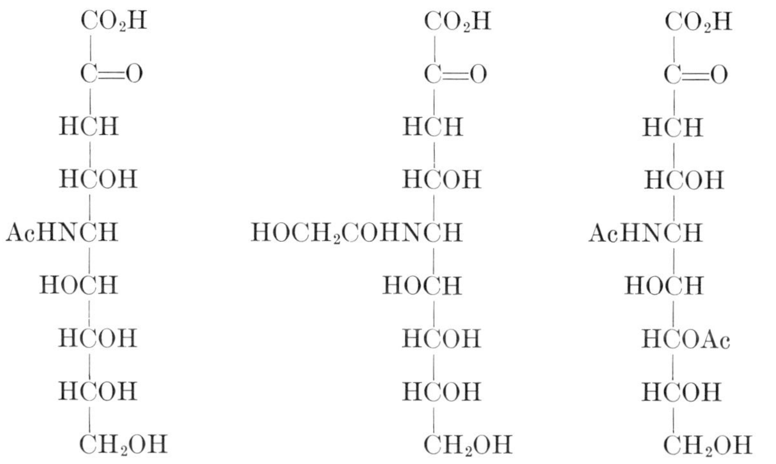

N-Acetylneuraminic acid | *N*-Glycolylneuraminic acid | *N*-Acetyl-*O*-acetylneuraminic acid

In addition:
BSM: *N*-Acetyl-di-*O*-acetyl (C-7, 8 or 9) and another *N*-acetyl-*O*-acetyl.

Figure 2. Sialic acids of BSM

In the early work the crystalline sialic acids were known to be reducing sugars, had carbonyl groups, and gave indirect and direct Ehrlich reactions. The empirical formulas were relatively simple, and in retrospect the long delay in the establishment of the structures of these compounds seems surprising. Blix explained this delay on his first choice of the bovine submaxillary mucin as the source of the sialic acids (*19*). The ovine glycoprotein contains considerably fewer *O*-acetyl groups. The variable empirical analyses obtained upon different preparations (Table I) arose from the uncontrolled partial removal of *O*-acetyl groups during the purification processes which involved exposure to alkali, and the empirical formulas could not be consistently reproduced.

The two earliest structures (*14, 20*) shown in Figure 3 were based on the reactivity of these compounds to the Ehrlich reagent. The pyrrole-2-carboxylic acid of the second formula was presumed to be attached to protein by the hydroxyl group at the 4-position of the pyrrole ring.

Further developments are shown in Figure 4. On the basis that glucosamine reacted with pyruvic acid in the presence of alkali to yield pyrrole-2-carboxylic acid, in 1% yield, Gottschalk (*21*) proposed that sialic acid was formed by an aldol condensation reaction between *N*-acetylglucosamine and pyruvic acid. Kuhn and Brossmer (*15*) and Zilliken and Glick (*22*) showed that the reverse reaction also took place under alkaline conditions. Cornforth, Firth, and Gottschalk (*23*) synthesized crystalline *N*-acetylneuraminic acid (NANA) from *N*-acetylglucosamine and oxaloacetic acid (pH 11, 20°C). Under conditions less subject to misinterpretation, Heimer and Meyer (*24*) found that *Vibrio cholerae* enzymes cleaved NANA into an *N*-acetylhexosamine and pyruvic acid.

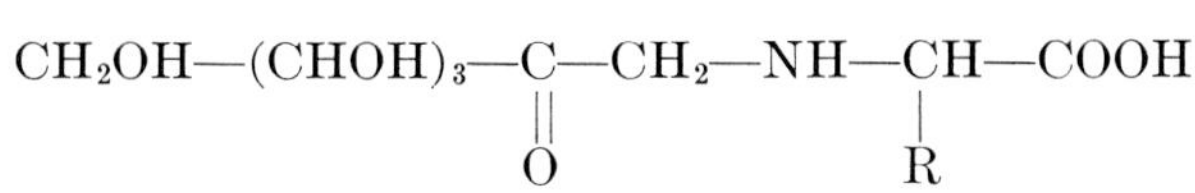

N-1-lysyl (?)-1-deoxyfructose gives indirect Ehrlich reaction

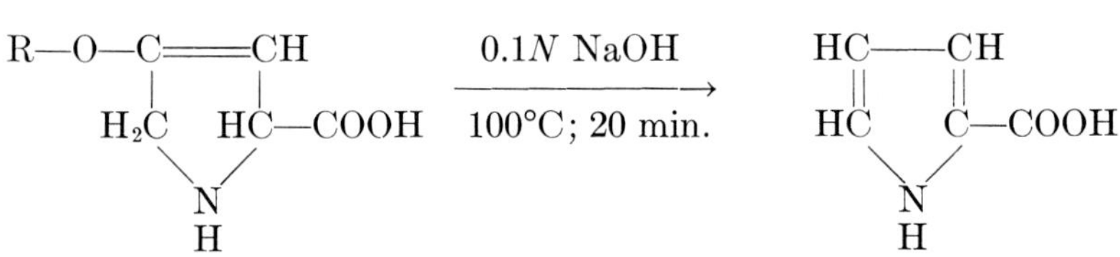

R = protein chain (Thr or Ser residue ?)

Figure 3. Early structures for sialic acids

The aldol condensation of the hexosamine and pyruvic acid was shown to have been preceded by a 2-epimerization of *N*-acetylglucosamine to *N*-acetylmannosamine when Roseman and Comb (*25*) identified the *N*-acetylmannosamine and showed that it reacted with pyruvic acid with NANA formation.

The final step in the structural work (*26*) was the determination of the configuration of the OH group at C-4 as trans to the acetamido group at C-5, correcting an earlier allocation of a cis configuration.

Gottschalk (*21*):

$$\text{D-glucosamine} + \text{pyruvic acid} \xrightarrow[\text{OH}^-]{\text{heat}} \text{pyrrole-2-carboxylic acid}$$

Zilliken and Glick (*22*); Kuhn and Brossmer (*15*):

$$\text{NANA} \xrightarrow{\text{OH}^-} N\text{-acetylglucosamine} + \text{pyruvic acid}$$
$$(+ \text{ degradation products})$$

Cornforth, Firth and Gottschalk (*23*):

$$N\text{-Acetylglucosamine} + \text{oxaloacetic acid} \xrightarrow[20°\text{C}]{\text{pH 11}} \text{NANA}$$

Heimer and Meyer (*24*):

$$\text{NANA(Me ester)} \xrightarrow[\text{enzyme}]{V.\ cholerae} N\text{-acetylhexosamine} + \text{pyruvic acid}$$

Roseman and Comb (*25*):

$$\text{NANA} \underset{\substack{\text{enzyme} \\ \text{pH 7.1}}}{\overset{Clos.\ perfringens}{\longleftrightarrow}} \text{pyruvic acid} + N\text{-acetylmannosamine}$$

Kuhn and Brossmer (*26*):

Unambiguous synthesis shows C-4 OH on right

Figure 4. Synthesis of N-*acetylneuraminic acid (NANA)*

Glycosyl Ester Linkages in Glycoproteins

A major contribution to the current concept of the structures of glycoproteins was their establishment as a protein core to which are attached many oligosaccharides (or small polysaccharides). This type of structure was first shown by Gottschalk and associates for bovine and ovine submaxillary glycoproteins (*27, 28*). This structure was made probable by treatment of bovine and ovine glycoproteins (BSM and OSM) with 0.01*N* $Ba(OH)_2$ at 80°C for 10–15 min. After neutralization and dialysis, the dialyzable materials were 6-*O*-(*N*-acetylneuraminyl)-2-acetamido-2-deoxy-D-galactose, and chromogens formed from the reducing galactosamine unit. The total recovery was about 80% of the sugar units initially present.

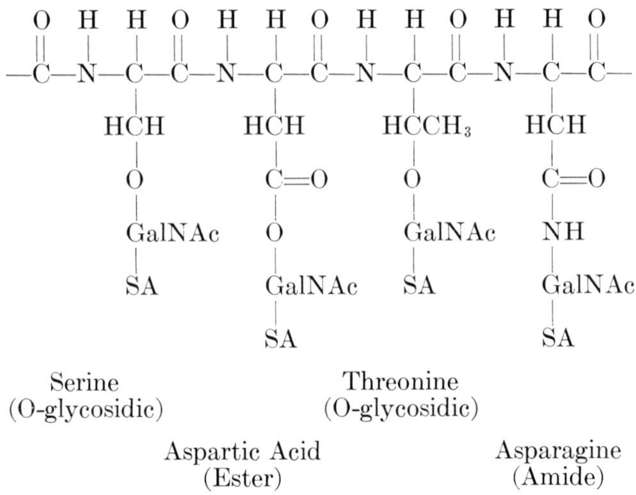

Figure 5. Possible linkages of oligosaccharides to the protein of mucus glycoproteins (SA, sialic acid; GalNAc, 2-acetamido-2-deoxy-D-galactose)

On the basis of the amino acids known to be present in substantial amounts in these glycoproteins, the only functional groups in the core amino acid chain that could form glycosidic linkages were carboxyl groups (esters or amides) and hydroxyl groups. The possible linkages are shown in Figure 5 (*29*). Recently, S-galactosyl peptides of cysteine have been found in urine (*30*), but S-amino acids are absent or present in only small amounts in mucus glycoproteins.

The removal of the oligosaccharide units by treatment with alkali was ascribed by Gottschalk (*31*) to labile glycosidic ester linkages between the carboxyl groups of dicarboxylic acids in the protein core and the *N*-acetylhexosamine units of the oligosaccharides. Later hydroxylamine was said to form hydroxamates. Reductions by lithium borohydride, which reduces esters but not amides or free carboxyl groups, gave small yields of the amino alcohols corresponding to glutamic and aspartic acids (*32, 33*).

Hashimoto and Pigman (*34*) pointed out that the number of dicarboxylic acids was grossly insufficient to account for *O*-esters as the principal glycosidic linkage, whereas the hydroxyamino acids were present in about equimolar proportions to the hexosamine and sialic acid units. As a result, an *O*-glycosyl linkage to the OH groups of serine and threonine was proposed as the principal linkage. Harbon *et al.* (*35*) also reported that hydroxamates were not formed and that the action of hydroxylamine was the same as that of sodium hydroxide under the same conditions of pH. The absence of ester linkages ($<5\%$) was shown finally by Bertolini and Pigman (*36*) who applied the quantitative hydroxylamine reaction of Hestrin to the bovine and ovine submaxillary glycoproteins (BSM and OSM) and other products and found that the only esters present were the *O*-acetyl groups of the sialic acids. Hydroxamate esters were formed but were dialyzable and behaved chromatographically as acetohydroxamic acid.

The proposal of the ester linkage was later withdrawn when it was found that the decomposition products of lithium borohydride (presumably dimethoxyborane) in acidic anhydrous methanol can reduce free carboxyl groups (*37*).

Glycosidic Linkages to the Hydroxyl Groups of Serine and Threonine Residues

On the basis of the close correlation between the number of hydroxyamino acids in bovine submaxillary mucin and the number of hexosamine and sialic acid residues, Hashimoto and Pigman (*34*) proposed that an *O*-glycosidic linkage was the only one that could be present in large amounts. Anderson *et al.* (*38*) reported a disappearance of much of the serine after incubation of the proteoglycan of chondroitin sulfate with alkali and suggested that this loss resulted from a β-elimination reaction.

Direct proof for the existence of *O*-glycosidic linkages involving the hydroxyl groups of serine and threonine was provided simultaneously from three laboratories. Anderson *et al.* (*39*) reported partial losses of serine and threonine after treatment with 0.5*N* NaOH (or 0.45*N* KOH) at 4°C or room temperature for about 20 hours. Subsequent reduction with platinum showed formation of some alanine and α-aminobutyric acid. Harbon *et al.* (*40*) treated ovine submaxillary glycoprotein at pH 12.8 for 45 minutes at 70°C. The serine and threonine content decreased by 78 and 60%, respectively. Treatment of this product with 0.1*M* sulfite, (pH 9, 24 hours, room temperature) caused a conversion of the dehydroserine residues to cysteic acid, but had no action on the dehydrothreonine residues. This reaction has been further studied by Simpson *et al.* (*41*).

The best procedure for most purposes was the treatment with alkali in the presence of sodium borohydride carried out on BSM (*42*); the dehydroserine linkages were converted to alanine. Since this procedure did not reduce all dehydrothreonine residues, the procedure was modified by a final reduction in the presence of palladium chloride and borohydride (*43, 44*). The reactions involved are shown in Figure 6.

Figure 7 shows that the best conditions for elimination of side chains from BSM and OSM are probably 0.3*M* sodium borohydride in 0.1*N* NaOH at 45°C for about 10–12 hours. Over 90% of the total hexosamine and sialic acid were dialyzable under these conditions (*36*). The reduction of dehydroserine and dehydrothreonine residues was not studied,

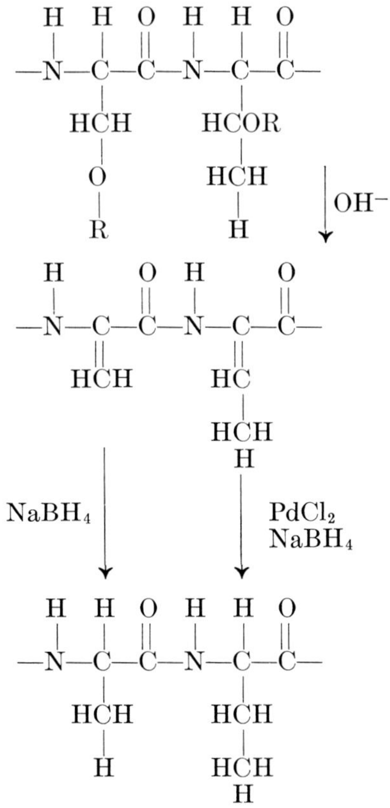

Figure 6. The β-elimination reaction applied to BSM and OSM in the presence of sodium borohydride. (R-glycosyl radical.)

but under similar conditions Downs and Pigman (*44*) found high recoveries as alanine and α-aminobutyric acid. With porcine submaxillary glycoprotein (PSM), the use of 0.6*M* sodium borohydride, 0.2*N* NaOH at 45°C for 6 hours, and later palladium reduction gave a 90% recovery of the serine which was lost and recovered as alanine; the corresponding recovery of threonine as α-aminobutyric acid was 75%. Some of the serine (13%) of the original PSM and threonine (16%) remained intact and presumably was originally free of side chains (*45*).

An extensive study of the alkaline–borohydride reactions was made by Weber and Winzler (*46*) of a number of different glycoproteins. The optimal conditions were suggested as: 0.4*M* sodium borohydride, 0.1 to 0.2*N* NaOH at 50°C for 16–32 hours. As will be shown later, their quantitative interpretations cannot be considered as valid, but the conditions are similar to those proposed by Bertolini and Pigman (*36*).

The optimal conditions for the β-elimination of *O*-glycosyl groups from the peptide core of glycoproteins and proteoglycans have not been well established. Such a study should involve quantitative determination of the loss of serine and threonine, recovery of alanine and α-aminobutyric acid, and recovery of reduced sugars originally attached to the hydroxyamino acids. The conditions may depend upon the material under study especially the glycoproteins in contrast to the proteoglycans. The

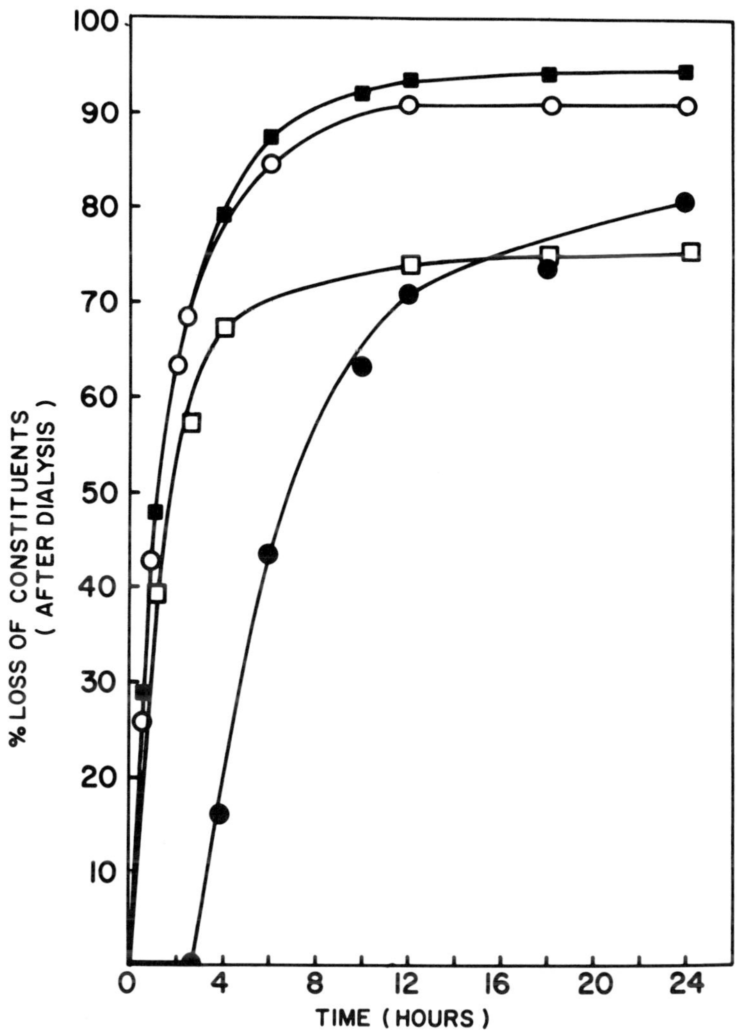

Figure 7. The rate of cleavage of the main components of bovine submaxillary mucin by alkaline sodium borohydride at 45°C. ○——○, *N-acetylneuraminic acid;* ■——■, *N-acetylhexosamine;* ●——●, *protein;* □——□, *N-acetylhexosamine (before dialysis).*

conditions for the latter types have been studied by Meyer, Danishefsky, and Rodén and associates (*38, 39, 47, 48*).

At the present time, the conditions used by Bertolini and Pigman (*36*) and the reductions as described by Downs and Pigman (*44*) seem to offer the most general utility.

Recovery of Oligosaccharide Chains

The β-elimination reaction as carried out within alkaline borohydride solutions provides a method of obtaining the oligosaccharide side chains, theoretically intact except for reduction of the "head" reducing group to an alditol.

For example, by alkaline β-elimination of BSM and OSM, Gottschalk and co-workers (*27, 28*) had obtained in addition to intact disaccharide chains considerable amounts of "chromogens" in which the reducing *N*-acetylgalactosamine had been converted to the anhydro derivative. Using the conditions of Bertolini and Pigman (*36*) involving both sodium hydroxide and borohydride, Murty and Horowitz (*49*) obtained about 80% yields of 6-*O*-(*N*-acetylneuraminyl)-2-acetamido-2-deoxy-D-galactitol.

As shown in Figure 7, Bertolini and Pigman found that after treatment of OSM and BSM with alkaline borohydride over 90% of the hexosamine and sialic acid became dialyzable but that 22 and 10%, respectively, of unreduced hexosamine was released after acid hydrolysis of the dialyzable fraction. The unreduced bound hexosamine was shown to be a part of tri- and tetrasaccharides containing several hexosamine units, some carrying free amino groups in the original BSM and OSM. BSM contained greater amounts of such oligosaccharides than OSM. Also, some of the BSM oligosaccharides had glucosamine units instead of galactosamine units.

Carlson (*6*) had shown earlier that porcine submaxillary glycoproteins (PSM) produced five reduced oligosaccharides after alkaline borohydride treatment (conditions: 0.05*M* KOH, *M* $NaBH_4$ at 45°C for 15 hours); removal of more than 90% of the side chains was reported. The oligosaccharides were separated and varied in size from mono to pentasaccharides. Katzman and Eylar (*50*) found similar but not identical oligosaccharides to be present in PSM, prepared by a different method.

Although incidently, the studies of the oligosaccharides present in BSM, OSM, and PSM indicated that the oligosaccharide chains in a single mucus glycoprotein molecule may vary in composition. This unexpected situation was later amplified by a study of porcine submaxillary glycoproteins separated first by human H(O), A, and ("-") blood-group types and then analyzed individually (*51, 52*). This microheterogeneity of oligosaccharides in glycoproteins has also been pointed out by Huang, Mayer, and Montgomery for ovalbumin (*53*).

Morgan (*54*) first showed the liberation of dialyzable oligosaccharide fragments from human blood-group A, B, and O substances by the action of alkali; these gave the direct Ehrlich reaction. Kabat and associates (*55, 56, 57, 58*) used sodium borohydride and alkali to reduce the amount of chromogen formation and obtained 60–80% of dialyzable oligosaccharides. Some of them had a hexosaminitol at the reducing end, but some contained a terminal 3-hexenetetrol, presumably derived by a secondary β-elimination of a sugar unit involved in a $1 \rightarrow 3$ glycosidic linkage to the sugar unit attached to the protein core. Since the original oligosaccharide chains were believed to be considerably larger than those isolated, these results suggested a series of secondary β-eliminations, the so-called peeling reaction. The conditions used were 0.1*N* NaOH, 0.26*M* $NaBH_4$ at room temperature for several days. Morgan and co-workers (*59, 60*) found similar oligosaccharides using 2.5% trimethylamine in 50% methanol at 50°C using blood-group substances of several types.

Mayo and Carlson (*61*) studied the action of alkaline borohydride on the model disaccharide, chondrosine:

O-β-D-glucopyranosyluronic acid-(1→3)-
2-acetamido-2-deoxygalactopyranose

Under the experimental conditions used earlier for blood-group substances (0.2*M* NaOH, 0.26*M* $NaBH_4$, room temperature, but for only 40 min) about 35% cleavage of the $1 \rightarrow 3$ bond occurred. With *M* $NaBH_4$ and 0.05*M* NaOH at 50°C for several minutes, complete reduction of the hexosamine residue occurred without cleavage of the disaccharide.

Mechanisms of β-Elimination of O-Glycosyl-Seryl and -Threonyl Linkages

For the *O*-glycosyl (Gl-O) derivatives of serine (I) and threonine (II), the structures are:

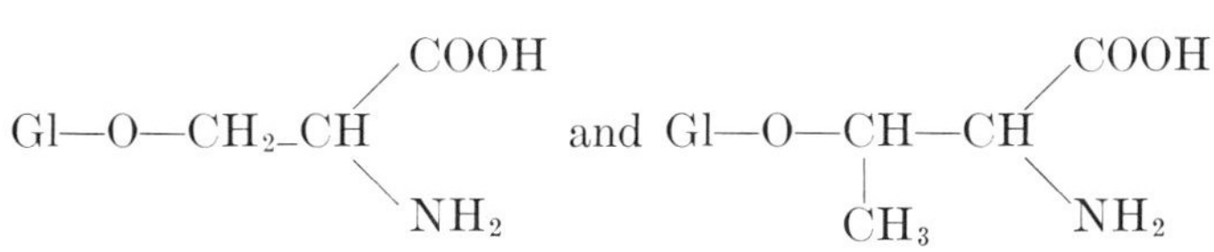

I II

These derivatives of 2-acetamido-2-deoxy-β-D-glucose were stable to 0.1*M* NaOH, 0.3*M* $NaBH_4$ at 20°C for 24 hours, but the corresponding amides were partially cleaved (*62*).

The lability to alkali (without sodium borohydride) of several substituted serine β-D-glucopyranosides was studied by Derevitskaya *et al.* (*63*) and some of the results are reported in Table II. The serine gly-

Table II. Effect of Substitution on β Elimination of Model Glucopyranosylserines (*63*)

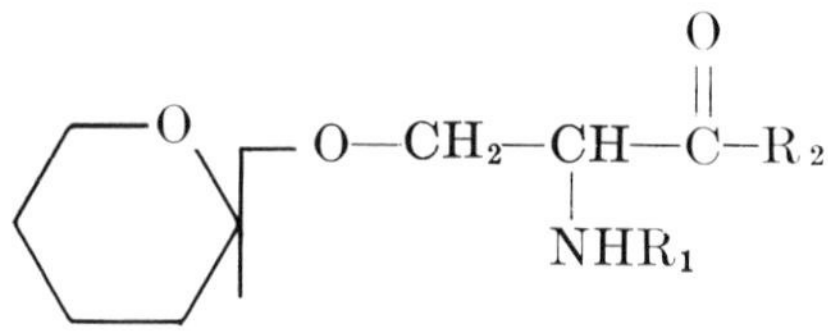

	R_1	R_2	*% Degradation* [a]
III	$-\overset{O}{\overset{\|}{C}}OCH_2Ph$	$-NHCH_3$	95
IV	$-\overset{O}{\overset{\|}{C}}OCH_2Ph$	$-OCH_3$	15
V	$-\overset{O}{\overset{\|}{C}}OCH_2Ph$	$-OH$	0
III	$-\overset{O}{\overset{\|}{C}}OCH_2Ph$	$-NHCH_3$	95
VI	$-\overset{O}{\overset{\|}{C}}CH_2NHCOCH_2Ph$	$-NHCH_3$	62
VII	$-\overset{O}{\overset{\|}{C}}CH_2NH_2$	$-NHCH_3$	0
VIII	$-H$	$-NHCH_3$	0
IX	$-H$	$-OH$	0 [b]

[a] Conditions used: pH 11, 37°, 24 hours.
[b] 0.1 *N* NaOH, 8 days.

coside (I) was stable as was its methylamide (VIII). Attachment of a glycyl group to the amino group of the sẹryl methylamide (VII) also produced a stable glycosidic linkage. Carbobenzoxy derivatives of the original compound or of its *N*-glycyl derivative were labile to alkali when the carboxyl group was present as the methylamide (III and VI). However, the corresponding methyl ester (IV) and the free carboxylate ion (V) were relatively stable, the former presumably as a result of prior alkaline hydrolysis.

In Figure 8 the stability of several glycosyl di- and tripeptides is shown (*47, 64*). The wavy line indicates cleavage by alkaline borohydride. For these compounds the existence of a free carboxyl or a free amino group in an *O*-glycosyl-hydroxyamino acid prevents β-elimination.

Vercellotti *et al.* (*65*) studied the rates of β-elimination for *O*-benzyl ethers of *O*-glycosyl-*N*-(2,4-dinitrophenyl)-L-serine and L-threonine methyl esters in non-aqueous solvents using various bases. The corresponding acetyl derivatives of 2-acetamido-2-deoxy-D-galactose were also included. Generally, these substances were highly reactive, but the nature of the solvent and the base type were influencing factors. Thus, no elimination occurred in benzene when trimethylamine was used as base.

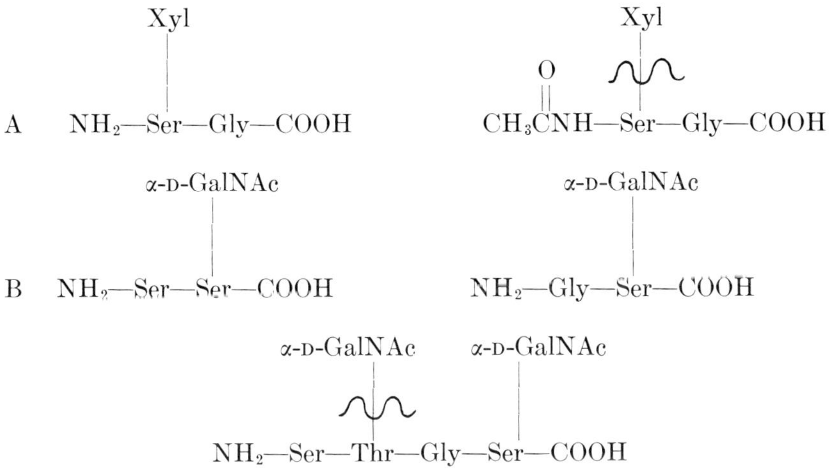

Figure 8. Susceptibility to alkaline β-elimination
A is from References 47 *and* 48; *B is from Reference* 64

Isbell (*66*) showed that many reactions of carbohydrates in the presence of bases could be explained by the formation of an enolic anion through removal of the protons at the α-position; the ion is stabilized by resonance between the two forms:

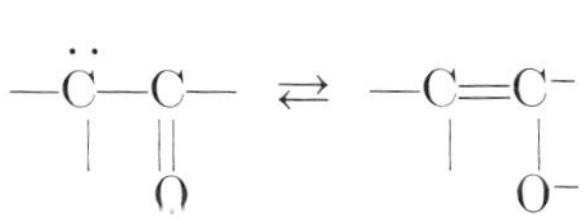

The presence of such enolic ions in alkaline solutions of sugars was shown later by Isbell *et al.* (*67*). The mechanism of the alkaline β-elimination reaction of substituted serine and threonine glycosidases based on this concept is shown in Figure 9.

The differences in alkaline lability of compounds I–IX results from the varying stability of their enolic intermediate. The alkaline stability of compounds I, II, and V results from resonance stabilization of the carboxylate ion. Esters and amides (R_2) do not show such resonance, and the formation of the intermediate enolic ion will result in alkaline lability.

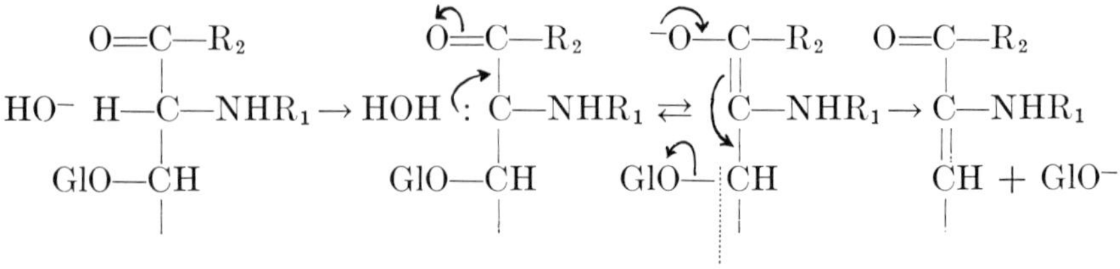

(GlO is the glycosyloxy group)

Figure 9. Mechanism of β-elimination of substituted serine and threonine glycosides

However, when a free amino group is present (R_1) in compounds VIII and IX, the unpaired electrons of the amino nitrogen atom may influence the electronegativity of the carbonyl carbon atom and thereby decrease its tendency to withdraw electrons from the adjacent α-carbon atom:

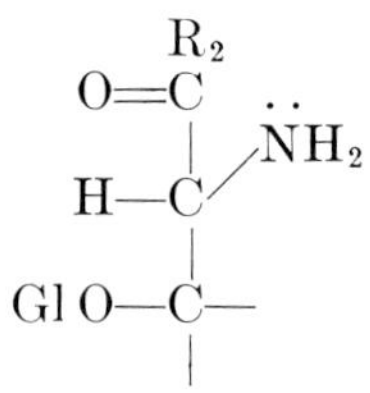

An *N*-acetyl group, however, would prevent this effect (Reaction A, Figure 8).

If R_2 is an ester or an amide group, the release of electrons from the enolic ion to the glycosyloxy linkage will be favored; also, resonance stabilization of the carboxylate ion would be eliminated.

These considerations indicate that the nature of R_1 and R_2 is an important factor in determining the alkaline lability of the *O*-glycosidic bond. β-Elimination usually occurs most readily when there are no free amino or carboxyl groups on the serine or threonine residue to which the glycosyl is attached. When both functional groups are unsubstituted β-elimination does not occur (I, II, IX). The *N*-2,4-dinitrophenyl derivatives of Vercellotti *et al.* represent an extreme case, possibly involving a different mechanism of β-elimination.

Dehydrothreonine Residue

Only minor differences have been observed in the rate of β-elimination between *O*-glycosylserine and *O*-glycosylthreonine residues (*42*).

Figure 6 shows that the dehydrothreonine residue is a derivative of 2-amino-2-butenoic acid. This residue is not easily reduced by sodium borohydride, and complete reduction requires a palladium catalyst (*43*, *44*). Also, the peptide bond involving the amino group of the dehydrothreonine residue is not hydrolyzed by dilute acids whereas that of the dehydroserine residue is labile under the same conditions (*44*), as will be discussed below.

These results suggest that the double bond shifts to the three position during β-elimination reaction:

$$\overset{|}{\underset{|}{C}}{=}CHCH_3 \rightleftarrows H\overset{|}{\underset{|}{C}}{-}CH{=}CH_2.$$

If a true equilibrium exists between the two forms, the shift must be very slow since extended times of reduction by borohydride do not improve the yield significantly.

A deactivation by the methyl group seems unlikely since compounds of the type $R_1R_2C = C(COOH)_2$ (R_1 and R_2, H or Me) were shown by Kadin (*67*) to exhibit no such effect and are reducible by sodium borohydride. However, the propriety of comparing these compounds with dehydrothreonine residues is questionable.

Further Applications of the β-Elimination Reaction

As described earlier, the alkaline β-elimination with or without sodium borohydride, has several valuable applications. First, it can be used to measure (within 5–10%) the number of *O*-glycosidic linkages to serine and threonine residues in glycoproteins. *O*-Glycosidic linkages to hydroxylysine and hydroxyproline are, in contrast, stable to treatment with alkalies (*29*). The other principal glycosidic linkage in glycoproteins, *N*-glycosylasparagine, is relatively stable to alkali but may be demonstrated semiquantitatively in alkaline borohydride (0.2*M* NaOH, *M* $NaBH_4$ at 100°C) by the formation of aspartic acid and corresponding amounts (60%) of reduced oligosaccharides (*69*).

A second major application is for the isolation of oligosaccharide and polysaccharide units attached by *O*-glycosidic linkages to serine and threonine residues. It seems that high concentrations of sodium borohydride relative to alkali are required for reduction of the reducing unit of the sugar chain freed by the β-elimination reaction and minimization of secondary reactions of alkali at this terminal reducing group. For the human blood-group substances and proteoglycans, further work may be required to establish optimal conditions for β-elimination, which may differ somewhat with each particular substance. If $1 \rightarrow 3$ glycosidic link ages are present in the sugar chain, especially between the reducing terminal group and the sugar unit in the second position, even stricter control of the conditions seems necessary to prevent β-elimination involving the second sugar residue.

Another important application is that β-elimination provides a new method for the selective cleavage of proteins and peptides into smaller peptides. This procedure may be used for the determination of the primary structure of the protein core of glycoproteins. It may also be used to establish the positions of the oligosaccharide chains on the protein core. This method involves β-elimination by alkali followed by acid hydrolysis at pH 2.2, 100°C for 1 hour. These hydrolytic conditions were first shown by Patchornik and Sokolovsky (*70*) for peptides containing dehydroserine. The cleavage removes the α-amino group of the unsaturated amino acid residue. The resulting product has pyruvic acid at the amino terminal end, whereas the carboxyl group at the cleavage point carries the amino group as the amide.

This approach was used qualitatively by Kotchetkov *et al.* (*71*), who showed that the blood-group A substance from pig stomach linings after treatment with alkali and subsequent acid hydrolysis was cleaved into smaller fragments which still carried the carbohydrate side chains.

Harbon *et al.* (*72*) hydrolyzed the alkaline β-elimination product from OSM with 3*N* HCl (90 min at 100°C). Pyruvic acid was found in the hydrolyzates to the extent of 83% of the serine lost, and 2-ketobutyric acid accounts for 63% of the threonine which had disappeared. Niederhiser *et al.* (*73*) reported quantitative conversions to these keto acids by alkaline treatment of bile mucins and showed that the elimination can be followed by an increase in the absorption at 240 nm.

Downs and Pigman (*44*) carried out this procedure (Figure 10) as quantitatively as possible for BSM and separated the products of mild acid hydrolysis on Sephadex G-25. (G-50 provides better separation.) The major fraction comprising more than 40% of the protein core, contained pyruvic acid as the amino terminal unit. The presence of pyruvic acid in this position was shown by conversion of the peptide to the phenylhydrazone, which was reduced to alanine by using sodium borohydride and palladium catalyst. Simultaneously, the dehydrothreonine units were quantitatively reduced to α-aminobutyric acid residues. These reactions are shown in Figure 10. Dansylation of the reduced peptide followed by acid hydrolysis showed alanine to be the sole amino terminal residue. The molecular weight of the reduced peptide was found to be 3000 by equilibrium ultracentrifugation. The peptide still contained one disaccharide residue, as the β-elimination was carried out under conditions milder than those reported as optimal in this laboratory to minimize alkaline cleavage of the protein core. In unpublished work Downs and Pig-

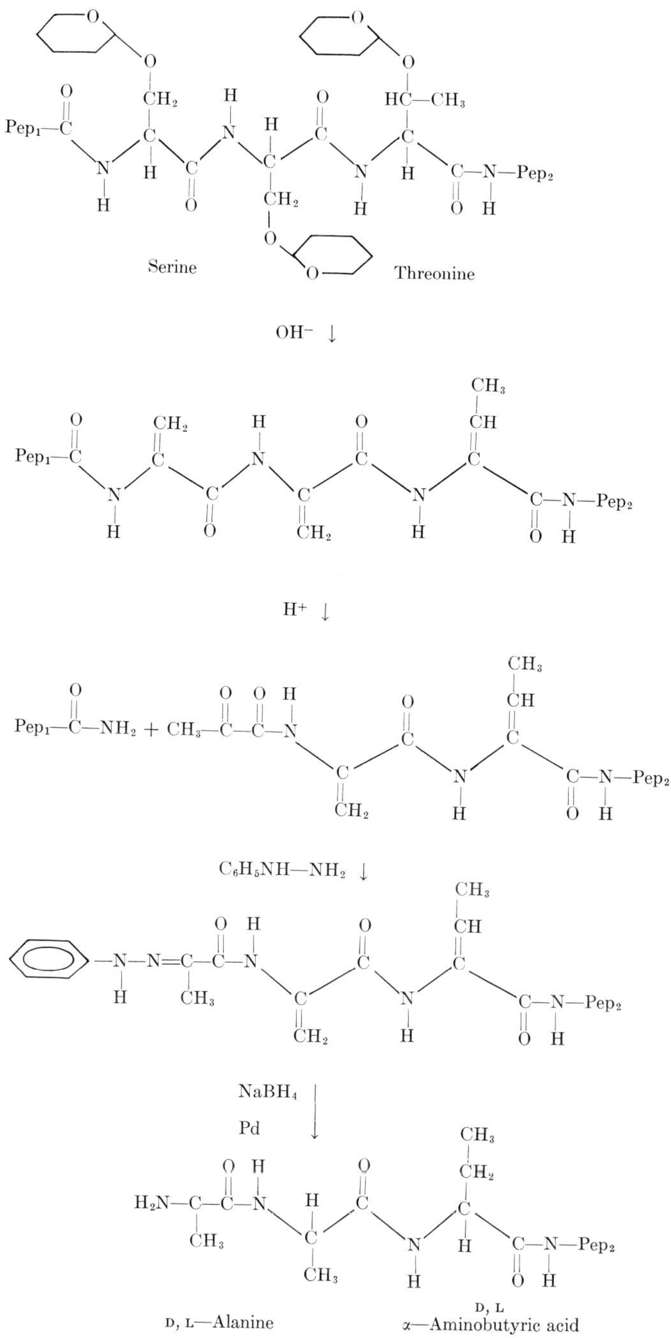

Figure 10. Conversion of BSM to glycopeptides by β-elimination followed by mild acid hydrolysis and hydrogenation

man found that OSM gives three peptide fractions after separation on Sephadex G-25, which were closely similar to those reported for BSM.

Under alkaline conditions similar to those used above for BSM, some arginine residues are converted to ornithine which under the usual conditions for amino acid analysis appears as apparent lysine (*74*).

To compare the composition of the principal peptide with that of the original BSM on a molar basis, proline was selected as 3.0, and the other components were expressed as molar ratios. On this basis, the calculated molecular weight was 2800 for the main glycopeptide. The composition of the reduced glycopeptide on this basis is given in Table III in the third column. In the next column the composition is given after correction for the loss of arginine, serine, and threonine, as discussed earlier. The composition of the corrected peptide is closely similar to that of the original BSM on a molar ratio basis, the principal exception being the loss of two serine residues. These were present as dehydroserine residues in the original β-eliminated BSM after mild acid hydrolysis and reaction with 2,4-dinitrophenylhydrazine using the conditions of Harbon *et al.* (*72*); most of the missing serine residues were recovered as free pyruvic acid in the eluates from the Sephadex G-50 column (*74*).

Table III also contains data for a glycopeptide obtained after repeated trypsin digestion of desialyzed BSM, fractionation on Sephadex

Table III. Molar Composition of Core Protein (Peptide) of BSM, Trypsin and β-Elim. Glycopeptides (Pro = 3)

AA	*BSM*	*β-Elim Glycopeptide (B) (reduced)*	*β-Elim Glycopeptide (B) (corrected)*	*Trypsin Glycopeptide (B)*
Ser	5.4	1.8	3.5[a,c]	6.2
Thr	3.9	1.0	3.7[a]	4.2
Gly	4.9	5.2	5.2	5.2
Ala	3.4	5.1	3.4[a]	3.4
Glu	1.8	2.0	2.0	2.1
Asp	0.7	0.6	0.6	0.35
Val	1.9	2.0	2.0	1.7
Leu	1.1	1.2	1.2	1.3
Ile	0.45	0.4[e]	0.4[e]	0.3
Pro	(3)	(3)	(3)	(3)
Arg	1.1	0.8	1.0[b]	1.1
α-Aminobutyric	—	2.7	—	—
Total	27.7		25.8[c]	28.8
MW (calc)	(2540)		2350 (2540)[d]	2640

[a] Corrected for conversion of serine to alanine and threonine to α-aminobutyric acid.
[b] Corrected for conversion of arginine to ornithine under these conditions.
[c] Two dehydroserine units were lost in the acid hydrolysis step.
[d] Corrected for loss of two dehydroserine units.
[e] The value of 0.9 in the original paper should have been 0.4.

G-50, and purification on Dowex 50. Again, the main fraction B corresponded to more than 40% of the original protein. Its molar composition is given in Table III on the basis of proline as 3.0 (also Arg, 1), and its molecular weight corresponded to this composition (75).

The two major peptides obtained from BSM by the two different processes (β-elimination and trypsin digestion, respectively) show amino acid compositions that closely resemble that of the original BSM, except for some deviation in the serine content.

Literature Cited

1. Pigman, W., Anet, E. F. L. J., "The Carbohydrates," p. 165, Vol. IA, W. Pigman and D. Horton, Eds., Academic, New York, 1972.
2. Overend, W. G., "The Carbohydrates," p. 332, Vol. IA, W. Pigman and D. Horton, Eds., Academic, New York, 1972.
3. Aminoff, D., "The Carbohydrates," p. 740, Vol. IIB, W. Pigman and D. Horton, Eds., Academic, New York, 1970.
4. Pigman, W., Moschera, J., "Biology of the Cervix," Chapt. 9, R. J. Blandau, Ed., 1972.
5. Montgomery, R., "The Carbohydrates," p. 628, Vol. IIB, W. Pigman and D. Horton, Eds., Academic, New York, 1970.
6. Carlson, D. M., *J. Biol. Chem.* (1968) **243,** 616.
7. Bertolini, M., Pigman, W., *Carbohyd. Res.* (1970) **14,** 53.
8. Blix, G., *Hoppe Seyler's Z. Physiol. Chem.* (1936) **240,** 43.
9. Blix, G., *Skand. Arch. Physiol.* (1938) **80,** 46.
10. Blix, G., Lindberg, E., Odin, L., Werner, I., *Nature* (1955) **175,** 340; *Acta Soc. Med. Upsaliensis* (1956) **61,** 1.
11. Klenk, E., *Hoppe-Seyler's Z. Physiol. Chem..* (1941) **268,** 50; (1942) **273,** 76.
12. Klenk, E., Lauenstein, K., *Hoppe-Seyler's Z. Physiol. Chem.* (1952) **291,** 147.
13. Gottschalk, A., Lind, P. E., *Nature* (1949) **164,** 232.
14. Gottschalk, A., *Nature* (1951) **167,** 845.
15. Kuhn, R., Brossmer, R., *Ber.* (1956) **89,** 2471.
16. Heimer, R., Meyer, K., *Biochim. Biophys. Acta* (1958) **27,** 490.
17. Blix, F. G., Gottschalk, A., Klenk, E., *Nature* (1957) **179,** 1088.
18. Blix, G., Lindberg, E., *Acta Chem. Scand.* (1960) **14,** 1809.
19. Blix, G., personal communication.
20. Gottschalk, A., *Nature* (1954) **174,** 652.
21. Gottschalk, A., *Biochem. J.* (1955) **61,** 298; *Nature* (1955) **176,** 881; *Arch. Biochem. Biophys.* (1957) **69,** 37.
22. Zilliken, F., Glick, M. C., *Naturwissenschaften* (1956) **43,** 536.
23. Cornforth, J. W., Firth, M. E., Gottschalk, A., *Biochem. J.* (1958) **68,** 57.
24. Heimer, R., Meyer, K., *Proc. Nat. Acad. Sci.* (1956) **42,** 728.
25. Roseman, S., Comb, D. G., *J. Amer. Chem. Soc.* (1958) **80,** 3166; Comb, D. G., Roseman, S., *ibid.* (1958) **80,** 497.
26. Kuhn, R., Brossmer, R., *Angew. Chem.* (1962) **1,** 218.
27. Gottschalk, A., Graham, E. R. B., *Biochem. Biophys. Acta* (1959) **34,** 380.
28. Graham, E. R. B., Gottschalk, A., *Biochim. Biophys. Acta* (1960) **38,** 513.
29. Spiro, R. G., *Ann. Rev. Biochem.* (1970) **39,** 608.
30. Lote, C. J., Weiss, J. B., *FEBS Letters* (1971) **16,** 81.
31. Gottschalk, A., *Nature* (1960) **186,** 949.

32. Gottschalk, A., Murphy, W. H., Graham, E. R. B. *Nature* (1962) **194,** 1051.
33. Gottschalk, A., Murphy, W. H., *Biochim. Biophys. Acta* (1961) **46,** 81.
34. Hashimoto, Y., Pigman, W., *Ann. N. Y. Acad. Sci.* (1962) **93,** 541.
35. Harbon, S., Herman-Boussier, G., Clauser, H., *Bull. Soc. Chim. Biol.* (1963) **45,** 1279.
36. Bertolini, M., Pigman, W., *J. Biol. Chem.* (1967) **242,** 3776.
37. Gottschalk, A., König, W., *Biochim. Biophys. Acta* (1968) **158,** 358.
38. Anderson, B., Hoffman, P., Meyer, K., *Biochim. Biophys. Acta* (1963) **74,** 309.
39. Anderson, B., Seno, H., Sampson, P., Riley, J. G., Hoffman, P., Meyer, K., *J. Biol. Chem.* (1964) **239,** PC2716.
40. Harbon, S., Herman, G., Rossignol, B., Jollès, P., Clauser, H., *Biochem. Biophys. Res. Commun.* (1964) **17,** 57.
41. Simpson, D. L., Hranisavljevic, J., Davidson, E. A., *Fed. Proc.* (1971) **30,** 1315.
42. Tanaka, K., Bertolini, M., Pigman, W., *Biochem. Biophys. Res. Commun.* (1964) **16,** 404.
43. Tanaka, K., Pigman, W., *J. Biol. Chem.* (1965) **240,** PC1488.
44. Downs, F., Pigman, W., *Biochemistry* (1969) **8,** 1760.
45. Payza, N., Rizvi, S., Pigman, W., *Arch. Biochem. Biophys.* (1969) **129,** 68.
46. Weber, P., Winzler, R. J., *Arch. Biochem. Biophys.* (1969) **129,** 534.
47. Bella, A., Jr., Danishefsky, I., *J. Biol. Chem.* (1968) **243,** 2660.
48. Stern, E. L., Lindahl, B., Rodén, L., *J. Biol. Chem.* (1971) **246,** 5707.
49. Murty, V. L. N., Horowitz, M. I., *Carbohyd. Res.* (1968) **6,** 266.
50. Katzman, R. L., Eylar, E. H., *Arch. Biochem. Biophys.* (1966) **117,** 623.
51. Pigman, W., Downs, F., Moschera, J., Weiss, M., "Blood and Tissue Antigens," p. 205, D. Aminoff, Ed., Academic, New York, 1970.
52. Payza, N., Martinez, L., Pigman, W., *Animal Blood Groups Biochem. Gen.* (1970) **1,** 195.
53. Huang, C., Mayer, H. E., Jr., Montgomery, R., *Carbohyd. Res.* (1970) **13,** 127.
54. Morgan, W. T. J., *Biochem. J.* (1946) **40, XV.**
55. Schiffman, G., Kabat, E. A., Thompson, W., *J. Amer. Chem. Soc.* (1962) **84,** 73; *Biochemistry* (1964) **3,** 113.
56. Lloyd, K. O., Kabat, E. A., *Carbohyd. Res.* (1967) **4,** 165.
57. Lloyd, K. O., Kabat, E. A., Licerio, E., *Biochemistry* (1968) **7,** 2976.
58. Kabat, E. A., "Blood and Tissue Antigens," p. 187, D. Aminoff, Ed., Academic, New York, 1970.
59. Rege, V. P., Painter, T. J., Watkins, W. M., Morgan, W. T. J., *Nature* (1964) **203,** 360; (1964) **204,** 740.
60. Painter, T. J., Watkins, W. M., Morgan, W. T. J., *Nature* (1965) **206,** 594.
61. Mayo, J. W., Carlson, D. M., *Carbohyd. Res.* (1970) **15,** 300.
62. Montreuil, J., Monsigny, M., Buchet, M.-T., *Compt. Rend.* (1967) **264,** 2068.
63. Derevitskaya, V. A., Vafina, M. G., Kotchetkov, N. K., *Carbohyd. Res.* (1967) **3,** 377.
64. Ozeki, T., Yosizawa, Z., *Arch. Biochem. Biophys.* (1971) **142,** 177.
65. Vercellotti, J. R., Nienaber, N., Chang, C. J., *Carbohyd. Res.* (1970) **13,** 63.
66. Isbell, H. S., *J. Res. Natl. Bur. Std.* (1944) **32,** 45.
67. Isbell, H. S., *Carbohyd. Res.* (1971) **19,** 319.
68. Kadin, S. B., *J. Org. Chem.* (1966) **31,** 620.
69. Austen, B. M., Marshall, R. D., *Biochem. J.* (1971) **124,** 14P.
70. Patchornik, A., Sokolovsky, M., *J. Amer. Chem. Soc.* (1964) **86,** 1206.
71. Kotchetkov, N. J., Derevitskaya, V. A., Kara-Murza, S. G., *Carbohyd. Res.* (1967) **3,** 403.
72. Harbon, S., Herman, G., Clauser, H., *European J. Biochem.* (1968) **4,** 265.

73. Neiderhiser, D. H., Plantner, J. J., Carlson, D. M., *Arch. Biochem. Biophys.* (1971) **145,** 155.
74. Moschera, J., Pigman, W., unpublished results.
75. Pigman, W., Moschera, J., Weiss, M., Tettamanti, G., *European J. Biochem.* (1973), in press.

RECEIVED January 14, 1972.

14

Solubility of Polysaccharides and Their Behavior in Solution

ROY L. WHISTLER

Department of Biochemistry, Purdue University, Lafayette, Ind. 47907

Polysaccharides consisting of one type of sugar unit uniformly linked in linear chains are usually water insoluble even when the molecules have a low molecular weight with degrees of polymerization (DP) 20–30. Insolubility results from the fit of molecules and their preference for partial crystallization. An exception to the rule is in (1 → 6)-linked homoglycans, which because of the extra degrees of freedom provided by the rotation about the C-5 to C-6 bonds gives higher solution entropy values. Homoglycans with two types of sugar linkages or heteroglycans composed of two types of sugars are more soluble than purely homogeneous polymers. Ionized linear homoglycans are soluble but like all soluble linear polymers easily form gels because of segmental association which sometimes may be in a double helix formation. As these junction zones develop a stronger tertiary structure, gel hardness increases.

As one examines the 300 or more natural polysaccharides that have been studied, it is seen that each polymer shows selected physical properties valuable to the biologic system which gives it eclectic origin. Besides physical service, some polysaccharides also serve as food reserves. Where service is entirely physical, it may be as a structural element, lubricant, emollient, gelation or water holding agent, complexant, or adhesive. All functionality of the polysaccharides rests to some degree on physical properties which derive from the macromolecularity of the polysaccharides and their inherent secondary molecular forces. Among the most significant of the latter are hydration characteristics which may be so extensive as to allow gelation or dissolution.

Not only are water–polysaccharide relations important effectors in biologic systems, but the physical properties of polysaccharides in aqueous dispersions are widely used industrially to give desired behavior in food and non-food products. Here, also, the useful effects are derived from the interaction of polysaccharide molecules with themselves and with molecules of their environment. Environmental molecules always include water molecules and sometimes proteins, lipids, additional polysaccharides, and other molecules used industrially. By interactions with these various environmental molecules, polysaccharides perform their useful practical functions which are mainly to give viscosity, solution stability, suspendability, emulsifying action, gelation without syneresis, and compatibility with proteins, other polysaccharides, or other ingredients present.

Solubility of Polysaccharides

Polysaccharides dissolve in water by continuous hydration with a transfer of interpolysaccharide binding to polysaccharide–water binding and with the action assisted to more or less degree by entropy as the molecules assume lower energy conformations. All polysaccharides have affinity for water; in the dry state their affinity for initial levels of water is as great as that of phosphorus pentaoxide. At normal humidities polysaccharides contain 8–10% water as water of hydration. In the solid state all polysaccharides have regions where molecules or chain segments are in a jumbled or disorganized arrangement with intermolecular forces and intermolecular hydrogen bonding only partially satisfied because of the random spatial arrangement. These amorphous regions, therefore, have numerous unsatisfied hydrogen bonding positions which can hydrate. For some polysaccharides and particularly for cellulose and starch, the relation of hydration to ambient relative humidity has been intensively examined, and hysteresis effects from moisture absorption and desorption are explained. Initial water molecules of absorption occupy hydrogen bonding positions not otherwise involved in intermolecular bonding of the polysaccharide molecules. When a soluble polysaccharide is placed in water, the abundant water molecules quickly penetrate amorphous regions and surround available polymer sites, competing for and eventually reducing to negligible numbers still other intermolecular bonds. Segments of a polysaccharide chain become fully solvated and by kinetic action move toward solution, tearing apart more interpolysaccharide bonds which are immediately solvated. Soon many sections of the polysaccharide chains are fully solvated and are solubilized while a lessening number of segments are still attached to other polysaccharide chains which are not yet completely solvated. This intermediate stage in the dissolution of a polymer molecule represents a transient gel state and represents a universal stage in the dissolution of all polysaccharides. As hydration continues, polysaccharide molecules become completely surrounded by partially immobilized water molecules which represent the solvation layer, and molecules move into solution where they may remain monodispersed and assume preferred low energy shapes and conformations or where they may become involved with other polysaccharide molecules to develop various degrees of gel structure.

These structures and conformations of polysaccharide molecules and their intermolecular associations give polysaccharide solutions, polysaccharide dispersions, and polysaccharide gels their special properties in biologic systems and in industrial usage. The ease with which polysaccharides dissolve is related to their conformation and structure in many the same ways as is their solution behavior.

Linear polysaccharides of uniform structure can fit together for strong intermolecular binding and usually can form perfectly ordered arrays to develop crystalline regions that firmly tie molecular segments together in effective cross-links. Such molecules are difficult to dissolve and once in solution can easily segmentally recombine during crystal development. Thus, they form particles that upon continued accretion reach precipitating size.

Molecules which most easily associate and form crystals have regular, extended, ribbon-like structures. Glycosidic linkages which do not form ribbon structures will lead to molecules that are more easily dissolved and to solutions of greater stability. Any structures which contain especially flexible units such as $1 \rightarrow 6$ linkages will lead to easier solubility and more stable solutions because of more favorable entropy of solution. Irregularities in the polysaccharide structure will also produce greater solubility and solution stability. Branching greatly reduces the possibility of intermolecular association and usually leads to easily soluble polysaccharides which form stable solutions. Also, introducing formal ionic charges, usually anionic charges resulting from carboxyl, sulfate, or phosphate groups, develop polysaccharides that are readily soluble and which form stable solutions so long as the formal charges exist. Formal charges do not eliminate, however, the possibility of interchain association which may, occasionally, result in a degree of cross-linking and consequent gel formation.

Generally, reinforcing, cell-wall polysaccharides are least soluble while emollients, mucilaginous, and food reserve polysaccharides represent the most soluble group. Exceptions to the generalization that reserve food polysaccharides are easily soluble occur in starch amylose and seed mannan. Starch amylose is readily dispersible in most of its natural forms since it occurs mixed with easily soluble amylopectin which facilitates the dissolution of the amylose.

Almost all low molecular weight polysaccharides—degrees of polymerization (DP) less than 15–20—are soluble in water. Solubility decreases with increase in the ease with which molecules can associate and with narrowness of molecular weight distribution.

Some polysaccharides classified by degree of water solubility are given in Table I. The list includes only a few of the known polysaccharides. Those listed have a reasonably high molecular weight, having DP of 100 to several thousand. Highly branched polysaccharides, constituting the majority of polysaccharides, are almost always very soluble in water.

Table I. Examples of Polysaccharides in Different Solubility Groups

Solubility Group	*Linkage*	*Polysaccharide*	*Sugar units*
Least			
	1→4	Cellulose	β-D-glucopyranosyl
		Chitin	2-acetamido-2-deoxy-β-D-glucopyranosyl
		Xylan	β-D-xylopyranosyl
		Amylose	α-D-glucopyranosyl
		Mannan (ivory nut)	β-D-mannopyranosyl
		Pectic acid	α-D-galacturonopyranosyl
		Alginic acid	α-L-guluronopyranosyl β-D-mannuronopyranosyl
	1→3	Callose (callan)	β-D-glucopyranosyl
		Paramylan	β-D-glucopyranosyl
		Pachyman	β-D-glucopyranosyl
	1→3 (mainly, 1→6 (minor	Laminaran	β-D-glucopyranosyl
Intermediate			
	1→6	Pustulan	β-D-glucopyranosyl
	1→3, 1→4 (1:1)	Nigeran	α-D-glucopyranosyl
	(1:3)	Lichenan	β-D-glucopyranosyl
Easily Soluble			
	1→4, 1→6 (2:1)	Pullulan	α-D-glucopyranosyl
	1→4 main chain 1→6 branches		
		Amylopectin	α-D-glucopyranosyl
		Glycogen	α-D-glucopyranosyl
		Guaran	α-D-galactopyranosyl β-D-mannopyranosyl
	1→4, 1→3 (2:1)	Cereal glucan (branched?)	β-D-glucopyranosyl
		Anionic polysaccharides	

Least Soluble Polysaccharides. As indicated, the most insoluble polysaccharides are those which are reinforcing structural polymers in cell walls. The best example is cellulose, the universal reinforcing structural component of all higher plants. Cellulose owes its insolubility to its regular structure, consisting of β-D-glucopyranosyl units in Reeves Cl conformation joined by $(1 \rightarrow 4)$-linkages; the high uniformity of the arrangement presents a ribbon-like shape. When the molecules are biosynthesized in plant cell walls, they deposit with some amorphous regions but with many regions, constituting 50% or more of the cellulose, of high crystalline order. Such regions have been extensively described by diffractionists and in particular by Meyer and Misch (*1*) and by Hermans (*2*, *3*). The fit is so perfect, and consequently the intermolecular binding is so strong that cellulose is insoluble in water and even in strongly alkaline solutions. Strongly alkaline solutions, such as 18% sodium hydroxide solution, swell cellulose, perhaps in the manner conceived by Warwicker and Wright (*4*). It is believed that the cellulose chains in ribbon-like conformation with equatorial bonds nearly in the plane of the ribbon are laid on top of each other in stacks which are joined in cellulose I and II by hydrogen bonding. Alkaline solutions may cause rupture of these hydrogen bonds, allowing the stacks or sheets, which remain more or less intact to separate. Reagents which complex with cellulose molecules—*e.g.*, cuprammonium solution—can lead to eventual dissolution of the cellulose molecules. However, the normal fit of cellulose molecules to each other is so extensive that only molecules of DP up to about 15–80 can be dissolved in 15–17% sodium hydroxide solution, and only molecules of less than 15 DP remain soluble in neutral aqueous solutions.

A near relative of cellulose is chitin, differing only in that the hydroxyl at C-2 of each D-glucopyranosyl unit is replaced by an acetamido group. Chitin is the reinforcing structural element in the shells and mantles of *Crustacea,* such as shrimp, lobster, and crabs, and in the exoskeleton, shells, and wings of insects. Chitin is also fairly abundant in fungi and some green algae. Like cellulose it is extensively crystalline and insoluble in water and alkaline solutions.

Another near relative of cellulose is xylan, differing in its lack of a hydroxymethyl group attached to C-5 but having instead only a hydrogen atom. High molecular weight, linear, homogeneous xylan molecules have not been examined and may not exist in nature. X-ray and infrared analysis suggests longer intramolecular hydrogen bonding of sugar units than occurs in cellulose (*5*, *6*). Even though almost all xylans have branched chains and the majority have other sugar units as chain substituents, most xylans are water insoluble although all are soluble in alkaline solutions. As expected, solubility decreases when side chain substituent sugar groups are removed.

Pure amylose, a linear glucan with α-D-$(1 \rightarrow 4)$ linkages, is normally insoluble in cold water but can be dissolved in hot water or in alkaline solutions. On cooling, partial crystallization (retrogradation) occurs as the temperature is lowered below about 65°. The precipitation rate depends also upon molecular weight (*7*). The glycosidic linkages cause the amylose molecule in solution to form a statistical coil without identifiable helical character (*8*). When complexed with agents such as iodine and butanol, the amylose molecule forms a helical structure containing the complexing agent. In this form the regularly shaped helices associate at temperatures depending upon the nature of the complexant but usually lower than 60° and form precipitating crystals.

Glucans that are β-D-(1 → 3) linked are insoluble in water. Laminarabiose is readily soluble in cold water, but as the oligosaccharide series is ascended, the solubility decreases so that laminarapentaose has a solubility of less than 1%.

Ivory nut mannan such as that from palm seed is the only identified water insoluble food reserve polysaccharide which occurs in nearly pure form. It dissolves in 5% alkaline solution, and its food reserve nature is indicated by its disappearance when the seed germinates. Perhaps its lack of solubility permits the palm seeds to withstand rather long soaking periods before germination. Most other mannans found in other biological sources are branched and more readily water soluble.

Thus, polysaccharide molecules with sugar units uniformly 1 → 4 or 1 → 3 linked seem to produce structures which can align to form strong intermolecular associations and even crystalline regions. Such structures are insoluble in water. Even when the molecules are soluble in alkaline solutions, neutralization leads to reassociation and precipitation.

No conclusion can be drawn for the 1 → 2 linked β-D-glucopyranosyl polymers since the one so far known from *Agrobacteria* (*10*), though water soluble, has a DP of only 22.

Glucomannans (*11*) are essentially linear polysaccharides with consistent 1 → 4 linkages joining β-D-glucopyranosyl and β-D-mannopyranosyl units (1:2 ratio in deciduous wood) in a rather random distribution. These diheteroglycans are soluble in alkaline solution but are insoluble in water; thus, the chains are capable of strong association. Replacement of D-mannopyranosyl units by some D-glucopyranosyl units, both presumably in Cl conformation, does not especially disrupt the chain regularity or its ability for intermolecular association. Alteration of the type of glycosidic linkages, such as replacement of 1 → 4 by 1 → 3 and especially by 1 → 6, has a much greater effect on solubility.

Polysaccharides of Intermediate Solubility. Pustulan, having a uniform 1 → 6 linkage between β-D-glucopyranosyl units and with one 3-O-acetyl group for every ten sugar units, is insoluble in cold water but is soluble in hot water. It is inferred that greater solubility results from the greater freedom of rotation of sugar units provided by the low energy requirement for rotation around the equatorial bond from carbon C-5 to the glycosidically bonded hydroxymethyl group (Figure 1). The low energy requirement for rotation of sugar units relative to each other should confer lower free energy of solution to the polymer. Removing the acetyl groups should lower the solubility, but the deacetylated polymer remains soluble in hot water.

Nigeran with a DP of 300–350 is also soluble in hot water but is insoluble in cold water. The irregularities in the chain produced by combining 1 → 4 and 1 → 3 linkages is expected to reduce intermolecular fit so that molecules can be more easily solvated and dissolved.

Lichenan with a DP usually reported above 100 is soluble in hot water but precipitates when its solutions are cooled.

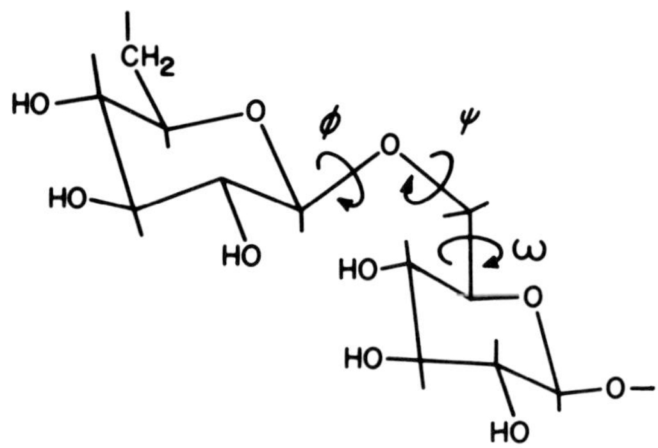

Figure 1. Rotations possible in a (1 → 6)-linked glucan

Easily Soluble Polysaccharides. Pullulan with a DP of 250 α-D-glucopyranosyl units joined by 1 → 4 and 1 → 6 linkages in a linear chain is water soluble. Again the variability in chain linkages plus the exceptional ease of unit rotation provided by 1 → 6 links give the polysaccharide molecule greater numbers of low energy conformations. The poor fit of one chain to another and the low free energy of solution contribute to developing solubility solution stability.

Introduction into a polysaccharide chain of substituents, which by their bulk decrease the fit of one polymer molecule to another, greatly improves water solubility and solution stability. A perfect example is the excellent solubility of guaran. This (1 → 4)-linked chain of β-D-mannopyranosyl units has a single α-D-galactopyranosyl side chain joined by 1 → 6 linkages at every second unit of the main chain (*12, 13*). The substituent D-galactosyl units inhibit strong intermolecular associations between the chains so that guaran solutions are stable; also the chains remain extended in solution and produce high viscosity. Removal of the side chain produces insoluble mannan of the type seen in palm seeds.

Solubilization of Polysaccharides by Chemical Modification

From the examples given above, it is apparent that water solubility of a polysaccharide can be instilled or improved by placing substituents on the linear structure which reduce the fit of one polysaccharide molecule to another or by providing anionic groups which can improve hydration and present ionic charges which by coulombic repulsion aid in molecular separation. Cellulose is a good example for illustrating such effects; it has been extensively studied because of its industrial availability at low cost (*14*).

Hydroxyalkylcellulose. Reaction of cellulose with ethylene or propylene oxides produces hydroxyethyl or hydroxypropyl derivatives. By forming the hydroxyethyl derivative about the same ratio of hydrogen bonding sites to carbon atoms is provided as in the underivatized cellulose, but the substituent groups reduce the fit between polymer chains so that the derivative can be dissolved in water to produce stable solutions. The cellulose derivative has many of the solution properties of guaran.

Methylcellulose. The placement of methyl groups on some of the hydroxyl groups of cellulose reduces the steric fit of the cellulose chains so much that they become water soluble. Certain acetylated celluloses also develop water solubility as a consequence of lowered interpolysaccharide associative forces. The solubility of partially methylated cellulose shows the powerful extensive effect of disrupting or reducing intermolecular associations. As each methyl ether grouping is formed, a strongly hydratable and effective solubilizing group is replaced by an ether linkage of equivalent hydrogen bonding capacity, and a methyl group with low hydration properties is inserted. Therefore even though some of the solubilizing hydroxyl groups are replaced by less solubilizing methyl groups, the loss in solubility is more than made up by the decreased association between cellulose chains, and consequently the cellulose ether dissolves. The overall solubility of the individual cellulose derivative has been reduced by the methylation. This is evident from the behavior of methylated cellulose, which, although it dissolves in cold water, precipitates when the solution is heated. Thus, the molecules can sufficiently solvate to dissolve, but they are just barely soluble, and any decrease in the solvation—*e.g.*, that occurring at the higher thermal energies of increased temperature—cause the molecules to aggregate and precipitate. Precipitation of methylcellulose with the formation of gels when its solutions are heated may also be influenced by changes in the structure of the water surrounding the dissolved molecules. Aqueous solutions of certain types of solutes known as hydrophobic solutes cause a structuring of liquid water in their vicinity. Such structured water would collapse when the solution is heated leaving the methylcellulose molecules free to associate to produce gels and even to form enlarged micelles which precipitate. These effects are not confined to the introduction of methyl groups but to any groups which reduce the fit between polysaccharide molecules and simultaneously lower the solvation of all or parts of the polymer molecule. Substituents can produce kinks in the polysaccharide chains which also reduce their ease of alignment.

Carboxylated Cellulose. Uniform introduction of carboxymethyl groups into cellulose to a degree of solubility (DS) of 0.45 or more produces derivatives whose alkali salts are water soluble. Solubility is a consequence of the ionic repulsion and of the disruptive effect of the substituent groups on molecular fit. If a carboxylated cellulose is prepared by the oxidation of cellulose with dinitrogen tetraoxide and under conditions where depolymerization is minimal, a carboxyl DS of 0.8 is required to produce solubility in the form of the sodium salt (*15*). Thus, introduction of ionic charge by converting the equatorial hydroxymethyl group to a carboxyl group does not extensively interfere with interchain steric fit, and absence of bulky substituents requires a greater number of coulombic repulsive forces to cause solubilization.

Other sodium salts of linear polyglycuronic acids such as sodium pectate and sodium alginate are water soluble. However these, as well as soluble sodium celluronate, become insoluble and form a gel when the pH of the solution is lowered to 2.5–3 where the carboxyl group is in the acid form and ionization is repressed. Insolubilization of alginic acid occurs regardless of its existence as a diheteroglycan in the form of a block polymer composed of alternating stretches of D-mannuronic acids and L-guluronic acids.

Cellulose Sulfates and Phosphates. Conversion of cellulose to sulfate or phosphate monoesters produces soluble derivatives. These ester groups are highly hydrated, offer steric interference to molecular fit, and are ionized at all pH levels so they continually produce coulombic repulsion.

Insolubilization of Polysaccharides by Chemical Modification

In reverse to solubilization, soluble polysaccharides are made less soluble by removing branches or substituents to produce a more uniformly linear polysaccharide with improved possibility of intermolecular fit, by removing formal charges, by lessening the number of strong hydration sites, or by completely overcoming hydration effects by introducing hydrophobic substituents.

Examples of these effects are common; most xylans decrease in solubility when L-arabinofuranosyl side-chains are removed by mild acid hydrolysis. Guaran decreases in solubility as its D-galactopyranosyl single unit side chains are hydrolyzed off; amylopectin develops an insoluble fraction when its α-D-(1 → 6) branch points are hydrolyzed by pululanase. As indicated above, polyglycuronates become insoluble as the molecular charge is eliminated by repressing ionization of the carboxyl group. Here the lower hydration of a carboxyl group compared with the carboxylic anion is important and may facilitate insolubility.

Working with cellulose and starch sulfates (*16*), we have shown that progressive introduction of 3,6-anhydro rings, through alkali induced displacement of the sulfate group at carbon C-6 by attack of the oxygen at C-3, gradually leads to lower solubility and eventually to insolubility. Insolubility results from excessive loss of hydration capacity because of the sulfate groups and of the hydroxyls at C-3 positions and their replacement by a single ether linkage.

Introduction of hydrophobic groups decrease the water solubility of a polysaccharide. With large hydrophobic groups, such as benzyl or stearoyl, insolubility is produced at low levels of derivatization.

One might properly conclude that neutral polysaccharides, when soluble are often just barely so and that withdrawal of a small degree of hydration can lead to insolubilization.

Solution Behavior of Polysaccharides

Depending upon chemical structure and the conformations that are possible, polysaccharides in solution may develop secondary structures such as helices, tertiary structures formed from junction zones or by double helix or triple helix unions and even quaternary structures from the cross linking of tertiary structures. Polysaccharides thus mimic proteins and nucleic acids, which are specific types of sugar–phosphoric acid copolymers.

Methods used to obtain conformational information and establish secondary, tertiary, and quaternary structures involve electron microscopy, x-ray diffraction, refractive index, nuclear magnetic resonance, infrared radiation, optical rotation, and anisotropy, as well as a variety of rheological procedures and molecular weight measurements. Extrapolation of solid state conformations to likely solution conformations has also helped. The general principles of macromolecules in solution has been reviewed by Morawetz (*17*), and investigative methods are discussed by Bovey (*18*). Several workers have recently reexamined the conformations of the backbone chain of xylans (*19, 20, 21*). Evidence seems to favor a left-handed chain chirality with the chains entwined perhaps in a two fold screw axis. Solution conformations are more disordered than those in crystallites (*22*). However, even with the disordering effects of branching and fewer intermolecular hydrogen bonds per sugar unit in xylan, compared with a glycohexan, the polymers associate in solution. Thus, Blake, Murphy, and Richards (*23, 24*) show that upon cooling a heated aqueous dispersion of hemicellulose, changes occur in optical rotation, viscosity, and light transmission which indicate molecular aggregation.

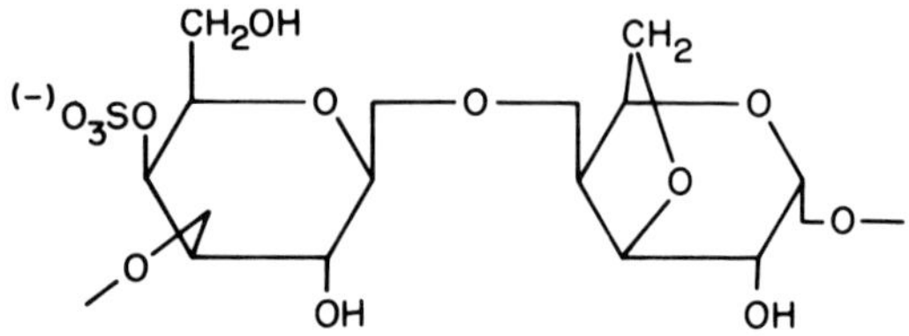

Figure 2. κ-Carrageenan

Rees *et al.* (*25*) have extensively examined the solution conformation and interactions of a number of polysaccharides, especially carrageenan fractions. Kappa-carrageenan is unusual as it forms gels when a solution of its potassium salt is cooled. Rees conceives that double helix formation occurs in solution and, interlocking helices develop in the gel state.

It is possible that gel formation is abetted by insolubility factors within the polysaccharide make up. Kappa-carrageenan is thought to be a barely soluble polysaccharide. It contains enough 3,6-anhydro rings to insolubilize a neutral polysaccharide (Figure 2). Solubility is maintained by the presence of the 4-O-sulfate groups. Yet it appears that

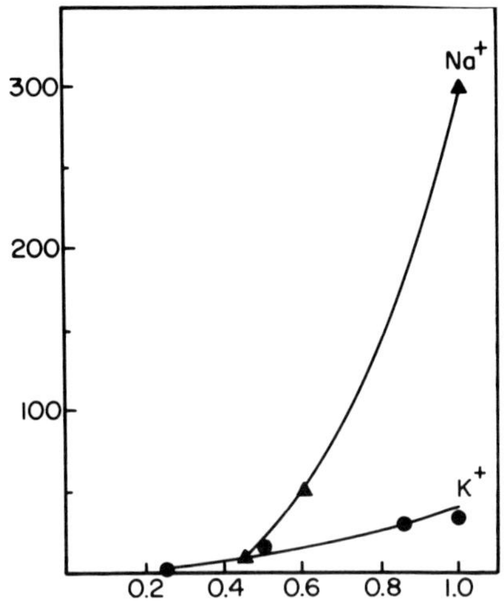

Figure 3. Viscosity of sodium and potassium salts of waxy corn starch sulfate (DS 0.5)

replacement of the counter ion sodium with potassium is enough to insolubilize the structure. The insolubilization is influenced by the much weaker electronic fields surrounding potassium ions as compared with sodium ions and their consequent reduced structuring effect on adjacent water molecules. For example, the large difference between the potassium and sodium salts of a waxy corn starch sulfate are evident in the viscosities of their solutions (*26*) (Figure 3).

Rees and coworkers (*27*, *28*) have suggested that optical rotatory changes during solution cooling of κ- and ι-carrageenans can be attributed to the formation of helices. With kappa-carrageenan they obtain optical rotatory changes on cooling the solution. We have obtained similar results as shown in Figure 4. Because of the value of gelling agents in various applications we are examining specifically modified starch and cellulose

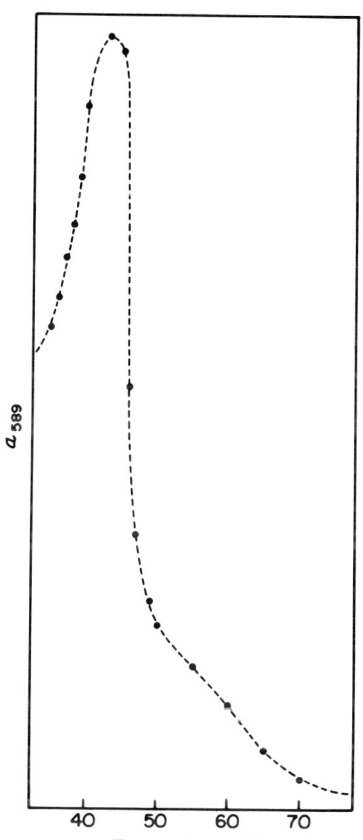

Figure 4. Changes in optical rotation on cooling a 1% solution of carrageenan

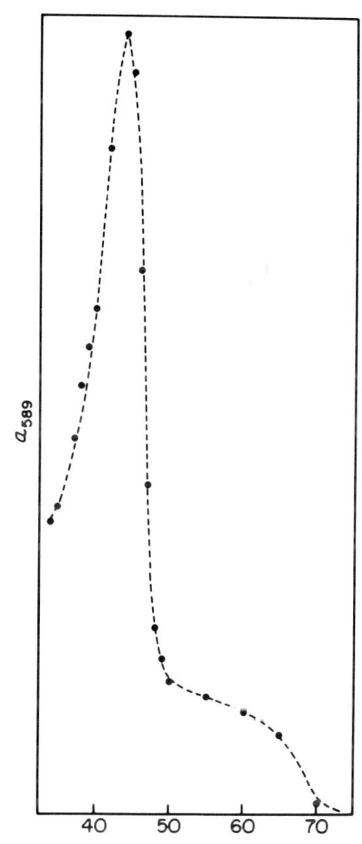

Figure 5. Changes in optical rotation on cooling a 1% solution of cellulose sulfate (DS 0.8)

sulfates. Many of these produce gels when their solutions are cooled (*29*). It is common even for the sodium salts of many cellulose sulfates to gel when their aqueous solutions are cooled. An example is shown in Figure 5, and resemblance to the cooling curve for κ-carrageenan is evident. Although starch amylose easily forms helices (*30*), it has generally been assumed, based on known solid state structures, that cellulose derivatives in solution exist in extended ribbon-like conformations. It is doubtful, but possible, that cellulose sulfate can form a double helix. Formation of such secondary structures might be aided by the presence of sulfate groups. There is already evidence that some 3,6-anhydro-β-D-glucopyranosyl units in the structure facilitate gel formation. Presence of this type of unit in the polysaccharide chain may be influential in determining chain conformation. These structures and the canonical cellulose 6-sulfate are presently under x-ray examinations.

Literature Cited

1. Meyer, K. H., Misch, L., *Helv. Chim. Acta* (1937) **20**, 232.
2. Hermans, P. H., "Physics and Chemistry of Cellulose Fibers," Elsevier, New York, 1949.
3. Liang, C. Y., Merchessault, R. H., *J. Polymer Sci.* (1959) **37**, 385.
4. Warwicker, J. O., Wright, A., *J. Appl. Polymer Sci.* (1967) **11**, 659.
5. Settineri, W. J., Marchessault, R. H., *J. Polymer Sci.* (1966) **11(C)**, 253.
6. Marchessault, R. H., Liang, C. Y., *J. Polymer Sci.* (1962) **59**, 357.
7. Whistler, R. L., Johnson, C., *Cereal Chem.* (1948) **25**, 418.
8. Brant, D. A., Dimpfl, W. L., *Macromol.* (1970) **3**, 655.
9. Clarke, A. E., Stone, B. A., *Rev. Pure and Applied Chem.* (1963) **13**, 134.
10. Gorin, P. A., Spencer, J. F. T., Westlake, D. W. S., *Can. J. Chem.* (1961) **39**, 1067.
11. Timell, T. E., *Adv. Carbohydrate Chem.* (1964) **19**, 247; (1965) **20**, 409.
12. Ahmed, Z. F., Whistler, R. L., *J. Amer. Chem. Soc.* (1950) **72**, 2524.
13. Whistler, R. L., Durso, D. F., *J. Amer. Chem. Soc.* (1952) **74**, 5140.
14. Whistler, R. L., Ed., "Industrial Gums," Revised Ed., Academic, New York, in press.
15. Whistler, R. L., Singh, P. P., in preparation.
16. Whistler, R. L., Hirase, S., *J. Org. Chem.* (1961) **26**, 4600.
17. Morawetz, H., "Macromolecules in Solution," Interscience, New York, 1965.
18. Bovey, F. A., "Polymer Conformation and Configuration," Academic, New York, 1969.
19. Settineri, W. J., Marchessault, R. H., *J. Polymer Sci.* (1965) **11**, 253.
20. Marchessault, R. H., Liang, C. Y., *J. Polymer Sci.* (1962) **59**, 357.
21. Rees, D. A., Skerrett, R. J., *Carbohydrate Res.* (1968) **7**, 334.
22. Flory, P. J., "Principals of Polymer Chemistry," Cornell University, Ithaca, 1953.
23. Blake, J. D., Murphy, P. T., Richards, G. N., *J. Macromol. Sci.-Chem.* (1970) **A4**, 1247; *Carbohydrate Res.* (1971) **16**, 49.
24. Blake, J. D., Richards, G. N., *Carbohydrate Res.* (1971) **18**, 11.
25. Rees, D. A., Scott, W. E., *J. Chem. Soc.* (1971) (b) 469.
26. Whistler, R. L., Unrau, D. G., Ruffini, G., *Arch. Biochem. Biophys.* (1968) **126**, 647.

27. Rees, D. A., Steele, I. W., Williamson, F. B., *J. Polymer Sci.* (1969) **C28,** 261.
28. McKinnon, A. A., Rees, D. A., Williamson, F. B., *Chem. Commun.* (1969) 701.
29. Baker, C. W., Whistler, R. L., unpublished.
30. Foster, J. F., "Starch:Chemistry and Technology," p. 349, R. L. Whistler, Ed., Academic, New York, 1965.

RECEIVED December 2, 1971.

15

Sugar Structure and Taste

R. S. SHALLENBERGER

New York State Agricultural Experiment Station, Cornell University, Geneva, N. Y. 14456

The saporous unit for sweet taste in sugars is the α-glycol moiety in the gauche conformation. Sugar sweetness is diminished when the glycol unit has the eclipsed conformation and an intramolecular hydrogen bond. Also if an OH group is disposed to hydrogen bond elsewhere in the molecule (the ring oxygen atom), the ability of an α-glycol moiety to elicit sweet taste is diminished. When α-glycol OH groups are in the anti conformation, they are apparently too far apart to cause sweet taste. Evidence for the above conclusions is derived from studies of various sugars, model compounds, and the mutarotation reaction and leads to a general concerted hydrogen bond model for the initial chemistry of sweet taste.

The relation between sugar structure and taste has interested investigators for many years. With the development of a more complete stereochemical description of the sugars, reasons for their varying sweetness seem to become apparent (*1, 2*). One saporous unit in the sugar is the α-glycol moiety. It seems that varying sugar sweetness is caused by:

1. Eclipsing of vicinal hydroxyl groups permitting them to participate in an intramolecular hydrogen bond

2. Positioning of an hydroxyl group, which is also part of an α-glycol pair of hydroxyl groups, so that it may hydrogen bond elsewhere in the molecule (for example to the ring oxygen atom)

3. A pyranose or furanose ring structure and an α-glycol configuration leading to the anti glycol conformation

Among the most interesting examples of varying sugar sweetness is the fact that crystalline or freshly dissolved β-D-fructopyranose is about twice as sweet as sucrose, but after mutarotation or during thermal mutarotation, sweetness diminishes markedly (*3*). α-D-Glucopyranose is about two-thirds as sweet as sucrose, but the mutarotated solution is even less sweet. Yet crystalline β-D-glucopyranose is sweeter than the α-anomer, and the mutarotated solution is again less sweet. In the first one would conclude that the α-anomer is sweeter than the β-anomer, but in the second the reverse conclusion could be reached. α-D-Galactopyranose is the 4-epimer of α-D-glucopyranose, but it is only one-half as sweet as glucose. α-D-Mannopyranose, the 2-epimer of glucose is also only one-half as sweet, but the β-D-anomer is bitter. Finally, because of their varying biological activity, it is believed that optical isomers have different sweetness. While this is true of the amino acids, the D- and L-series of sugars are equally sweet (*2*).

Several examples are given to indicate that the eclipsing of OH groups results in reduced sweetness, but the best evidence comes from the nature of the fructose mutarotation reaction.

When crystalline β-D-fructopyranose is newly dissolved in water, it is twice as sweet as sucrose, but shortly thereafter it is only slightly sweeter. Fructose mutarotates rapidly, and such phenomena have been associated by Isbell (*4*) with the formation of furanose forms of the sugars. Using a gas chromatographic procedure (*5*), we have shown (*6*) that the mutarotation primarily results from the formation of that isomer present in the sucrose molecule or β-D-fructofuranose.

At equilibrium in water at 20°, gas–liquid chromatography indicates that there is 76% β-D-fructopyranose, 20% β-D-fructofuranose, and 4% of an unknown compound, which has a specific rotation of about +122° (if the value of +17° assigned by Hudson (*7*) to β-D-furanose is correct). We deduced that the furanose form is void of sweetness for at least two reasons. As an example of hydrogen bonded hydroxyl groups, both hydroxy-methyl substituents are so dispersed as to be (perhaps) completely bonded to the ring oxygen atom (*8*).

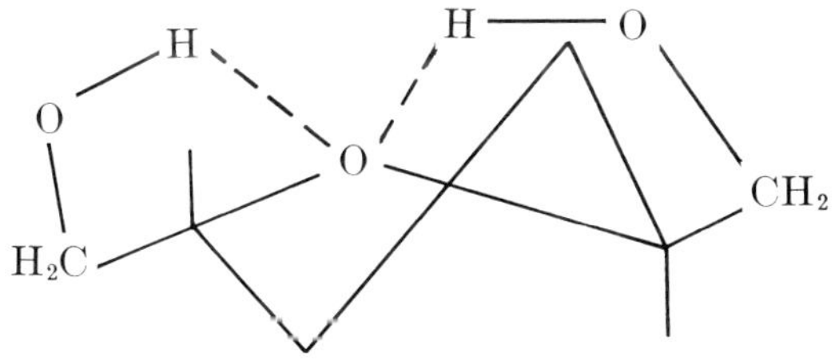

This is one example of the second criterion mentioned previously. However, the other OH substituents, depending upon the furanose ring conformation, are either eclipsed or in the anti conformation. In the former they are disposed to form a strong intramolecular hydrogen bond; in the latter they are incapable of such bonding. Further evidence to support the contention that free β-D-fructofuranose is nearly tasteless is seen in the thermal mutarotation (*3*) of D-fructose. As the temperature of D-fructose solutions is increased, the relative sweetness of the solution, measured under comparable conditions, drops markedly. The optical rotation increases, suggesting the formation of a more dextrarotatory isomer, at the expense of the sweet β-D-pyranose. These phenomena are shown in Figure 1. (The sweetness data are those of Tsuzuki and Yamazaki (3).) At 60° we find 58% β-D-fructopyranose, 30% β-D-fructofuranose, and 12% of the unknown compound, tentatively assigned to α-D-fructopyranose based on furanose conformation principles. Two cis bulky hydroxy methylene substituents on a furanose ring are unlikely.

Based upon this, we predicted that if the configuration and conformation of di-D-fructose anhydride I established by Lemieux and Najarajan (9) was correct, the compound should be nearly tasteless (the dihedral angle between α-glycol groups is 150° and 75° in the α- and β-furanoside rings, respectively). This was true.

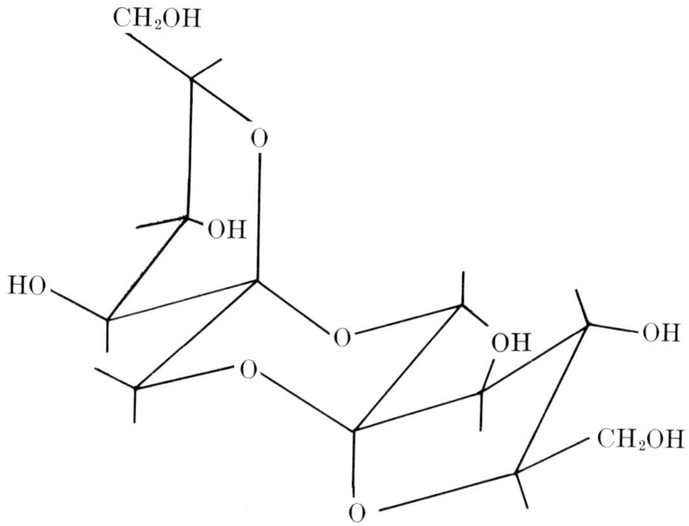

Thus, it is assumed that foods containing free fructose which have, or are in, an environment for anhydride formation will lose sweetness.

Another example of where reduced sweetness can be associated with intramolecular hydrogen bonding is the sugar galactose. Only one-half as sweet as glucose, the difference must be in the fact that the OH group on position-4 in galactose is axially disposed and capable of bonding to the ring oxygen atom. The same situation is found in α-D-mannose. Why β-D-mannose tastes bitter is not known, but it posssesses within its structure the key to structural requirements for bitter taste.

The glucose anomers have interesting taste characteristics. As mentioned before, by their behavior in solution, one could conclude that either the α-D or the β-D-anomer is sweeter. What this information suggests is that perhaps the sugars have gone into equilibrium with an unknown, but a significant proportion of conformers whose vicinal OH groups are eclipsed, anti, or disposed to bond the ring oxygen atom, such as

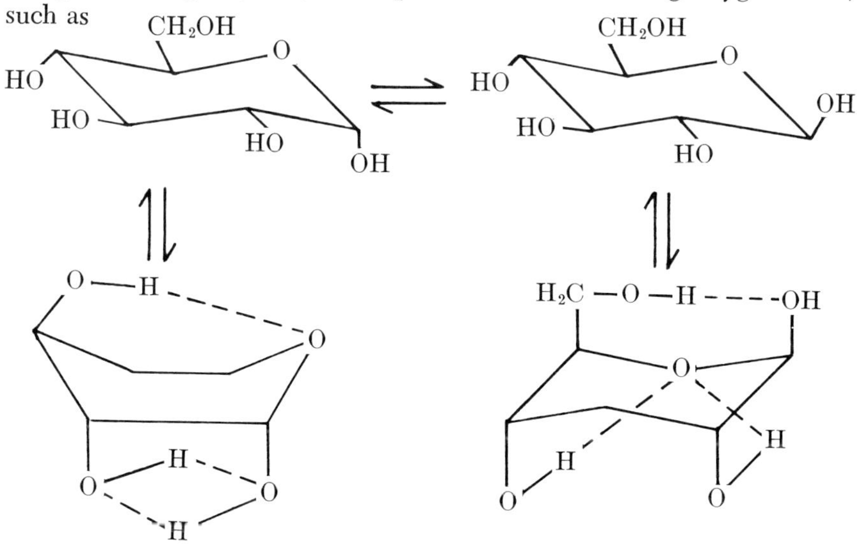

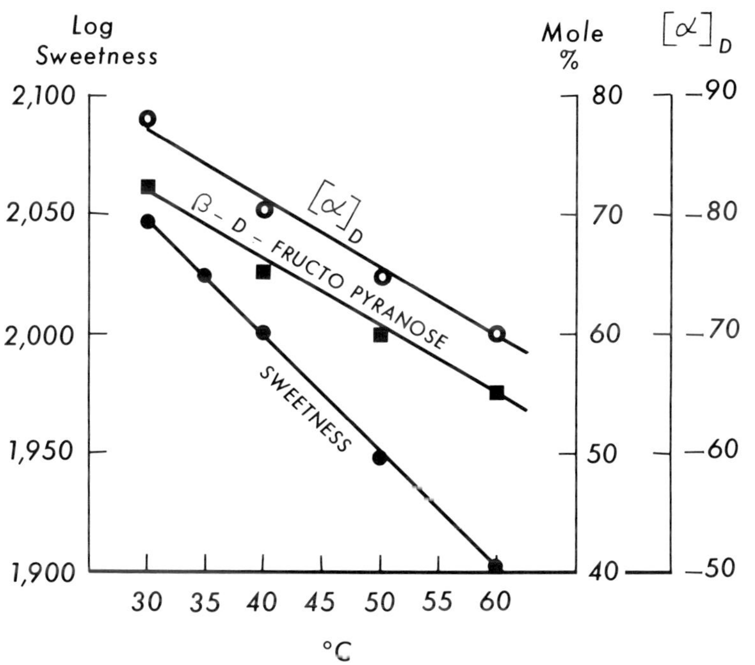

Figure 1. Change in the sweetness and concentration of D*-fructo pyranose during thermal mutarotation*

There is little evidence for such phenomena. Energetically, boat forms of the sugars at present are, highly unfavored, but the B3 conformer shown for α-D-glucopyranose may explain the reaction of that sugar with borate. The 1C conformer shown for β-D-glucopyranose is equally unfavorable but a possible candidate for marked decrease in sweetness of this compound in solution. The stabilizing H-bonds shown should be severed at elevated temperatures, and if sugar sweetness does vary inversely with the degree of intramolecular H-bonding, the sweetness of glucose in solution should increase with temperature. This is an established fact with important practical application.

The strongest evidence for the notion that anti α-glycol conformations, such as persist in β-D-fructofuranose, are devoid of sweet taste is found in levoglucosan.

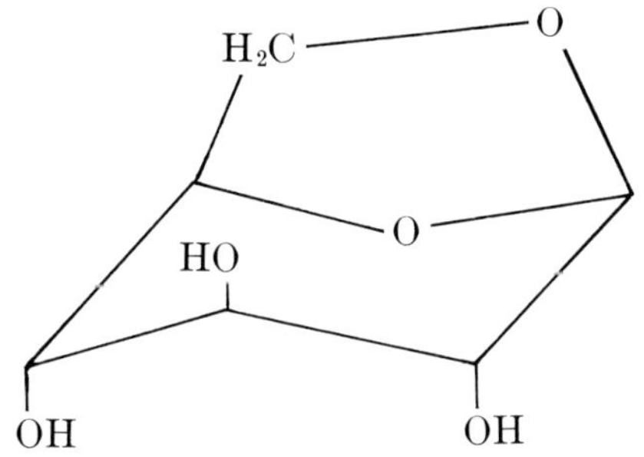

When crystals of this material are placed upon the tongue, something happens, but a taste panel that was established cannot say just what. They agreed, however, that the compound is tasteless. Possibly the panel responds to a sense of coldness because of the heat of dissolution of the compound.

From studies such as these, it was deduced that the ideal sugar moiety eliciting sweet taste was the α-glycol unit in the gauche conformation, regardless of whether or not the OH groups were cis or trans. A diagrammatic way of representing the relation between glycol conformation and sweetness is shown in Figure 2. With an O · · · O distance of near 2.5 A, the eclipsed OH groups are so strongly H-bonded, they cannot elicit sweet taste, and when they are anti with an O · · · O distance of 3.71 A, they are too far apart. The maximum for sweet taste seems to be about 3 A.

Many more examples of the relation between sugar structure and taste could be given. Birch *et al.* (*10*) have studied a number of free sugars, sugar derivatives, and substituted sugars.

As these studies developed, it became apparent that if the sugar sweetness varied inversely with the degree of intramolecular hydrogen bonding, then perhaps the initial chemistry of sweet taste resulted from

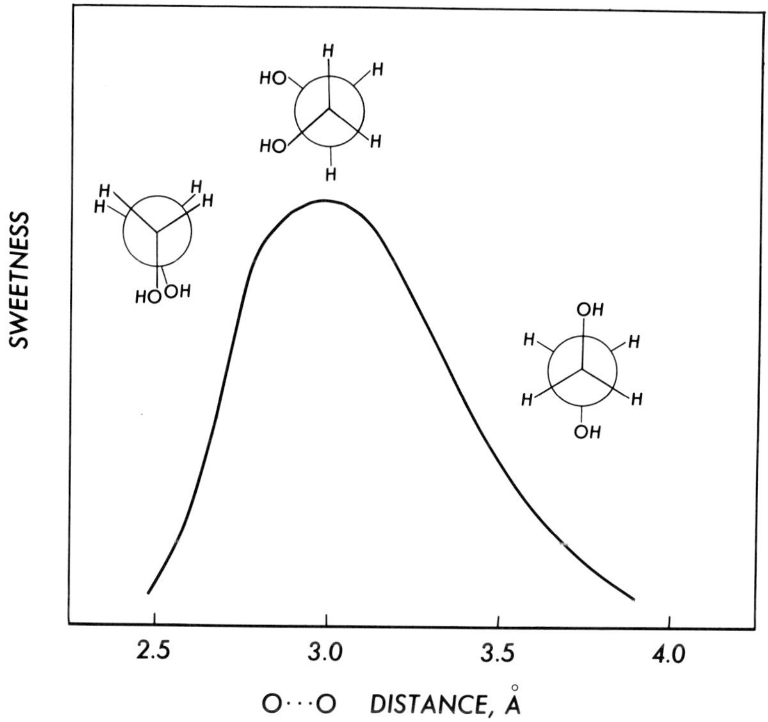

Figure 2. Diagram of the potential of sugar α-glycol groups to elicit sweet taste

intermolecular H-bonding (*11*). Thus, we began to view the α-glycol unit as an AH,B system conventionally used (*12*) to describe and define the hydrogen bond. Based upon our conclusion that the α-glycol unit when in the gauche conformation was the most favorable for causing sweet taste, we decided that the receptor site could also be an AH,B unit and the AH proton distance to B must also be about 3 A although the identity of the receptor AH,B unit remained uncertain. Viewing the sweet unit of a sugar as an AH,B unit led us to look for such a unit in the various sweet-tasting compounds, particularly with an AH, proton to B distance of about 3 A. With varying chemical identity, the unit is present in all compounds with sweet taste (*1*) and probably is a prerequisite for sweet taste. Thus, the initial chemistry of the sweet taste response seemed to be a concerted intermolecular hydrogen bond interaction between the saporous unit of a sweet substance and a commensurate chemical site on the receptor.

Sweet Compound	— A — H . . . B —	Receptor Site
	— B . . . . HA —	

It was with the sweet taste of enantiomorphs that we encountered the strongest criticism to our explanation of why sugars vary in their sweetness and also as to the identification of the saporous unit as an AH,B system in all sweet compounds.

Amino acids which belong to the D-series are generally sweet tasting, but those in the L-series are tasteless or bitter when R in $RCNCNH_2CO_2H$ is larger than the ethyl radical.

Equally, the naturally occurring D-series of sugars usually taste sweet, while it has generally and tacitly been assumed that the synthetic L-series is tasteless. To support this assumption, it has been reported that L-glucose is tasteless, and D-mannose is sweeter than L-mannose (*13*).

The D and L-sugars and amino acids are enantiomorphs and differ in absolute configuration about every asymmetric carbon atom. The pentose L-arabinose is structurally related to the sweet-tasting hexose, D-galactose, and, to recognize this, we predicted (*2*) that L-arabinose would probably taste sweet.

L-Arabinose did taste just about as sweet as D-galactose, and L-sorbose, the 5-epimer of D-fructose, also tasted sweet although it was only about one-fifth as sweet as D-fructose.

To extend these findings, a series of seven pair of D- and L-sugars was submitted to the taste panel to compare the sweet taste characteristics of each enantiomorphic pair. The method used was that of a paired comparison technique where sweetness scores are assigned to the taste of each sugar. To prevent the need to identify the anomeric and crystalline form of each sugar (the supply of certain rare enantiomorphs was small), 10% solutions of the sugars were allowed to come to mutarotational equilibrium before they were tasted. Thus, the tautomeric or conformational composition of a reducing sugar solution would not be important so long as only enantiomorphic sugars are compared. We found (*2*) that there was no statistically significant difference between the sweet taste of the enantiomorphic sugars. D-glucose was just about as sweet as L-glucose. The critical enantiomorphic pair to be compared were D- and L-fructose and M. L. Wolfrom confirmed that L-fructose tastes very sweet (*14*).

We conclude that the strange stereochemical features, which students have been assigning to the taste bud for years, many be more myth than fact. All that really needs to be accounted for is the fact that the one amino acid (alanine) which tastes sweet in its D- and L-forms has a side chain smaller than the ethyl radical. If one constructs a spatial barrier about 3 A from the postulated AH,B site, it becomes a simple matter of whether or not an amino acid in its L-form can be positioned over the site. A sugar with vicinal OH groups as AH,B can make any approach to the receptor site to elicit sweet taste, and, therefore, there should be

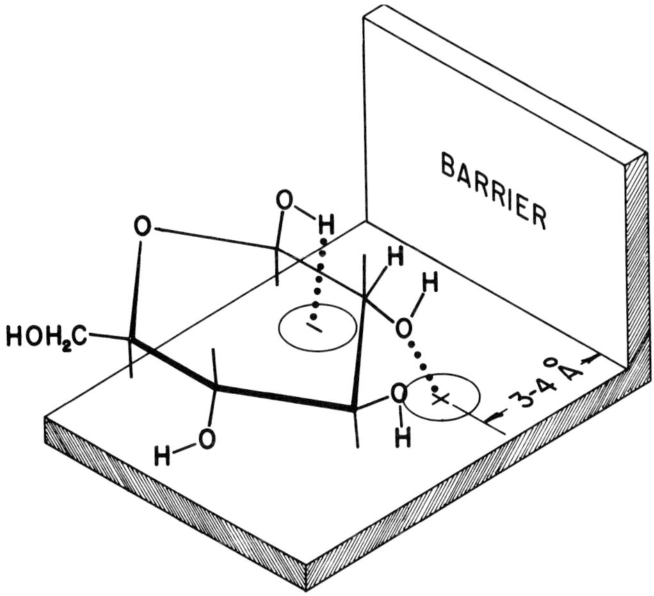

Figure 3. Positioning of β-D-glucose over the proposed taste receptor site to initiate sweet taste response

no difference in the ability of D- and L-forms to elicit sweet taste. The model with β-D-glucose as the tastant is shown in Figure 3. To demonstrate the ability of β-L-glucose to be positioned over the site, the anomeric hydroxyl group and the $-CH_2OH$ group need to be interchanged and the molecule turned over.

Literature Cited

1. Shallenberger, R. S., Acree, T. E., *Nature (London)* (1967) **216**, 480.
2. *Nature (London)* (1969) **221**, 555.
3. Tsuzuki, Y., Yamazaki, J., *Biochem. Z.* (1953) **323**, 525.
4. Isbell, H., Pigman, W. W., *J. Res. Natl. Bur. Std.* (1938) **20**, 773.
5. Shallenberger, R. S., Acree, T. E., *Carbohydrate Res.* (1966) **1**, 495.
6. Shallenberger, R. S., "Frontiers in Food Research," p. 40, Cornell University, 1968.
7. Hudson, C. S., *J. Amer. Chem. Soc.* (1908) **30**, 1564; (1909) **31**, 655.
8. Barker, S. A., Brimacombe, J. S., Foster, A. B., Whiffen, D. H., Zweifel, G., *Tetrahedron* (1959) **7**, 10.
9. Lemieux, R. U., Nagarajan, R., *Can. J. Chem.* (1964) **42**, 1270.
10. Birch, G. G., Lee, C. K., Rolfe, E. J., *J. Sci. Food Agr.* (1970) **21**, 650.
11. Shallenberger, R. S., *New Scientist* (1964) **23**, 569.
12. Pimentel, G. C., McClellan, A. L., "The Hydrogen Bond," Freeman, 1960.
13. Boyd, W. C., Matsubara, S., *Science* (1962) **137**, 669.
14. Wolfrom, M. L., personal communication.

RECEIVED December 2, 1971

16

Sugar Transport Systems and the Evolution of Mutarotases

J. MARTYN BAILEY, PETER G. PENTCHEV, PETER H. FISHMAN, SALLY A. MULHERN, and JOHN W. KUSIAK

Biochemistry Department, George Washington University School of Medicine, Washington, D. C. 20005

Diverse transport mechanisms for sugars have evolved in living organisms, including the phosphotransferase and permease systems in bacteria. In mammals several different sugar transport processes seem to have evolved to satisfy the more complex requirements for control and regulation in multicellular organisms. The enzyme mutarotase is found in all tissues which transport glucose, and the embryological and evolutionary development of sugar transport mechanisms in organs of different species correlates well with tissue levels of the enzyme. The inhibition pattern of transport and mutarotase by phloretin and a series of estrogenic compounds are essentially identical. The evidence suggests that the enzyme mutarotase contains some glucose-binding function particularly suitable, which led to its retention through a long evolutionary history in recognizable form in various transport systems for glucose.

This review outlines the current knowledge of the enzyme mutarotase (aldose-1-epimerase) and evaluates the evidence that it may have evolved from an origin in primitive bacteria into an important transport system for sugars in higher organisms.

It is assumed that transport processes for sugars arose early in the evolution of living systems since synthesis of simple sugars by the formol condensation mechanism would occur at an early stage on the primtive earth. At the level of the unicellular organism, development of transport systems for nutrients present in the environment would give competitive advantages when the supply became limited. It is presumed that development of specific and efficient transport mechanisms has been under considerable selective pressure and that different mechanisms for solving the problem most efficiently may have evolved in response to different environmental conditions.

A brief description of sugar transport in bacteria and mammals is given principally to illustrate general principles and to outline the diversity of the processes which have evolved, particularly in mammals. The selection of material for this section is of necessity, therefore, somewhat arbitrary, and more comprehensive surveys of sugar transport may be found in several recent reviews (*1, 2, 3, 4, 5, 6, 7, 8, 9, 10, 11, 12, 13, 14*).

The evolutionary history and properties of mutarotase is mainly studied here. The enzyme catalyzes the interconversion of the anomeric forms of a number of aldopyranose sugars related configurationally to D-glucose. The specificity and the mechanism of the enzyme catalysis are of considerable basic interest to understanding the mechanism of the mutarotation of sugars. The study of the enzyme also has significance however since it has many of the biological properties attributed to the sugar carrier or permease molecule thought to be involved in active transport of sugars. Since no biological requirement for a catalyst of mutarotation of sugars has been established in higher organisms, there is a distinct possibility that the catalysis of mutarotation is a coincidental consequence of the sugar-binding function of the protein (*15, 16*). The biological and catalytic properties of mutarotase, therefore, are a subject of considerable theoretical interest for carbohydrate chemistry and for biological transport.

Sugar Transport in Bacteria

Transport processes are often classified into two types: active transport in which the transported species is transferred across a membrane from a region of lower, to one of higher concentration (*i.e.*, against a thermodynamic potential gradient), and passive transport in which the compound does not accumulate against a concentration gradient but, nevertheless, crosses the membrane at a rate greater than expected by simple diffusion. This latter process is often termed facilitated diffusion. It is generally believed that the active transport process involves at least two subsystems: 1. a component in the membrance capable of combining with the substrate and accomplishing its translocation and 2. a system for coupling the translocation process to a source of energy to establish a concentration gradient. If the energy-coupling mechanism is inactive or blocked, the system then may show the characteristics of passive transport, facilitating transport without the establishment of a concentration gradient. One of the central outstanding questions in the study of membrane transport is the exact mechanism of the energy-coupling process.

It seems that in bacteria at least two different types of active sugar transport have evolved. In the first of these the phosphotransferase system discovered by Kundig, Ghosh, and Roseman (*17, 18, 19*) the sugar undergoes a covalent transformation to the sugar phosphate as an intermediate stage in the transport process. In the second type of transport the so called permease system, no covalent transformation of the sugar is apparently necessary (*20, 21, 22, 23, 24*).

For the phosphotransferase system there is no problem in identifying the energy source for the establishment of the thermodynamic potential gradient. It is apparent that a high-energy phosphate bond in phosphoenol-pyruvate is the immediate source of the energy responsible for the vectorial translocation of the sugar molecule across the membrane (*17*).

For the permease-type systems, however, since no covalent intermediate has been identified, the source of energy and the type of coupling to the process of translocation have not been defined. The bulk of the evidence favors a coupling of translocation to hydrolysis of high-energy phosphate or to redox processes of the cell (*2*). Current opinion favors the idea that energy-coupling is mediated *via* some conformational transformation of the transport proteins which accomplishes the vectorial translocation. A useful *in vitro* criterion for identifying the transport protein might be such a conformational change in response to some type of concentration gradient. Some evidence that the mutarotase protein undergoes such a transformation in response to osmotic gradients, a transformation which can be reversed by substrate sugars, is presented below.

The molecular and enzymatic basis of the permease and phosphotransferase types of transport has been under extensive study for several years. The main features of both and the present state of our knowledge concerning the molecular components are outlined schematically in Figures 1 and 2.

$$\text{H.Pr} + \text{P-Enolpyr.} \xrightarrow{\text{Enzyme I}} \text{P-H.Pr}$$

$$\text{Sugar} + \text{P-H.Pr} \xrightarrow{\text{Enzyme II}} \text{Sugar-P}$$

$$\text{Sugar-P} \xrightarrow{\text{Phosphohydrolase}} \text{Sugar}$$

VECTORIAL PHOSPHORYLATION BY MEMBRANE-BOUND ENZYME II
RESULTS IN TRANSLOCATION OF SUGAR ACROSS MEMBRANE.

Figure 1. The bacterial phosphotransferase sugar transport system

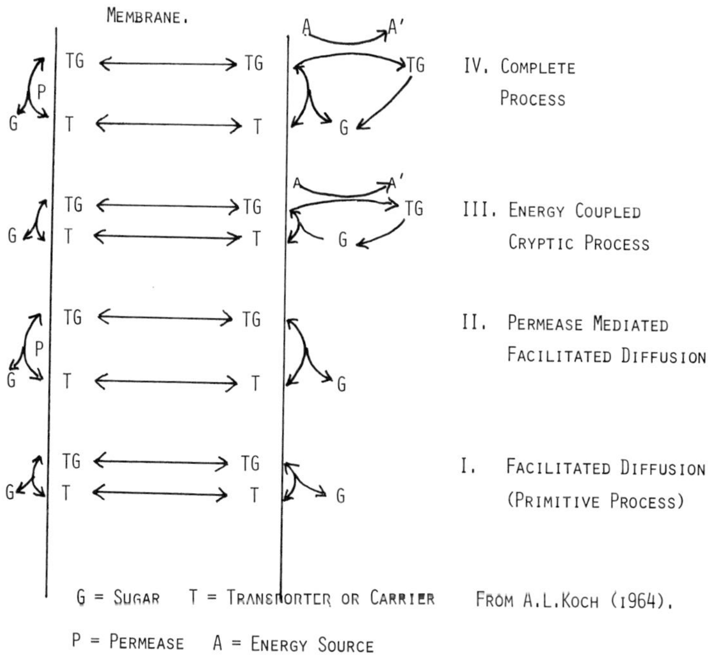

Figure 2. Proposed model for a bacterial permease system

The Phosphotransferase Systems of Bacteria. The main features of the phosphotransferase system, which are outlined in Figure 1, include a heat-stable protein HPr, which accepts a phosphate group from phospho-enol-pyruvate as donor, in a reaction catalyzed by a soluble Enzyme I. The phosphorylated HPr then transfers the phosphate group to the sugar in a reaction catalyzed by a membrane-bound Enzyme II. This latter enzyme determines the sugar specificity of the system and is frequently inducible (*18*). The phosphorylated sugar may then be dephosphorylated intracellularly, the net result being an accumulation of free sugar inside the cell. This general process may be classified as a group translocation mechanism. It seems that it has evolved in a number of bacterial species, including *E. coli*, *S. typhimurium*, *B. subtilis*, *A. aerogenes*, and *S. aureus*, specifically to carry out sugar transport (*19*).

The Permease Systems of Bacteria. The best defined of these is the galactoside permease of *E. coli*. This transport system mediates the active accumulation of galactosides in the presence of metabolic energy and the facilitated diffusion of these compounds when the energy system is blocked (*8*). A specific galactoside-binding protein has been implicated, but it seems clear that the system is different from the phosphotransferase system described above since no covalent intermediates of the transported sugars appear to be involved. Various procedures have been used attempting to isolate the components of the permease system. A sugar-binding (M) protein was isolated by Kennedy and co-workers by a specific labelling technique (*25, 26, 27*), and the technique of osmotically shocking bacteria also releases a galactose-binding protein into the medium from *E. coli* (*28*). Addition of the purified binding protein has been shown in some cases to restore the ability of the cells to transport galactose. It seems from these experiments in bacteria that the sugar-binding protein may be located in a periplasmic space between the bacterial cell wall and the plasma membrane and is not necessarily an integral part of the membrane itself (*29*).

One of the most comprehensive early models outlining the nature of a specific permease mechanism is that of Koch (*30*) (illustrated in Figure 2). The main features of this model imply three principal components:

1. Externally, the highly stereospecific binding or permease step P,
2. The carrier or transport mechanism T which bridges the permeability barrier of the cell, and
3. Internally, the mechanism which couples transport to energy metabolism.

Depending upon the loss of one or more components, the system can be shown to mimic the four principal types of sugar transport observed in various mutants of *E. coli*.

Koch points out the reasonableness of the model from the point of view of evolution since a membrane barrier capable of retaining important metabolites but of selectively allowing the passage of certain classes of nutrients would be an early development. The addition of the specific permease and energy-coupling features to the primitive mechanism illustrates the manner in which a complex transport process may evolve.

The various proposed components of the permease system are based upon the response of the transport system to genetic or environmental changes. The complex nature postulated for the intact permease system is necessary to account for the various observed phenomena such as facilitated diffusion, active concentration, facilitated efflux, exchange diffusion, and counter transport of one compound driven by the downhill efflux of a second (*2*).

Much of the work defining the galactoside permease systems of *E. coli* was done before the discovery of the ubiquitous phosphotransferases of bacteria. There is now much discussion as to the relative importance of the permease and phosphotransferase systems and on the possible re-interpretation of some of the models proposed for permease systems, because of the new information on the phosphotransferases.

Sugar Transport in Mammals

Diversity of the Transport Systems. In multicellular organisms, which have complex physiological and regulatory requirements for sugar metabolism, it might be expected that considerably more variety and complexity would be associated with the processes of sugar transport. The mammalian organism is composed of many different organs and organelles, each having specific requirements for energy supply and metabolism. Also certain tissues—*e.g.*, intestinal epithelium and kidney tubule—have evolved specific, highly efficient mechanisms for glucose absorption against concentration gradients. In these tissues the glucose fluxes far exceed the metabolic requirements of the individual cells. In certain other tissues glucose uptake and metabolism are under hormonal control, chiefly by insulin, and specific changes in the permeability of membranes to glucose are believed to occur under the influence of this hormone. Some of the various sugar transport systems which have been identified or postulated in mammals include carrier-mediated facilitated diffusion in red blood cells (*31, 32, 33, 34*), insulin-sensitive passive transport in muscle and adipose tissue (*35*), phospho–enol–pyruvate-linked phosphotransferases in intestine (*36*), pyrophosphate phosphotransferases in kidney and liver (*37*), sodium-dependent active transport in kidney and intestine (*38*), mutarotase-linked active transport in kidney and intestine (*39*), sodium-independent active transport in kidney and intestine (*40*), feedback-regulated active transport in certain tumor tissues (*41*), and trehalase-linked active transport in kidney (*42*). The list is not exhaustive or necessarily exclusive, and a number of the systems may be interrelated or have certain components in common.

The definition on a molecular basis of these different sugar transport systems has not progressed to the same extent as with bacteria, partly because of the greater convenience of bacteria as experimental organisms, and also because of the greater complexity of the mammalian systems themselves.

It seems that processes related to the bacterial phosphotransferases and permeases probably also operate in mammals, although their relative importance and function have not been exactly defined. A phospho–enol–pyruvate–phosphotransferase for amino sugars has been identified in intestine (*36*), and a different phosphotransferase system has been identified in kidney tubules (*37*). In the latter case because of the high K_m for glucose, it has been proposed that the process is not of great importance under normal physiological conditions but may operate in certain pathological states such as diabetes when urine glucose levels are much higher than normal.

The predominant active sugar transport systems of intestine and kidney are specific for sugars which have the pyranose structure with a free equatorial hydroxyl group at C-2. It is doubtful that phosphorylation is significant in this transport system since sugars such as 1-deoxy glucose, 6-deoxy-glucose and α-methyl glucoside, which are not phosphorylated, are nevertheless actively transported (*43, 44*). It has been shown that active sugar transport in these tissues is linked in some way to active sodium transport. This sodium pump is also responsible for most of the water reabsorptive capacity of kidney and intestine. Besides the sodium-dependent process, there is in kidney and intestine a sodium-independent active transport system for certain sugars (*40*).

The Role of Phospholipid-Sugar Complexes in Membrane Transport. Because of the high lipid content of cell membranes, it has seemed necessary to find some additional mechanism to render the hydrophylic substrates of the transport process more compatible with a lipid environment. It has been shown that sugars and amino acids can form complexes with certain phospholipids which make them soluble in lipid solvents (*45, 46, 47*). These phospholipid-sugar complexes are believed by some to be attractive candidates for the intermediate form of the sugar during transit across the membrane. LeFevre has studied the properties of the complexes extensively and has demonstrated that addition of phospholipids accelerates the transfer of sugar through chloroform into an aqueous phase (*48*).

Formation of the phospholipid-sugar complexes can be followed in similar experiments where transfer of sugars from glass fiber strips into a hexane phase is measured in glass roller tubes as shown in Figure 3. (Glass-filter paper strips loaded with the individual C^{14}-labelled sugars were incubated with hexane solutions of the intestinal phospholipid fractions (A, B, or C) in glass roller tubes. Phospholipids were isolated from intestinal lipid extracts by chromatography on silicic acid. Fractions A, B, and C represent phospholipids eluted by 25%, 50%, and 75% methanol-in-ether, respectively (*104*). Progress of complex formation was followed by determining the uptake of radioactive sugar as a function of time.) Marked differences in the affinities of the sugars are found. Uptake of glucose, galactose, and mannose is close to 1 mole of sugar per mole of phospholipid phosphorus for the less polar (A and B) fractions and 2 to 3 moles per mole for the most polar C fraction. The complexes with pentoses form more rapidly than hexoses but are also less stable. The subsequent decrease in pentose content of the complexes noted in Figure 3 results from dissociation and subsequent deposition of the sugars on the glass walls of the roller tube. This greater rate of dissociation of pentoses is also evident when release of the radioactive sugar from the hexane layer into an aqueous layer is measured. When excess sugar is present, the rate of formation of the complexes approximates a first-order reaction, and the relative rates of formation may then

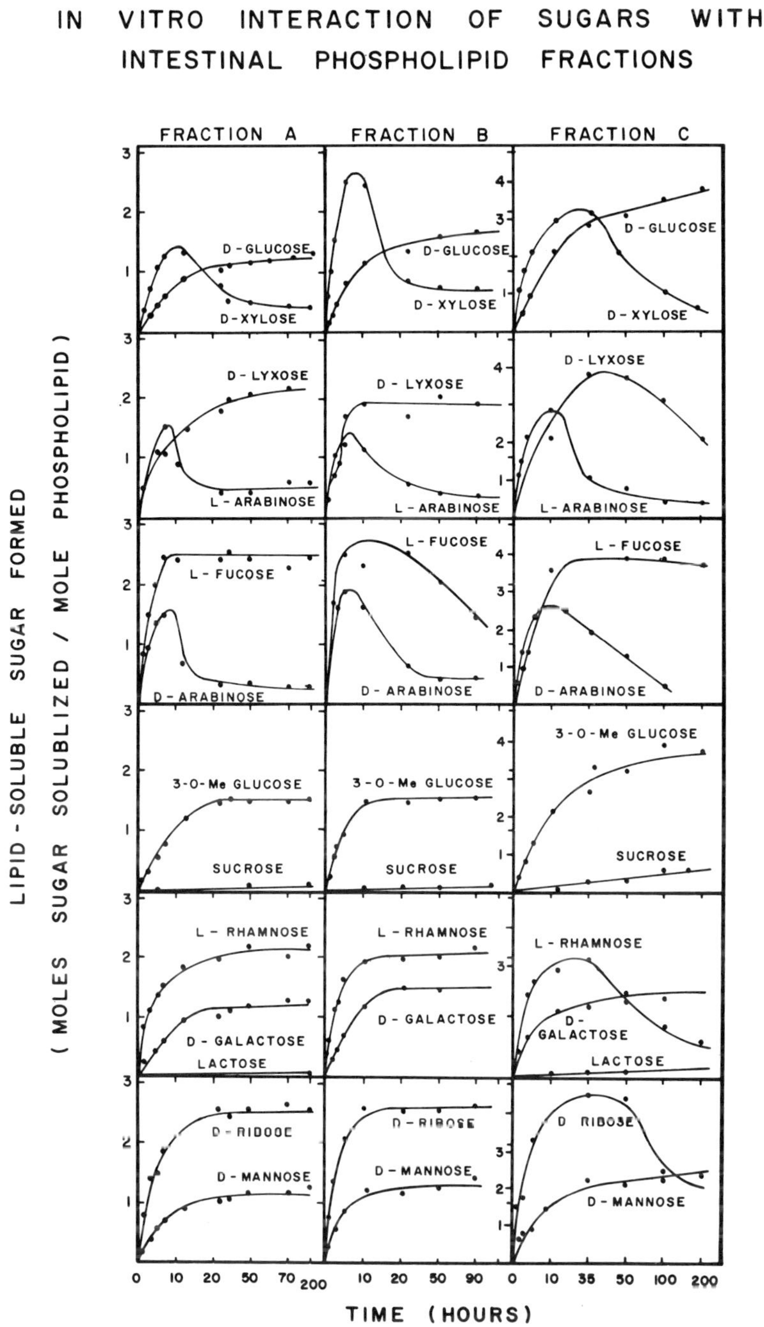

Figure 3. Formation of phospholipid–sugar complexes

be compared on a quantitative basis. It is found that the rate constants for formation of the hexose-phospholipids are some 1/4 to 1/5 of those for the pentoses or deoxy-hexoses. A similar relative rate of dissociation is also obtained when the rate of exchange of the radioactive sugar complexes with unlabelled sugars is measured in a non-aqueous assay system.

The significance, if any, of these complexes in sugar transport is not yet understood. The specificity pattern however has some suggestive correlations with those observed for transport, and the complexes may have some secondary role in determining the overall specificity (similar to that perhaps played by the hypothetical transporter or T substance of Figure 2) in the overall proposed scheme for the permease system. Considered in this sense the primary specificity of the system would be determined by the permease protein (P) in accelerating the formation of the substrate-transporter complex, but the overall specificity of the system would reflect the properties of all components.

Self-Regulation in Sugar Transport Systems. The principal sugar reabsorption process in kidney and intestine does not appear from present evidence to have any inherent self-regulatory features. For example, with normal levels of blood glucose, reabsorption in the kidney continues until the tubular filtrate is essentially free of glucose. Increasing levels of glucose in the filtrate—*e.g.*, as in diabetes—finally saturate the transport mechanism, and excess glucose is then spilled into the urine. In man the transport maximum, *Tm,* for glucose is about 300 mg per minute.

The insulin-sensitive glucose transport of muscle and adipose tissue is externally regulated in the sense that increasing blood glucose stimulates secretion of insulin by the pancreas. Metabolism of glucose is also subject to the feedback controls of the glycolytic enzymes (*49*), and release of glucose from liver into blood may also be self-regulated (*50*).

There are however indications that an internally self-regulating process operates in many species of bacteria, which rigorously controls the entry and accumulation of sugars *via* the phosphotransferase system (*51*). In certain bacteria glucose and α-methyl glucoside transport and phosphorylation are non-competitively inhibited by glucose-6-PO_4 (*52, 53, 54*). The inhibitory sites for the two esters are distinct and accessible from both sides of the membrane, and the inhibition of glucose transport by glucose-1-PO_4 is antagonized by glucose-6-PO_4, and vice-versa. It seems that in the bacteria the sugar phosphates are centrally important in the regulation of carbohydrate transport (*55*).

It has been found that a similar type of self-regulatory active transport also operates in mammalian cells, in this case in the Ehrlich ascites tumor cell (*41*). These cells show an active concentrative ability for glucose at low ambient glucose concentrations. This enables the cell to

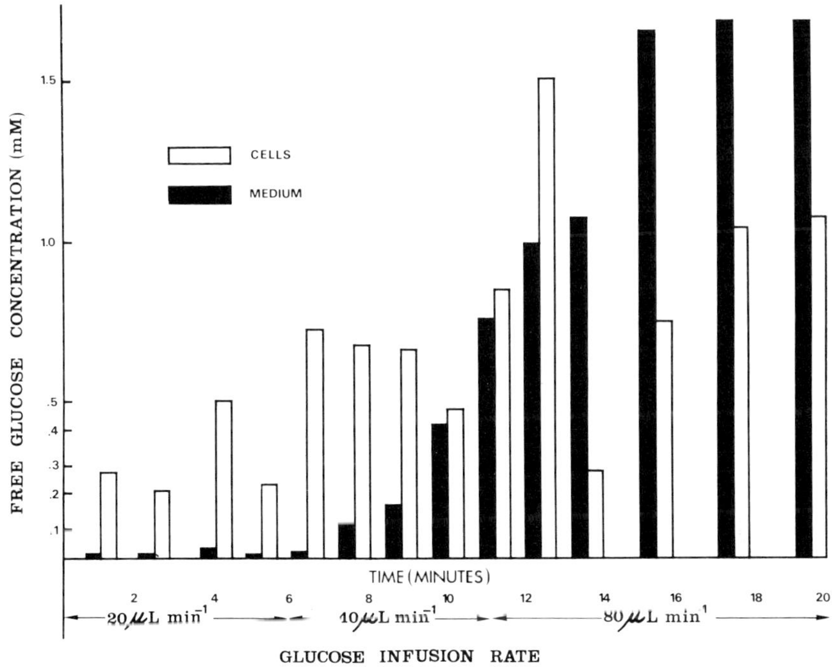

Figure 4. Accumulation of glucose by ascites tumor cells under semi-steady-state infusion conditions

maintain an internal glucose concentration which is higher than the extremely low levels found in peritoneal fluid of the host animal. It was first noted by Coe (*56, 57, 58*) that the cells showed a rapid initial uptake of glucose to a concentration which often exceeded that of the medium, but that the initial rapid concentrative phase was followed within 1 minute by an apparent decrease in permeability of the membrane and a fall in internal glucose to levels less than that in the medium.

By using a steady-state infusion technique to keep external glucose at low levels it is possible to maintain the concentrative phase of uptake for many minutes (Figure 4). (A suspension of Ehrlich ascites tumor cells in Krebs–Ringer buffer (pH 7.4) was oxygenated with 95% O_2/5% CO_2 at 15°. A 1% solution of D-glucose was infused into the suspension at a rate of 20 μliters per minute, and samples were removed for determination of glucose content of cells and medium using glucose oxidase. At 6 minutes the glucose infusion rate was increased to 40 μliters per minute and then to 80 μliters per minute at 11 minutes. A sharp decrease in permeability to glucose occurred at between 12 and 14 minutes (*41*).) Under these conditions of low ambient glucose, the concentration in the cell can attain levels 10–20 times those of the medium. As the rate of infusion is increased, the external (and consequently the internal) glucose levels increase until (at 14 minutes) a point is reached at which a sharp decrease in permeability to glucose occurs. The influx of glucose is then reduced, becoming equal to that of glucose metabolism and thus maintaining a constant internal glucose concentration. It is apparent that a self-regulatory system of this type for glucose transport in normal mammalian cells could be of considerable physiological significance.

Discovery of Mutarotase

Evidence of an enzyme preparation catalyzing the mutarotation of glucose was first noted by Bentley and Neuberger in 1948 in their studies of glucose oxidase (notatin) of *Penicillium notatum* (*59*). In 1951 Keilin and Hartree demonstrated that glucose oxidase preparations indeed contained a second enzyme, which was named mutarotase. This enzyme accelerated the oxidation of glucose by converting α-glucose to the β-form for which glucose oxidase is specific (*60*).

Keston in 1954 demonstrated a similar enzyme in mammalian tissues (*61*). Because the enzyme was present in particularly high concentration in kidney and intestine and because the pattern of substrate and inhibitor specificity resembled that for sugar absorption by these tissues, Keston proposed that the enzyme was involved in active transport of sugars. According to the postulated mechanism, mutarotase accelerated conversion of sugars to a preferentially absorbed form, which then because of a concentration gradient diffused back into the blood. The original hypothesis however was subjected to considerable criticism by Crane and others (*38*) when it was shown that compounds such as 1-deoxy-D-glucose (*43*) and α-methyl glucoside (*44*), which are inherently incapable of mutarotation, are substrates for the transport system of intestine and also competitively inhibit absorption of other actively transported sugars.

With the establishment of the permease hypothesis, however, it was apparent that the mere formation of a complex with the mutarotase protein may be the necessary interaction in transport (*15*). The subsequent mutarotation could be considered to be a coincidental consequence of the complex formation. To support this idea, it was found that 1-deoxy glucose and α-methyl glucoside are excellent competitive inhibitors of the enzyme (*16, 61*). Keston also showed that a number of cataractogenic sugars were inhibitors of lens mutarotase (*62*). It has since been shown that in all cases where a sugar is a substrate for the mammalian intestinal transport system it is also a competitive inhibitor of mutarotase.

Purification of the enzyme from kidney has been reported for hog (*68, 71, 74*), beef (*72*), sheep, lamb, rabbit (*70*) and human (*76*).

Bentley and Bhate (*66*) described the properties of the enzyme isolated from *P. notatum*, and Wallenfels and Herrman have reported the isolation and substrate specificity of a mutarotase from *E. coli* (*63*). The widespread occurrence of the enzyme in higher plants (*53*), fish (*69*), birds (*70*) and amphibia (*69*) has been noted. Sacks has recently described the properties of the enzyme in human red blood cells (*75*).

The possible involvement of mutarotases in sugar transport has been examined from a number of viewpoints, and the principal evidence which has emerged is summarized in Table I.

Table I. Summary of Evidence for Involvement of Mutarotases in Sugar Transport

Evolutionary History
- A) Increased amounts in higher organisms
- B) Stable catalytic and physical properties
- C) Tissue distribution matches glucose fluxes
- D) Correlates with evolution of kidney function in marine and fresh-water fish

Non-Involvement in Metabolism
- A) G-6 P has internal mutarotation mechanism
- B) Hexokinases and phosphatases have no anomeric specificity

Developmental History
- A) Correlates with development of transport in embryonic kidney and intestine

Substrate and Inhibitor Specificity Match Transport
- A) All interacting sugars are transported
- B) Inhibition by phloridzin and estrogens identical
- C) Osmotic transformations suggest energy-coupling mechanism

Comparative Properties in Different Species

The evolutionary history and function of a genetically determined constituent such as an enzyme can be studied in several ways. One approach is to examine the distribution and properties of the enzyme in different species and to relate changes to the known evolutionary sequence based upon phylogenetic relationships. A second method is based on the principle that ontogeny recapitulates embryonic phylogeny. Thus the embryological abundance of a protein in relation to morphological and functional characteristics can give information relative to the evolutionary development of a function. This type of approach was successfully used by Foa to investigate the development of the glucose permeability barrier and insulin sensitivity in embryonic chick hearts (*77*).

Both approaches have been used for mutarotase and have given a considerable amount of information on the phylogenetic distribution of the enzyme and its relationship to developing sugar transport systems.

Table II. Species Distribution of Mutarotases

Source [a]	*Mutarotase Content, units/gram* [b]
E.coli (*63*)	54
Mushroom (*62*)	39
Green pepper (*62*)	19
Catfish kidney (*87*)	226
Bullfrog kidney (*87*)	240
Chicken kidney (*87*)	800
Human kidney (*76*)	1300
Hog kidney (*71*)	4200

[a] Numbers in parentheses are the source references.

[b] 1 unit of enzyme catalyzes the conversion of 1 μmole of α-glucose to β per minute at 25°C. Since mutarotase catalyes the anomerization of its substrates—*i.e.*, converts α-D-glucose to the β-form, the resulting change in optical rotation can be followed with a polarimeter. This gives a convenient assay procedure for measuring mutarotase activity from various sources (*69,62,109*).

Table III. Mutarotase Levels in Organs and Tissues of the Rat

Tissue	*Mutarotase* [a], *units/gram*
Kidney (whole)	1617 ± 87
Kidney (cortex)	3160
Liver	605 ± 102
Upper small intestine (whole)	237 ± 48
Upper small intestine (mucosa)	421 —
Lower small intestine	162 ± 15
Stomach	142 ± 51
Lung	141 ± 10
Spleen	128 ± 51
Heart	88 ± 31
Brain	37 ± 12
Diaphragm	15 ± 7
Skeletal muscle	7 ± 5

[a] Data from Bailey and Pentchev (*119*).

The relative amounts of the enzyme in the various species and in the various organs of the mammals are listed in Tables II and III. The amounts of the enzyme tend to increase with increasing evolutionary development of the species. This, together with the widespread distribution, suggests evidence of a significant function. In mammals the enzyme has become concentrated in those tissues such as kidney, liver, and intestine where glucose fluxes are greatest. These are remarkable concentrations for an enzyme. For example, in some mammalian kidneys the amount is sufficient to convert twice the tissue weight of α- to β-glucose per minute under physiological conditions. From the measured specific activity of crystalline mutarotase (*72*), it can be calculated that the enzyme comprises about 0.3% of the total soluble protein in the kidney.

Despite the broad spectrum of distribution, the catalytic properties of the enzyme in the different species are remarkably similar. The same four sugars (D-glucose, D-galactose, D-xylose, and L-arabinose) are substrates for the enzyme from all sources (Table VI). Some changes, however, have taken place, and the marginal activity for maltose found in the lower species (*67*) has been lost in the mammals, whereas the relative activity towards the pentoses has become enhanced. The K_m values and turnover numbers for the substrates of purified beef kidney mutarotase are given in Table V.

The addition of a second sugar to the assay system often markedly reduces the enzyme-catalyzed mutarotation of D-glucose. Of the 50 sugars which have been tested, 30 are inhibitors of the enzyme in varying degrees (Table VII). The inhibition was measured for each sugar over a range of concentrations from which it was shown that the interactions in each case followed classical Michaelis kinetics for competitive inhibition (Figure 5) and from which the corresponding K_I values for each sugar were derived as described in the figure. (In Figure 5 the mutarotation rate of 16.67 *mM* α-D-glucose with 40 units of purified mutarotase was measured in the presence and absence of varying concentrations of an equilibrium mixture of a second sugar. The enzyme-inhibitor-complex dissociation constants (K_I) were obtained from the equation

$$V/V_I = \frac{K_m}{K_m + [S]} \cdot \frac{[I]}{K_I} + 1$$

where V and V_I represent the enzymatic velocities in the presence and absence of the inhibitor (I). When V/V_I is plotted against $[I]$, the intercept of this line representing $V/V_I = K_M/([K_M + S]) + 1$ gives a point on the $[I]$ axis equal to K_I. That the inhibition shown by this

Table IV. Substrate Specificities of Mutarotases from Different Sources

	Relative Velocities				
Sugar	*E. coli* [a]	*P. notatum* [b]	*Green Pepper* [c]	*Beef Kidney* [d]	*Mouse Kidney* [e]
D-Glucose	100	100	100	100	100
D-Galactose	108	107	151	215	230
L-Arabinose	25	128	49	190	179
D-Xylose	171	43	18	250	272
Maltose	2	3	6	0	0

[a] Calculated from data of Wallenfels and Herrmann (*63*).
[b] From Bentley and Bhate (*65*).
[c] Reference (*67*).
[d] Reference (*72*).
[e] Reference (*87*).

Table V. Kinetic Parameters of Crystalline Beef Kidney Mutarotase

Substrate	K_M *mM*	*Turnover Nos.* $\times 10^6$ min^{-1}
α-D-Glucose	19	1.0
α-D-Galactose	6.5	1.2
α-D-Xylose	13.2	1.2
α-L-Arabinose	8.3	1.9
α-D-Fucose	2.0	0.35

Table VI. Substrate Specificity for Catalyzed Mutarotation[a]

	Mutarotation Rate		*Relative*[a] *Increase*
	Spontaneous Ksp	*Plus Enzyme* K	ΔK/Ksp
D-Arabinose	0.138	0.136	0
L-Arabinose	0.132	0.900	5.8
D-Fucose	0.095	0.185	0.95
L-Fucose	0.098	0.090	0
D-Galactose	0.042	0.640	14.3
D-Glucose	0.028	0.345	11.3
L-Glucose	0.028	0.028	0
D-Glucosamine	0.117	0.117	0
2-Deoxy-D-glucose	0.176	0.181	0
3-*O*-Methyl-D-glucose	0.026	0.027	0
D-Lyxose	0.470	0.445	0
D-Mannose	0.042	0.042	0
L-Mannose	0.042	0.042	0
L-Rhamnose	0.262	0.275	0
D-Talose	0.200	0.201	0
D-Xylose	0.094	0.540	4.7
L-Xylose	0.092	0.088	0
D-Ribose	0.263	0.267	0

[a] Mutarotation of 0.3% solutions of the freshly dissolved sugars in 12 ml of 5 *mM* EDTA, pH 7.4 was followed. Significant differences in mutarotation rates (ΔK) in the presence and absence of 100 units of bovine kidney enzyme were expressed as the ratio $\Delta K/Ksp$. Differences of less than 5% in these rate constants were not considered significant. Of the 18 sugars listed, nine have been tested previously as substrates for other mammalian mutarotases with essentially the same pattern as described here. The pattern of specificity indicates that a 3-point attachment of enzyme and substrate is necessary for catalysis of mutarotation.

[b] Data from (*72*).

method is competitive in nature has also been shown for several sugars by conventional Lineweaver–Burk plots and by demonstration of reversal of inhibition at higher sugar concentrations (*72*).) When sugars which are also substrates were used as inhibitors, the K_I values as expected are the same as the respective K_M values. Many sugars which are not substrates are nevertheless extremely potent competitive inhibitors. The affinity of two of these inhibitors (L-fucose and L-xylose) was about 10 times that of substrate glucose.

Table VII. Competitive Inhibitors of Mutarotase[a]

Inhibitor Sugar	K_1 [b] mM	*Inhibitor Sugar*	K_1 [b] mM
D-Fucose	2.0	L-Arabitol	27.7
L-Fucose	2.6	D-Cellobiose	31.5
L-Xylose	3.7	D-Maltose	32.5
2-Deoxy-D-ribose	5.5	3-*O*-Methyl-D-glucose	35.5
D-Galactose	6.3	D-Xylose	38.0
D-Allose	6.4	L-Glucose	44.5
L-Arabinose	8.0	D-Erythrose	47.7
D-Ribose	9.8	D-Sorbitol	67.5
α-Methyl-D-glucoside	12.6	D-Xylitol	68.5
D-Xylose	14.0	D-Inositol	69.5
D-Galacturonic acid	18.4	D-Arabinose	77
D-Glucuronic acid	21.0	β-Methyl-D-xyloside	82.5
2-Deoxy-D-glucose	24.0	D-Melibiose	110
Galactitol	25.4	2-Deoxy-D-galactose	149
Erythritol	25.5	D-Mannitol	149

[a] Ability of the various sugars to bind to the enzyme was tested by measuring the inhibition of the enzyme catalyzed mutarotation of D-glucose. The mutarotation rate of 16.67 *mM* α-D-glucose was determined in the presence and absence of a second sugar added in its anomeric equilibrium form, using pure bovine kidney mutarotase. K_I for the sugars was determined by plotting V/V_I against $[I]$ where V and V_I represent the enzymatic velocities in the presence and absence of inhibitor and $(I]$ the concentration of the competitive inhibitor. The intercept of this line with a line representing $V/V_I = K_m/(K_m + [S]) + 1$ corresponds to a point on the $[I]$ axis equal to K_I (The K_m for glucose is 19 *mM*.). The following sugars were tested and were not inhibitors (a non-inhibitor being defined for practical purposes as a sugar with a K_I of greater than 150 *mM*): D-arabitol, D-fructose, *N*-acetyl-D-galactosamine, D-galactosamine, D-galactose-6-phosphate D-glucosamine, *N*-acetyl-D-glucosamine, D-glucose-6-phosphate, D-lactose, D-mannose, L-mannose, α-methyl-D-mannoside, D-raffffinose D-sedoheptulose, D-sucrose, D-trehalose, D-turanose, and D-rhamnose. The pattern of specificity indicates that a 2-point attachment of enzyme and substrate is sufficient for efficient binding.
[b] Data taken from Reference (*72*).

It seems that for catalysis of mutarotation to take place, the sugar must make a three-point attachment to the enzyme, one point being through the anomeric hydroxyl group, and the other two involving the hydroxyl groups at C-2 and C-3 of the sugar ring. It is presumed that the three-point attachment is necessary for correct alignment of the anomeric hydroxyl at the active center. For a competitive inhibitor, however, the pattern of specificity (Table VII) indicates that a two-point attachment suffices in the correct ring conformation. These points can be any two of the three which are required for substrate capability.

The molecular weight of the enzyme has remained remarkably constant from the bacteria to the higher mammals (Table VIII). The molecular weight from all species averages 36,500, and the deviations from this value fall within the experimental range of the methods used, in this case sedimentation equilibrium in the ultracentrifuge or electrophoresis of the SDS–enzyme-complexes on polyacrylamide gel.

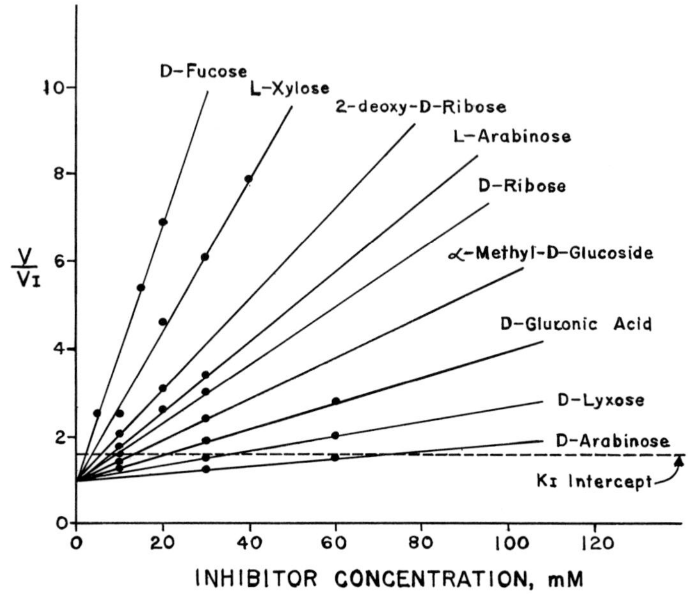

Figure 5. Competitive inhibitors of bovine kidney mutarotase

Table VIII. Molecular Weights of Mutarotase from Different Species[a]

	Molecular Weight		
Species	*SDS Gel Electrophoresis*	*Gel Permeation Analysis*	
		*5*mM	*155*mM
Human	37,000	45,800	37,300
Calf	36,800	51,000	37,000
Ox	37,800	49,500	37,000
Green Pepper	36,200	51,200	—
Lamb	36,000	51,000	—
Embryo Calf	—	50,800	—
Rabbit	36,500	49,000	37,500
Chicken	—	48,000	—
Yellow Perch	37,000	45,800	37,400
E. coli	35,000	—	—

[a] The molecular weights determined by gel permeation analysis in high osmolarity (155 *mM*), agree with the particle weights obtained by SDS gel electrophoresis or sedmentation equilibrium in the ultracentrifuge. The ability to transform to a less compact orm in low osmolarity (5 *mM*) appears to be a common property of all mutarotases.

A detailed examination of the comparative kinetic properties of the enzyme from different species has been made (*87*). It appears that there are three kinetically distinct forms of mutarotase. The form with a K_m of 12 *mM* is found in fish and some mammals, whereas the type with a

K_m of 19 *mM* is found in most other mammals (Figure 6). In the calf all three forms of the enzyme are found at different stages of development with a 5 *mM* type predominating in the embryo, and the 12 *mM* form being found in the newborn animal up to the age of about 6 months (*87*). Recently the situation in humans has been found to be unusual since the purified human kidney enzyme is the 12 *mM* type, whereas the liver enzyme has a K_m of 19 *mM* (*76*).

The physiological or evolutionary significance of these various forms of the enzyme is not understood. It is apparent since most of their other chemical and physical properties are identical (*88*) that they are closely related.

Embryological Development of Mutarotase and Sugar Transport in Kidney and Intestine

There is considerable evidence indicating that the embryonic kidney does not develop complete functional capacity until sometime after the birth of the animal (*85, 86*).

The morphological and functional development of the kidney has extensively been studied in rats. There is a rapid differentiation of the renal cortex after birth (*83*). The proximal segments of the tubules grow most rapidly, and micropuncture techniques have shown that glucose is reabsorbed solely by this segment of the tubule. Baxter and Yoffey have reported that in the newborn rat the peripheral region of the renal cortex is characterized by a neogenic zone of undifferentiated tissue, which cannot store the vital dye, trypan blue. By 28 days after birth, the tubules

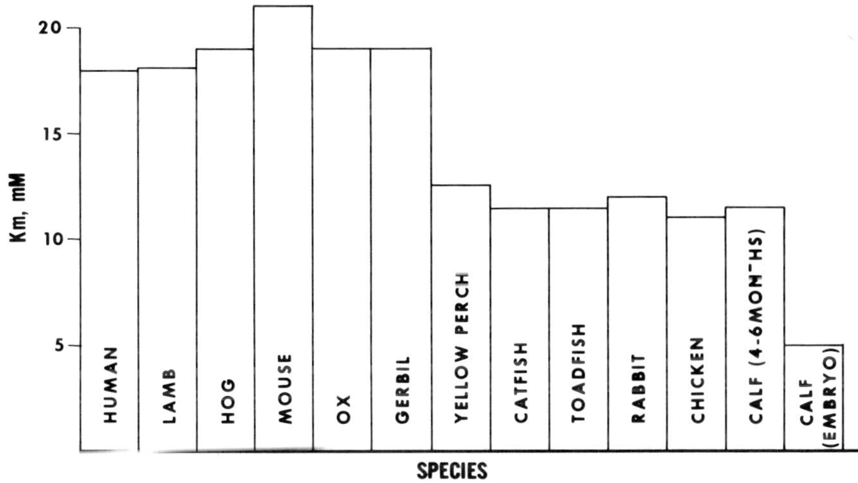

Figure 6. K_m Values for mutarotases from different species using D-glucose as substrate

in this region are fully developed, as judged by the ability to concentrate the dye. Renal function does not seem to be critical during the gestational period if the placenta is functional since fetuses without kidneys have been shown to survive to term (*83*).

The time of appearance of mutarotase in various organs of the embryonic and newborn rat has been compared with the development of functional ability of the tissue to transport sugars (*102*). The mutarotase content of the embryonic kidney was marginal compared with that of the mature organ (Figure 7). (Numbers of embryonic, newborn (0–3 days), suckling (3–20 days), and infant rats (20–30 days) were sacrificed, and the mutarotase and lactic acid dehydrogenase contents of the kidneys were measured. One unit of mutarotase catalyzes the conversion of 1 μmole of α-glucose per minute. One lactic acid dehydrogenase unit catalyzes the oxidation of 1 μmole of lactic acid per minute (*102*).) Low levels of the enzyme (10 to 20 units per gram) appear in the kidney on day 17. The enzyme content rises gradually to about 60–70 units per gram before birth. A rapid and progressive increase in the mutarotase levels begins immediately following birth. Adult levels of the enzyme (1100–1200 units/gram) were not attained until between the 20th and 30th days following birth. Lactate dehydrogenase, which was also assayed in the same tissue extracts, showed no striking changes in the immediate postnatal period. The studies on sugar uptake indicated a

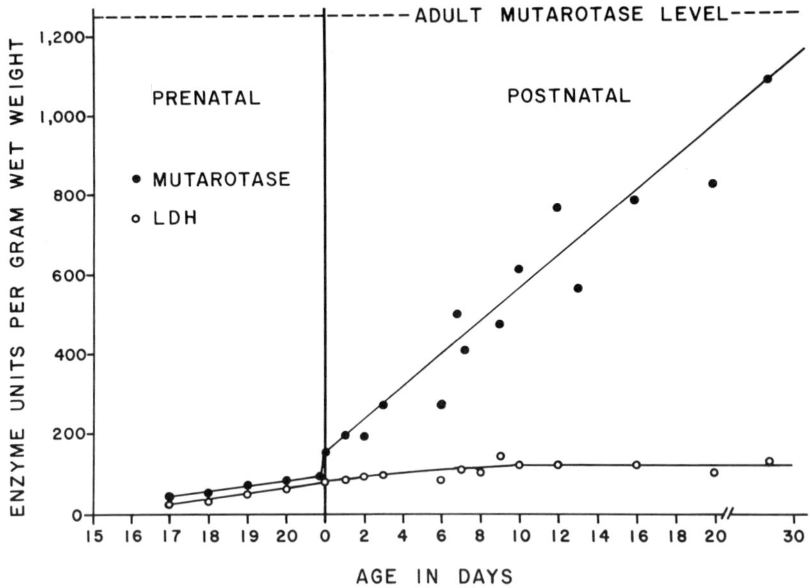

Journal of Biological Chemistry

Figure 7. Mutarotase and LDH content of developing rat kidney (102)

functional correlation also (*102*). Kidney slices from adult or immature animals transported galactose against a concentration gradient by a process sensitive to dinitrophenol. The neonatal kidney, which has low levels of mutarotase, has not yet developed an active transport mechanism. Concentration gradients of sugar and inhibition of uptake by dinitrophenol were not observed (Figure 8). (Segments of adult and newborn rat kidney were incubated separately in buffer containing D-galactose-1-^{3}H, with and without dinitrophenol (DNP) (7×10^{-5} *M*). Galactose uptake and concentration gradients were measured as described in Ref. *102*.)

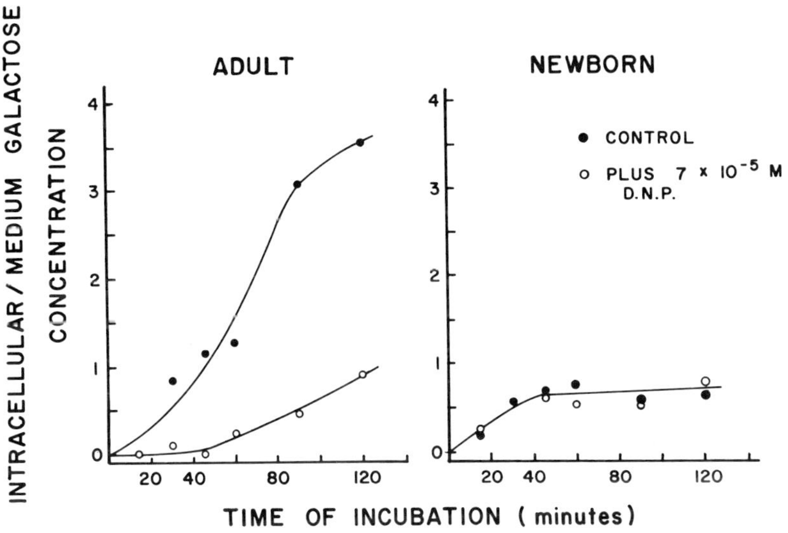

Figure 8. Galactose transport by kidney slices from adult vs. *newborn rats* (102)

Measurable amounts of mutarotase appear earlier in embryonic intestine, and the levels are considerably higher than in the kidney of embryos of corresponding ages (Figure 9). From a value of about 40–50 units per gram in the 15-day-old embryo, the enzyme increases by day 20 to about 200 units per gram. The values peaked at about twice the adult level during the 24 hours immediately preceding and the 36 hours following birth. There was a second maximum (800 units per gram) between the 7th and 14th days postnatal. Enzyme levels returned to normal at the end of the suckling period. In contrast, no corresponding fluctuations of similar amplitude were noted in the intestinal content of lactate dehydrogenase during the same period of development.

Studies on intestinal absorption were carried out *in vivo* (*102*). It was found that in contrast to the situation in kidney the galactose transport mechanism was well developed in the small intestine of newborn

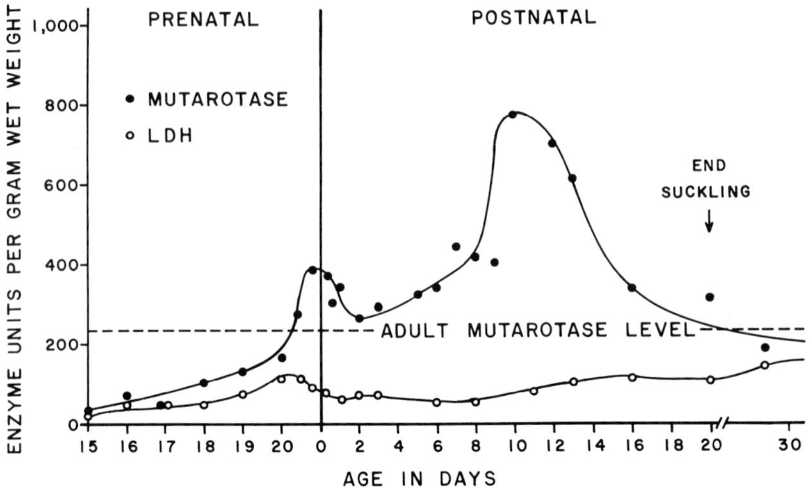

Journal of Biological Chemistry

Figure 9. Mutarotase and LDH content of developing rat intestine
For details see Figure 7 (102)

and early post-natal rats (Figure 10). (Left: Newborn rats were fed by stomach tube a mixture containing 200 μgrams each of ^{3}H-galactose and inulin-^{14}C in 0.2 ml of NaCl solution with and without addition of phloridzin (0.3 *mM*). After 30 minutes the tissues were analyzed. Galactose absorption is expressed on the ordinate as R_1:R_2 where R_1 is the initial ^{3}H:^{14}C ratio in the mixture and R_2 the final ^{3}H:^{14}C ratio of the material remaining in the excised intestine. Right: 4-day-old rats were fed by stomach tube with an NaCl solution containing 1.1 μmoles each of D-galactose-1-^{3}H and D-mannose-1-^{14}C and with and without 0.3 *mH* phloridzin. The relative rates of absorption were measured after 30 minutes from the change in the ^{3}H-galactose to ^{14}C-mannose ratio (*88*).) Preferential uptake of ^{3}H-D-galactose was observed with ^{14}C-inulin and ^{14}C-D-mannose as reference compounds. This preferential uptake was abolished by phoridzin.

In the kidney the number of nephrons is more or less well established at birth, and the increase in cortex size is predominantly a result of increase in tubular volume. The data of Arataki (*84*) on the number of rat nephrons per kidney, taken along with the observed increase in size of the kidney, have been used to calculate the average increase in the volume of the individual tubule. This increase in tubular volume results primarily from growth of the proximal segment of the tubule, which is the site of sugar reabsorption. A close correlation between tubular volume and mutarotase content was observed over the entire postnatal growth period (*102*).

Similar studies have been reported by Segal *et al.* (*85*) on the late development of the sugar transport processes in kidney, as measured by the ability of kidney slices from newborn animals to accumulate α-methyl glucoside. The concentrative ability is first observed 5 days following birth and does not reach the adult levels until 25 days of age.

The embryonic and newborn animals, however, do not spill glucose into the urine, and bladder samples taken following glucose injection indicate that the sugar is completely reabsorbed. Segal *et al.* (*85*) report that a rudimentary transport system is present in the newborn, as indicated by a decrease in accumulation in the presence of phloridzin. Thus, it seems possible that some more primitive mechanism for glucose reabsorption may operate in the embryonic kidney.

Enzyme Levels in Marine and Freshwater Fish

The evolutionary development of the kidney can also be followed in fish (*105*). The protovertebrates evolved in freshwater rivers. These species developed the glomerulus as an appendage to the more primitive nephrostomatous tubule. Since freshwater fish are osmotically superior to their environment, the need arose for a kidney which like the mammalian kidney filters off excess water but retains salts, sugars, and amino acids. The glomerular kidney of these fish accomplishes this by allowing waste products to be filtered from the blood while permitting glucose, salts, and amino acids to be reabsorbed in the tubules. Conversely, those

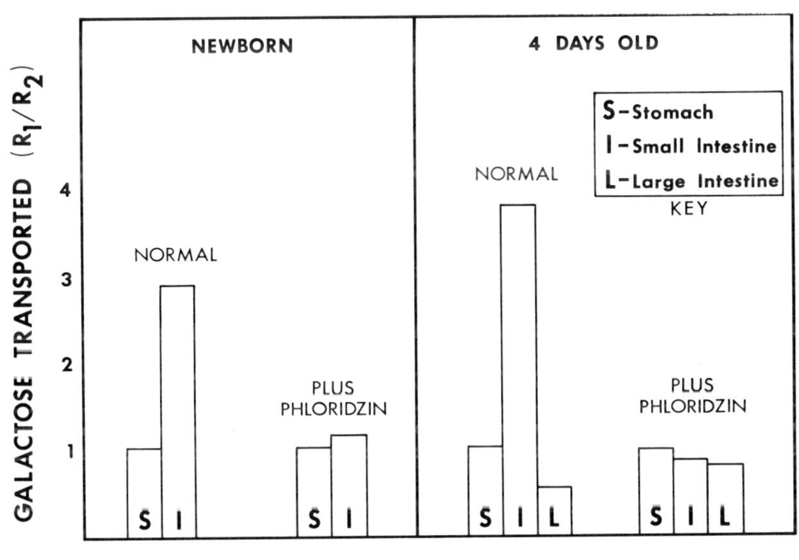

Figure 10. Galactose transport by new-born rat intestine (102)

fish which migrated to the sea must retain water and reject salt and have, therefore, evolved a glomerulus which is reduced in size and activity. Since only a minimum fluid is excreted in marine fish, there is a minimum requirement for glucose reabsorption in the tubules.

The relative amounts of the enzyme in kidneys of fish from different habitats have been measured (*105*) (Table IX). Particularly interesting is the finding that the mutarotase levels in kidneys of freshwater fish are always higher than those in saltwater fish, averaging six times greater. This extends an earlier observation by Keston that the enzyme was present in low levels in the aglomerular toadfish kidney (*68*).

Table IX. Mutarotase Content of Marine *vs.* Freshwater Fish Kidneys

Species	*Enzyme units/gram*[a] *wet weight*
Freshwater fish	
Catfish (*Ictalurus punctatus*)	226
Carp (*Carpiodes*)	187
Rock bass (*Ambloplites rupestris*)	103
Largemouthed bass (*Micropterus salmoides*)	143
Crappie (*Pomoxis annularis*)	132
Pumpkin seed (*Lepomis gibbosus*)	182
Warmouth (*Crypticanthodes maculatus*)	384
Saltwater fish	
Croaker (*Micropogon undulatus*)	26
Flounder (*Paralichtys*)	10
Sea trout (*Cynoscian nebulosus*)	13
Sheepshead (*Archosargus probayocephalus*)	27
Toadfish (*Opsanus*)	54
Blow fish (*Sphaeroides maculatus*)	51
Sea robin (*Prionotus evolans*)	40
Ling (*Molva molva*)	20
Sand shark (*Carcharus taurus*) (*Elasmobranch*)	63

[a] Data from Reference (*105*).

The mutarotase levels in other tissues and organs were measured for comparative purposes. The higher levels in freshwater *vs.* saltwater fish are confined to kidney tissue. The amounts of enzyme in liver, spleen, heart, and skeletal muscle are about the same for the two groups. In Table X enzyme levels in tissues of the rat are compared with those in fish. Most mammalian tissues contain from two to five times more enzyme than the corresponding tissue in the fish, including freshwater fish kidney. Mutarotase levels in the mammalian kidney however are over 60 times higher than those in marine fish. The evidence is convincing that the difference is related to the altered functional requirement for sugar reabsorption.

Absence of Rate-Limiting Anomerization Reactions in Glucose Metabolism

The only known metabolic function for mutarotase is in certain bacteria where a primitive oxidative transformation of β-glucose occurs, catalyzed by the enzyme glucose oxidase (*59, 60*). The anomeric specificities of the various enzymes involved in glucose metabolism in mammals have only recently been defined. Glucose-6-phosphate is an interesting compound in this respect since it is at the crossroads of several important metabolic pathways for which anomeric specificities have recently been reported by Salas, Vinuela, and Sols (*80*). It was shown that phosphoglucomutase interacts specifically with α-D-glucose-6-phosphate to give α-D-glucose-1-phosphate, whereas phosphohexose

Table X. Comparative Mutarotase Levels in Tissues of Mammals and Marine Fish

Tissue	*Enzyme units/gram fresh wt.*[a] *Rat*	*Fish*	*Ratio (rat/fish)*
Kidney	1617	26	64
Liver	605	185	3
Intestine	237	45	5
Spleen	128	82	2
Heart	88	76	1
Muscle	7	15	0.5

[a] Data from Reference (*105*).

isomerase apparently requires the open chain form of glucose-6-phosphate. Entry into the pentose pathway occurs *via* β-glucose-6-phosphate since glucose-6-phosphate dehydrogenase is specific for the β-anomer.

The anomeric preferences of these enzymes suggested that rate-controlling anomerizations may occur during glucose metabolism, which could involve mutarotase.

The anomeric specificities of a number of glycolytic enzymes have been determined by using the reduced triphosphopyridine nucleotide (TPN)-coupled rate of oxidation with glucose-6- phosphate dehydrogenase, or by the enzyme glucose oxidase, which is also specific for the β-anomer (*106*).

The hexokinases from yeast or ascites tumor cells showed only a marginal preference for α glucose. The K_m values for α and β glucose (58 and 66 *mM* for the yeast enzyme and 77 and 82 *mM* for the tumor enzyme) were essentially the same, and the turnover numbers were only 10–20% higher with α-glucose.

By analysis of the products with glucose oxidase, it was shown that the anomeric composition of the glucose liberated from glucose-6-phosphate by the enzymes acid or alkaline phosphatase or by glucose-6-phosphatase from rat liver was essentially the same as that of the substrate, thus indicating a lack of anomeric specificity for these enzymes also (*106*).

The mutarotation of glucose-6-phosphate is too rapid to be measured by conventional polarimetric methods (*80, 81*). Use was made of the specificity of glucose-6-phosphate dehydrogenase for the β-anomer. By monitoring NADP reduction spectrophotometrically, mutarotation of the α-anomer to the oxidizable β- form was measured. The absolute mutarotation rate of glucose-6-phosphate was thus shown to be about 240 times that of D-glucose under the same conditions (*106*). The activation energy for the reaction (21.8 kcal), however, is not significantly different from that of D-glucose.

The inherent rapid mutarotation of glucose-6-phosphate, thus, seems to eliminate the possibility that a rate-limiting anomerization could have a significant regulatory role in mammalian carbohydrate metabolism (*81, 106*). A requirement for mutarotase at other points in the metabolic pathways also seems unlikely, as illustrated by the overall summary of the reactions given in Figure 11. (The mutarotational half-life of either α- or β- glucose is about 7 minutes under physiological conditions. Both anomers are phosphorylated at about the same rate by hexokinases, glucokinase, or pyrophosphate phosphotransferase. The glucose-6-phosphate, formed from either anomer exclusively, rapidly mutarotates with a half-life of only 1.5 seconds to an equilibrium mixture containing about 61% of the β-anomer. This mutarotation is not catalyzed by mutarotase. The α-anomer of glucose-6-phosphate is required by phosphoglucomutase and phosphohexose isomerase whereas glucose-6-phosphate dehydrogenase is specific for the β form. Both anomers are dephosphorylated equally readily by acid or alkaline phosphatase and by glucose-6-phosphatase from liver. The mutarotational half-life of glucose is about 6 seconds in the presence of the average content (700 units per gram) of mutarotase in liver. Because of this lack of anomeric specificity of the enzymes phosphorylating and dephosphorylating glucose and the rapid spontaneous mutarotation of glucose-6-PO_4, it seems unlikely that anomerization reactions could become rate-limiting in glucose metabolism (*106*).)

Further information on the lack of requirement for mutarotase in glucose metabolism has come from a study of the metabolism of a rapidly growing tumor. Ehrlich ascites cells were found to contain less than 0.5 unit of mutarotase per gram, wet weight (this figure should be compared with up to 3000 units per gram for kidney and 400 units per gram for intestinal mucosa). These cells metabolized glucose at rates up to

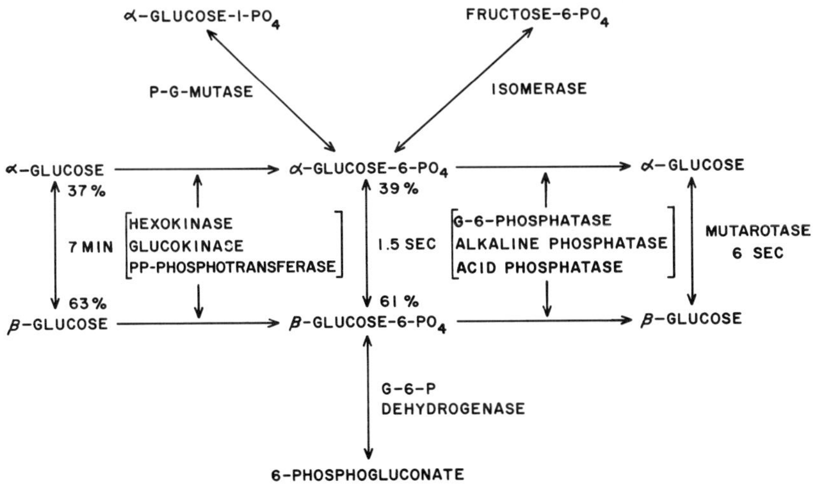

Journal of Biological Chemistry

Figure 11. Anomeric specificities of glucose-metabolizing enzymes (106)

5 μmoles/gram/minute—*i.e.*, 10 times the maximum velocity of the enzyme-catalyzed mutarotation reaction (*107*).

The rapid spontaneous mutarotation of glucose-6-phosphate has been shown to result from an intramolecular catalysis of the reaction by the phosphate group at carbon 6 (*81*). The cleavage of glucose into three carbon fragments, which is essentially a reversal of the aldol condensation reaction, requires the ketohexose as substrate. The necessary isomerization reaction to form the ketohexose then uses the open-chain form intermediate of the mutarotation reaction. Salas *et al.* (*80*) have speculated that the enhanced mutarotation of glucose-6-phosphate may thus have been the key requirement which led to the evolution of the phosphorolytic pathway for glucose metabolism.

Nature of the Catalytic Site of Mutarotase

The chemical properties of the crystalline enzyme obtained from beef kidney cortex, have been studied extensively (*72, 88*). Based upon a molecular weight of 37,000, and a specific activity of 26,500 units/mg, the catalytic coefficient for glucose is calculated to be 0.98×10^6 (*72*). Because of this high efficiency, the nature of the active site is of considerable interest.

The enzyme contains four sulfhydryl groups, and all four may be required for activity since inactivation parallels the decrease in SH content over the entire rate range (Figure 12). (A two-fold molar excess of *p*-mercuri-benzate was added to a sample of crystallized beef kidney

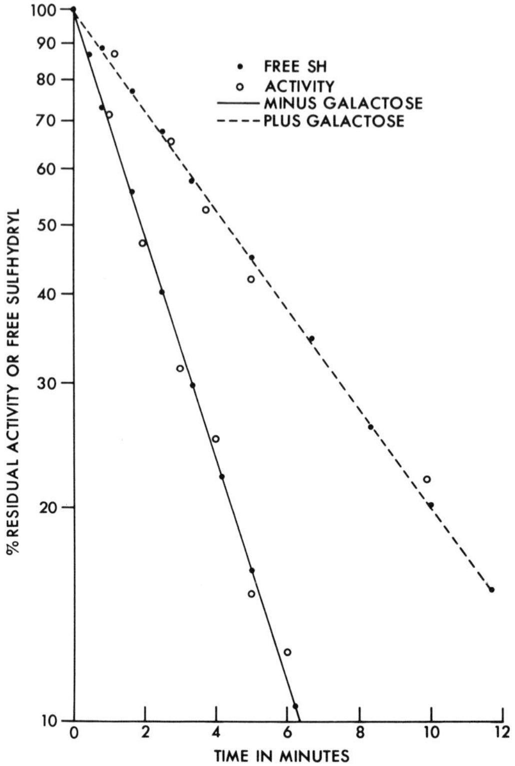

Figure 12. Sulfhydryl-dependent inactivation of mutarotase and protection by substrate

mutarotase at zero time. Decrease in the free sulfhydryl content was followed in a recording spectrophotometer at 250 mμ, and samples were removed at intervals for assay of enzyme activity in an EDTA-mercaptoethanol buffer. The first-order rate constants for inactivation and for complexing of the four SH groups in the molecule were the same (0.363 minute^{-1}) and were both reduced equally (0.160 minute^{-1}) when substrate galactose was added.) All four SH groups are also protected equally by substrate galactose, which indicates that they are close to, if not actually a part of, the catalytic site.

Wallenfels (*64*) has proposed that proton abstraction and addition by the concerted action of a histidine imidazole nitrogen and the SH group could form the basis for the enzyme-catalyzed mutarotation. Amino acid analysis indicates the presence of ten histidine residues per molecule of enzyme (*88*).

Other evidence indicates however that it is unlikely that histidine is part of the catalytic site. The enzyme can be photo-inactivated in the presence of oxygen using visible light and either methylene blue or rose bengal as sensitising dyes (Figure 13). (Solutions of pure bovine kidney mutarotase were photo-oxidized at 12.5°C with 0.2% methylene blue as photosensitising dye, and the residual enzyme activity was determined at intervals. The rate of inactivation was first-order down to less than 2% residual activity and was reduced by addition of substrates (about 40% by D-glucose (shown), and 65% by α-methyl-D-glucoside), but was unaffected by addition of a non-substrate (α-methyl-D-mannoside), indicating that the photosensitive group is probably part of the active center of the enzyme.) The presence of substrate glucose decreases the rate of photo-inactivation, whereas the non-substrate mannose does not, indicating that the photo-oxidizable group is near, or part of, the active center. The pH-profile of the photosensitivity does not match that of either histidine or imidazole and indicates fairly conclusively that histidine is not involved (Figure 14). (The first-order rate constants were measured for photoinactivation of pure bovine kidney mutarotase, in the presence of rose bengal (10 μgrams/ml) as photosensitizing dye in solutions of different pH values.) By measuring the photo-oxidation of

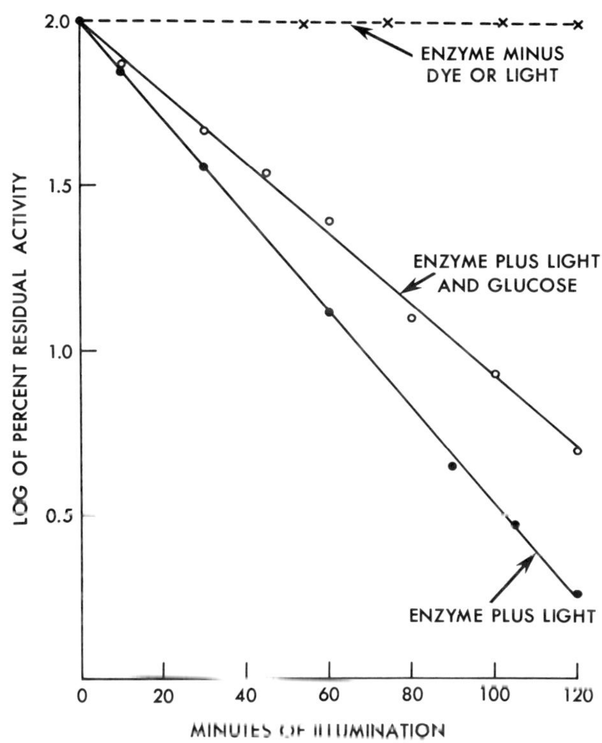

Figure 13. Photochemical inactivation of mutarotase

model amino acids in the same system, it was also possible to eliminate SH, methionine, lysine, or tyrosine as the photosensitive group. The pH profile of the photooxidation most closely matched that of the amino acid tryptophan (Figure 14). Tryptophan probably functions (as it does in lysozyme) as part of the binding site for the sugar. By analogy with the similar reaction in the hydrolysis catalyzed by lysozyme, it seems conceivable that binding of the sugar, perhaps accompanied by a conformational change in the enzyme molecule, may favor formation of the half-chair form of the sugar as a reactive intermediate in the mutarotation reaction. Tipson and Isbell have obtained evidence of the half-chair form as an intermediate in the spontaneous mutarotation of sugars (*109*). The transition to the half-chair form is believed to involve a free-energy change of about 10 kcal. The activation energy of the enzyme-catalyzed mutarotation has been shown to be about 10 kcal less than the spontaneous (*78*). Such an enhancement would be an example of the rack and strain theory of enzyme catalysis (*90*) and has considerable appeal since the favorable changes in free-energy would reduce the requirement for potent acids or nucleophiles in the active center and may explain why none is apparently present.

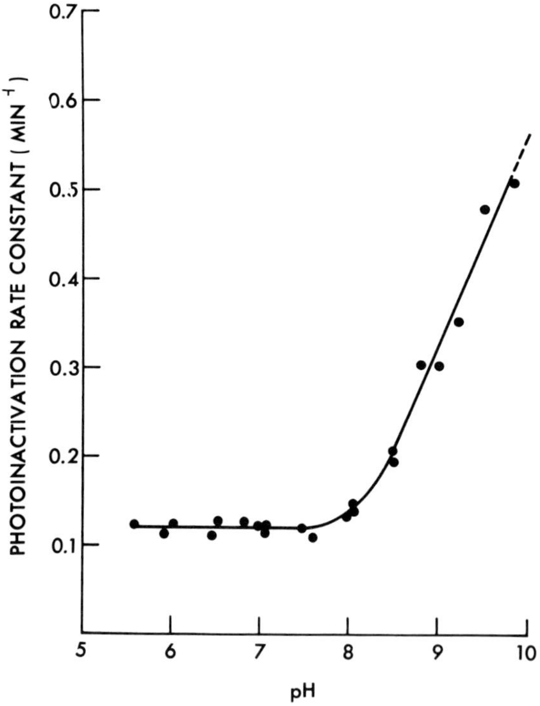

Figure 14. Influence of pH on the photoinactivation of mutarotase

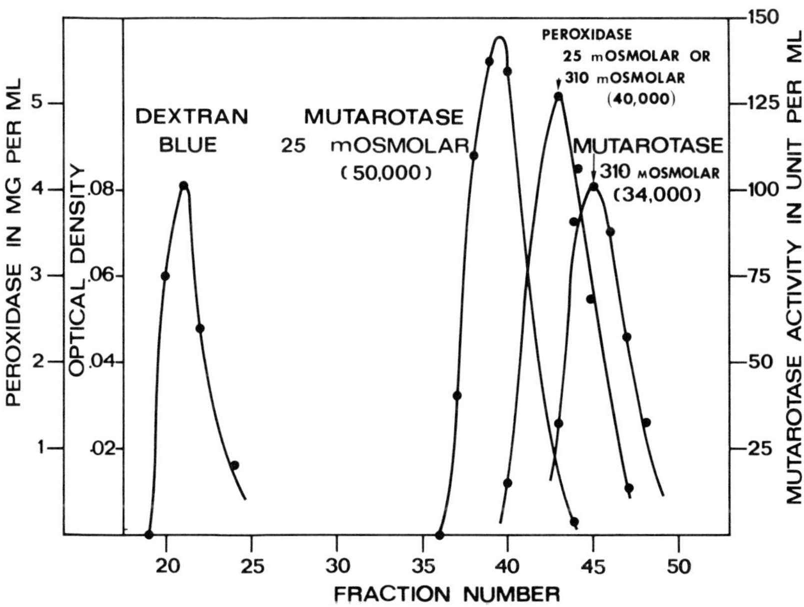

Figure 15. Osmotically induced transformations of mutarotase monitored by gel-filtration chromatography

Osmotically Induced Transformations in the Mutarotase Molecule. When measured by techniques such as gel-filtration analysis, the apparent molecular weight of the pure enzyme is sensitive to the osmotic environment (*110*). As indicated by the values given in Table VIII, the enzyme has an apparent molecular weight of about 50,000 when measured in buffers of low osmolarity (5 *mM*), whereas in buffers of high osmolarity (155 *mM*) the apparent molecular weight is 37,000. Since gel-filtration is a measure of the Stokes radius, it is apparent that the enzyme is converted to a less compact form in buffers of low osmolarity. The molecular weight of 37,000 agrees well with the particle-weight measured by sedimentation equilibrium in the ultracentrifuge or by electrophoresis in SDS gels (*111*). The sensitivity to the osmotic environment is an inherent property of the enzyme rather than a change in properties of the gel since filtration of standard proteins is essentially the same under both conditions (Figure 15). (Elution of pure bovine kidney mutarotase was monitored from the same Biogel P100 chromatographic column in buffers of two different molarities (25 *m* osmolar EDTA or 310 *m* osmolar Krebs-Ringer pH 7.4). Mutarotase eluted as a compact form with an apparent molecular weight of 34,000 at the high osmolarity but with an apparent molecular weight of 50,000 in buffers of low osmolarity. The elution pattern of control proteins, peroxidase (shown) and carbonic anhydrase or myoglobin added as internal standards, was unaffected by the change in osmolarity. Dextran-blue was used as a standard dye for determining the void volume. Similar transformations induced by change in the osmotic environment were found for mutarotases from human, calf, beef, green pepper, lamb, embryo calf, rabbit, chicken, and yellow perch.) The apparent transformation to a more compact form is undergone by mutarotases from all species which have been tested (Table VIII).

The selectivity of membrane filters of graded pore size has also been used to show the transformation (*11*). The filtration through Diaflo XM50 membrane increased from 8% in low osmolarity to about 35% in Krebs–Ringer buffer. Similar results were obtained with 155 millosmolar NaCl or KCl (Table XI). An interesting observation is the finding that addition of glucose partially reverses the effect of high osmolarity (Figure 16). (Pure bovine kidney mutarotase (200 units) was filtered with an XM50 filter (approximate 50,000 molecular weight cutoff), and the relative concentrations of enzyme in the upper cell and in the filtrate were measured in response to the solutions added to the upper cell. Increase in the osmolarity transformed the enzyme to a more compact and filterable form. The transformation was partially reversed by addition of glucose and entirely reversed on returning to low osmolarity (10*mM* tris buffer).) This effect is specific for glucose and other substrate sugars. It was not produced by nonsubstrates such as D-mannose or L-fucose. (Table II). More work is required to determine if these transformations have any relationship to a transport function of the protein.

Comparative Studies on Substrate and Inhibitor Specificities of Sugar Transport

The increasingly sophisticated studies in recent years on the specificity of sugar transport have enlarged the list of possible stereoconfigurations which may be transported by kidney or intestine under various experimental conditions. The relatively simple requirements originally outlined by Crane (*3*) have tended to become somewhat less definite. The direct comparison of the specificity of sugar transport with that of an enzyme such as mutarotase may therefore be misleading in that the spectrum of specificities showed by the transport system is not as clearly definable. As noted previously, it is also possible that the overall process of transport may require a number of individual components. Thus the specificity of the overall process will reflect an average or compromise specificity of all components. It should also be remembered that the functional transport specificity observed may represent the simultaneous superposition of more than one mechanism.

Table XI. Species and Protein Specificity of Osmotically Induced Transformations of Kidney Mutarotases[a]

Type of Protein	*Percent Protein Filtered* 5 mM *Tris buffer*	155 mM *buffer*	155 mM *KCl*
Bovine Kidney Mutarotase	15	35	35
Rabbit Kidney Mutarotase	16	37	—
Human Kidney Mutarotase	44	73	—
Hog Kidney Mutarotase	17	47	—
Ovalbumin	8,9	10,10	—
Bovine Serum Albumin	20	16	—
Myoglobin	63,59	65,67	—

[a] Purified mutarotases and other proteins used as controls were filtered through a 50 ml Diaflo filtration apparatus under the conditions indicated. Results are expressed in terms of the ratio of protein in the filtrate to that in the upper chamber. For the human and hog enzymes a different batch of XM 50 membranes was used.

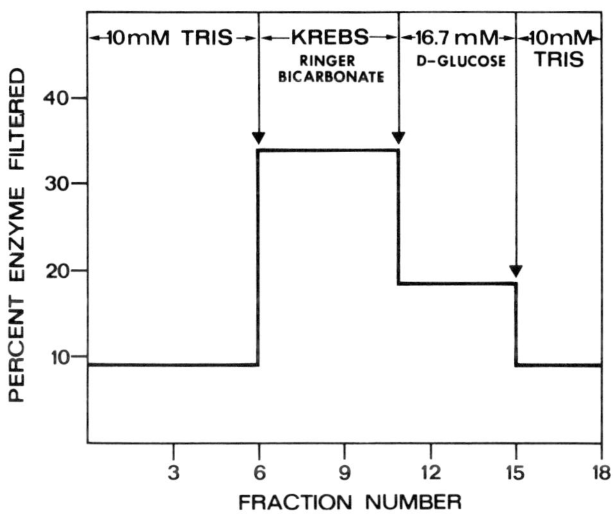

Figure 16. Osmotically induced transformations of mutarotase monitored by membrane filtration analysis

However, if these limitations are remembered, it can be instructive to compare the patterns of interaction which have been variously reported for substrates and inhibitors of mutarotase, with similar patterns for the kidney and intestinal sugar transport processes. In Table XIII the reported specificities for mammalian intestine are compared. In cases where comparative data are available, all sugars which are actively transported or which are passively transported but share the same carrier as glucose also interact with the active center of mutarotase. Particularly interesting is the observation that L-fucose, the most potent sugar inhibi-

Table XII. Sugar Specificity for Induced Transformation of Mutarotase

Sugar Tested	*Reverses Transformation*	*Substrate for Enzyme*	*Competitive Inhibitor*
D-Glucose	+	+	+
D-Xylose	+	+	+
L-Arabinose	+	+	+
D-Galactose	+	+	+
D-Fucose	+	+	+
L-Fucose	–	–	+
D-Ribose	–	–	+
2-Deoxy-D-glucose	–	–	+
α-Methyl-D-glucoside	–	–	+
D-Mannose	–	–	–
Sucrose	–	–	–

tor of glucose transport (*38*), is also the most potent (nonsubstrate) inhibitor of the mutarotation reaction.

Inhibition by Phloridzin and Certain Estrogens. Phloridzin has been shown to be a potent inhibitor of sugar transport *in vivo* and *in vitro* in many tissues and cell types (*112*). It was originally shown by Keston (*68*) that mutarotase was inhibited by low concentrations (10^{-4}–10^{-5} *M*) of phloridzin similar to those which block glucose absorption in kidney and intestine. Compounds related in structure to phloridzin, including diethylstilbestrol and certain other estrogens, have also been found to be potent inhibitors (*16*). Many of these compounds were shown by LeFevre to block glucose transport in erythrocytes (*98*). In a recent, comprehensive study, Diedrich and Stringham have compared the ability of phloridzin, phloretin, and 30 structurally related compounds to inhibit the enzyme with their efficiency as transport inhibitors in intestine and red blood cells (*94*). For the red-cell transport system there was a close correlation between changes which reduced the effectiveness of a compound as an inhibitor of mutarotase and those which decreased its ability to inhibit transport (Table XIV).

Diedrich has also measured by indirect means the content of the phloridzin-sensitive transport site in dog kidney and obtained a figure of 0.68 μmole per 100 grams of tissue (*113*). Dog kidney mutarotase has not yet been purified, but the content in those species where the turnover number is known (*58*) is calculated to be about 0.5 μmole per 100 grams of cortex, in fair agreement with Diedrich's estimate for the transport site.

It was also found as a result of this work that mutarotase appears to have two binding sites for phloridzin, one for the aglycone portion and one for the sugar. The lesser inhibition by phloridzin is thought to reflect reversal of the inhibition produced by the aglycone portion of the

Table XIII. Comparison of Ability of Sugars to Inhibit Mutarotase with Literature Data on Active Transport[a]

Sugar	*Inhibits*[b] *Mutarotase*	*Transported by Intestine*
D-Galactose	Yes	Yes
D-Mannose	No	No
3-*O*-Methyl-D-glucose	Yes	Yes
1-Deoxy-D-glucose	Yes	Yes
2-Deoxy-D-galactose	No	No
D-Xylose	Yes	Yes
α-Methyl-D-glucoside	Yes	Yes
L-Glucose	Yes	Yes
L-Mannose	No	No
6-Deoxy-D-glucose	Yes	Yes
D-Fructose	No	No
D-Glucosamine	No	No
D-Fucose	Yes	Yes

[a] Data from (*15*).
[b] Some sugars such as L-arabinose and 2-deoxy-glucose do not fit the pattern since neither is reportedly transported by intestine, but both interact with mutarotase (Table VII). However, 2-deoxy-glucose is reportedly transported by kidney cells (*100*), and L-arabinose also inhibits transport of D-xylose in the intestine (*101*).

Table XIV. Comparative Inhibition of Mutarotase and Red Blood Cell Glucose Transport by Phloretin and Related Compounds

	Relative Potency[a]	
Inhibitor	*Glucose Transport*	*Mutarotase*
Phloretin	100	100
Diethylstilbestrol	60	70
Hexestrol	70	43
Dienestrol	15	16
Naringenin	7	11
Phloroacetophenone	—	0.2
Phloroglucinaldehyde	0.2	—

[a] Inhibition of red-cell glucose transport is from LeFevre (*98*) and Forsling and Widdas (*99*); mutarotase data are from Diedrich (*93*).

molecule. Diedrich points out the interesting parallelism with the work of Alvarado, who has demonstrated that the intestinal glucose-binding protein also seems to have two binding sites for phloridzin, one for the aglycone and one for the sugar portion (*96*). Alvarado makes a similar suggestion—*i.e.*, that binding of glucose by the intestinal glucose-binding site is inhibited allosterically by phloretin—as does Diedrich for the mutarotase binding site.

It seems reasonable to conclude from these studies that if mutarotase and the sugar-binding site in intestine and the red blood cell are not identical at least they are structurally very similar.

Arguments against the Hypothesis

The bulk of the evidence obtained to date indicates only a strong circumstantial association of mutarotase with sugar transport. Evidence of this kind by its nature cannot provide definite proof that the enzyme is essential for transport. The similarities in substrate specificity could be coincidental, and the common inhibition by phloridzin and related compounds could merely be related to the fact that the transport-system and mutarotase are glucose-specific although for different functional reasons. Similar objections can also be made against the evidence that levels of the enzyme in embryonic kidney and intestine parallel the capacity of the tissues to transport sugars. It is known that the functional capacity for transport accompanies the development of the proximal tubules in the kidney. There may be other enzymes which are preferentially localized in tubules, whose levels would be found to increase in the same way as mutarotase. It must be admitted however that the relatively high content of mutarotase compared with other enzymes (up to 5000 units per gram in cortex) and the correlation between the molar content of the enzyme with that measured for the glucose transport binding-site by Diedrich (*113*) makes this objection less valid.

It may be that the enzyme has no function in mammals. As suggested by Chase (*114*), it may represent some vestige of an evolutionary history beginning with the primitive bacteria for which anomer-specific oxidative processes for glucose (*cf.* glucose oxidase) were important (*60*). It does seem however unlikely if this were the case, that selective pressures during a long evolutionary history not only failed to eliminate an unnecessary enzyme but actually led to some hundredfold increase in its activity in mammalian kidneys compared with the bacteria. Since the enzyme is apparently not required at any stage of the known metabolic pathways for glucose, it seems by default that it must be related to some specific function of the kidney such as sugar transport.

One type of argument which is of doubtful validity is to assume that all sugar transport necessarily results from the operation of a single biological mechanism and to claim that one mechanism alone can account for all the observations. The concept that no covalent transformation of the sugar was necessary in the bacterial permease system certainly obscured for a considerable time the separate existence of the omnipresent phosphotransfrases. The discovery of this latter system by Kundig, Ghosh, and Roseman forced a complete reevaluation of our ideas about bacterial sugar transport (*17*). It seems that there is now some danger of attempting to account for *all* the properties of bacterial sugar transport based on the phosphotransferase mechanism alone.

One of the strongest arguments against the involvement of mutarotase in transport is the fact that it is not present in any membrane structure, at least as isolated by presently available techniques. In intestine the enzyme is localized in the mucosa. When brush borders were isolated using EDTA treatment, the enzyme was found exclusively in the supernatant fraction (*115*). In kidney the enzyme is concentrated in the cortex. Isolated kidney tubules also contain high concentrations, but single-cell preparations made from the tubules by brief treatment with trypsin contain only small amounts. The bulk of the enzyme is again found in the supernatant fraction (*107*). Since it is widely believed that the key reactions of transport must occur in the cell membrane, it is difficult to envision how the enzyme could participate in establishing a concentration gradient for sugars. Similar objections can also be made against the sugar-binding proteins which are liberated from bacteria by osmotic shock. Many of these seem to be localized not in the bacterial membrane but in a periplasmic space between the bacterial cell wall and the plasma cell membrane (*116*).

Whatever the site of the enzyme may be, Keston *et al.* have recently produced fairly conclusive evidence that glucose, which is reabsorbed by the kidney, is exposed to mutarotase at some stage of the process (*117*). Glucose infused into the renal artery spills into urine when the renal threshold is exceeded in the same anomeric form as that administered, whereas reabsorbed glucose in the renal vein is mutarotated. Hill has also shown that the anomer infused in excess is excreted in excess (*73*).

Some recent developments in the physiology of water transport may eventually explain the apparent extracellular location of the enzyme. It seems that water transport in intestinal and kidney epithelium and in certain other organ structures in mammals is accomplished *via* standing osmotic gradients, which are established in the extracellular spaces. This concept has been extensively developed by Diamond (*118*), who has produced convincing evidence on the role of these extracellular spaces in water transport. Using a quantitative, predictive model, Diamond has shown how the phenomenon of standing osmotic gradients can explain the hitherto puzzling fact that water reabsorption in kidney and intestine always appears to deliver isoosmolar fluid to the blood, whereas in other structures such as gall bladder and stomach the process can lead to extreme concentrative phenomena.

If the tissue locations of mutarotase were extracellular, it would be possible to speculate on a role for the enzyme in the transport of sugar *via* these water-transporting capillary spaces. We are presently attempting to develop fluorescent antibody to the enzyme to use in visualizing the location of the enzyme in tissues which transport glucose.

Intracellular uptake of glucose itself raises certain problems. How is metabolic breakdown of glucose prevented if passage of the sugar from the lumen to blood proceeds *via* an intracellular route? Do transporting cells contain diffusion barriers to prevent access of glucose to the metabolic machinery? The metabolism-linked membrane permeability mechanism, used by certain other cells (*2*, *41*), would be self-defeating in this case.

Another objection, which has been made to the mutarotase hypothesis, concerns the fact that as originally stated—*i.e.*, that an intermediate of the catalyzed mutarotation was the preferentially absorbed form of the sugar—there was no provision for coupling the reaction to an energy source. The process therefore could not lead to a concentration gradient for sugar, such as is observed *in vitro* and *in vivo*. If binding of a sugar to the mutarotase protein is the key reaction in transport and catalyzed mutarotation a side-effect, however, then several possibilities for coupling to an energy source become apparent.

It is believed by several workers that transport may involve a direct coupling to the redox processes of the cell (*2*). Mutarotase has four free sulfhydryl groups. As noted above, all are apparently needed for substrate binding, and all four are protected equally from oxidation by the presence of a substrate sugar. Recent work has shown that oxidation of the sulfhydryl groups by mercuric ions also blocks the conformational transformation which the enzyme undergoes in response to changes in ionic strength. Reduction to the free sulfhydryl group by reaction with cysteine restores this ability (*87*).

The apparent transformations which the enzyme undergoes in response to changes in its osmotic environment could also be related to a transport function. The ability to reverse these changes specifically with glucose and other substrate sugars suggests a mechanism whereby the energy required to pump glucose uphill against a concentration gradient could be supplied by the standing osmotic gradients for sodium chloride. Such a scheme would be a novel and somewhat radical departure from present concepts of transport. It must be admitted however that these osmotic gradients are capable of doing work. They could theoretically be used to establish concentration gradients for non-ionic transported species if suitable mechanisms for coupling the two processes evolved. The ultimate proof of involvement in transport would be to reconstitute *in vitro* using the purified enzyme, a similar system showing concentrative ability for substrate sugars.

Also the concentration of the enzyme in specific organs correlates with high glucose fluxes, rather than the active transport capacity of a tissue. In liver high fluxes of glucose can occur, but active transport of glucose does not take place. However, in most species the mutarotase content of liver is usually intermediate between that of kidney and small intestine (*119*). Water reabsorption, particularly in the proximal tubule of the kidney, can produce concentrated solutions of non-electrolytes such as glucose. Some mechanism for increasing the permeability of the tubule to glucose would thus be a useful development in sugar transport even without a capability for direct coupling to an energy-yielding process. It seems that such a possibility should prove amenable to direct test by measuring penetration rates of glucose through biological membranes, following addition of purified mutarotase.

The objection has also been made that although in most tissues the aglycone phloretin is a more potent inhibitor of glucose transport than phloridzin, thus paralleling their relative effectiveness as inhibitors of mutarotase, in the intestine the situation is reversed (*38*). Diedrich has recently resolved this apparent conflict, however, by demonstrating that phloretin is probably the active inhibitor in intestine also but is prevented by a diffusion barrier from reaching the transport site. Phloridzin, however, is efficiently bound and releases phloretin at the site of transport under the action of an active phloridzin hydrolase. The transport site in intestine is thus seen as having the same enhanced sensitivity to the aglycone phloretin, as do transport systems in other tissues (*95*).

The objection that phloridzin is not specific for mutarotase but is an inhibitor of other glucose-metabolizing enzymes such as hexokinase is however a valid one. Information using inhibitors of a more specific nature would be desirable. The effect of specific antibodies to mutarotase on the active transport of sugar in an *in vitro* system—*e.g.*, intestinal sacs or kidney slices—would be interesting. Apparently, because of the close structural similarities of the enzyme in all species, the native enzyme is not a particularly good antigen (*88*). Experiments with the pure enzyme, covalently linked to hemacyanin, are now in progress, attempting to improve the antigenicity.

One of the principal arguments used previously against the hypothesis that mutarotase is involved in transport was that certain sugars such as 1-deoxy-D-glucose and α-methyl-D-glucoside, which inherently can not mutarotate, are nevertheless actively transported. This argument of course is no longer valid since it has been shown that these sugars do interact strongly with the enzyme and are efficient competitive inhibitors of the catalyzed mutarotation of glucose (Table VII) (*15*). The necessary interaction in transport is the binding of the sugar, and the mutarotation is considered to be a coincidental side-effect. It has been shown that a three-point atachment to the enzyme is required for catalyzed mutarotation, but that a two-point attachment will suffice for efficient binding (*107*). L-Glucose, which is not a substrate, binds to the enzyme with an affinity which is about half that of D-glucose (Table VII). It has recently been shown by Crane *et al.* that L-glucose is also actively transported by the sodium-dependent process in intestine (*38*). Since demonstration of active transport of a marginally active sugar such as L-glucose often requires careful control of the experimental conditions (*38*), it seems that the list of sugars given in Table VII might have a useful predictive function for such studies.

Conclusions

It seems in the course of evolution of the higher species that a diversity of transport processes for sugars has arisen. This is a reasonable and an inevitable consequence of the varied requirements for regulation and control in a complex multicellular organism. Sugar transport systems, each with distinct functional characteristics, are found in skeletal muscle, diaphragm, heart, adipose tissue, lens, red blood cells, kidney, and intestine. These varied processes however have two major points in common. The first is that they all interact with glucose, which is their only substrate of any physiological importance. The second is that they are all inhibited by micromolar concentrations of phloretin and closely related compounds such as diethylstilbestrol. This implies that the glucose-receptors in these varied transport systems must have a common structural identity, at least in one portion of the molecule. As we have seen, these structural features are also shared by the enzyme mutarotase. While this enzyme may have played a role in direct oxidative use of glucose in primitive organisms, with the evolution of the phosphorolytic pathways of glucose metabolism, the need for a mutarotation catalyst has been eliminated.

Similar or identical structural features in two proteins fulfilling the same function may arise during evolution in one of two ways. In the first usually referred to as divergent evolution, the genetic information which has arisen responding to the need for a structure to bind glucose—*e.g.*, the active center of a glucose-metabolizing enzyme—is retained and used when a requirement arises for a glucose-binding site—*i.e.*, a transport system. Examples of such homologies amongst enzymes in different metabolic pathways are becoming increasingly common (*103*). The second way in which similar structures may arise is by convergent evolution. In this mechanism two unrelated species can evolve by separate evolutionary pathways similar structures for two proteins—*e.g.*, two proteolytic enzymes (*103*) which fulfill the same function.

The evidence suggests that the various transport systems and the mutarotase protein must contain some glucose-binding structure of peculiar suitability, which has led to its retention through a long evolutionary history in recognizable form. There is persuasive evidence although so far mainly circumstantial in nature that the mutarotase protein in kidney and intestine is a component of a major transport system for glucose. With the recent isolation of the enzyme in crystalline form (79, 88), more direct testing of this hypothesis can be expected in the near future.

Literature Cited

1. Whittam, R., Wheeler, K. P., "Transport Across Cell Membranes," *Ann. Rev. Physiol.* (1970) **32,** 21.
2. Kaback, H. R., "Transport," *Ann. Rev. Biochem.* (1970) **39,** 561.
3. Crane, R. K., "Intestinal Absorption of Sugars," *Physiol. Rev.* (1960) **40,** 789.
4. Thurau, K., Valtin, H., Schnermann, J., "Kidney," *Ann. Rev. Physiol.* (1968) **30,** 441.
5. Heinz, E., "Transport through Biological Membranes," *Ann. Rev. Physiol.* (1967) **29,** 21.
6. Hoffman, J. F., "The Cellular Functions of Membrane Transport," pp. 1–291, Prentice–Hall, Englewood Cliffs, 1963.
7. Albers, W., "Biochemical Aspects of Active Transport," *Ann. Rev. Biochem.* (1967) **36,** 727.
8. Kepes, A., "Etudes Cinetiques sur La Galactoside Permease D'*Escherichia coli*," *Biochem. Biophys. Acta* (1960) **40,** 70.
9. Hokin, L. E., Hokin, M. R., "Biological Transport," *Ann. Rev. Biochem.* (1963) **32,** 553.
10. Csaky, T. Z., "Transport Through Biological Membranes," *Ann. Rev. Physiol.* (1965) **27,** 415.
11. Judah, J. D., Ahmed, K., "Biochemistry of Sodium Transport," *Biol. Rev.* (1964) **39,** 160.
12. Cirillo, V. P., "Sugar Transport in Microorganisms," *Ann. Rev. Microbiol.* (1961) **15,** 197.
13. Stein, W. C., "The Movement of Molecules Across Cell Membranes," Academic, New York, 1967.
14. Wilbrandt, W., "Active Transport and Secretion," A. Kleinzeller and A. Kotyk, Eds., Academic, New York, 1960.
15. Bailey, J. M., Pentchev, P. G., "Inhibition of Kidney and Intestinal Mutarotase by Actively Absorbed Sugars," *Biochem. Biophys. Res. Commun.* (1964) **14,** 161.
16. Bailey, J. M., Pentchev, P. G., "Inhibition of Rat Intestinal and Rat Kidney Mutarotases by Actively Transported Sugars," *Amer. J. Physiol.* (1965) **208,** 385.
17. Kundig, W., Ghosh, S., Roseman, S., "Phosphate Bound to Histidine in a Novel Phosphotransferase System," *Proc. Nat. Acad. Sci.* (1964) **52,** 1067.
18. Kundig, W., Kundig, F. D., Anderson, B., Roseman, S., "Restoration of Active Transport of Glycosides in *E. coli* by a Component of a Phosphotransferase System," *J. Biol. Chem.* (1966) **241,** 3243.
19. Kundig, W., Roseman, S., *Methods Enzymol.* (1966) **9,** 396.
20. Beckwith, J. R., "Regulation of the Lac Operon," *Science* (1967) **156,** 597.
21. Halpern, Y. S., Lupo, M., "Effect of Glucose and Other Compounds on Transport of Alpha Methyl Glucoside in *E. coli* K12," *Biochim. Biophys. Acta* (1960) **126,** 63.
22. Kepes, A., *Colloq. Ges. Physiol. Chem.* (1961) **12,** 100.
23. Kennedy, E. P., Jones, T. D., "Characterization of the Membrane Pro-

tein Component of the Lactose Transport System of *E. coli*," *Federation Proc.* (1971) **30**, 644.
24. Kaback, H. R., "The Molecular Basis of Membrane Transport," D. Tosteson, Ed., Prentice–Hall, Englewood Cliffs, 1969.
25. Fox, C. F., Kennedy, E. P., "Specific Labeling and Partial Purification of the M Protein, a Component of the β-Galactoside Transport System of *E. coli*," *Proc. Nat. Acad. Sci.* (1965) **54**, 891.
26. Fox, C. F., Carter, J. R., Kennedy, E. P., "Genetic Control of the Membrane Protein Component of the Lactose Transport System of *E. coli*," *Proc. Nat. Acad. Sci.* (1967) **57**, 698.
27. Jones, T. D. M., Kennedy, E. P., "Characterization of the Membrane Protein Component of the Lactose Transport System of *E. coli*," *Federation Proc.* (1968) **27**, 644.
28. Anraku, J., "The Reduction and Restoration of Galactose Transport in Osmotically Shocked Cells of *E. coli*," *J. Biol. Chem.* (1967) **242**, 793.
29. Heppel, L., "Selective Release of Enzymes from Bacteria," *Science* (1967) **156**, 1451.
30. Koch, A. L., "The Role of Permease in Transport," *Biochim. Biophys. Acta* (1964) **79**, 177.
31. LeFevre, P. G., "Sugar Transport in the Red Blood Cell: Structure-Activity Relationships in Substrates and Antagonists," *Pharmacol. Rev.* (1961) **13**, 39.
32. LeFevre, P. G., "Rate and Affinity in Human Red Blood Cell Sugar Transport," *Amer. J. Physiol.* (1962) **203**, 286.
33. LeFevre, P. G., "The Dimer Hypothesis for Sugar Permeation through Red Cell Membrane. Reinvestigation of Original Evidence," *Biochim. Biophys. Acta* (1966) **120**, 395.
34. Faust, R. G., "Monosaccharide Penetration into Human Red Blood Cells by an Altered Diffusion Mechanism," *J. Cell Comp. Physiol.* (1960) **56**, 103.
35. Hall, J. C., "The Effect of Insulin on Intact Muscle from Normal and Diabetic Rats," *J. Biol. Chem.* (1960) **235**, 1.
36. Weiser, M. M., Isselbacher, K. J., "Phosphorylation of Amino Sugars by Intestinal Mucosa: Role of Phosphoenolpyruvate vs. Adenosinetriphosphate," *Federation Proc.* (1969) **28**, 463.
37. Nordlie, R. C., Soodsma, J. F., "Phosphotransferase Activities of Kidney Glucose-6-Phosphatase," *J. Biol. Chem.* (1966) **241**, 1719.
38. Crane, R. K., "Absorption of Sugars," Handbook of Physiology III—American Physiological Society, Washington, D. C. (1968).
39. Keston, A. S., "Occurrence of Mutarotase in Animals: Its Proposed Relationship to Transport and Reabsorption of Sugars and Insulin," *Science* (1954) **120**, 355.
40. Kleinzeller, A., Lotspeich, W. D., "Active Transport of 2-Deoxy Sugars in Kidney Cortex Cells," *Federation Proc.* (1967) **26**, 660.
41. Fishman, P. H., Bailey, J. M., "Metabolic Suppression of Glucose Penetration in Ascites Tumor Cells," *Federation Proc.* (1969) **28**, 847.
42. Grossman, I. W., Sacktor, B., "Histochemical Localization of Renal Trehalase. Demonstration of a Tubular Site," *Science* (1968) **161**, 571.
43. Crane, R. K., "Studies on Transmural Potentials in Relation to Intestinal Absorption," *Biochim. Biophys. Acta* (1967) **135**, 61.
44. Landau, B. R., Bernstein, L., Wilson, T. H., "Hexose Transport by Hamster Intestine in Vitro," *Amer. J. Physiol.* (1962) **203**, 237.
45. LeFevre, P. G., Jung, C. Y., Chaney, J. E., "Glucose. Transfer by Red Cell Membrane, Phospholipids in ($H_2O/CHCl_3/H_2O$) Three-Layer Systems," *Arch. Biochem. Biophys.* (1968) **126**, 677.
46. Bailey, J. M., Woodson, M., "Some Properties of Phospholipid-Amino

Acid Complexes from Ascites Tumor Cells," *Federation Proc.* (1963) **22**, 417.
47. Reiser, S., Christiansen, P. A., "Formation of a Complex Between Valine and Intestinal Mucosal Lipid: Its Possible Role in Valine Absorption," *J. Lipid Res.* (1968) **9**, 606.
48. LeFevre, P. G., Jung, C. Y., Chaney, J. E., "Enhanced Migration of Glucose from Water into Chloroform in Presence of Phospholipids," *Arch. Biochem. Biophys.* (1968) **126**, 664.
49. Weber, G., Lea, M. A., Stamm, N. B., "Advances in Enzyme Regulation," Vol. 6, p. 101, G. Weber, Ed., Pergamon, 1968.
50. Bailey, J. M., Whelan, W. J., "The Role of Glucose and AMP in Regulating the Conversion of Phosphorylase a to Phosphorylase b," *Biochim. Biophys. Res. Commun.* (1971) **46**, 191.
51. Monod, J., "Recherches sur la Croissance des Cultures Bacteriennes," Hermann, Paris, 1942.
52. Kaback, H. R., "Uptake of Amino Acids by 'Ghosts' of Mutant Strains of *E. coli*," *Federation Proc.* (1960) **19**, 130.
53. Kaback, H. R., "Current Topics in Membrane Biochemistry," A. Kleinzeller and F. Bronner, Eds., Academic, New York, 1970.
54. Kaback, H. R., "Regulation of Sugar Transport in Isolated Bacterial Membrane Preparations from *E. coli*," *Proc. Nat. Acad. Sci.* (1969) **63**, 724.
55. Kotyk, A., Kleinzeller, A., "Affinity of the Yeast Membrane Carrier for Glucose and its Role in the Pasteur Effect," *Biochim. Biophys. Acta* (1967) **135**, 166.
56. Coe, E. L., Saha, J., "A Rapid Filtration Method for Estimating the Rate of Glucose Penetration into Ehrlich Ascites Carcinoma Cells," *Arch. Biochem. Biophys.* (1966) **114**, 282.
57. Coe, E. L., Saha, J., "Evidence Indicating the Existence of Two Modes of Glucose Uptake in Ehrlich Ascites Tumor Cells," *Biochem. Biophys. Res. Commun.* (1967) **26**, 441.
58. Coe, E. L., Ibsen, K. H., Dixon, M., McKee, R. W., "Glycolysis of Small Amounts of Glucose by Ehrlich Ascites Carcinoma Cells," *Cancer Res.* (1966) **26**, 276.
59. Bentley, R., Neuberger, A., "The Mechanism of the Action of Notatin," *Biochim. J.* (1948) **45**, 584.
60. Keilin, D., Hartree, E. F., "Biological Catalysis of Mutarotation of Glucose," *Biochem. J.* (1952) **50**, 341.
61. Keston, A. S., "Mutarotase Inhibition by 1-deoxy-Glucose," *Science* (1964) **143**, 698.
62. Bailey, J. M., Fishman, P. H., Pentchev, P. G., "Studies on Mutarotases I. Purification and Properties of a Mutarotase from Higher Plants," *J. Biol. Chem.* (1967) **242**, 4263.
63. Wallenfels, K., Herrman, K., "Aldose-1-Epimerase (mutarotase) and *E. coli*. Reiningung and vorlaufige molekulare Charakterisierung," *Biochem Z.* (1965) **343**, 294.
64. Wallenfels, K., Hueho, F., Herrman, K., "On the Enzymatic Catalysis of the Mutarotation of Aldoses. Properties of the Mutarotase from *E. coli*," *Biochem. Z.* (1965) **343**, 307.
65. Bentley, R., Bhate, D. S., "Mutarotase from *P. notatum* I. Purification, Assay and General Properties of the Enzyme," *J. Biol. Chem.* (1959) **235**, 1219.
66. Bentley, R., Bhate, D. S., "Mutarotase from *P. notatum*. The Mechanism of the Mutarotation Reaction," *J. Biol. Chem.* (1960) **235**, 1225.
67. Bailey, J. M., Fishman, P. H., Pentchev, P. G., "Mutarotase in Higher Plants. Inhibition and Properties," *Science* (1966) **152**, 1270.
68. Keston, A. S., "Purification of Kidney Mutarotase, Evidence for Proposed Unitary Transport Theory," *Federation Proc.* (1955) **14**, 234.

69. Keston, A. S., "Kinetics and Distribution of Mutarotases and Their Relationship to Sugar Transport," *J. Biol. Chem.* (1964) **239,** 3241.
70. Mulhern, S., Fishman, P. H., Bailey, J. M., "Studies on Conformational Changes in Different Mutarotases," *Federation Proc.* (1970) **29,** 910.
71. Chase, A. M., Lapedes, S. M., Lu-Ku, Li, "The Molecular Weight and Other Properties of a Mutarotase (Aldose-1-Epimerase) from Hog Kidney," *Proc. VIIth Int. Congr. Biochem.* (1967) **F,** 141.
72. Bailey, J. M., Fishman, P. H., Pentchev, P. G., "Studies on Mutarotase III. Isolation and Characterisation of a Mutarotase from Bovine Kidney Cortex," *J. Biol. Chem.* (1969) **244,** 781.
73. Hill, J. B., "Mutarotase and Glucose Anomers," Parts I and II, *Biochem. Med.* (1967) **1,** 62.
74. Lapedes, S. M., Chase, A. M., "Aldose-1-Epimerase from Hog Kidney. Isolation and Evidence of Purity and Inhibition Kinetics," *Biochem. Biophys. Res. Commun.* (1968) **31,** 967.
75. Sacks, W., "Mutarotase in Erythrocytes-Isolation and Properties," *Science* (1967) **158,** 498.
76. Kusiak, J., "Purification and Properties of Human Mutarotase," M.S. Thesis, George Washington University (January, 1972).
77. Foa, P. P., "Glucose Uptake in the Developing Chick Embryo Heart," *Amer. J. Phys.* (1966) **211,** 981.
78. Chase, A. M., "Concerning the Role of Mutarotase," *J. Cell. Biol.* (1968) **39,** 165.
79. Fishman, P. H., Mulhern, S., Bailey, J. M., "Photochemical Oxidation Studies of a Crystalline Mutarotase," *Federation Proc.* (1970) **29,** 876.
80. Salas, M., Vinuela, E., Sols, A., "Spontaneous and Enzymatically Catalyzed Anomerization of Glucose-6-Phosphate and Anomeric Specificity of Related Enzymes," *J. Biol. Chem.* (1965) **240,** 561.
81. Bailey, J. M., Fishman, P. H., Pentchev, P. G., "Anomalous Mutarotation of Glucose-6-Phosphate. An Example of Intramolecular Catalysis," *Biochemistry* (1970) **9,** 1189.
82. Oliver, J., "New Directions in Renal Morphology. A Method, its Results and its Future," *Harvey Lectures* (1944) **40,** 102.
83. Baxter, J., Yoffey, S., "The Post-Natal Development of Renal Tubules in the Rat," *Amer. J. Anat.* (1948) **82,** 189.
84. Arataki, M., "On the Post-Natal Growth of the Kidney with Special Reference to the Number and Size of the Glomeruli in the Albino Rat," *Amer. J. Anat.* (1926) **36,** 399.
85. Segal, S., Rea, C., Smith, I., "Separate Sugar Transport Systems for Sugars and Amino Acids in Developing Rat Kidney Cortex," *Proc. Nat. Acad. Sci.* (1971) **68,** 372.
86. MacNamara, P., Rea, C., Segal, S., "Sugar Transport. Effect of Temperature on Concentrative Uptake of α-Methyl Glucoside by Kidney Cortex Slices," *Science* (1971) **172,** 1033.
87. Mulhern, S., "Molecular and Kinetic Properties of Mutarotases," Ph.D. Thesis, George Washington University (1971).
88. Fishman, P. H., "The Chemistry and Biology of Mutarotases," Ph.D. Thesis, George Washington University (1970).
89. Chipman, D. M., Sharon, N. Y., "The Mechanism of Lysozyme Action," *Science* (1969) **165,** 453.
90. Koshland, D. E., Neet, K. E., "The Catalytic and Regulatory Properties of Enzymes," *Ann. Rev. Biochem.* (1968) **37,** 359.
91. Crane, R. K., Forstner, G., Eicholtz, A., "An Effect of Na+ on the Apparent Michaelis Constants for Intestinal Sugar Transport *in Vitro,*" *Biochim. Biophys. Acta* (1965) **109,** 467.
92. Crane, R. K., "Gradient Coupling and the Membrane-Transport of Water

Soluble Compounds. A General Biological Mechanism," *Protides Biol. Fluids* (1967) **15,** 227.

93. Diedrich, D. F., Stringham, C. H., "Mutarotase Still a Candidate for the Role of Membrane Glucose Transport," *Arch. Biochem. Biophys.* (1970) **138,** 493.
94. Diedrich, D. F., Stringham, C. H., "Active-Site Comparison of Mutarotase with the Glucose Carrier in Human Erythrocytes," *Arch. Biochem. Biophys.* (1970) **138,** 497.
95. Diedrich, D. F., "Is Phloretin the Sugar Transport Inhibitor in Intestine?" *Arch. Biochem. Biophys.* (1968) **127,** 803.
96. Alvarado, F., "Hypothesis for the Interaction of Phloridzin and Phloretin with Membrane Carriers for Sugars," *J. Biol. Chem.* (1959) **234,** 3022.
97. Keston, A. S., "Inhibition of Action of Lens Mutarotases on Glucose by Cataractogenic Sugars and Corresponding Polyols," *Arch. Biochem. Biophys.* (1963) **102,** 306.
98. LeFevre, P. G., Marshall, J. K., "The Attachment of Phloretin and Analogues to Human Erythrocytes in Connection with Inhibition of Sugar Transport," *J. Biol. Chem.* (1959) **234,** 3022.
99. Forsling, M. L., Widdas, W. F., "The Effect of Temperature on the Competitive Inhibition of Glucose Transfer in Human Erythrocytes by Phenolphthalein, Phloretin and Stilboestrol," *J. Physiol.* (1968) **194,** 545.
100. Kleinzeller, A., "Intestinal Transport of Electrolytes, Amino Acids and Sugars," p. 232, W. M. Armstrong and A. S. Nunn, Eds., Charles C. Thomas, Illinois, 1971.
101. Bihler, I., "Intestinal Sugar Transport. Ionic Activation and Chemical Specificity," *ibid.* p. 144.
102. Bailey, J. M., Fishman, P. H., Pentchev, P. G., "Studies on Mutarotases VI. Enzyme Levels and Sugar Reabsorption in Developing Rat Kidney and Intestine," *J. Biol. Chem.* (1970) **245,** 559.
103. Whelan, W. J., Schultz, J., Eds., "Homologies in Enzymes and Metabolic Pathways and Metabolic Alterations in Cancer," North-Holland, 1970.
104. Bailey, J. M., "Lipid Metabolism in Cultured Cells VI. Lipid Biosynthesis in Serum and Synthetic Growth Media," *Biochim. Biophys. Acta* (1966) **125,** 226.
105. Bailey, J. M., Fishman, P. H., Mulhern, S., "Studies on Mutarotases IV. Mutarotase in Kidneys of Saltwater versus Freshwater Fish," *Proc. Soc. Exper. Biol. Med.* (1969) **131,** 861.
106. Bailey, J. M., Fishman, P. H., Pentchev, P. G., "Studies on Mutarotases II. Investigations of Possible Rate-Limiting Anomerizations in Glucose Metabolism," *J. Biol. Chem.* (1968) **243,** 4827.
107. Pentchev, P. G., "Studies of Mammalian Mutarotase," Ph.D. Thesis, George Washington University (1967).
108. Tipson, R. S., Isbell, H. S., "Infrared Absorption Spectra in the Study of Mutarotational Equilibria of Monosaccharides," *J. Res. Nat. Bur. Stand.* (1962) **66A,** 31.
109. Bailey, J. M., Fishman, P. H., Mulhern, S., Murray, R., "Studies on Mutarotases V. Serum Mutarotase Levels in Renal Disease," *Clin. Biochem.* (1970) **3,** 11.
110. Kusiak, J. W., Bailey, J. M., "Chemi-Osmotic Gradients and Molecular Transformations of Mutarotases," *Federation Proc.* (1972) **31,** 2712.
111. Mulhern, S., Fishman, P. H., Bailey, J. M., "Studies on Conformational Changes in Different Mutarotases," *Federation Proc.* (1970) **29,** 910.
112. Diedrich, D. F., "*In Vitro* Evaluation of Relative Inhibitory Potency of Phloridzin and its Congeners," *Amer. J. Physiol.* (1965) **200,** 621.
113. Diedrich, D. F., "Glucose Transport Carrier in Dog Kidney, Its Concentration and Turnover Number," *Amer. J. Physiol.* (1966) **211,** 581.

114. Chase, A. M., "Distribution and Function of Mutarotases," *Amer. Zoologist* (1969) **9,** 581.
115. Bailey, J. M., Pentchev, P. G., "Distribution of Mutarotase in Subfractions of Rat Intestinal Mucosa and Rat Kidney," *Proc. Soc. Exper. Biol. Med.* (1964) **115,** 796.
116. Heppel, L. A., "Selective Release of Enzymes from Bacteria," *Science* (1967) **156,** 1451.
117. Keston, A. S., Brandt, R., Barach, I. M., "Quantitative Relationships between Glucose Reabsorption and Mutarotation by Dog Kidney *in vivo,*" *Biochem. Biophys. Res. Commun.* (1972) **46,** 610.
118. Diamond, J. M., "Standing Gradient Osmotic Flow. A Mechanism for Coupling of Water and Solute Transport in Epithelia," *J. Gen. Physiol.* (1967) **50,** 2061.
119. Bailey, J. M., Pentchev, P. G., Woo, J., "Distribution of a Mutarotase Activity in Rat Tissues and Possible Function in Active Transport of Sugars," *Biochim. Biophys. Acta* (1965) **94,** 124.

RECEIVED December 2, 1971.

17

Glycosylation as the Paradigm of Carbohydrase Action

Evidence from the Actions of Amylases

EDWARD J. HEHRE, GENTARO OKADA, and DOROTHY S. GENGHOF

Albert Einstein College of Medicine, New York, N. Y. 10461 and Shizuoka University, Shizuoka, Japan

Evidence is presented which supports the thesis that glycosyl-proton interchange defines the underlying unity of all actions catalyzed by the so-called glycoside hydrolases and glycosyl transferases. These enzymes form a single class of glycosylases, effecting the formation of one glycosylic bond at the expense of another. Recent findings of α-glucosaccharide synthesis from the β-anomer of D*-glucopyranose and β-anomer of maltose by glucoamylase and β-amylase, respectively, confirm this model and show that glycosyl transfer is more general than the term (which implies biochemical action on glycosidic bonds) suggests. The capacity of α-amylases as a group to synthesize maltosaccharides from α-*D*-glucosyl fluoride is reviewed, and new findings are presented of similar syntheses from α-maltosyl fluoride. The significance of the emergence of the glycosylation paradigm for the field is discussed.*

This paper advances the thesis that enzymically catalyzed glycosyl-proton interchange is the reaction responsible for the great array of syntheses and degradations brought about by the so-called hydrolases acting on glycosyl compounds and the glycosyl transferases and, thus, is one of the major reactions of living matter. According to this concept, all hydrolase and transferase actions on carbohydrates effect the glycosylation of one compound at the expense of another (in exchange for a proton) in a process of glycosylic bond redistribution. The gain of unity and generality has a further advantage. That is, a glycosylic bond which may be defined as the union joining the anomeric carbon of a cyclic form of a sugar to the anomeric hydroxyl or to any group replacing it has an unambiguous meaning in contrast to glycosidic bond traditionally used in connection with glycosyl transfer. The latter, though generally understood by biochemists as the bond between the anomeric carbon and the bridge atom of glycosides, saccharides, sugar phosphates, nucleosides, and sugar nucleotides, is blurred by its inconsistency with the long standing view of carbohydrate chemists that not all of these compounds are glycosides and by confusion with a glycosidic linkage (which contains two bonds).

With features of complete generality and ability to be articulated in simple clear terms as the redistribution of glycosylic bonds, the glycosylation model seems to represent a truly unifying paradigm for the field in the same sense as the original hydrolase concept was held to be. Evidence for the validity of the model has been obtained from recent experimental studies on the scope of catalytic action of amylases; however, strong support comes from a very wide range of contributions to carbohydrate enzymology made by many investigators, including recent demonstrations of glycosyl-enzyme transition states in hydrolytic and nonhydrolytic reactions.

Breakdown of the once universally held belief that all carbohydrates are hydrolases began some 30 years ago with the discovery of several enzymic syntheses, recognized as other than reversals of hydrolysis. These included not only the well-known initial example of formation of glycogen-like and amylaceous polysaccharides from glucose 1-phosphate, but also the synthesis of dextran (*1, 2, 3*), levan (*4, 5*), and methyl β-D-glucoside (*6*) from more familiar glycosidic substrates (sucrose and phenyl β-D-glucoside). Early pioneer workers had attempted to explain reactions of the latter two types on the basis of the then unassailable hydrolase paradigm. Thus, Beijerinck (*7*), who had observed levulan formation from sucrose (but not D-fructose) by bacterial viscosaccharase, assumed a process of sucrose hydrolysis plus reversal of hydrolysis. Since ordinary D-fructose would not provide for the latter step, *levulose im status nascens* was postulated as doing so. Rabate (*8, 9*) years later accounted for observations of enzymic formation of aliphatic β-D-glucosides from aryl β-D-glucosides (but not from D-glucose) by a similar two-step process—*e.g.*, hydrolysis of *p*-acetylphenyl β-D-glucoside (picein), plus reversal of hydrolysis to form methyl β-D-glucoside by the reaction,

$$\textit{Glucose naissant}\ C_6H_{12}O_6 + CH_3OH \rightleftharpoons C_6H_{11}O_5\text{—}O.CH_3 + H_2O$$

Saccharide syntheses recognized as other than reversals of hydrolysis were, for a time, widely viewed as resulting from the exchange of P-ester bonds for glycosidic linkages. However, recognition that an α-D-glucoside radical is common to sucrose and glucose 1-phosphate (*2, 3*) and proofs of the absence of a phosphate ester requirement for dextran and levan syntheses (*10, 11*) and for disaccharide synthesis by sucrose phosphorylase (*12*) led to the more rational concepts of mixed acetal precursor structure and transglycosidation; Japanese scientists arrived independently at Umglucosidierung in work (*6*) that long remained unknown.

In 1951 the term transglycosylation was introduced (*13*) to indicate reactivity of the glycosyl rather than the glycosyloxy radical in saccharide synthesis, on the basis that a change of the glycosidic bridge atom necessarily would occur in the enzymic conversion (*14*) of ribose 1-phosphate to inosine and had been proved by using ^{18}O for reactions catalyzed by glycogen and sucrose phosphorylases (*15*).

That the same enzyme might catalyze transfer and hydrolytic reactions was suspected early and ultimately shown (*16*) for levansucrase; however, the first clear demonstrations of this capacity in 1950 involved two well known hydrolases, yeast invertase (*17, 18*) and β-glucosidase (*19*).

The great range of glycosyl transfer reactions now recognized (*20, 21, 22, 23, 24, 25, 26, 27, 28, 29, 30*), including many catalyzed by enzymes known for their hydrolytic activity, has confirmed glycosyl (*vs.* glycosyloxy) group activity (*13*), as well as the insufficiency of the hydrolase paradigm. Nevertheless, in the situation that has prevailed up to the present, the hydrolase model of action has been kept alive by the formal internationally adopted system of enzyme nomenclature and classification (*21, 31, 32, 33*). Carbohydrases having little or no known activity but hydrolysis are often described without alluding to the transfer concept or else as having little or no glycosyl transfer (transferase) activity (*34, 35, 36, 37, 38, 39, 40, 41*)—illustrating the common meaning of the latter as a process of rearranging pre-existing glycosidic bonds.

Glycosyl transfer has long been formally recognized as embracing transfer to water, and overwhelming evidence exists that hydrolysis is the glycosylation of water. However, a particular inconsistency is in the dissociation of glycosyl transfer from the *de novo* synthesis of glycosidic linkages. The enzymic formation of glycosidic linkages where none existed before has been assigned to phosphoryl transfers by mutases and kinases or to reversal of hydrolysis by glycoside hydrolases. Some authors have applied the glycosyl transfer concept to saccharide synthesis from glycoses by the latter enzymes, but most authors have not. Rather, because of thermodynamic and structural implications, the concept of transfer has usually been restricted to reactions with substrates already possessing a glycosidic linkage. Over the years this restriction has been repeatedly stated e.g., in 1956 (*23*), "although enzymic condensation of free sugars ('reversion') has been reported, the donor in transfer reactions appears always to be a glycoside, the nonreducing part of the molecule providing the transferred radical"; in 1970 (*29*), "The synthesis of complex saccharides (oligosaccharides, glycosides, and polysaccharides) involves the process of transglycosylation. In this process, the glycosyl donor may be sugar phosphate, sugar nucleotide, polysaccharide, or oligosaccharide."

In any event a model valid for enzymic saccharide syntheses as far apart in behavior as those stemming from substrates with mixed acetal and hemiacetal structures might be expected to carry certain assumptions and implications for the entire discipline. This is the case with the type of reaction mentioned at the outset,

$$\text{Glycosyl-}X + \text{H-}X' \rightleftharpoons \text{Glycosyl-}X' + \text{H-}X \qquad (1)$$

expressed as Equation 1, which was reported in 1960 (*42*) as a compact and more general modification of earlier overall glycosyl transfer expressions (*20, 22, 23*). This reduces all carbohydrase-catalyzed reactions to an interchange between glycosyl and hydrogen about sites -X and -X' of various potential kinds,

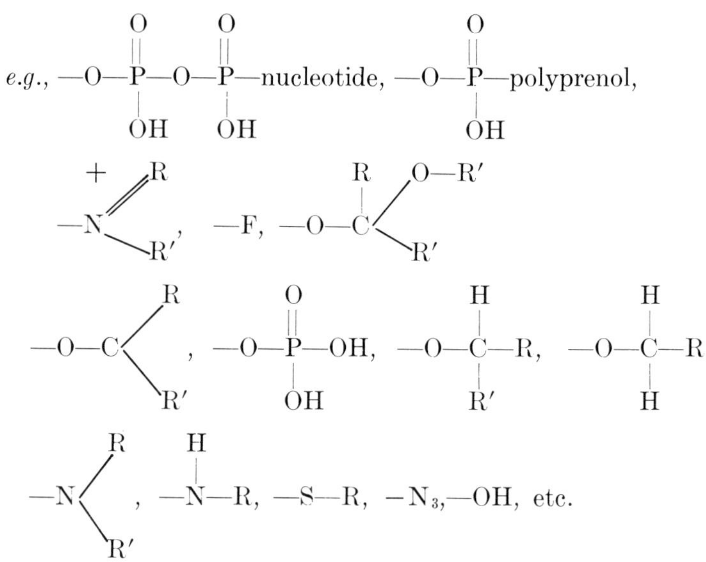

This model makes no assumption as to detailed reaction mechanism, but it does allow for glycosyl-proton interchange with enzyme (as H-X') to form glycosyl–enzyme intermediates, whether covalently linked or of carbonium–oxonium type (*42*). In the past few years such glycosyl–enzyme intermediates have been demonstrated for a number of carbohydrases. Among enzymes providing for retention of configuration, a covalently linked β-D-glucosyl enzyme intermediate has been isolated (*43*) for sucrose phosphorylase, confirming theoretical (*44*) and experimental (*45*) predictions of a double displacement mechanism. On substantial but less direct evidence, β-glucosidase has been reported to catalyze hydrolysis by this same mechanism (*46, 47*). In contrast, avian lysozyme has been found from model building studies (*48*) and secondary isotope effect measurements during hydrolysis (*46*) to form a stabilized oxycarbonium ion–enzyme complex. Likewise, *E. coli* maltodextrin phosphorylase and muscle phosphorylase b are reported (*49*) to have an oxonium ion type of glucosyl–enzyme transition complex accounting for their glucosyl transfer reactions. From all published results (*37, 40, 41, 50*), hydrolytic reactions leading to products with inversion of configuration, whether catalyzed by β-amylase, papaya lysozyme, or exo-β-1,3-glucanase, seem as likely to proceed by an oxycarbonium ion–enzyme transition (*51*) as by a single nucleophilic displacement reaction (*44*). These findings highly support glycosyl-proton interchange (*42*) as the unifying type reaction for so-called hydrolase and transferase actions.

Equation 1 does postulate that the glycosyl transfer process is concerned with the redistribution of glycosylic rather than glycosidic bonds and is satisfied by glycosyl donors that have simpler structure than usually assumed to be required. For example, the model includes -OH among potential -*X* sites and places syntheses of glycosidic linkages by the condensation of free sugars among the glycosyl transfer reactions rather than separate from them. Until recently, no direct experimental evidence was available to prove or disprove this point, despite the fact that many investigators have studied such reactions.

Previous literature contains nearly 200 reports of reversion syntheses of glycosides and oligosaccharides by the action of glycosidases on monosaccharides, alone or with alcohols (*52, 53, 54, 55, 56*). Bourquelot and his associates (*52, 53, 54*) in particular prepared gentiobiose and other disaccharides and more than 40 crystalline glycosides by this method, and their contributions formed one of the pillars of the hydrolase concept. More recent observations of free sugar condensations, involving purified glycosidases which catalyze hydrolysis and glycosyl transfer from saccharides to organic acceptors, have sometimes been interpreted as proceeding through a glycosyl–enzyme intermediate (*57, 58, 59, 60, 61, 62*); but, equally often, these have not been attributed to (or have been explicitly dissociated from) glycosyl transfer (*63, 64, 65, 66, 67, 68, 69, 70*). The failure of investigators to apply the glycosyl transfer concept to free sugar condensations catalyzed by glucoamylase (*71, 72, 73, 74, 75, 76, 77*) or β-amylase (*69*) also illustrates the persistence of the hydrolase concept for these inverting amylases (*35*).

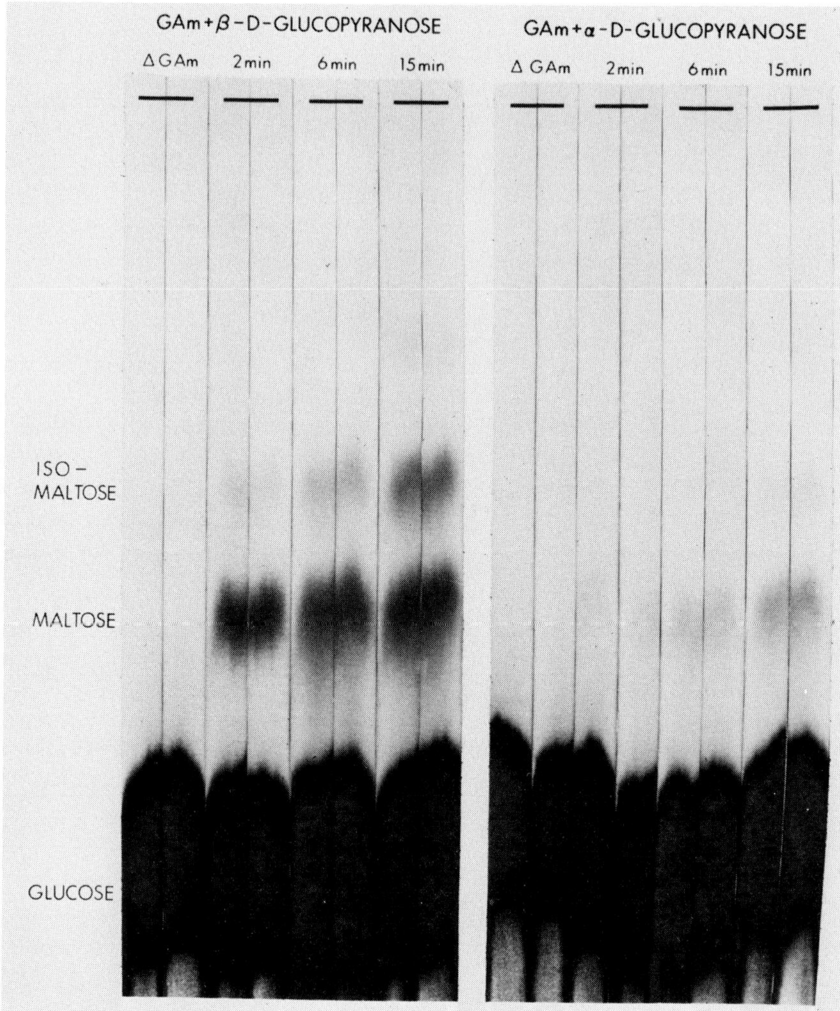

Figure 1. Chromatograms from mixtures of crystalline glucoamylase. (GAm), incubated with β- or α-D-glucopyranose at 30°C for 2, 6, or 15 minutes and from controls of inactivated enzyme (ΔGAm) *plus the β- or α-D-glucose* (78).

Evidence for the Generality of the Glycosyl Transfer Process

The standard technique used to obtain condensation reactions invariably involved enzymes acting on high concentrations of sugars (or sugars plus alcohols) for long periods—usually days or weeks—guaranteeing that the D-glucose, D-galactose, or other so-called substrate used in the past could never have represented a single chemical species but always a mixture of isomers.

By using the pure anomeric forms of sugars separately with very brief incubation to limit mutarotation, we obtained evidence (*78*) for the rapid condensation of sugars of appropriate anomeric configuration by several amylases. Crystalline *Rhizopus niveus* glucoamylase, which hydrolyzes starch and maltosaccharides to β-D-glucopyranose, was the first amylase examined. Previous investigators had tested long-incubated mixtures of purified glycoamylase and D-glucose and had found isomaltose synthesized as the sole or predominant product, rather than maltose; whereas, the latter was known to be the much preferred substrate for hydrolysis (*76*).

As shown in Figure 1 (*78*), we observed that when glucoamylase is incubated with pure β-D-glucopyranose, maltose is synthesized most rapidly—within 2 min. Then isomaltose is formed more slowly, even though the energy required to glucosylate the C6 carbinol is less than that required to glucosylate the carbinol at C4. With longer time maltose does not increase, but isomaltose becomes very abundant. α-D-Glucopyranose begins to function as a substrate only after it begins to mutarotate into the β-anomer.

Figure 2 (*78*) illustrates the correspondence found for the first time between condensation and hydrolysis for these glucoamylase-catalyzed reactions. Most important is that the condensations of β-D-glucopyranose to maltose and isomaltose involve configurational inversion from which we infer that the glucosylic bond is split and the D-glucopyranosyl group transferred.

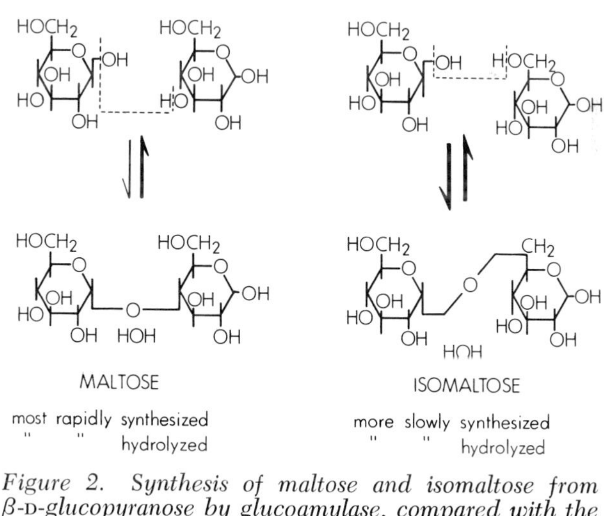

Figure 2. Synthesis of maltose and isomaltose from β-D-glucopyranose by glucoamylase, compared with the known hydrolytic reactions. The configurational inversion observed in the condensations indicates that the D-glucopyranosyl part of β-D-glucopyranose is interchanged with hydrogen at the C_4 or C_6 site of a second D-glucose molecule (78).

Using the pure anomers of maltose in the same way, we found that crystalline hog pancreatic α-amylase causes the very rapid synthesis of maltotetraose from α-maltose but not from β-maltose; whereas, crystalline sweet potato β-amylase causes the very rapid synthesis of the same compound, specifically from β-maltose. Configurational inversion again marks this latter condensation as glycosyl-hydrogen interchange, or glycosyl transfer.

It probably would be best to continue to reserve the latter term, which is tied to the group transfer concept, for glycosylations in which water is not a reactant, using hydrolysis and condensation for glycosylation reactions involving water. However, the above final point of evidence that the glycosyl radical is transferred during condensation, plus all that is already in the literature on the reactivity of the glycosyl group, shows the complete generality of glycosylation as defined by Equation 1. And this allows a clear issue to be raised over the use of separate hydrolase and transferase models for grouping, naming, and describing enzymes that potentially can (and often do) catalyze hydrolysis and transfer reactions. A single, completely general model should contain more information on the possible catalytic activities of the enzymes of glycosylation than separate, less general models; the principle is Occam's Razor.

To show the potency of the glycosylation model by an experimental example, a study of the catalytic potential of α-amylases was undertaken early in 1970. These enzymes were purposely selected because they head the Enzyme Commission's list of glycoside hydrolases (*32*) and because a generation of work under the hydrolase notion had provided a long familiar picture of their action as hydrolytic and specific for the glycosidic linkages in α-D-($1 \rightarrow 4$)-glucans (*35, 36, 79, 80, 81, 82, 83, 84, 85, 86*). When viewed from the glycosyl-hydrogen interchange model (Equation 1), the potential range of α-amylase action was not settled, but it invited exploration—not only beyond hydrolysis and condensation, but beyond glycosyl transfer from pre-formed glycosidic linkages.

A few studies (*86*) had indicated that certain α-amylases can catalyze glycosyl transfers to organic acceptors. In 1961 Matsubara (*87*) observed that crystalline *Aspergillus oryzae* Takaamylase A brings about transmaltosidation from aryl α-maltosides to alcohols. As far as we know, such transfers between sites other than the C_4-carbinols of D-glucose have not been observed for any other α-amylase. Other investigators (*88, 89, 90, 91*), more recently, had reported that the unusual saccharifying α-amylase from *Bacillus subtilis* var. *amylosacchariticus* Fukumoto catalyzes α-D-glucopyranosyl (monomeric) transfer reactions, a capacity not known for other α-amylases, though long known for certain α-D-transglucosylases that hydrolyze starch (*92, 93, 94, 95*). Still other workers had reported (*96, 97, 98, 99, 100*) that α-amylase preparations cause some conversion of maltose or maltosaccharides to higher homologs; such conversions, however, may be interpreted as combinations of free sugar condensation and hydrolysis (*78, 100*) and, hence, do not provide evidence for a widened catalytic range for α-amylases. In some instances (*96, 97*), the authors hesitated to attribute these reactions to the α-amylase, and Pazur and S. Okada (*102, 103*) reported that a separate glucanosyl transferase accompanies various crude α-amylase preparations. (In a paper published during our study, Robyt and French (*101*) reported that hog pancreatic α-amylase can catalyze maltotriosyl transfer from maltotetraose in addition to causing condensation and hydrolysis of this substrate. Maltoheptaose synthesis was not directly observed but was proposed on the basis of calculations involving the ratios of ^{14}C-labeled D-glucose, maltose, and maltotriose produced from reducing-end labeled maltotetraose.) No study to our knowledge had shown that α-amylases as a group catalyze glycosylation reactions beyond hydrolysis and condensation, and no study had shown any α-amylase capable of cleaving a glycosylic bond other than that of the α-D-glucopyranosyloxy radical.

Catalytic Capacities of α-Amylases Not Explained by Their Representation as Hydrolases

Our experimental approach avoided the maltosaccharides generally used as substrates in studying hydrolytic reactions and which may undergo condensation reactions. Instead, by parallelling our observation (*78*) that pancreatic α-amylase catalyzes very rapid maltosaccharide synthesis from a hydrolytic end-product, α-maltose, we have found that α-amylases as a group catalyze similar glycosylations on a very much greater scale from certain structural analogs of their end-products.

The first such substrate examined (*104*) was α-D-glucopyranosyl fluoride, a stereoanalog of the simplest α-amylase product, α-D-glucopyranose. Features of this compound are the small dimensions of the F atom (covalent radius, 0.72 A) that replaces the anomeric hydroxyl (O = 0.74 A, H = 0.30 A) of the parent sugar, meaning likely access of the analog to enzyme sites where α-D-glucopyranose fits; also, the high lability of its glycosylic (C_1 to F) bond; finally, the high heat of hydration, tight binding of water, and exceptionally strong hydrogen bonding ability of the liberated fluoride ion (*105*).

Table I (*104*) shows the yields of products from maltose to maltopentaose recovered from digests of 0.2 *M* crystalline α-D-glucosyl fluoride with crystalline α-amylase preparations from six different biological sources. The digests were incubated at 30°C for 10 minutes, heat inactivated, and chromatographed for product isolation and analysis.

Maltose and maltotriose were synthesized from α-glucosyl fluoride by all the α-amylases. Maltotetraose and sizeable amounts of the pentaose

Table I. Maltosaccharide Synthesis from α-D-Glucosyl Fluoride by α-Amylases[a]

	Yield from 20 μmoles Substrate				
α-Amylase[b]	*Maltose*[c]	*Maltotriose*[c]	*Maltotetraose*[c]	*Maltopentaose*[c]	*Total Maltosaccharides*[d]
B. subtilis, liquifying	101	110	70	248	14.7
B. subtilis, thermostable	93	82	37	231	12.3
B. stearothermophilus	143	61	30	90	8.9
Hog pancreas	56	68	29	0	4.2
Aspergillus oryzae	25	14	0	0	1.1
B. subtilis, saccharifying	376	222	5	0	16.8

[a] Reference *104*.
[b] Crystalline enzymes, 1 mg/ml.
[c] Micrograms recovered from digests held 10 minutes at 30°C.
[d] % of α-D-glucosyl fluoride substrate.

(which actually includes some maltohexaose which was not well resolved) were formed by several of the enzymes. The total yields of saccharides ranged from 1% of the α-D-glucosyl fluoride substrate for *Aspergillus oryzae* α-amylase to 12–17% for the different *B. subtilis* α-amylases in these 10-minute digests.

Though the data are not included in Table I, concurrent tests were also made of the action of all the α-amylases on crystalline α-maltose (*106*) and β-maltose. The unique saccharifying α-amylase of *B. subtilis* var. *amylosaccharitICUS* showed α-D-glucosyl transferring activity with both maltose anomers, confirming the conclusion of Japanese authors (*88, 89, 90, 91*). The remaining α-amylases showed no detectable action on either maltose anomer under these conditions.

Table II (*104*) shows the results of an experiment to learn the proportion of α-D-glucosyl fluoride that is hydrolyzed *vs.* that used for saccharide synthesis by the different crystalline α-amylases and by diluted saliva. The substrate was again 0.2 *M* and the enzymes at one-tenth to one-one hundredth the concentrations of the preceding experiment. These digests were analyzed directly for amounts of fluoride (*104, 107*) and D-glucose (glucose oxidase method) liberated during 2 hours at 30°C.

All the amylases caused release of more fluoride than D-glucose, both expressed as percent of substrate. The difference between them gives the percent substrate converted to saccharides. The main point is that with six of the seven α-amylases two to twenty times more α-D-glucosyl fluoride was converted to oligosaccharides than hydrolyzed to D-glucose. The great discrepancy was readily confirmed by direct visualization of the products by chromatography. In this experiment, preparations of so-called α-1,4-glucan 4-glucanohydrolase have not only brought about glycosylation of C_4 carbinol sites but they have done so in preference to hydrolyzing the substrate, even though the molar concentration of water was some 275 times greater than that of the α-D-glucosyl fluoride.

Many so-called glycoside hydrolases catalyze some glycosyl transfer to acceptors other than water. However, it has always been assumed that carbohydrases which have strictly or predominantly hydrolytic action on one substrate would necessarily prove solely or predominantly hydrolytic with all utilizable (suitably diluted) substrates. This is the essential assumption of the hydrolase concept; it is not made by the glycosylation model, and it is this assumption which our experimental findings, as well as theoretical base, deny.

Table II. Saccharide Synthesis *vs.* α-D-Glucosyl Fluoride Hydrolysis by Different α-Amylases[a]

α-Amylase	*Yield (% Substrate)* HF[b]	D-*Glucose*[b]	*Saccharides*[c]
B. subtilis, liquifying[d]	15.8	0.8	15.0
B. subtilis, thermostable[d]	13.3	2.7	10.6
B. stearothermophilus[d]	5.0	1.6	3.4
Hog Pancreas[d]	9.9	2.2	7.7
Human saliva (1:10)	7.8	0.7	7.1
Aspergillus oryzae[d]	8.4	6.9	1.5
B. subtilis, saccharifying[d]	23.7	8.5	15.2

[a] Reference *104*.
[b] Released on 2 hours incubation at 30°C.
[c] Fluoride released, less D-glucose released.
[d] Crystalline enzyme, 100 μgrams/ml.

Maltosaccharide Synthesis from Dilute Solutions of α-Maltosyl Fluoride

The foregoing observations of a α-amylase action on α-D-glucosyl fluoride all involved the use of high enzyme concentrations—at least 100 μg/ml. At 10 μg/ml, none but the *B. subtilis* saccharifying α-amylase rapidly used this substrate. This result is not surprising in view of the very limited ability of ordinary α-amylases to cleave α-D-glycosyl units from the nonreducing ends of maltosaccharide chains. However, the importance of investigating end-product analogs that might be better substrates than α-D-glucosyl fluoride encouraged us to try the next higher homolog, α-maltosyl fluoride.

As detailed under Experimental, we have obtained this compound *via* the known hepta-*O*-acetyl α-maltosyl fluoride (*108*), and although the preparations we have made and studied so far are only partly pure, we believe the results obtained with them are valid. A final chromatographic step is used to prepare the compound, but one by-product co-migrates with α-maltosyl fluoride, and a large amount is present in our preparations. This substance seems to be an α-D-glucopyranosyl-1,6-anhydro-β-D-glucopyranose; it is hydrolyzed to D-glucose and levoglucosan by the purified α-glucosidase of *Candida tropicalis*. None of the α-amylases shows any detectable action on this accompanying substance.

The action of different crystalline α-amylases on α-maltosyl fluoride may be illustrated by a typical experiment in which digests (18 m*M* in this substrate) were incubated 30 min at 30°C and then chromatographed. Figure 3 shows that digests representing the seven α-amylases tested contain products migrating as maltosaccharides, and five of the digests show saccharides migrating at least to the hexasaccharide level. Several also contain nonreducing products which have not yet been examined, but which very likely are members of the maltosaccharide 1-fluoride series observed previously in some digests with α-D-glucosyl fluoride (*104*).

The most important features are, first, that some of the α-amylases—*i.e.*, those in digests Numbers 4, 5, 6, and 7 have catalyzed glycosylations from α-maltosyl fluoride at one-tenth to one-one hundredth the enzyme concentrations required for action on α-D-glucosyl fluoride. The *Aspergillus oryzae* α-amylase, which had the weakest synthetic action on α-D-glucosyl fluoride of the enzymes tested at 100 μg/ml, has produced under conditions of this experiment an extended series of maltosaccharides at 1 μg/ml.

Secondly, although all of the digests contain more maltose than the control, several of them show obviously greater amounts of higher saccharides than of maltose. This was confirmed, for example, in digest Number 1 with *B. subtilis* liquifying α-amylase. When the individual products of this digest were recovered by chromatography, digested with crystalline *Rhizopus nivens* glucoamylase, and the liberated D-glucose measured, 6.9% of the α-maltosyl fluoride substrate was accounted for as maltose and D-glucose, whereas, 31.5% (4½ times as much) was recovered as higher saccharides.

In this case the predominance of glycosylation of C_4-carbinols over substrate hydrolysis is in a mixture where the molar concentration of water was more than 3000 times that of the α-maltosyl fluoride. In other experiments the substrate concentration could be further reduced at least 5-fold without altering the fact of a substantially greater extent of maltosidic linkage synthesis than hydrolysis. We now are referring to a very dilute 4 m*M*, solution of substrate with which, if the hydrolase concept were valid, hydrolysis ought to have been predominant.

Glycosylation as the Paradigm of Carbohydrase Action

We turn from these simple experiments with amylases, which we believe place their capacities in a perspective different from the past, to

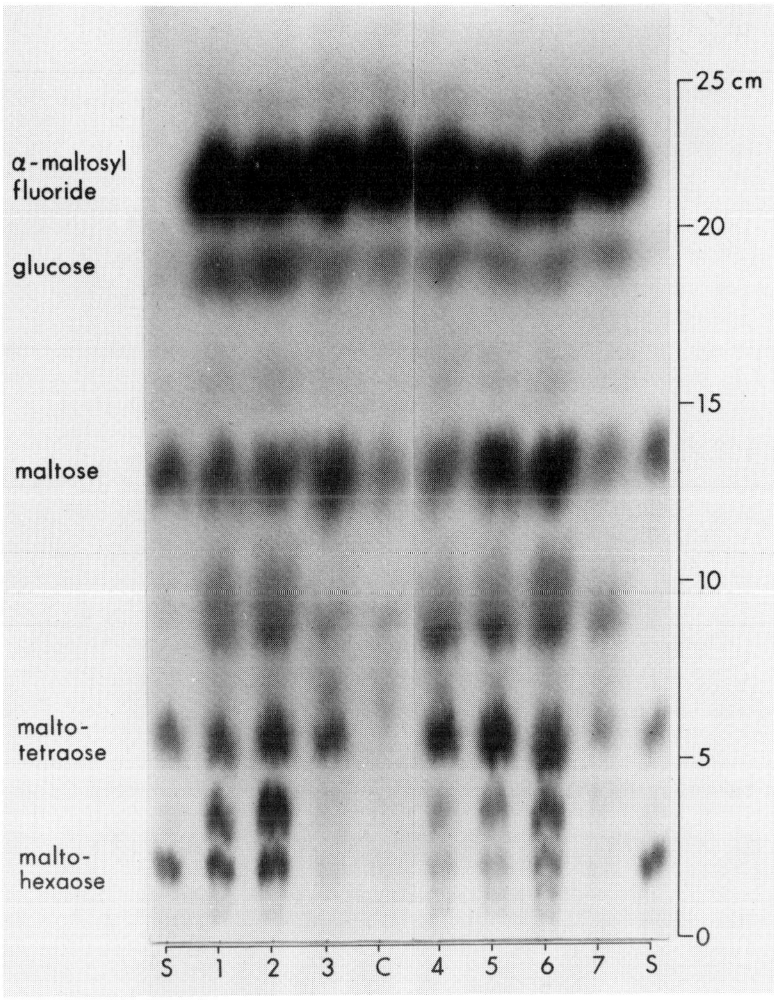

Figure 3. Maltosaccharide synthesis from α-maltosyl fluoride by α-amylases. Mixtures (100 μliters) contained 1.4 mg substrate (α-maltosyl fluoride, 1.8 μmoles) and α-amylase in 0.05 M pH 5.5 acetate buffer (or buffer alone). After incubation (30°C, 30 min) 15 μliters of each mixture was chromatographed.

1 = *cryst.* B. subtilis *liquifying α-amylase, 100 μg/ml.*
2 = *cryst.* B. subtilis *heat-stable α-amylase, 100 μg/ml.*
3 = *cryst.* B. stearothermophilus *α-amylase, 100 μg/ml.*
4 = *cryst. hog pancreatic α-amylase, 10 μg/ml.*
5 = *human saliva, 1:250 dilution.*
6 = *cryst.* Asp. oryzae *α-amylase, 1 μg/ml.*
7 = *cryst.* B. subtilis *saccharifying α amylase, 1 μg/ml.*
C = *control of α-maltosyl fluoride incubated without enzyme.*
S = *authentic maltohexaose, maltotetraose, maltose standards.*

the thesis that prompted their undertaking—*i.e.*, that the single type reaction derived by considering glycosyl transfer as glycosyl-proton interchange (*42*) underlies all reactions that carbohydrates in solution undergo under the influence of carbohydrases presently dissociated into hydrolases and transferases.

Table III. Evidence That the Actions of "Glycoside Hydrolases" (like One Glycosylic Bond at the Expense

Glycoside Hydrolase [E.C. 3.2.1]	*Alteration of Glycosylic Bond in Reactions of Hydrolysis*	
	Inversion of Configuration of Anomeric Carbon	*Change in the Atom Linked to the Anomeric Carbon*[a]
α-Amylase		O to ^{18}O (*51,109*) F to O (*104*), present work
β-Amylase	+ (*110,111,112*)	O to ^{18}O (*51,109*) F to O, present work
Glucoamylase, mold	+ (*112,113,114 115,116*)	F to O (*104,116,117, 118,119*)
Exo-β-1,3-glucanase	+ (*41,112,115,120*)	
Exo-β-1,4-glucanase	+ (*121*)	
Dextranase, *Penicillium*	+ (*122*)	
Cycloglucanase, mold	+ (*123,124,125*)	
Cycloglucanase, *B. polymyxa*	+ (*126*)	
Hyaluronidase, testis		O to ^{18}O (*127*)
Lysozyme, hen egg		O to ^{18}O (*128*)
Lysozyme, papaya	+ (*40*)	
α-D-Glucosidase		O to ^{18}O (*129*) F to O (*117,118,119*)
β-D-Glucosidase		O to ^{18}O (*129,130*) S to O (*131,132,133*) F to O (*116,119*)
α-D-Galactosidase		F to O (*119,134*)
β-D-Galactosidase		O to ^{18}O (*58*) S to O (*58*) F to O (*119,134*) N to O (*135*)
α-D-Mannosidase		F to O (*119*)
β-D-Fructofuranosidase		O to ^{18}O (*136*)
β-D-Glucuronidase		O to ^{18}O (*138*)

[a] O replaces N on hydrolysis of nucleosides by nucleosidases (*139,140*) and replaces S on hydrolysis of thioglucosides by thioglucosidase (*141,142*).

[b] Glycosidic linkage synthesis from free sugars (present as anomeric mixtures) is known for most of the listed enzymes (*52,53,54,55,56,57,58,59,60,61,62,63,64,65,66,67, 68,69,70,71,72,73,74,75,76,77*). Though less conclusive than the tabulated examples, these observations indicate that condensations are glycosylations on the basis that the enzymes cleave the glycosylic bond on hydrolysis.

The process of glycosyl transfer, defined in this mechanistic way, extends to hydrolysis and its reverse (*i.e.,* condensation synthesis from glycoses with expulsion of water) as well as to reactions in which water is not a reactant. However, the terms glycosyl transfer and transglycosylation are bound to the group transfer concept and inevitably carry a

those of "Glycosyl Transferases") Represent the Formation of of Another—*i.e.*, Glycosylation

Alteration of Glycosylic Bond in Reactions of Condensation and Transfer	
Inversion of Configuration of Anomeric Carbon [b]	*Change in the Atom Linked to the Anomeric Carbon* [c, d]
	F to O (*104*), present work
+ (*78*)	
+ (*78*)	
	F to O (*118*)
	O to N (*137*)

[c] Transfer reactions from saccharide or glycoside donors are well known for many so-called hydrolases (*20,21,22,23,24,25,26,27,28,29,30*). Like the tabulated examples which illustrate the point directly, such transfer reactions are glycosylations since the same enzymes cleave the glycosylic bond on hydrolysis.

[d] N' replaces N in transfer reactions catalyzed by certain nucleosidases, as in the synthesis of DPN analogs by DPNase (*140,143,144*).

sense of reactions not involving water and of substrates with a glycosidic bond. We therefore propose a less encumbered and maximally general term, glycosylation, to cover the full range of the reactions.

The evidence that glycosylation accounts for reactions variously seen as hydrolysis, condensation, and transfer is summarized in Table III for enzymes currently designated as glycoside hydrolases. It is abundantly clear that hydrolysis, catalyzed by these enzymes, is the exchange of one glycosylic bond for another since inversion of configuration of the anomeric carbon atom and/or a change in the atom linked to the anomeric carbon have been documented for a wide range of hydrolytic reactions. As for condensations, which involve the hemiacetal group of sugars and lead to oligosaccharide synthesis, these also represent the formation of one glycosylic bond from another; the proof is provided by our examples of α-glucosaccharide synthesis from the β-anomer of D-glucopyranose by glucoamylase and from β-maltose by β-amylase (*78*), supported by abundant evidence of a less direct nature (footnote *b* in Table III). Finally, glycosyl transfers, which do not involve water yet which are commonly observed with enzymes called hydrolases, also represent the formation of one glycosylic bond at the expense of another (Table III, right hand column and footnotes *c* and *d*). The reactions of the glycosyl transferases, a great number of which proceed with inversion of configuration and/or a change in the atom linked to the anomeric carbon, clearly reduce to the same expression.

One of the merits of this simple formulation is that it provides an unambiguous language for reactions involving glycosyl compounds that are not saccharides or glycosides. Thus, glycosyl fluorides are hydrolyzed by so-called glycoside hydrolases including α- and β-amylases (*see* Table III for references), and are also used as substrates for transfer reactions (to a small extent by α-D-glucosidase (*118*), preponderantly by α-amylases (*104,* Figure 3)). α-D-Glucosyl fluoride also serves as a donor for sucrose synthesis by sucrose phosphorylase, dextransucrase and amylosucrase (*145, 146, 147*), and for high polymer dextran and amylopolysaccharide synthesis by the latter two glycosyl transferases (*104, 146, 147*). A clear meaning (and expression of unity) is conveyed by describing all of these reactions as redistributions of glycosylic bonds; whereas, it would seem inappropriate to speak of the foregoing transfer reactions as redistributions of glycosidic bonds. Glycosyl fluorides contain a mixed acetal-fluoride function, but they are not glycosides, and they lack the glycosyloxy (glycosido) grouping present in all other known substrates for the several enzymes that are now found to synthesize oligo- and polysaccharides from glycosyl fluorides.

This is not to deny the possible usefulness of distinguishing glycosylation reactions in which water enters as a reactant from those in which it does not. Such a subdivision of reactions does not, however, affect the fundamental unity of the glycosylation process which may be considered as leading in every case to the formation of a new glycosylic bond having (usually) a different free energy of hydrolysis from that of the bond from which it arose. The positions of final equilibrium of different glycosylation reactions seem to be distributed over an extremely wide range, with the equilibrium position of each representing a function of the overall thermodynamics. However, thermodynamic factors do not determine whether a glycosylation reaction will, for example, be hydrolytic or nonhydrolytic. That is a matter determined especially by the structure of the particular enzyme, but also by the kinds and concentrations of donor and acceptor substrates and possibly other active compounds available to interact with the enzyme. It would seem (*29, 148*) that certain enzymes may have only hydrolytic activity (though they can utilize organic acceptors for condensation), that others have practically no hydrolytic activity, and that between these extremes there are very many enzymes which catalyze hydrolytic and nonhydrolytic reactions to differing degrees. Many of the latter group have previously been classified as hydrolases either for historical reasons or because in their reactions the hydrolytic process often ultimately extends to the cleavage of products of concurrent transfer reactions and, therefore, supervenes as final equilibrium approaches.

The glycosylation model accounts for such sequences perfectly well, and altogether displaces the familiar hydrolase concept which is defunct and should be abandoned. Only the former can account for the catalysis by enzymes with hydrolytic capacity of transfers that do not involve water as a reactant, for the identity of the glycosyl group(s) transferred in hydrolytic and nonhydrolytic reactions by a particular enzyme, for the glycosyl-specific inhibition of both reaction types by the aldonolactones (*149*). Moreover, the designation of an enzyme as a hydrolase implies that it has an overwhelming affinity for water and that no acceptor in low concentration can complete with 55 *M* water effectively enough to divert the major reaction from hydrolysis. This assumption is contradicted by the present evidence of abundant oligosaccharide synthesis by α-amylases acting on dilute solutions of α-maltosyl fluoride and by a recent finding (*150*) that hydrolysis of starch-type α-glycan by *B. subtilis* α-amylase is strongly diverted to glycoside formation in the presence of low concentrations of *p*-nitrophenyl β-D-glucoside. A further striking example of nonhydrolytic glycosylation from a highly diluted substrate, involving β-D-galactoside galactohydrolase, illustrates the significance that such reactions may have in cellular physiology. Jobe and Bourgeois (*151*) find that when lactose is supplied to *lac*⁺ *E. coli*, the low basal level of β-D-galactosidase hydrolyzes only part of the minute amount of lactose capable of entering the cell. A significant fraction of lactose (as a 3×10^{-4} *M* solution) is converted by the enzyme to allolactose, *O*-β-D-galactopyranosyl-(1 → 6)-D-glucopyranose, in a reaction in which the D-glucose moiety split from a lactose molecule is re-glycosylated at the C_6 carbinol before water or any other acceptor available in the cell can approach the active galactosyl–enzyme complex. The allolactose so produced is the inducer of the *lac* operon, thereby accounting for the synthesis of β-D-galactosidase when *E. coli* is grown in the presence of lactose.

We believe that the glycosylation paradigm provides significant alternatives to certain currently accepted premises and ways carried over from another era. For example, it confronts the present classification of the enzyme catalysts as hydrolases acting on glycosyl compounds and glycosyl transferases (*31*) with the alternative that all these enzymes belong together forming the great natural class of glycosylases (*104*), precisely defined and delineated from all other enzymes by their action as catalysts of glycosyl-proton interchange. There is nothing to prevent the division of this class into two subgroups having the same members as the present dissociated groups, on the basis of whether an enzyme generally uses water as a reactant or keeps it out. However, the important advantage of having the parental class of glycosylases as a starting point is that it allows for the development of alternative schemes of subdivision using other bases, the merits of which may then be compared with each other. The glycosylation model offers a rational approach to finding the best system of classifying and naming these intimately interrelated enzymes.

Likewise, it challenges the widespread custom of describing especially hydrolytic reactions in terms of splitting particular compounds or glycosidic linkages (an approach already known to be inadequate in the case of β-glycanases (*152, 153, 154*), with the alternative of specifying in the first place the glycosyl groups transferred to water. One consequence would be that knowledge of hydrolytic reactions could more directly influence and be influenced by the knowledge of other types of glycosylation reactions than at present. Another influence would lead the operational language away from such blurred expressions as breaking or forming a glycosidic bond found increasingly in the biochemical literature. A failure to focus on the reactive (glycosylic) bond of the glycosidic linkage is detected even among the precise Rules of Carbohydrate Nomenclature. Thus, "A glycoside is a mixed acetal resulting from the exchange of an alkyl or aryl radical for the hydrogen atom of the hemiacetal of a cyclic form of an aldose or ketose" (*155*); or "Mixed acetals, resulting from the replacement of the hydrogen atom of the anomeric or glycosidic hydroxyl group by group *X*, derived from an alcohol or phenol (XOH), are named 'glycosides'" (*156*). These are exact as geometric constructions, but it is doubted if any glycoside in nature (apart from glycosyl glycosides) has ever resulted from such an exchange or replacement. Presumably, all have arisen by enzymic interchange of a glycosyl radical with hydrogen from the carbinol of an alcohol or phenol. Surely, none has arisen by the interchange of a glycosyl radical with an hydroxyl group—*i.e.*,

$$X\text{-glycosyl} + Y\text{-OH} \rightleftharpoons Y\text{-glycosyl} + X\text{-OH},$$

as suggested in a recent review on glycosides (*157*).

Perhaps, it should be suggested speculatively that the open-ended type reaction of glycosylation has likely served the life process throughout evolutionary development for saccharide syntheses and degradations beyond enumeration. And possibly one of the fundamental reasons why cyclic sugar units exist as components of contemporary living matter has to do with the fact that these complex ring structures provide glycosyl radicals which can be interchanged with hydrogen and transferred to new chemical sites under the influence of specific enzyme catalysts. That sugars in solution have the tendency to dissociate into glycosyl and (anomeric) OH radicals may be inferred from reports (*158*) of the spontaneous (and acid–base catalyzed) exchange of the oxygen of the anomeric hydroxyl with that of water.

That the carbohydrases are glycosylases, and therefore are able to utilize substrates of simpler structure than customarily assumed to be required, is supported by new evidence. Lehmann and Schröter (*167*) report the hydration of D-glucal and D-galactal to the corresponding 2-deoxy-D-hexoses by β-D-glucosidase and β-D-galactosidase, respectively; also the synthesis of glyceryl 2-deoxy-β-D-glycosides from these glycals in the presence of glycerol as acceptor. These reactions may be viewed as instances of glycosyl–proton interchange (about C_2), with the double bond that is opened by the enzymes between C_1 and C_2 of the glycals having the character of a glycosylic bond.

Experimental

Reaction Components. The α-amylase preparations examined included twice crystallized hog pancreatic α-amylase (Worthington Biochemical Corp.) and 3-times crystallized *Aspergillus oryzae* Takaamylase A (Sankyo Co., Ltd., Tokyo) specified as homogeneous on electrophoresis and ultracentrifugation. Temperature-resistant α-amylase from *Bacillus subtilis* was the twice recrystallized product of Daiwa Kasei Co., Ltd., Osaka. Twice crystallized saccharogenic α-amylase from *B. subtilis* var *amylosacchariticus* Fukumoto and 3-times crystallized liquifying α-amylase from *B. subtilis* var *amyloliquefaciens* Fukumoto were products of Seikagaku Kogyo Co., Tokyo. Twice crystallized thermophilic α-amylase from *B. stearothermophilus Donk.* strain BS-1 (homogeneous by sedimentation, electrophoresis, and gel filtration) was contributed by Kyoko Ogasahara (*159*). Weighed amounts of the microbial enzymes were used to prepare solutions in 0.1 *M* pH 5.5 acetate buffer; the pancreatic α-amylase (in 0.5 saturated NaCl containing 0.003 *M* $CaCl_2$) was first separated from the suspending medium by high speed centrifugation then dissolved in the 0.1 *M* pH 5.5 acetate buffer. Crude salivary α-amylase (saliva diluted 1:250 in the same buffer and clarified by centrifugation) was also examined.

The α-maltosyl fluoride used as substrate was obtained as follows. Recrystallized β-maltose octaacetate (*160*), α_D^{26} + 63.1° ($c = 3$, $CHCl_3$), was dissolved in liquid anhydrous HF (10 grams in 40 ml.) at −15°C. After reaction for 20 min between −15°C and −10°C and 10 min at 20°C, the hepta-*O*-acetyl-α-maltosyl fluoride was extracted into ice-cold $CHCl_3$. This solution was then repeatedly washed with ice-cold water, dried with Na_2SO_4, and concentrated to a syrup at 30°C. Treatment with petroleum ether gave a bulky white precipitate which on vacuum drying at 25°C yielded 6.55 grams of a hard granular solid, α_D^{26} + 106° ($c = 2$, $CHCl_3$); lit. (*108*), α_D^{20} + 111.1° ($CHCl_3$). Examination under the polarizing microscope showed the presence of some needle-shaped highly birefringent crystals embedded in an amorphous matrix.

For deacetylation, 2 grams of the α-maltosyl fluoride heptaacetate in 50 ml of methanol, cooled in an ice-salt mixture, was treated with 1.6 ml of 0.5 *N* barium methoxide (*161*). After 16 hours at 0°C barium ion was removed with moist Amberlite 1R-120-H, and the filtered solution was evaporated to a syrup at 35°C. Final drying from solution in methanol gave 0.87 gram of a glassy product. This contained α-maltosyl fluoride (hydrolyzable to maltose and HF by β-amylase or mild acid treatment) admixed with a large quantity of maltose, some D-glucose, and traces of several unidentified carbohydrates. For final purification, the material (730 mg) was chromatographed on sheets of Whatman No. 1 paper, and the α-maltosyl fluoride, migrating slightly faster than D-glucose in the solvent used, was recovered by elution with methanol; yield, 118 mg of an amorphous powder, α_D^{31} + 105.8° ($c = 1.3$, methanol).

Chromatographic examination (0.2 mg/spot) showed the presence of a single major component which slowly reduced $AgNO_3$ plus traces of D-glucose and maltose and faint traces of two unidentified saccharides (Figure 3). Chromatograms of material treated with 0.1 *N* H_2SO_4 (100°C, 10 min), or with crystalline sweet potato β-amylase (30°C, pH 5.5, 4 hours) showed abundant maltose but also an appreciable amount of residual material at the original location. Analysis of these digests for liberated fluoride (*104, 107*) gave a value corresponding to 44% of that expected for pure α-maltosyl fluoride (assuming complete hydrolysis). The accompanying substance that co-migrated with α-maltosyl fluoride was recovered by subjecting 80 mg of the mixture to acid hydrolysis (0.1 *N* H_2SO_4, 10 min, 100°C) followed by chromatography and elution.

This acid resistant substance (16.3 mg) had α_D + 93° ($c = 0.4$, H_2O) and was hydrolyzed by purified *Candida tropicalis* α-glucosidase (*162*) to D-glucose and a substance identified as levoglucosan on the basis of its nonreducing nature and chromatographic mobility, indistinguishable from levoglucosan (a preparation of which was obtained by catalytic deacetylation of an authentic sample of 2,3-*O*-diacetyl-1,6-anhydro-β-D-glucopyranose, kindly supplied by Donald M. Marcus). It thus seems to be an α-D-glucopyranosyl-1,6-anhydro-β-D-glucopyranose, possibly maltosan (lit. (*163*), $\alpha_D + 79°$ ($c = 4$, H_2O)) arising with maltose from alkaline degradation of α-maltosyl fluoride in the same way as levoglucosan arises from α-D-glucosyl fluoride (*164, 165*). The resistant accompanying material was not detectably attacked by the various α-amylases under conditions similar to those used in the experiment of Figure 3.

Experimental Memoranda. Concentrations were carried out under reduced pressure in a rotary evaporator at or below 35°C. All chromatography was carried out using Whatman No. 1 paper and two 22-hour ascents in 1-butanol–pyridine–water (6:4:3). Staining was by a $AgNO_3$ dipping technique (*95, 166*) with papers hung in air for exactly 7 min following application of the NaOH reagent. Fluoride was determined by the method of Megregian and Maier (*107*) modified (*104*) to permit analysis of small samples.

Acknowledgment

Support of the Corn Refiners Association, Washington, D. C., is gratefully acknowledged.

Literature Cited

1. Hehre, E. J., *Science* (1941) **93,** 237.
2. Hehre, E. J., Sugg, J. Y., *J. Exper. Med.* (1942) **75,** 339.
3. Stacey, M., *Nature (London)* (1942) **149,** 639.
4. Aschner, M., Avineri-Shapiro, S., Hestrin, S., *Nature (London)* (1942) **149,** 527.
5. Hestrin, S., Avineri-Shapiro, S., *Nature (London)* (1943) **152,** 49.
6. Miwa, T., Mafune, K., Furuya, T., *Igaku to Seibutsugaku (Japan)* (1942) **1,** 80.
7. Beijerinck, M. W., *Folia Microbiol.* (1912) **1,** 377.
8. Rabaté, J., Thesis, Fac. Sci. Paris, p. 190, Declume, Lons-Le-Saunier (France), 1934.
9. Rabaté, J., "Die Methoden der Fermentforschung," p. 1818, Vol. 2, Lief. 5, E. Baumann and K. Myrbäck, Eds., Thieme, Leipzig, 1940.
10. Hehre, E. J., *Proc. Soc. Exper. Biol. Med.* (1943) **54,** 240.
11. Hestrin, S., Avineri-Shapiro, S., *Biochem. J.* (1944) **38,** 2.
12. Doudoroff, M., Barker, H. A., Hassid, W. Z., *J. Biol. Chem.* (1947) **168,** 725.
13. Hehre, E. J., *Advan. Enzymol.* (1951) **11,** 297.
14. Kalckar, H. M., *J. Biol. Chem.* (1945) **158,** 723.
15. Cohn, M., *J. Biol. Chem.* (1949) **180,** 771.
16. Hestrin, S., Feingold, D. S., Avigad, G., *Biochem. J.* (1956) **64,** 340.
17. Bacon, J. S. D., Edelman, J., *Arch. Biochem.* (1950) **28,** 467.
18. Blanchard, P. H., Albon, N., *Arch. Biochem.* (1950) **29,** 220.
19. Takano, K., Miwa, T., *J. Biochem. (Japan)* (1950) **37,** 435.
20. Barker, S. A., Bourne, E. J., *Quart. Rev. Chem. Soc.* (1953) **7,** 56.
21. Hoffmann-Ostenhof, O., *Advan. Enzymol.* (1953) **14,** 219.
22. Stacey, M., *Advan. Enzymol.* (1954) **15,** 301
23. Edelman, J., *Advan. Enzymol.* (1956) **17,** 189, 225.

24. Leloir, L. F., Cardini, C. E., Cabib, E., "Comparative Biochemistry," p. 97, Vol. 2, M. Florkin and H. S. Mason, Eds., Academic, New York, 1960.
25. Dedonder, R. A., *Ann. Rev. Biochem.* (1961) **30,** 347.
26. Hassid, W. Z., Neufeld, E. F., "Enzymes," p. 277, Vol. 6, P. D. Boyer, H. Lardy, and K. Myrbäck, Eds., Academic, New York, 1962.
27. Tsuiki, S., Hashimoto, Y., Pigman, W., "Comprehensive Biochemistry," pp. 153, 172, Vol. 5, M. Florkin and E. H. Stotz, Eds., Elsevier, New York, 1963.
28. Glaser, L., "Comprehensive Biochemistry," p. 93, Vol. 15, M. Florkin and E. H. Stotz, Eds., Elsevier, New York, 1964.
29. Hassid, W. Z., "The Carbohydrates," pp. 302, 312, Vol. IIA, 2nd ed., W. Pigman and D. Horton, Eds., Academic, New York, 1970.
30. Nisizawa, K., Hashimoto, Y., "The Carbohydrates," p. 241, Vol. IIA, 2nd ed., W. Pigman and D. Horton, Eds., Academic, New York, 1970.
31. Hoffmann-Ostenhof, O., *Ann. Rev. Biochem.* (1960) **29,** 73.
32. International Union of Biochemistry, "Enzyme Nomenclature," pp. 98, 136, Elsevier, Amsterdam, 1965.
33. Dixon, M., Webb, E. C., "Enzymes," pp. 740–746, 2nd ed., Academic, New York, 1964.
34. Tsujisaka, Y., Fukumoto, J., Yamamoto, T., *Nature* (1958) **181,** 770.
35. Jermyn, M. A., *Rev. Pure Applied Chem.* (1961) **11,** 92.
36. French, D., "Control of Glycogen Metabolism," p. 25, W. J. Whelan and M. P. Cameron, Eds., Little, Brown, Boston, 1964.
37. Thoma, J. A., Spradlin, J. E., Dygert, S., "The Enzymes," p. 115, Vol. V, P. D. Boyer, Ed., Academic, New York, 1971.
38. Dahlqvist, A., *Acta Chem. Scand.* (1960) **14,** 9.
39. Avigad, G., *Biochem. J.* (1965) **97,** 715.
40. Dahlqvist, F. W., Borders, C. L., Jr., Jacobson, G., Raferty, M. A., *Biochemistry* (1969) **8,** 694.
41. Nelson, T. E., *J. Biol. Chem.* (1970) **245,** 869.
42. Hehre, E. J., *Bull. Soc. Chim. Biol.* (1960) **42,** 1713.
43. Abeles, R. H., Voet, J. G., *J. Biol. Chem.* (1970) **245,** 1020.
44. Koshland, D. E., Jr., *Biol. Rev. Cambridge Phil. Soc.* (1953) **28,** 416.
45. Silverstein, R., Voet, J., Reed, D., Abeles, R. H., *J. Biol. Chem.* (1967) **242,** 1338.
46. Dahlqvist, F. W., Rand-Meir, T., Raferty, M. A., *Biochemistry* (1969) **8,** 4214.
47. Legler, G., *Biochim. Biophys. Acta* (1968) **151,** 728.
48. Blake, C. C. F., Johnson, L. N., Mair, G. A., North, A. C. T., Phillips, D. C., Sarma, V. R., *Proc. Roy. Soc. (London)* (1967) **B167,** 378.
49. Tu, J. I., Jacobson, G. R., Graves, D. J., *Biochemistry* (1971) **10,** 1229.
50. Thoma, J. A., *J. Theor. Biol.* (1968) **19,** 297.
51. Mayer, F. C., Larner, J., *J. Amer. Chem. Soc.* (1959) **81,** 188.
52. Bourquelot, E., *Ann. Chim.* (1913) **29,** 145.
53. Bourquelot, E., *J. Pharm. Chim.* (1914) **10,** 361.
54. Oppenheimer, C., "Die Fermente und Ihre Wirkungen," pp. 532–540, Vol. 1, 5th ed., Thieme, Leipzig, 1925.
55. Gottshalk, A., "The Enzymes," p. 551, Vol. 1, J. B. Sumner and K. Myrbäck, Eds., Academic, New York, 1950.
56. Veibel, S., "The Enzymes," pp. 583, 621, Vol. 1, J. B. Sumner and K. Myrbäck, Eds., Academic, New York, 1950.
57. Jermyn, M. A., *Science* (1957) **125,** 12.
58. Wallenfels, K., *Bull. Soc. Chim. Biol.* (1960) **42,** 1715.
59. Chiba, S., Shimomura, T., *Agr. Biol. Chem.* (1966) **30,** 536.
60. Clancy, M. J., Whelan, W. J., *Arch. Biochem. Biophys.* (1967) **118,** 724.
61. Andersen, B., *Acta Chem. Scand.* (1967) **21,** 828.
62. Nelson, T. E., Larner, J., *Biochim. Biophys. Acta* (1970) **198,** 538.

63. Peat, S., Whelan, W. J., Hinson, K. A., *Nature* (1952) **170,** 1056.
64. Stephen, A. M., Kirkwood, S., Smith, F., *Can. J. Chem.* (1962) **40,** 151.
65. Stark, J. R., Thambyrajah, V., *Biochem. J.* (1970) **120,** 17P.
66. French, D., *Advan. Carbohydrate Chem.* (1954) **9,** 149.
67. Lukomskaya, I. S., *Enzymologia* (1962) **24,** 329.
68. Chesters, C. G. C., Bull, A. T., *Nature* (1964) **202,** 454.
69. Abdullah, M., French, D., *Nature* (1966) **210,** 200.
70. Abdullah, M., French, D., *Arch. Biochem. Biophys.* (1970) **137,** 483.
71. Pfannmuller, J., Noë, A., *Science* (1952) **115,** 240.
72. Tamura, T., Shibata, S., Kawase, N., Saeki, F., Suzuki, S., *Denpunto Gijutsu Kenkyu Kaiho (Japan)* (1961) **24,** 45; *Chem. Abstr.* (1964) **61,** 817.
73. Kuroiwa, S., Nakamura, M., *Kogyo Kagaku Zasshi (Japan)* (1961) **64,** 942; *Chem. Abstr.* (1962) **57,** 5015.
74. Fukumoto, J., Tsujisaka, Y., Kimoto, K., *Symp. Enzyme Chemistry, Fukuoka (Japan) 14th* (1962) p. 13; *Abstr. Sixth Internat. Congr. Biochem.* (1964) **VI,** 529.
75. Underkofler, L. A., Denault, L. J., Hou, E. F., *Die Stärke* (1965) **17,** 179.
76. Pazur, J., Okada, S., *Carbohyd. Res.* (1967) **4,** 371.
77. Watanabe, T., Kawamura, S., Sasaki, H., Matsuda, K., *Die Stärke* (1969) **21,** 18, 42.
78. Hehre, E. J., Okada, G., Genghof, D. S., *Arch. Biochem. Biophys.* (1969) **135,** 75.
79. Oppenheimer, C., "Die Fermente und Ihre Wirkungen," pp. 640, 667, Vol. 1, 5th ed., Thieme, Leipzig, 1925.
80. Myrbäck, K., Neumüller, G., "The Enzymes," p. 653, Vol. 1, J. B. Sumner and K. Myrbäck, Eds., Academic, New York, 1950.
81. Fisher, E. H., Stein, E. A., "The Enzymes," p. 313, Vol. 4, 2nd ed., P. D. Boyer, H. Lardy, and K. Myrbäck, Eds., Academic, New York, 1960.
82. Manners, D. J., *Advan. Carbohyd. Chem.* (1962) **17,** 371.
83. Bernfeld, P., "Comparative Biochemistry," p. 355, Vol. 3, M. Florkin and H. S. Mason, Eds., Academic, New York, 1962.
84. Whelan, W. J., "Methods of Carbohydrate Chemistry," p. 252, Vol. 4, R. L. Whistler, Ed., Academic, New York, 1964.
85. Courtois, J. E., Perles, R., "Traite de Biochimie Generale," p. 287, Fasc. 2, Vol. 2, M. Javillier *et al.*, Eds., Masson et Cie, Paris, 1964.
86. Greenwood, C. T., Milne, E. A., *Advan. Carbohyd. Chem.* (1968) **23,** 282, 305.
87. Matsubara, S., *J. Biochem. Tokyo* (1961) **49,** 226.
88. Mantani, S., Yamamoto, T., Fukumoto, J., "Proceedings of the Symposium on Amylases," p. 95, Amylase Research Group, Eds., Osaka Technological Research Society, Osaka, 1967.
89. Yoshida, H., Hiromi, K., Ono, S., *J. Biochem. Tokyo* (1967) **62,** 439.
90. Yoshida, H., Hiromi, K., Ono, S., *J. Biochem. Tokyo* (1969) **65,** 741.
91. Yoshida, H., Hiromi, K., Ono, S., *J. Biochem. Tokyo* (1969) **66,** 183.
92. Miller, K. D., Copeland, W. H., *J. Biol. Chem.* (1958) **231,** 987, 997.
93. Sawai, T., *J. Biochem. Tokyo* (1959) **45,** 49.
94. Sawai, T., *J. Biochem. Tokyo* (1960) **48,** 382.
95. Sawai, T., Hehre, E. J., *J. Biol. Chem.* (1962) **237,** 2047.
96. Whelan, W. J., "Control of Glycogen Metabolism," p. 26, W. J. Whelan and M. P. Cameron, Eds., Little, Brown, Boston, 1964.
97. French, D., Abdullah, M., *Cereal Chem.* (1966) **43,** 555.
98. Greenwood, C. T., Milne, E. A., Ross, G. R., *Arch. Biochem. Biophys.* (1968) **126,** 244.
99. Okada, S., Kitahata, S., Higashihara, M., Fukumoto, J., *Agr. Biol. Chem.* (1969) **33,** 900.

100. Robyt, J. F., French, D., *Fed. Proc. Fed. Amer. Soc. Exp. Biol.* (1969) **28,** 345.
101. Robyt, J. F., French, D., *J. Biol. Chem.* (1970) **245,** 3917.
102. Pazur, J. H., Okada, S., *J. Biol. Chem.* (1966) **241,** 4146.
103. Pazur, J. H., Okada, S., *J. Biol. Chem.* (1968) **243,** 4732.
104. Hehre, E. J., Genghof, D. S., Okada, G., *Arch. Biochem. Biophys.* (1971) **142,** 382.
105. Sheppard, W. A., Sharts, C. M., "Organic Fluorine Chemistry," pp. 14–17, Benjamin, New Yolrk, 1969.
106. Hodge, J. E., Rendleman, J. A., Baker, B. G., Zobel, H. F., *Abstr. 153rd Meeting Amer. Chem. Soc.* (1967) **C,** 7.
107. Megregian, S., Maier, F. J., "Fluoride Drinking Waters," F. J. McClure, Ed., p. 566, Public Health Service Publication No. 825, Bethesda, 1962.
108. Brauns, D. H., *J. Amer. Chem. Soc.* (1929) **51,** 1920.
109. Halpern, M., Leibowitz, J., *Biochim. Biophys. Acta* (1959) **36,** 29.
110. Freeman, G. G., Hopkins, R. H., *Biochem. J.* (1936) **30,** 451.
111. Thoma, J. A., Koshland, D. E., Jr., *J. Biol. Chem.* (1960) **235,** 2511.
112. Eveleigh, D. E., Perlin, A. S., *Carbohyd. Res.* (1969) **10,** 87.
113. Fuwa, H., Nikuni, Z., *Mem. Inst. Sci. Indust. Res. (Osaka)* (1952) **9,** 194.
114. Weill, C. E., Burch, R. J., Van Dyke, J. W., *Cereal Chem.* (1954) **31,** 159.
115. Parrish, F., Reese, E. T., *Carbohyd. Res.* (1967) **3,** 424.
116. Barnett, J. E. G., Jarvis, W. T. S., Munday, K. A., *Biochem. J.* (1967) **105,** 9P.
117. Barnett, J. E. G., Jarvis, W. T. S., Munday, K. A., *Biochem. J.* (1966) **100,** 41P.
118. Barnett, J. E. G., Jarvis, W. T. S., Munday, K. A., *Biochem. J.* (1967) **103,** 699.
119. Barnett, J. E. G., Jarvis, W. T. S., Munday, K. A., *Biochem. J.* (1967) **105,** 669.
120. Barras, D. R., Stone, B. A., *Biochim. Biophys. Acta* (1969) **191,** 342.
121. Storvick, W. O., Cole, F. E., King, K. W., *Biochemistry* (1963) **2,** 1106.
122. Kobayashi, T., *Nippon Nogei Kagaku Kaishi (Japan)* (1954) **28,** 352.
123. Nikuni, Z., Fuwa, H., *Proc. Japan Acad.* (1952) **28,** 139.
124. Ben-Gershom, E., *Nature* (1955) **175,** 593.
125. Ben-Gershom, E., Leibowitz, J., *Enzymologia* (1958) **20,** 133, 148.
126. Robyt, J., French, D., *Arch. Biochem. Biophys.* (1964) **104,** 338.
127. Ludowieg, J., Vennesland, B., Dorfman, A., *J. Biol. Chem.* (1961) **236,** 333.
128. Rupley, J. A., Gates, V., *Proc. Nat. Acad. Sci. (U.S.)* (1967) **51,** 496.
129. Bunton, C. A., Lewis, T. A., Llewellyn, D. R., Tristram, H., Vernon, C. A., *Nature* (1954) **174,** 560.
130. Springhorn, S. S., Koshland, D. E., Jr., *Abstr. Amer. Chem. Soc. 128th Meet.* (1955) **37C.**
131. Jermyn, M. A., *Aust. J. Biol. Sci.* (1955) **8,** 577.
132. Jermyn, M. A., *Aust. J. Biol. Sci.* (1958) **11,** 114.
133. Wagner, G., Metzner, R., *Naturwissenschaften* (1965) **52,** 83.
134. v. Hofsten, B., *Biochim. Biophys. Acta* (1961) **48,** 159.
135. Sinott, M. L., *Biochem. J.* (1971) **125,** 717.
136. Koshland, D. E., Jr., Stein, S. S., *J. Biol. Chem.* (1954) **208,** 139.
137. Miwa, T., Takeshita, M., Nakamura, S., *Biochim. Biophys. Acta* (1960) **37,** 541.
138. Eisenberg, F., *Fed. Proc. Amer. Soc. Exp. Biol.* (1959) **18,** 221.
139. Carter, C. E., *J. Amer. Chem. Soc.* (1951) **73,** 1508.
140. Zatman, L. J., Kaplan, N. O., Colowick, S. P., *J. Biol. Chem.* (1953) **200,** 197.
141. Gadamer, J., *Arch. Pharmaz.* (1897) **235,** 44.

142. Ettlinger, M. G., Dateo, G. P., Jr., Harrison, B. W., Mabry, T. J., Thompson, C. P., *Proc. Nat. Acad. Sci. (U.S.)* (1961) **47,** 1875.
143. Zatman, L. J., Kaplan, N. O., Colowick, S. P., Ciotti, M. M., *J. Biol. Chem.* (1954) **209,** 467; *J. Amer. Chem. Soc.* (1954) **75,** 3293.
144. Alivisatos, S. G. A., Wooley, D. W., *J. Amer. Chem. Soc.* (1955) **77,** 1065; *J. Biol. Chem.* (1956) **219,** 823.
145. Gold, A. M., Osber, M. P., *Biochem. Biophys. Res. Comm.* (1971) **42,** 469.
146. Genghof, D. S., Hehre, E. J., *Proc. Soc. Exp. Biol. Med.* (1972) **140,** 1298.
147. Okada, G., Hehre, E. J., *Carbohyd. Res.* (1972), in press.
148. Miwa, T., "Proceedings of the International Symposium on Enzyme Chemistry Tokyo and Kyoto 1957," p. 162, Internat. Union of Biochem., Maruzen, Tokyo, 1958.
149. Levvy, G. A., Conchie, J., "Methods in Enzymology," p. 571, Vol. 8, S. P. Colowick and N. O. Kaplan, Eds., Academic, New York, 1966.
150. Takeshita, M., Hehre, E. J., unpublished experiments.
151. Jobe, A., Bourgeois, S., *J. Mol. Biol.* (1972) **69,** 397; also, personal communication, Alan Jobe.
152. Parrish, F. W., Perlin, A. S., *Nature (London)* (1960) **187,** 1110.
153. Perlin, A. S., "Advances in Enzymic Hydrolysis of Celluloses and Related Materials," p. 185, E. T. Reese, Ed., Macmillan, New York, 1963.
154. Reese, E. T., Mandels, M., "Advances in Enzymic Hydrolysis of Celluloses and Related Materials," pp. 197, 231, E. T. Reese, Ed., Macmillan, New York, 1963.
155. Rules of Carbohydrate Nomenclature, *J. Org. Chem.* (1963) **28,** 281, 285.
156. IUPAC Commission on the Nomenclature of Organic Chemistry (CNOC), IUPAC-IUB Commission on Biochemical Nomenclature (CBN), *Eur. J. Biochem.* (1971) **21,** 455, 464.
157. Courtois, J. E., Percheron, F., "The Carbohydrates," p. 213, Vol. IIA, 2nd ed., W. Pigman and D. Horton, Eds., Academic, New York, 1970.
158. Rittenberg, D., Graff, C., *J. Amer. Chem. Soc.* (1958) **80,** 3370.
159. Ogasahara, K., Imanishi, A., Isemura, T., *J. Biochem. Tokyo* (1970) **67,** 65, 77.
160. Wolfrom, M. L., Thompson, A., "Methods of Carbohydrate Chemistry," p. 334, Vol. 1, R. L. Whistler, Ed., Academic, New York, 1962.
161. Isbell, H. S., *Bur. Std. J. Res.* (1930) **5,** 1179.
162. Sawai, T., "Proceedings of the Symposium on Amylase, 1967," p. 111, Society of Amylase Researchers, Eds., Osaka, Japan, 1967.
163. Karrer, P., Kamienski, L., *Helv. Chim. Acta* (1932) **15,** 739.
164. Micheel, F., Klemer, A., Baum, G., Ristic, P., Zumbülte, F., *Chem. Ber.* (1955) **88,** 475.
165. Micheel, F., Klemer, A., *Advan. Carbohydrate Chem.* (1961) **16,** 85.
166. Trevelyan, W. E., Procter, D. P., Harrison, J. S., *Nature (London)* (1950) **166,** 444.
167. Lehmann, J., Schroter, E., *Carbohyd. Res.* (1972) **23,** 359.

RECEIVED December 7, 1971.

18

Immunochemical Studies on the Carbohydrate Moiety of Water Soluble Blood Group A, B, H, Le^a, and Le^b Substances and Their Precursor I Antigens

ELVIN A. KABAT

Departments of Microbiology, Neurology, and Human Genetics and Development, Columbia University and the Neurological Institute, Presbyterian Hospital, New York, N.Y. 10032

The carbohydrate moiety of blood group glycoproteins is reviewed and analyzed. Current work in the immunochemistry of the ABH and Lewis blood group determinants is described, and other antigenic determinants on the megalosaccharide are discussed. Blood group genetics and basic immunochemical concepts are explained.

Blood group substances are complex macromolecules possessing carbohydrate antigenic determinants which are recognized by their reaction with various kinds of specific antisera (An antigenic determinant is defined as the portion of the macromolecule which reacts with the antibody combining site.). Since Landsteiner used the antibodies to the A and B determinants normally found in the serum of group B and of group A individuals to identify the A, B, and O blood groups and to place blood transfusion on a rational basis, there has been much interest in the chemistry and immunochemistry of these substances (*1*). The A and B determinants are only two of many specific antigenic groupings or structures intimately related to one another and synthesized by a set of enzymes under genetic control. Other recognized carbohydrate determinants closely involved structurally with the A and B antigens are the H, Le^a, Le^b, and I blood groups.

The genetics of the human blood groups has been shown to follow classical Mendelian principles. The four major blood groups A, B, O and AB of human erythrocytes are generally considered to be controlled by three genes A, B, and O at a single locus. Several variants of the A gene actually exist, the most important of which are those designated A_1 and A_2. The A and B genes code for enzymes synthesizing products, antigenic determinants recognizable by their reactions with specific antisera, A_2 erythrocytes generally giving weaker reactions than A_1 erythrocytes. The O gene does not give rise to a distinctive antigenic determinant but leaves unaltered the H determinant which resulted from biosynthetic steps before those producing A or B determinants and is thus thought of as a precursor of the A and B determinants. The H determinant is controlled by another independent gene locus, Hh. Two other independent gene loci, the Lele and the Ii loci, control the addition of their respective antigenic determinants on to the same core structures which may bear A, B, and H determinants.

In secretions—e.g., saliva, gastric juice, ovarian cyst fluid, etc.—the presence of soluble glycoproteins bearing the A, B, and H determinants is also controlled by an independent gene locus Sese. Eighty % of A, B, O, and AB individuals secrete glycoproteins bearing determinants with A, B, and H specificity, depending on their genetic composition; the 20% who do not secrete A, B, or H substances are termed non-secretors; they secrete a glycoprotein with Le^a determinants. Rare individuals do not secrete A, B, H, or Le^a substances. The Le^b determinant results from the combined action of the H and Le genes when both occur in secretors. The Ii system also involves antigenic determinants synthesized at earlier stages than the A, B, H, and Le^a glycoproteins. Its complexity and relation to the other determinants is considered subsequently.

During the past quarter of a century, work on the structural chemistry of the blood group A, B, H, Le^a and Le^b substances from human, animal, plant and microbial sources has been actively pursued in various parts of the world (*1, 2, 3, 4, 5, 6, 7, 8, 9*). A genetic scheme for the biosynthesis of human blood group substances has been described by Ceppellini (*10*) and by Watkins and Morgan (*11*), and studies in various laboratories (*7*) on the glycosyltransferases involved have given convincing support for this concept.

The blood group substances are generally considered to be of two types, glycolipid and glycoprotein. The glycoproteins have high molecular weight and are found in secretions of mucosal tissues—notably, in saliva, gastric juice, pseudomucinous ovarian cyst fluids, etc. (*1*). The glycolipids occur in capillary endothelium and may also be found in serum. Blood group determinants on erythrocytes have generally been considered to be glycolipid (*12*), but recently Zahler (*13*), Whittemore *et al.* (*14*), and Gardas and Koscielak (*15*) have indicated that the amounts of glycolipid present are insufficient to account for the blood group activity of erythrocytes and concluded that some of the blood group substance on erythrocytes is also glycoprotein. Glycolipids with certain blood group activities, notably Le^a and Le^b, are absorbed on to erythrocytes from plasma and may thus confer a particular specificity (*16, 17*); these findings have limited our understanding of the nature of erythrocyte blood group substances. Using fluorescent anti-A and anti-B, the histochemical localization of blood group antigenic determinants in various cells and tissues has been described (*18*).

Oligosaccharides with blood group activity are found in human urine (*7*) and in human milk (*3*). Materials with blood group activity are widely distributed in nature (*1, 2, 3, 4, 5, 6, 7, 8, 9*) as are various proteins, plant (*19, 20*), and animal hemagglutinins, (*21, 22, 23*) which show specificity for blood group determinants.

Most of our knowledge of the structure and specificity of blood group substances has come from studies on the water soluble A, B, and H substances of secretions. About 1 in 100 individuals do not secrete A, B, H, or Le^a substances but secrete a precursor type of substance (*2, 25*) which has recently been shown to possess I and i specificity (*26, 27*). To secrete a particular blood group substance an individual must possess not only the gene which determines his capacity to synthesize the antigenic determinant but also the secretor (Se) gene, which controls whether or not the soluble glycoproteins in secretions will contain the A, B, or H antigenic determinants (*1, 2, 3, 4*). From tissues with large amounts of glandular mucosal cells—*e.g.*, stomach, intestine, etc. (*18, 28*)—substantial amounts of water soluble substances are also obtained after autolysis or peptic digestion (*1*). Human milk has proved a good source for a glycoprotein reacting with certain anti-I sera (*26, 27, 29*).

The presence of one antigenic determinant may reduce the extent of expression of another. Thus, in an A glycoprotein of a secretor, the expression of the H determinant is largely inhibited, and by using enzymes (*30*) which destroy A or B activity, H activity is greatly increased (*3, 7*). Also, gene interactions may occur as, for example, when the H and Le genes are both present, a new specificity termed Le^b arises (*10, 11*).

Over the past 15 years numerous oligosaccharides from blood group glycoproteins have been isolated in various laboratories (*3, 7*) after mild acid hydrolysis or by alkaline degradation followed by fractionation by various chromatographic methods. The most widely used source is the fluid from women with pseudomucinous ovarian cysts (*31*) which may contain extremely large quantities of these blood group substances and from which almost all of the oligosaccharides have been obtained. Blood group glycoproteins from various tissues have recently been shown to differ in structure. Thus, blood group A and H substances from hog and human gastric mucosa possess unique carbohydrate determinants absent

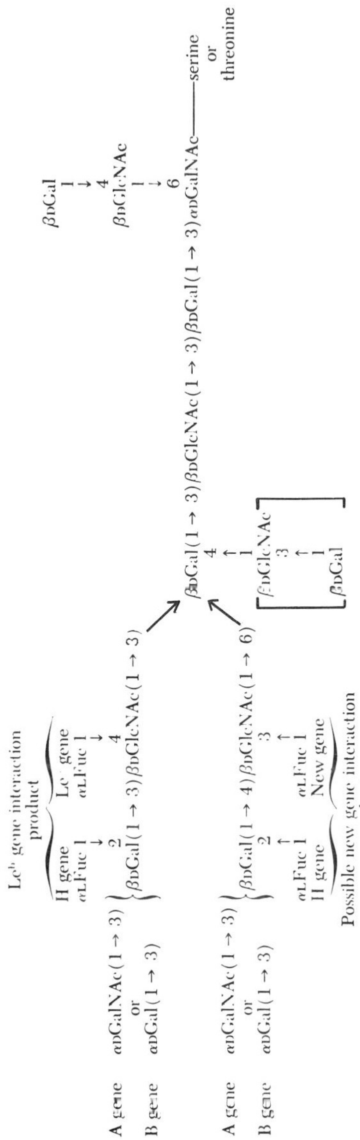

Figure 1. The structure presented is based on the earlier composite structure (37, 38) for the A, B, H, Le^a, and Le^b substances and shows the relationships of the type 1 and type 2 determinants upon which these antigens are built. It is subject to all of the limitations considered earlier. As in the earlier studies, incomplete chains are present and can result from incomplete biosynthesis or by degradation in the cyst cavity. The bracketed substitution on carbon-4 of the 3, 4, 6-linked galactose could be galactose; whatever the residue, it must be a sequence capable of giving galactitol on peeling. From the analytical composition one to two additional fucoses are present which have not been located (27).

in human ovarian cyst blood group substances (*32, 33, 34, 35*). However, identical highly active blood group A monofucosyl reduced pentasaccharide, as well as an identical H active blood group monofucosyl reduced tetrasaccharide, have been isolated from substances from hog gastric mucosa and from human ovarian cysts (*36*).

From the structures of the various oligosaccharides isolated after alkaline degradation with NaOD–$NaBD_4$ and based primarily on two branched oligosaccharides obtained from human ovarian cyst Le[a] substance and on the mechanism of the alkaline borohydride degradation, a composite structure for the blood group A, B, H, Le[a], and Le[b] substances has been advanced (Figure 1) (*37, 38*).

On the assumption that the two branched reduced oligosaccharides (*37*)

$$\begin{array}{l} \beta\text{DGal}(1 \rightarrow 4)\beta\text{DGlcNAc} \\ \qquad\qquad\qquad\qquad 1 \\ \qquad\qquad\qquad\qquad \downarrow \\ \qquad\qquad\qquad\qquad 6 \\ \qquad\qquad\qquad\quad \text{D-galactitol} \\ \qquad\qquad\qquad\qquad 3 \\ \qquad\qquad\qquad\qquad \uparrow \\ \qquad\qquad\qquad\qquad 1 \\ \beta\text{DGal}(1 \rightarrow 3)\beta\text{DGlcNAc} \end{array} \qquad [1] \text{ and}$$

$$\begin{array}{c} \beta\text{DGal} \\ 1 \\ \downarrow \\ 4 \\ \beta\text{DGlcNAc} \\ 1 \\ \downarrow \\ 6 \end{array}$$

$$\beta\text{DGal}(1 \rightarrow 3)\text{2acetamido2deoxyDgalactitol} \qquad [2]$$

and two oligosaccharides

$$\beta\text{DGal}(1 \rightarrow 4)\beta\text{DGlcNAc}(1 \rightarrow 6)\beta\text{DGal}(1 \rightarrow 3)\text{DGlcNAc} \quad [3]$$

isolated by Aston *et al.* (*39*) and

$$\beta\text{DGlcNAc}(1 \rightarrow 3)\beta\text{DGal}(1 \rightarrow 3)\text{DGalNAc} \qquad [4]$$

(*40*) which partially overlapped with them could be assembled to give a core megalosaccharide (Figure 2) upon which the various genes A, B, H, and Le would determine the addition of the other sugars to form the antigenic determinants (Figure 1). Almost all of the oligosaccharides with A, B, H, Le[a], Le[b], and I activity (Table I) could be accounted for by this structure (*7, 37*).

As originally proposed it was not clear that the reducing DGlcNAc of compound [3] was joined to the DGal of compound [2] as shown in Figures 1 and 2. However, from stepwise periodate oxidation and Smith degradation studies of a blood group H (Figure 3 (1)) made with Kenneth O. Lloyd, two such degradation steps (*38*) yielded a glycopeptide with an average composition of two DGal, one DGlcNAc, and one DGalNAc residues (Figure 3 (3)). Byron Anderson (*42*) has carried this degradation one step further by stepwise N-acetylation after each degradative step and found a ratio of 1 DGlcNAc, 1DGal, and 1DGalNAc.

The isolation following alkaline borodeuteride degradation of substantial amounts of free *N*-acetyl-Dgalactosaminitol indicated that a substantial portion of terminal nonreducing DGalNAc residues were linked directly to serine and/or threonine, and it was estimated that about one-half of the chains were of this type (Figure 3 (1)). It is not clear why those terminal DGalNAc residues resist the first stage of periodate oxidation and Smith degradation and are only destroyed in the second step; perhaps, they are inaccessible to periodate or are O-acetylated since the acetyl content of the intact blood group substances generally is substantially higher than required for complete N-acetylation of the hexosamines.

Also, free 3-hexenetetrols (R) were isolated following the alkaline degradation as were various oligosaccharides with 3-hexenetetrols at the reducing end. These would arise by alkaline elimination of chains originally substituted on carbons-3 and -4 of the galactose branch. The nature of the substitution on carbon-4 is not established, but large amounts of free galactitol isolated indicate that it is probably a DGal residue or an oligosaccharide chain capable of giving galactitol on continued peeling in alkaline borodeuteride; it is shown in brackets as a

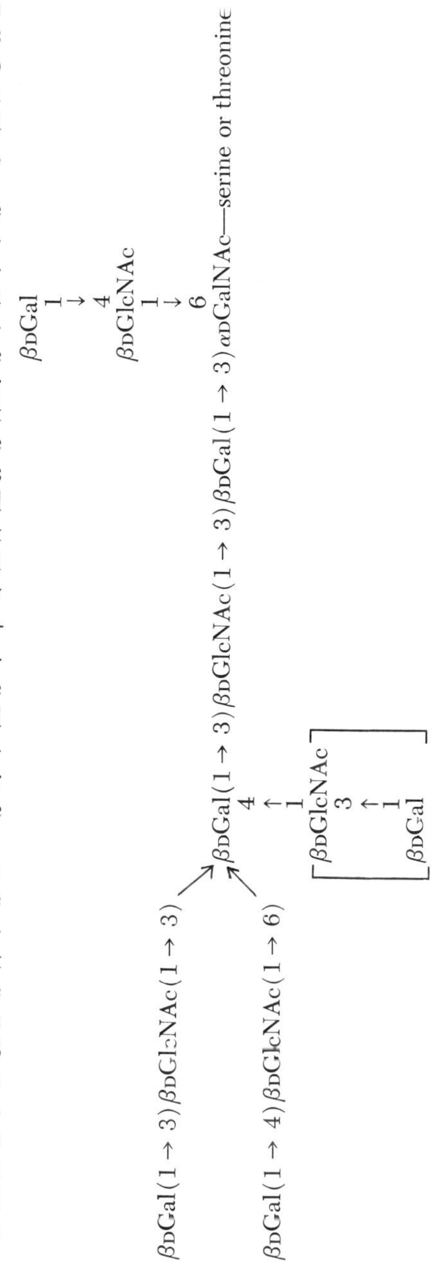

Figure 2. Proposed overall composite structure for OG oligosaccharide moiety (41).

Table I. Oligosaccharides with A, B, H, Le[a], Le[b] and I Specificity

Blood Group Activity	*Structure*
A(A_1 and A_2)	αLFuc 1 ↓ 2 αDGalNAc(1 → 3)βDGal(1 → 4)βDGlcNAc ¦[a,b] (1 → 6?)R[c, d]
A(A_1 and A_2)	αLFuc αLFuc 1 1 ↓ ↓ 2 3 αDGalNAc(1 → 3)βDGal(1 → 4)βDGlcNAc(1 → 6?)R
A(A_1)	αLFuc 1 ↓ 2 αDGalNAc(1 → 3)βDGal(1 → 3)βDGlcNAc(1 → 3)Dgalactitol
B	αLFuc 1 ↓ 2 αDGal(1 → 3)βDGal(1 → 4)βDGlcNAc ¦[a,b] (1 → 6?)R
B	αLFuc αLFuc 1 1 ↓ ↓ 2 3 αDGal(1 → 3)βDGal(1 → 4)βDGlcNAc(1 → 6?)R
B	αLFuc 1 ↓ 2 αDGal(1 → 3)βDGal(1 → 3)βDGlcNAc(1 → 3)Dgalactitol[e]

Isolated from Human Ovarian Cyst Blood Group Substances

Type	*Reference*
2	*36,46,47*
2	*36*
1	*36*
2	*36,47,48*
2	*36*
1	*48*

Table I.

Blood Group Activity	*Structure*
H	αLFuc 1 ↓ 2 βDGal(1 → 4) βDGlcNAc ¦ (1 → 6?)R
H	αLFuc αLFuc 1 1 ↓ ↓ 2 3 βDGal(1 → 4) βDGlcNAc(1 → 6?)R
H	αLFuc 1 ↓ 2 βDGal(1 → 3)DGlcNAc
Le^a	αLFuc 1 ↓ 4 βDGal(1 → 3) βDGlcNAc ¦ (1 → 3 or 4)Dgalactitol
?[f]	αLFuc 1 ↓ 3 βDGal(1 → 4) βDGlcNAc (1 → 6?)R
?[f, g]	αLFuc 1 ↓ 3 βDGal(1 → 4) βDGlcNAc(1 → 3)DGal
Le^b	αLFuc αLFuc 1 1 ↓ ↓ 2 4 βDGal(1 → 3) βDGlcNAc ¦ (1 → 3)DGal

Continued

Type	*Reference*
2	*36,50*
2	*36*
1	*50*
1	*37,51*
2	*37*
2	*44*
1	*52*

Table I.

Blood Group Activity	*Structure*
I[h]	βDGal(1 → 4)βDGlcNAc(1 → 6?)R
	βDGal(1 → 4)βDGlcNAc(1 → 6)hexane-1, 2, 4, 5, 6 pentols
	βDGal(1 → 4)βDGlcNAc 1 ↓ 6 Dgalactitol 3 ↑ 1 βDGal(1 → 3)βDGlcNAc
	βDGal(1 → 4)βDGlcNAc(1 → 6)2acetamido2deoxyDgalactosaminitol 3 ↑ 1 βDGal

[a] A dashed line throughout the table indicates that the shorter oligosaccharide was also isolated, usually by triethylamine hydrolysis (*47, 50, 51*).

[b] The constituent trisaccharides and disaccharides of the main chain without fucose and R have been isolated after mild acid hydrolysis and their structures determined.

[c] R = 3 hexenetetrol(s).

[d] The same oligosaccharide was also isolated from hog gastric mucin A + H substance (*36*).

[e] This oligosaccharide without fucose was also isolated (*48*) as was the oligosaccharide without fucose and with DGal instead of Dgalactitol (*49*).

βDGal(1 → 3)βDGlcNAc-residue. Aston *et al.* (*43*) have isolated

βDGal(1 → 3)βDGlcNAc(1 → 4)DGal

but there is no evidence that it originated from this portion of the molecule.

The second branched structure is supported since Marr *et al.* (*44*) have isolated βDGal(1 → 4)βDGlcNAc(1 → 6)DGalNAc; in compound [2] above, it could not be determined whether the linkage of the βDGal to the DGlcNAc was (1 → 3) or (1 → 4), but we have accepted the (1 → 4) linkage.

Oligosaccharides terminated by 3-hexenetetrols include βDGlcNAc-(1 → 6)R (*45*) for which the structure was established by hydroxylation and periodate oxidation. For the other compounds containing a βDGlc-

Continued

Type	Reference
2	*37,41,46*
2	*37,41*
2,1	*37*
	37,41

[f] Isolated from Le[a] substance; not the Le[a] determinant. Addition of fucose controlled by a different glycosyl transferase from that which adds to carbon-4 of the type 1 chain and is controlled by the Le gene (*38*).

[g] The β(1 → 3) linkage to Gal at the reducing end differs from the structure proposed in the megalosaccharide (Figures 1 and 2); 4.3 mg was isolated from 21 grams Le[a] substance and contained non-carbohydrate material (*44*).

[h] All four oligosaccharides had the same activity on a molar basis with anti I serum Ma (*27*), indicating the I specificity of this antiserum to be specific for the βDGal(1 → 4)βDG*lc*NAc(1 → 6)-structure.

NAc-R at the reducing end, the (1 → 6) linkage has not been directly established. From the mechanism of alkaline borodeuteride degradation, the linkage, assuming a pyranose ring in the DGal which is converted to hexenetetrol, could only have been on C-2 or C-6. The structure of compound [3] is consistent with this. Compounds terminated by βDGlcNAc-R are given in Table I. Since they all possess the βDGlcNAc-R structure, we have assumed the linkage is always β(1 → 6) as in the βDGlcNAc(1 → 6)R.

In the oligosaccharides terminated by DGlcNAc-R from the mechanism of alkaline borodeuteride degradation, the R originated from galactose residues from which chains substituted on carbons-3 and -4 had been eliminated. That a whole series of compounds terminated by the -DGlc-NAc-R have been isolated from individual ovarian cysts suggests that the proposed structure in Figure 1 represents the maximum composite structure. The chains in any substance represent heterogeneous populations which may be incomplete at any stage perhaps because of unfinished biosynthesis or possibly resulting from partial degradation in the cyst cavity.

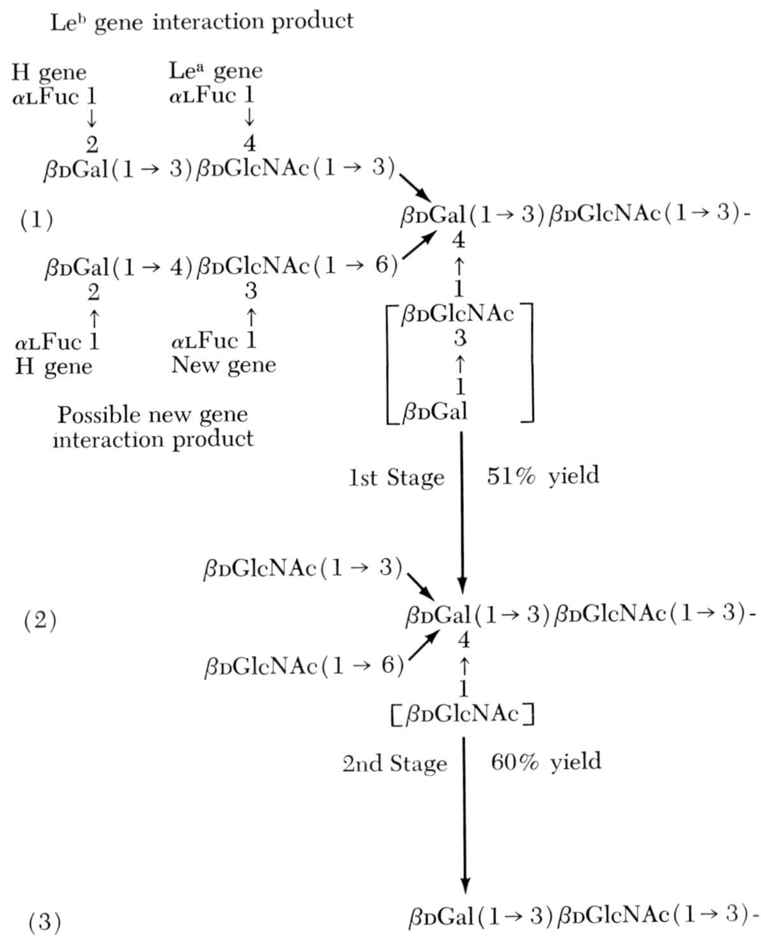

Figure 3. Proposed overall composite structure for Blood Group H substance

In addition to the DGlcNAc-R series of oligosaccharides, the free R would result from elimination of chains with the branched DGal substituted only on carbons-3 and -4 but not on carbon-6, and several oligosaccharides were isolated which are terminated by $\beta(1 \rightarrow 6)$ linked hexane-1, 2, 4, 5, 6-pentols which would have arisen from chains substi-

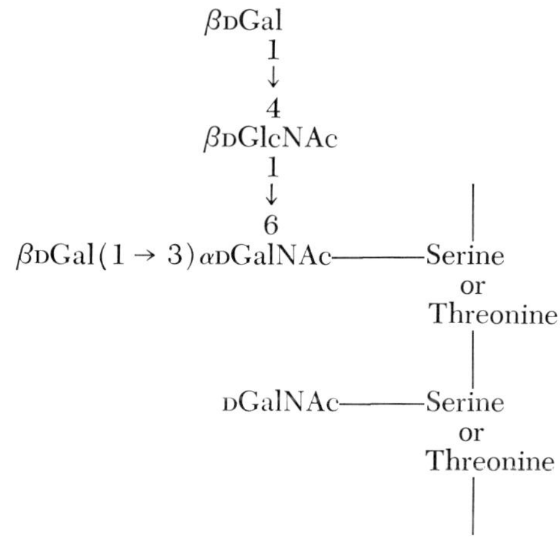

Sugar	Mole Ratios Found	Theory
Fuc	5.1 (19.3%)	4
Gal	6.8 (28.6%)	6
GlcNAc	4.1 (17.1%)	4
GalNAc	2 (8.4%)	2

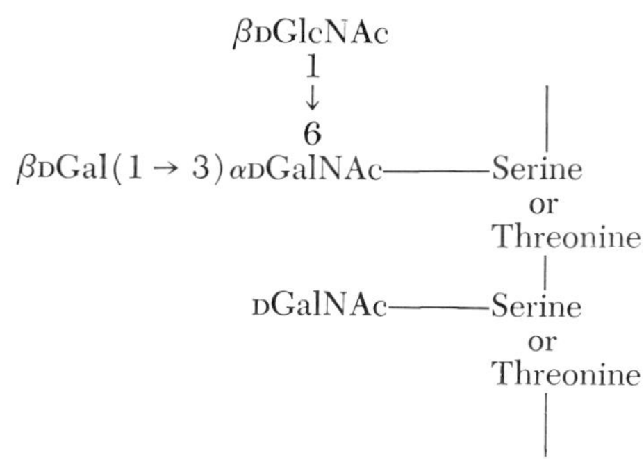

Sugar	Mole Ratios Found	Theory
Fuc	0 (1.1%)	0
Gal	2.2 (16.9%)	2
GlcNAc	3.9 (29.2%)	4
GalNAc	2 (14.8%)	2

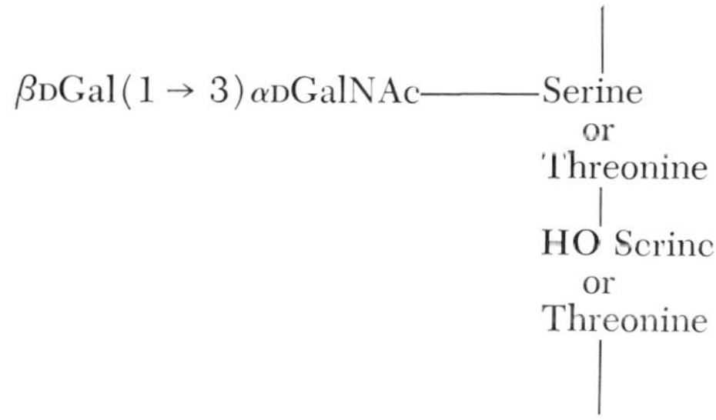

Sugar	Mole Ratios Found	Theory
Gal	2.3 (22.6%)	2
GlcNAc	1.1 (10.4%)	1
GalNAc	1 (9.5%)	1

(JS) and of products of two successive periodate oxidation and Smith degradation stages (38).

tuted on carbons-3 and -6 but not on carbon-4. This is an important indication of heterogeneity. The gross analytical compositions of the various blood group substances conform fairly well (*38*) to what would be expected from the overall structure, indicating that a substantial proportion of the chains are complete. This is also supported by the presence of considerable amounts of materials which behave as higher oligosaccharides. Heterogeneity in the intact blood group substances is also indicated by their distribution in several fractions on precipitation from 90% phenol by addition of ethanol (*1, 25, 37*). The overall structure also does not account for one to two moles of fucose which are calculated to be present from the composition (*38*). Alkaline borohydride degradation shows up to 80% destruction of serine, threonine, and *N*-acetylgalactosamine (*53*). The nature of the surviving carbohydrate protein linkages is being studied; the problem may be complicated by the presence of higher oligosaccharides which remain relatively non-dialyzable and of carbohydrate which remains linked to peptide.

Of special significance was the isolation (*25*) from ovarian cyst fluid of a Nigerian (OG) of a glycoprotein similar to the other blood group substances but lacking A, B, H, Le^a, and Le^b activity and having a very low fucose content. Similar materials have also been encountered previously (*2, 54, 55*). The OG material of which only a small amount was obtained showed a very low fucose content and high cross reactivity with Type XIV antipneumococcal horse serum and thus seemed to be a precursor type of substance (*25*). Degradation with alkaline borohydride of a smaller amount of material than used with the other substances yielded oligosaccharide [2] and two other oligosaccharides (*41*) identical to those isolated earlier from Le^a substance (*37*) βDGal(1 → 4)βDGlcNAc(1 → 6)hexane-1, 2, 4, 5, 6-pentol(s) and βDGal(1 → 4)βDGlcNAc-(1 → 6?)3 hexenetetrol(s) in addition to free 3 hexenetetrols, DGalactitol and *N*-acetyl-DGalactosaminitol. Comparison of the relative amounts of the oligosaccharides isolated from OG and from an Le^a substance indicated that OG gave many more type 2 chains with hexenetetrol(s) than hexanepentol(s) at the reducing end and thus which were originally substituted at C-3 and C-4 rather than only on C-3 and hence carried type 1 and type 2 determinants. Oligosaccharide, [1] which might have arisen from chains substituted on C-3 and C-6 but not on C-4 and which were reduced by borodeuteride before elimination could occur, was not isolated from OG. The amount expected by comparison with that obtained from Le^a substance would only have been 7 or 8 mg, and such a small quantity would have been difficult to isolate and purify. OG yielded considerably more *N*-acetyl-DGalactosaminitol than did the Le^a substance indicating that more unsubstituted chains of terminal DGalNAc are present in the former. Similarly, OG gave substantially less free hexenetetrols than did the Le^{a} substance indicating the presence of proportionately fewer chains linked on C-3 and C-4 but not on C-6. Thus the OG material (Figure 2) can readily be fitted into the composite structure in Figure 1 (*41*). Further evidence that it is a precursor was obtained by Jarkovsky *et al.* (*56*) who successfully used the OG substance as an acceptor on which an L-fucosyl residue has been transferred by a fucosyl transferase from milk to give an Le^{a} active product; a similar type of acceptor was also produced by a single periodate oxidation step from human A substance.

An unusual use for the stepwise periodate oxidation products of cyst H substance JS (Figure 3 (1)) was found when a mouse myeloma γ A

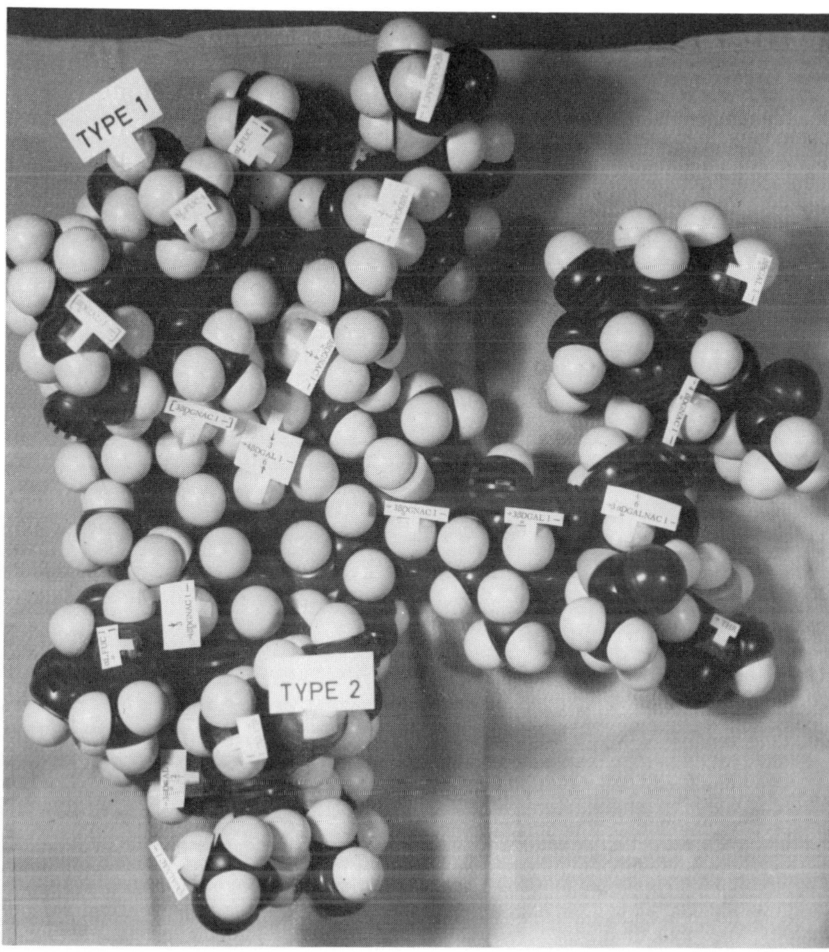

Figure 4. Composite close-packed model of the blood group A megalosaccharide (equivalent to Figure 1).

immunoglobulin was encountered which showed specificity for terminal β-linked DGlcNAc residues (*57*). In conformity with the composite structure (Figure 3 (1)), it did not react with the original JS or with the second, fourth, and fifth periodate oxidation stages but did react with the first (Figure 3 (2)) and third stages which would be the only ones containing terminal β-linked DGlcNAc residues. While the origin of the specific reactivity of this mouse immunoglobulin, as with other myeloma globulins with specific receptor sites, is still obscure, the finding of an immunoglobulin which reacts with materials related to the blood group substances may have broader implications.

The proposed structure is unusual (Figures 1 and 2) (*37, 38*) in that it places the two kinds of blood group determinant represented by βDGal(1 → 3)DGlcNAc (type 1) and βDGal(1 → 4)DGlcNAc (type 2) as branches on the same galactose [1] and would thus provide close association of the two types of determinants. A close packed molecular model of the blood group A megalosaccharide is shown in Figure 4. This branched structure has important implications for the biological and immunological activities of these molecules. When this branched structure was recognized, it was predicted (*7, 38*) that one might explain the difference between A_1 and A_2 substances in that A_1 substances might have the terminal αDGalNAc(1 → 3) residue on both chains while A_2 substance might only have it on one. This prediction has been completely confirmed by Moreno *et al* (*58*), who demonstrated the association of A_2 determinants with the type 2 chain and that A_2 substance from ovarian cyst and from saliva had very high Le^b activity. Since the terminal αDGalNAc residue on the type 1 A determinant blocks Le^b specificity, its absence on A_2 substance would account for the high Le^b specificity with the absence of the αDGalNAc on the type 1 chain. Moreover A oligosaccharides of the type 2 structure had been isolated from A_1 substance. Thus A_1 substance would have both types of A determinants while A_2 would only have A determinants of type 2. Independent evidence for the intimate association of A_1 with the Le^b and with the type 1 chain was the finding of an antibody specific for A_1Le^b erythrocytes (*59, 60*); it reacted only with A_1Le^b erythrocytes and not with A_2Le^b, OLe^b or A_1Le^a erythrocytes. Moreover, another antibody (*61*) reacted with O, I, Le(a − b+) or A_2ILe(a − b+) but not with A_1ILe(a − b+) erythrocytes, indicating the association of A_2 receptors on the type 2 chains and H, I, and Le^b activity associated with type 1 chains lacking the terminal αDGalNAc(1 → 3)-grouping (*58*). Recent findings by Schachter *et al.* (*62*) in Toronto strongly support this conclusion in that A_1 serum contains a galactosaminyl transferase which adds GalNAc to a type 1 oligosaccharide from milk, lacto-*N*-fucopentaose I, αLFuc(1 → 2)βDGal-

(1 → 3)βDGlcNAc(1 → 3)βDGal(1 → 4)DGlc, while serum from A_2 donors is much less effective.

Another important consequence of the branched structure may well be the ability of the type 1 and type 2 determinants on a single branch to combine with two sites on the same antibody molecule. Such a bivalent interaction would provide very strong binding as Karush (*63*) has shown for other systems.

An unusual development was the finding that the precursor OG cyst substance showed high blood group I and i activity (*26, 27*). The I antigen was recognized by the ability of human adult erythrocytes to react specifically with most human cold autohemagglutinins (*24, 64, 65*); these antisera however do not agglutinate human cord erythrocytes and erythrocytes from about 1 in 10,000 adults also are not agglutinated. The agglutinable adult erythrocytes have been termed I and the rare negatives i. The I antigen is weakly developed at birth but becomes well established on the erythrocyte during the first 18 months of life. Although cold agglutinins with I specificity from different individuals behave very differently, the I antigen was considered to be determined by a single gene which converted i into I (*24*). Earlier studies using enzymes had shown the I determinant on human erythrocytes to involve β-linked DGal and β-linked DGalNAc (*66*).

A source of soluble I substance was human milk and saliva, but various samples of milk reacted only with a small proportion of anti-I sera. A preparation obtained by Feizi *et al.* (*26*) from human milk was a glycoprotein with a composition similar to the blood group substances but having a very low fucose content as well as a very low GalNAc content. On testing a number of anti-I sera with the preparations from milk and from the OG precursor substance by quantitative precipitin assay, several patterns emerged (*27*). All anti-I sera precipitated with a fraction of the OG substance, which had been obtained after peptic digestion and ethanol precipitation, followed by extraction with 90% phenol and fractional precipitation twice between 10 and 20% ethanol in phenol (OG20%2×). Other OG fractions precipitable at lower ethanol concentrations as well as the milk preparation tended to react differently with various antisera. With one anti-I serum Ma, all OG fractions were of equal potency in precipitating the anti-I in the cold and milk precipitated about 90% of the antibody precipitated by OG fractions. With another antiserum, Ort, milk precipitated only about one-third of the antibody and with a third anti-I serum, Step, milk failed to react entirely; with these two anti-I sera the various OG fractions reacted less strongly than did OG20%2×. It was also significant that certain blood group substances prepared from cow stomach (abomasus) (*68*) also precipitated well with the various anti-I sera (*27*) while others precipitated

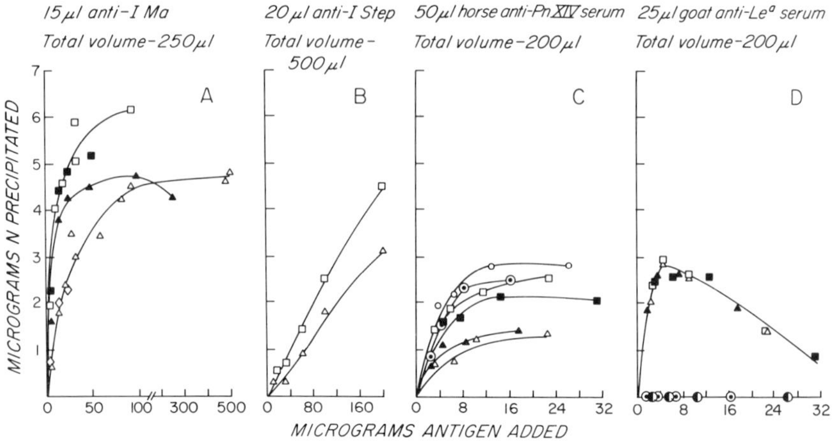

Figure 5. Quantitative precipitin curves of fractions from Le^a substance N-1 with anti-I sera Ma (A), Step (B), with type XIV antipneumococcal horse serum (C) and with goat anti-Le^a (D).

N-1 phenol insoluble, △; N-1 phenol insoluble of 1st 10%, ◇; N-1 10% 2×, ▲; N-1 10% of 1st 20%, □; N-1 20% 2×, ■; milk fraction C, ⊙; OG 10% 2×, ◐ (27).

poorly; with all three anti-I sera, differences among the various cow substances could be seen by hemagglutination inhibition as well.

In terms of the proposed structure for the OG substance (Figure 2) and the composite A or B blood group substance (Figure 1), it was possible by a one stage periodate oxidation and Smith degradation of A or B substances to produce I reactive determinants (*26, 27*). Thus, the I determinants were established as being in the interior of the blood group A, B, megalosaccharide.

Another important property of the isolated blood group substances revealed by the examination of I activity in relation to other immunochemical reactivities was the heterogeneity of the various blood group substances. Figure 5 shows the reactivity of the various fractions obtained (*27*) from a human ovarian cyst Le^a substance (N-1) by phenol–ethanol purification with two anti-I sera with type XIV horse antipneumococcal serum and with goat anti-Le^a. All four fractions of Le^a substance (Figure 5D) gave an identical quantitative precipitin curve with goat anti-Le^a. However, with the two anti-I sera and with type XIV antipneumococcal serum, the various fractions behaved quite differently; those with a lower solubility in phenol-ethanol were generally less reactive than those requiring a higher ethanol concentration to be precipitated. Thus, substances with approximately equal numbers of Le^a determinants per unit weight might vary substantially in the number of free I determinants. This is best accounted for by considering these fractions all to have about 80–95% complete Le[a] determinants with heterogeneity of the small proportion of non-Le[a] chains which show I reactivity or cross reactivity with type XIV serum resulting from incomplete biosynthesis or removal of blocking side chains, possibly because of degradation in the cyst cavity. The differences in solubility might be ascribed to the differences in the number of complete chains, the less soluble fractions having more complete chains.

It was possible by oligosaccharide inhibition studies to ascertain the

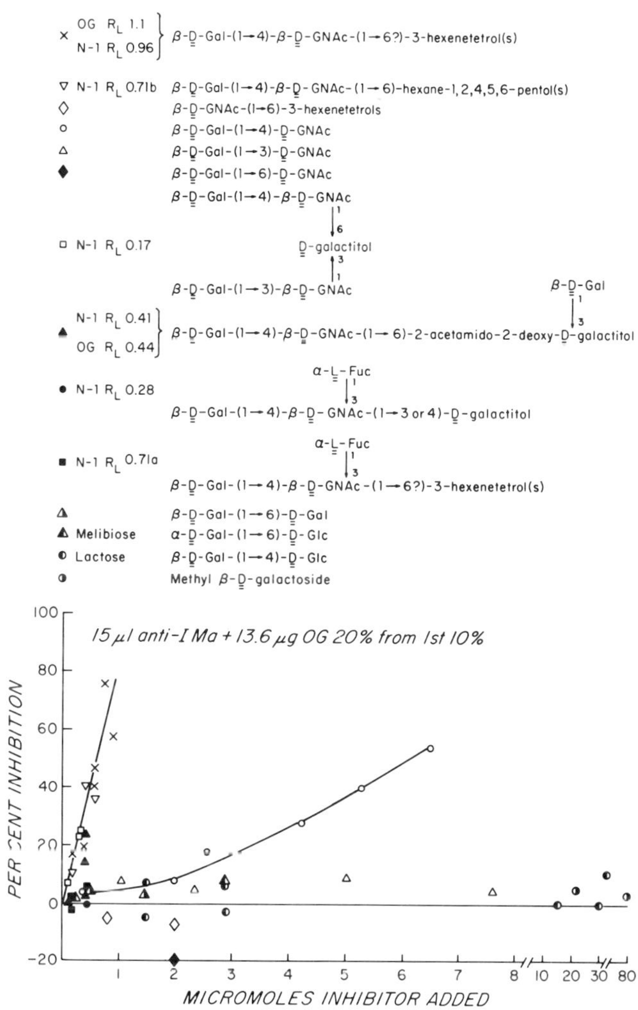

Figure 6. Quantitative inhibition assays of anti I serum Ma with oligosaccharides (27).

partial structure of the I determinant reacting with anti-I Ma (*27*). The only active disaccharide in inhibiting this system was βDGal(1 → 4)-DGlcNAc. With oligosaccharides possessing this structure plus a β(1 → 6) linkage, either to 3-hexenetetrol(s), to 1, 2, 4, 5, 6-hexanepentols, or to D-galactitol or N-acetyl-Dgalactosaminitol as in the branched reduced oligosaccharides [1] and [2], there was a substantial increase in inhibiting power, and all compounds were equally potent on a molar basis. Thus, the structure of the I determinant in OG reactive with anti-I Ma was at least βDGal(1 → 4)βDGlcNAc(1 → 6)- (*27*). The structure of determinants in OG reacting with the other anti-I sera and with anti-i remains to studied. The association of I activity with the type 1 chains mentioned above (*61*) would indicate that the sera used had a specificity distinct from that of anti-I Ma and that the genetics of the Ii system is much more complicated than is generally believed.

The two major sources of blood group A substances from mammalian tissues have been human ovarian cyst and hog gastric mucin (*1, 2, 3, 4, 5, 6, 7, 8*). Both yield very highly active materials, and their activity per unit weight is of the same order, but the cyst substances generally have slightly higher A activity than the A substances from individual hog stomachs per unit weight, and from Table I the same monofucosyl reduced A pentasaccharide has been isolated from both (*36*).

However stomach blood group substances, hog and human, must be quite different from cyst substances. The first structural indication of this came from the isolation from hog gastric mucin A + H substance of αDGlcNAc(1 → 3 or 4)Dgalactitol (*32*). This terminal non-reducing αDGlcNAc residue was responsible for the capacity of a portion of the purified hog stomach substance to be precipitated by concanavalin A while ovarian cyst substances did not precipitate. Antibody specific for the αDGlcNAc determinant was found in the serum of individuals who had been immunized with hog mucin A + H substance (*33*), and the oligosaccharide structures involved were αDGlcNAc(1 → 4)βDGal (1 → 3)DGalNAc and αDGlcNAc(1 → 4)βDGal(1 → 4)DGal NAc which were as active on a molar basis as αDGlcNAc(1 → 4)DGal or αDGlcNAc(1 → 3 or 4)Dgalactitol. All were more active than methyl αDGlcNAc (*34*). These oligosaccharides were unrelated to the A specificity of the hog mucin substances and did not react with the A specific lectin of *Dolichos biflorus*. They were responsible for the reactivity with concanavalin A since they were all good inhibitors of precipitation of concanavalin A by hog stomach blood group substance but were not quite as good as methyl αDGlcNAc.

Concanavalin A was found by Clarke and Denborough (*35*) to precipitate with blood group glycoproteins from human gastric juice from two group O non-secretors and 2 group O and 3 group A secretors; it did not precipitate with blood group substances from ovarian cyst fluid and saliva. These blood group glycoproteins had all been purified from concentrated fluid by density gradient centrifugation in cesium chloride (*71*) while the earlier preparations had been obtained from stomach or ovarian cyst fluid by peptic digestion and phenol-ethanol precipitation (*1*). Thus, the differences between blood group glycoproteins from various sources are independent of the method of preparation, and concanavalin A may prove useful in recognizing such differences. Their significance requires further study but is probably related to their function in various organs and tissues. There is substantial evidence that the αDGlcNAc determinants are on some of the hog stomach glycoprotein molecules carrying A determinants and not on a contaminating glycoprotein (*32, 70, 71, 72*).

In this connection Kochetkov and co-workers (*73*) have reported that hog stomach A + H substance contains several *N*-acetylhexosamine residues bound by (1 → 3) linkages to the *N*-acetyl-D-galactosamine residue which is linked to serine or threonine of the peptide backbone, but no oligosaccharides have yet been isolated.

Pig submaxillary mucin (*74*) exists in two forms in individual glands; one of these possesses A activity, and the other does not. Their gross compositions were identical. Five oligosaccharides were isolated, the largest having the structure

```
                  αL Fuc                 N-Glycolylneuraminyl
                    1                             1
                    ↓                             ↓
                    2                             6
αDGalNAc(1 → 3)βDGal(1 → 3)Nacetyl-D-galactosaminitol
```

Other oligosaccharides had a similar structure but lacked either the αDGalNAc, the *N*-glycolylneuraminyl residue, or both. A fourth oligosaccharide contained only *N*-glycolylneuraminyl-*N*-acetyl-D-galactosaminitol; free *N*-acetyl-D-galactosaminitol was also isolated. The heterogeneity of the pig submaxillary mucin substance thus parallels that of the other blood group substances except that the chains are much shorter.

The composite megalosaccharide structure accounts for all of the various blood group active oligosaccharides isolated with the single linkage exception in Table I and agrees with various biosynthetic schemes (*3, 7*). Moreover, the difucosyl A and B pentasaccharides from human urine (*7*) are as active as the difucosyl A and B oligosaccharides isolated from blood group substances (*75*) and have an identical structure except that glucose is substituted for GNAc-B (Table I). However, the megalosaccharide may not represent the totality of the carbohydrate structure in the blood group glycoproteins. Several oligosaccharides have been isolated in very small amounts from hydrolysates of ovarian cyst blood group substances which differ in one or another linkage from those in the megalosaccharide. They have been discussed in detail (*7*). In most instances they differ in a single linkage from those of the megalosaccharide, indicating either that at certain points in biosynthesis several possibilities exist or that the interior of various chains in the megalosaccharide may be different. The proposed megalosaccharide structure leaves one to two fucose residues unaccounted for, and certain of the isolated oligosaccharides have an αLFuc(1 → 6)DGal structure or two fucoses linked together (*76*). Transglycosylation is also a possibility for small amounts of oligosaccharides such as βDGal(1 → 3)βDGlcNAc(1 → 3)βDGal(1 → 4)-DGlcNAc (*34*) and certain of the fucosyl oligosaccharides (*76*) isolated after mild acid hydrolysis.

The linkage of the D-GalNAc at the reducing end to serine and threonine in the megalosaccharide (Figure 1) is shown by the correspondence between destruction of those three residues with alkaline borohydride (*53*). This was further established by Donald *et al.* (*77*) who isolated two glycopeptides after partial hydrolysis of cyst blood group B substance with acetic acid. These compounds each contained 25% of DGalNAc and accounted for 80% of the DGalNAc in the intact blood group substance. The two glycopeptides had only 6 and 3% of DGal and DGlcNAc and contained no fucose. Alkaline borohydride cleavage showed destruction of serine, threonine, and DGalNAc and formation of *N*-acetyl-Dgalactosaminitol. On a molar basis the DGalNAc destroyed was 96% of the serine plus threonine destroyed. Enzymatic studies showed the DGalNAc to be linked α to serine and threonine (*77*). Similar findings were obtained with H and Le^{a} substances, indicating the structural similarities in the region of the linkage of the carbohydrate to the protein.

Further structural studies on the blood group glycoproteins depends on the ability to isolate in a high state of purity, to characterize, and to establish the structures of oligosaccharides larger than the hexasaccharides and reduced hexasaccharides thus far obtained. The demonstration by Iyer and Carlson (*78*) that blood group glycoproteins at a concentration of 1 mg/ml in 0.05*N* NaOH and 1 *M* $NaBH_4$ at 50° C liberates oligosaccharides with a minimum of degradation offers a promising lead if a suitable methodology can be developed, and such studies are under way in this laboratory.

Literature Cited

1. Kabat, E. A., "Blood Group Substances," Academic, New York, 1956.
2. Morgan, W. T. J., "The Croonian Lecture. A Contribution to Human Bio-

chemical Genetics; the Chemical Basis of Blood Group Specificity," *Proc. Roy. Soc. (London) Ser. B.* (1960) **151,** 308.
3. Watkins, W. M., "Blood Group Specific Substances" in "Glycoproteins," p. 462, A. Gottschalk, Ed., Elsevier, Amsterdam (1966).
4. Marcus, D. M., "The ABO and Lewis Blood-Group System. Immunochemistry, Genetics, and Relation to Human Disease," *New Eng. J. Med.* (1969) **280,** 994.
5. Egami, F., Oshima, Y., Eds., "Biochemistry and Medicine of Mucopolysaccharides, Maruzen., Tokyo, 1962.
6. Prokop, O., Uhlenbruck, G., "Lehrbuch der Menschlichen Blut und Serumgruppen," 2nd ed., Leipzig, 1967.
7. Aminoff, D., Ed., "Symposium on Blood and Tissue Antigens," Academic, New York, 1970.
8. Morgan, W. T. J., "Molecular Aspects of Human Blood-Group Specificity, *Ann. N. Y. Acad. Sci.* (1970) **169,** 118.
9. Springer, G. F., "Importance of Blood-Group Substances in Interactions between Man and Microbes," *Ann. N. Y. Acad. Sci.* (1970) **169,** 134.
10. Ceppellini, R., "Physiological Genetics of Human Factors" in "Ciba Symposium on the Biochemistry of Human Genetics," p. 242, G. E. W. Wolstenholme and M. O'Connor, Eds., Churchill, London, 1959.
11. Watkins, W. M., Morgan, W. T. J., "Possible Genetical Pathways for the Biosynthesis of Blood Group Mucopolysaccharides, *Vox Sanguinis* (1959) **4,** 97.
12. Hakomori, S., Strycharz, G. O., "Investigations on Cellular Blood Group Substances I. Isolation and Chemical Composition of Blood-group ABH and Le^b Isoantigens of Glycolipid Nature," *Biochemistry* (1968) **7,** 1279.
13. Zahler, P., "Blood Group Antigens in Relation to Chemical and Structural Properties of the Red Cell Membrane," *Vox Sanguinis* (1968) **15,** 81.
14. Whittemore, N. B., Trabold, N. C., Reed, C. F., Weed, R. J., "Solubilized Glycoprotein from Human Erthyrocyte Membrane Possessing Blood Group A-, B-, and H- Activity," *Vox Sanguinis* (1969) **17,** 289.
15. Gardas, A., Kóscielak, J., "A, B, and H Blood Group Specificities in Glycoprotein and Glycolipid Fractions of Human Erythrocyte Membrane. Absence of Blood Group Active Glycoproteins in the Membranes of Nonsecretors," *Vox Sanguinis* (1971) **20,** 137.
16. Sneath, J. S., Sneath, P. H. A., "Transformation of Lewis Groups of Human Red Cells," *Nature* (1955) **176,** 172.
17. Marcus, D. M., Cass, L., "Glycolipids with Lewis Blood Group Activity," *Science* (1969) **164,** 553.
18. Szulman, A. E., "Chemistry, Distribution, and Function of Blood Group Substances," *Ann. Rev. Med.* (1966) **17,** 307.
19. Mäkelä, O., "Studies in Hemagglutinins of Leguminosae Seeds," *Ann. Med. Exp. Fenniae Supple. 11,* (1957) 35.
20. Etzler, M. E., Kabat, E. A., "Purification and Characterization of a Lectin (Plant Hemagglutinin) with Blood Group A Specificity from *Dolichos biflorus,*" *Biochemistry* (1970) **9,** 869.
21. Prokop, O., Uhlenbruck, G., Kohler, W., "A New Source of Antibody-Like Substances Having Anti-blood Group Specificity. A Discussion on the Specificity of Helix Agglutinins," *Vox Sanguinis* (1968) **14,** 321.
22. Hammarström, S., Kabat, E. A., "Purification and Characterization of a Blood-Group A Reactive Hemagglutinin from the Snail *Helix pomatia* and a Study of Its Combining Site," *Biochemistry* (1970) **8,** 2696.
23. Hammarström, S., Kabat, E. A., "Studies on Specificity and Binding Properties of the Blood Group A Reactive Hemagglutinin from *Helix pomatia,*" *Biochemistry* (1971) **10,** 1684.
24. Race, R. R., Sanger, R., "Blood Groups in Man, 5th ed., Blackwell, Oxford, 1968.

25. Vicari, G., Kabat, E. A., "Immunochemical Studies on Blood Groups XLII. Isolation and Characterization of a Substance from Human Ovarian Cyst Fluid of a Blood Group Substance Lacking A, B, H, Le[a], and Le[b] Specificity," *J. Immunol.* (1969) **102,** 821.
26. Feizi, T., Kabat, E. A., Vicari, G., Anderson, B., Marsh, W. L., "Immunochemical Studies on Blood Groups XLVII. The I Antigen Complex-Precursors in the A, B, H, Le[a], and Le[b] Blood Group System–Hemagglutination Studies," *J. Exp. Med.* (1971) **133,** 39.
27. Feizi, T., Kabat, E. A., Vicari, G., Anderson, B., Marsh, W. L., "Immunochemical Studies on Blood Groups XLIX. The I Antigen complex: Specificity Differences among Anti-I Sera Revealed by Quantitative Precipitin Studies; Partial Structure of the I Determinant Specific for One Anti-I Serum," *J. Immunol.* (1971) **106,** 1578.
28. Holborow, E. J., "Immunological Methods," p. 155, J. F. Ackroyd, Ed., Blackwell Oxford, 1964.
29. Marsh, W. L., Nichols, M. E., Allen F. H., "Inhibition of Anti-I Sera by Human Milk," *Vox Sanguinis* (1970) **18,** 149.
30. Iseki, S., Masaki, S., "Transformation of Blood Group Substance by Bacterial Enzyme," *Proc. Japan Acad.* (1953) **29,** 460.
31. Morgan, W. T. J., van Heyningen, R., "The Occurrence of A, B, and O Blood Group Substances in Pseudo-Mucinous Ovarian Cyst-Fluids," *Brit. J. Exp. Pathol.* (1944) **25,** 5.
32. Lloyd, K. O., Kabat, E. A., Beychok, S., "Immunochemical Studies on Blood Groups XLIII. The Interaction of Blood Group Substances from Various Sources with a Plant Lectin, Concanavalin A," *J. Immunol.* (1969) **102,** 1354.
33. Moreno, C., Kabat, E. A., "Immunochemical Studies on Blood Groups XLIV. Human Antibodies against a New Determinant Present in Blood Group Substances from Hog Gastric Mucosa," *J. Immunol.* (1969) **102,** 1363.
34. Etzler, M., Anderson, B., Beychok, S., Gruezo, F., Lloyd, K. O., Richardson, N. G., Kabat E. A., "Immunochemical Studies on Blood Groups XLVI. Oligosaccharides Isolated after Hydrolysis of Hog Gastric Mucin Blood Group A + H Substance Previously Treated with the Blood Group de-N-Acetylating Enzyme," *Arch. Biochem. Biophys.* (1970) **141,** 588.
35. Clarke, A. E., Denborough, M. A., "The Interaction of Concanavalin A with Blood-Group Glycoproteins from Human Secretions," *Biochem J.* (1971) **121,** 811.
36. Lloyd, K. O., Kabat, E. A., Layug, E. J., Gruezo, F., "Immunochemical Studies on Blood Groups XXXIV. Structures of Some Oligosaccharides Produced by Alkaline Degradation of Blood Group A, B, and H Substances," *Biochemistry* (1966) **5,** 1489.
37. Lloyd, K. O., Kabat, E. A., Licerio, E., "Immunochemical Studies on Blood Groups XXXVIII. Structures and Activities of Oligosaccharides Produced by Alkaline Degradation of Blood Group Lewis[a] Substance. Proposed structure of the Carbohydrate Chains of Human Blood Group A, B, H, Le[a], and Le[b] Substances," *Biochemistry* (1968) **7,** 2976.
38. Lloyd, K. O., Kabat, E. A., "Immunochemical Studies on Blood Groups XLI. Proposed Structures for the Carbohydrate Portions of Blood Group A, B, H, Lewis[a], and Lewis[b] Substances," *Proc. Nat. Acad. Sci.* (1968) **61,** 1470.
39. Aston, W. P., Donald, A. S. R., Morgan, W. T. J., "Oligosaccharides Containing a (1 → 6) Glycosidic Linkage Obtained from Human Blood Group Specific Glycoprotein," *Biochem. Biophys. Res. Commun.* (1968) **30,** 1.
40. Rege, V. P., Painter, T. J., Watkins, W. M., Morgan, W. T. J., "Three New Trisaccharides Obtained from Human Blood-Group A, B, H, and Le[a]

Substances: Possible Sugar Sequences in the Carbohydrate Chains," *Nature* (1963) **200,** 532.

41. Vicari, G., Kabat, E. A., "Immunochemical Studies on Blood Groups XLV. Structures and Activities of Oligosaccharides Produced by Alkaline Degradation of a Blood Group Substance Lacking A, B, H, Le^a, and Le^b Specificities, *Biochemistry* (1970) **9,** 3414.
42. Anderson, B., Kabat, E. A., unpublished data.
43. Aston, W. P., Donald, A. S. R., Morgan, W. T. J., A Trisaccharide *O*-β-D-Galactopyranosyl-(1 → 3)-*O*-*N*-acetyl-β-D-glucosaminopyranosyl-(1 → 4)-D-galactose Obtained from Human Blood Group H Substance," *Biochem. J.* (1968) **107,** 861.
44. Marr, A. M. S., Donald, A. S. R., Morgan, W. T. J., "Two New Oligosaccharides Obtained from an Le^a-active Glycoprotein," *Biochem. J.* (1968) **110,** 789.
45. Lloyd, K. O., Kabat, E. A., "Immunochemical Studies on Blood Groups XXXVI. The Nature of the Unsaturated Residue in Some Oligosaccharides Produced by the Action of Sodium Hydroxide–Sodium Borohydride on Blood Group A, B, and H Substances," *Carbohydrate Res.* (1967) **4,** 165.
46. Lloyd, K. O., Kabat, E. A., "Some Products of the Alkaline Degradation of Blood Group Substances by Alkaline Borohydride," *Biochem. Biophys. Res. Commun.* (1964) **16,** 385.
47. Painter, T. J., Watkins, W. M., Morgan, W. T. J., "Serologically Active Fucose Containing Oligosaccharides Isolated from Human Blood-Group A and B Substances," *Nature* (1965) **206,** 594.
48. Lundblad, A., Hammarström, S., Licerio, E., Kabat, E. A., Immunochemical Studies on Blood Groups LII. Structures and Activities of Oligosaccharides Produced by Alkaline Degradation of a Blood Group B Substance Previously Subjected to Partial Hydrolysis by Hydrochloric acid, *Arch. Biochem. Biophys.* (1972) **148,** 291.
49. Aston, W. P., Hague, G. M., Donald, A. S. R., Morgan, W. T. J., "A Serologically Active Tetrasaccharide Obtained from a Human Blood Group-B-Active Glycoprotein by Partial Acid Hydrolysis," *Biochem. J.* (1968) **110,** 157.
50. Rege, V. P., Painter, T. J., Watkins, W. M., Morgan, W. T. J., "Isolation of Serologically Active Fucose-Containing Oligosaccharides from Human H Substance," *Nature* (1964) **203,** 360.
51. Rege, V. P., Painter, T. J., Watkins, W. M., Morgan, W. T. J., "Isolation of a Serologically Active Fucose-Containing Trisaccharide from Human Blood-Group Le^a Substance," *Nature* (1964) **204,** 740.
52. Marr, A. M. S., Donald, A. S. R., Watkins, W. M., Morgan, W. T. J., "Molecular and Genetic Aspects of Human Le^b Specificity," *Nature* (1967) **215,** 1345.
53. Kabat, E. A., Bassett, E. W., Pryzwansky, K., Lloyd, K. O., Kaplan, M. E., Layug, E., "Immunochemical Studies on Blood Groups XXXIII. The Effects of Alkaline Borohydride and of Alkali on Blood Group A, B, and H substances," *Biochemistry* (1965) **4,** 1632.
54. Carlson, D. M., "Isolation of a "Sialic Acid-Rich" Glycoprotein from an Ovarian Cyst," *Federation Proc.* (1968) **27,** 530.
55. Baer, H., Bringaze, J. K., McNamee, M., "The Immunochemistry of Blood Group O. I. The Production of Precipitating and Agglutinating Antibodies to Various Human and Hog Blood Group Substances," *J. Immunol.* (1954) **73,** 67.
56. Jarkovsky, Z., Marcus, D. M., Grollman, A. P., "Fucosyltransferase Found in Milk. Product of the Lewis Blood Group Gene," *Biochemistry* (1970) **9,** 1123.
57. Vicari, G., Sher, A., Cohn, M., Kabat, E. A., "Immunochemical Studies on

a Mouse Myeloma Protein with Specificity for Certain β-Linked Terminal Residues of *N*-acetyl-D-Glucosamine," *Immunochemistry* (1970) **7,** 829.
58. Moreno, C., Lundblad, A., Kabat, E. A., "Immunochemical Studies on Blood Groups LI. A Comparative Study of the Reaction of A_1 and A_2 Blood Group Glycoproteins with Human Anti-A," *J. Exp. Med.* (1971) **134,** 439.
59. Crookston, M. C., Tilley, C. A., Crookston, J. H., "Human Blood Group Chimera with Seeming Breakdown of Immune Tolerance," *Lancet* (1970) **2,** 1110.
60. Swanson, J. Crookston, M. C., Yunis, E., Azar, M., Gatti, R. A., Good, R. A., *Lancet* (1971) **1,** 396.
61. Tegoli, J., Cortez, M., Jensen, L., Marsh, W. L., "A New Antibody Anti I Le^{bH} Specific for a Determinant Formed by the Combined Action of the I, Le, Se, and H Gene Products," *Vox Sanguinis* (1971) **21,** 397.
62. Schachter, H., Michaels, A., Crookston, M. C., Tilley, C. A., Crookston, J. H., "A quantitative Difference in the Activity of Group A-Specific-*N*-Acetylgalactosaminyl Transferase in Serum from A_1 and A_2 Human Subjects," *Biochem. Biophys. Res. Commun.* (1971) **45,** 1011.
63. Karush, F., "Affinity and the Immune Response," *Ann. N.Y. Acad. Sci.* (1970) **169,** 56.
64. Wiener, A. S., Unger, L. J., Cohen, L., Feldman, J., "Type Specific Cold Autoantibodies as a Cause of Acquired Hemolytic Anemia and Hemolytic Transfusion Reactions: Biologic Test with Bovine Red Cells," *Ann. Internal Med.* (1956) **44,** 221.
65. Jenkins, W. J., Marsh, W. L., Noades, J., Tippet, P., Sanger, R., Race, R. R., "The I Antigen and Antibody," *Vox Sanguinis* (1960) **5,** 97.
66. Marcus, D. M., Kabat, E. A., Rosenfield, R. E., "The Action of Enzymes from *Clostridium tertium* on the I Determinant of Human Erythrocytes," *J. Exp. Med.* (1963) **118,** 175.
67. Lloyd, K. O., Beychok, S., Kabat, E. A., "Immunochemical Studies on Blood Groups. XXXVII. The Structures of Difucosyl and Other Oligosaccharides Produced by Alkaline Degradation of Blood Group A, B, and H Substances. Optical Rotation and Circular Dichroism Spectra of These Oligosaccharides," *Biochemistry* (1967) **6,** 1448.
68. Beiser, S. M., Kabat, E. A., "Immunochemical Studies on Blood Groups. XII. Preparation of Blood Group Substances from Bovine Stomach Linings and a Comparison of Their Chemical and Immunochemical Properties with Those of Blood Group Substances from Other Species," *J. Immunol.* (1952) **68,** 19.
69. Marcus, D. M., Grollman, A. P., "Studies of Blood Group Substances I. Caprine Precipitating Antisera to Human Le^a and Le^b Blood Group Substances," *J. Immunol.* (1966) **97,** 867.
70. Marcus, D. M., Cass, L. E., "Studies of Blood Group Substances III. A Caprine Antiserum Containing Antibodies to Two Antigenic Determinants on Type H Hog Gastric Mucin," *J. Immunol.* (1967) **99,** 987.
71. Creeth, J. M., Denborough, M. A., "The Use of Equilibrium-Density-Gradient Methods for the Preparation and Characterization of Blood-Group-Specific Glycoproteins," *Biochem. J.* (1970) **117,** 879.
72. Lloyd, K. O., "The Preparation of Two Insoluble Forms of the Phytohemagglutinin, Concanavalin A, and Their Interactions with Polysaccharides and Glycoproteins," *Arch. Biochem. Biophys.* (1970) **137,** 460.
73. Kochetkov, N. K., Derevitskaya, V. A., Lichosherstov, L. M., Martynova, M. D., Senchenkova, S. N., "On the Structure of the Carbohydrate Chains of the Blood-Group Substance (A + H)," *Carbohydrate Res.* (1970) **12,** 437.
74. Carlson, D. M., "Structures and Immunochemical Properties of Oligo-

saccharides Isolated from Pig Submaxillary Mucins," *J. Biol. Chem.* (1968) **243,** 616.

75. Lundblad, A., Kabat, E. A., "Immunochemical Studies on Blood Groups XLVIII. Activity of Fucose-Containing Oligosaccharides Isolated from Human Urine and from Human Ovarian Cyst A and B Blood Group Substances," *J. Immunol.* (1971) **106,** 1572.
76. Schiffman, G., Kabat, E. A., Leskowitz, S., "Immunochemical Studies on Blood Groups XXIV. Some Oligosaccharides Isolated from Dialysates after Mild Acid Hydrolysis of Human Blood Group B Substances from Ovarian Cyst Fluid," *J. Amer. Chem. Soc.* (1960) **82,** 1122.
77. Donald, A. S. R., Creeth, J. M., Morgan, W. T. J., Watkins, W. M., "The Peptide Moiety of Human-Blood-Group Active Glycoproteins Associated with the ABO and Lewis Groups," *Biochem. J.* (1969) **115,** 125.
78. Iyer, R. N., Carlson, D. M., "Alkaline Borohydride Degradation of Blood Group H Substance," *Arch. Biochem. Biophys.* (1971) **142,** 101.

RECEIVED December 2, 1971. Work of the laboratories is supported by the National Science Foundation (NSF GB-25686 and 35243X) and a General Research Support Grant of the United States Public Health Service.

19

The Role of Sugar Phosphate in the Biosynthesis of Complex Saccharides

W. Z. HASSID

Department of Biochemistry, University of California, Berkeley, Calif. 94720

Nucleoside disphosphate sugars seem to be superior donors of the glycosyl moiety for complex sugar formation because they have a higher negative free energy of hydrolysis ($\Delta G°$) *than other glycosyl compounds. This seems to be the reason why most of the polysaccharides and oligosaccharides are synthesized* in vivo *from these sugar nucleotides. The conditions of the reactions and the factors by which they are affected in the production of the various complex saccharides will be discussed here. Synthesis of the cell wall polysaccharides, cellulose, xylan, pectin, and others will also be discussed.*

The synthesis of complex saccharides (oligosaccharides, glycosides, homo- and heteropolysaccharides) involves the process of transglycosylation. In this process the glycosyl donor may be sugar phosphate or sugar nucleotide oligosaccharide or polysaccharide. However, the most effective compounds to serve as donors for the glycosyl unit for enzymic synthesis of the complex carbohydrates have been shown to be phosphorylated sugars, especially the nucleoside diphosphate sugars. The reason for this is because they possess a high negative free energy of hydrolysis ($\Delta G°'$).

The Role of Sugar Phosphate in Biosynthesis of Complex Saccharides

Transglycosylation Reactions from Different Phosphate Containing Substrates. Although some oligosaccharides and polysaccharides can be synthesized *in vitro* from D-glucose-1-*P* (*1*), it is now believed that complex saccharide formation from this phosphorylated sugar by phosphorolysis is not a normal physiological process; these enzymes act only in a degradative capacity. Only six enzymes of this type are known:

glycogen phosphorylase (glycogen muscle phosphorylase) (*2*), starch phosphorylase (potato phosphorylase) (*3*), sucrose phosphorylase (*Pseudomonas saccharophila*) (*4*), maltose phosphorylase (*Neisseria meningitidis*) (*5*), cellobiose phosphorylase (*Ruminococcus flavefaciens*) (*6*), and laminaribiose phosphorylase (*Euglena gracilis*) (*7*). They are produced by the following reactions:

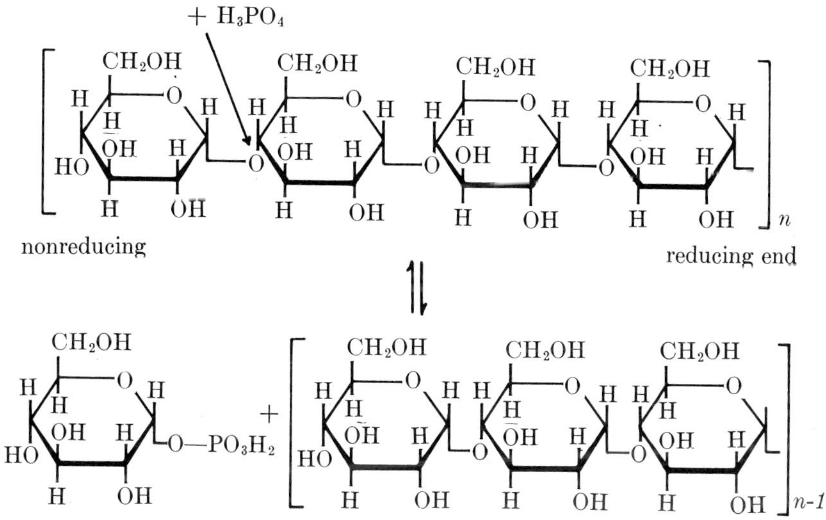

Formation of glycogen chains by glycogen muscle phosphorylase and starch amylose by potato phosphorylase.

(*Pseudomonas saccharophila*)

Sucrose + Pi ⇌ α-D-glucose 1-phosphate + D-fructose.

Sucrose formation by the sucrose phosphorylase reversible reaction.

(*Neisseria meningitidis*)

4-*O*-α-glucosyl-D-glucose + Pi ⇌ β-D-glucose 1-phosphate + D-glucose.
(maltose)

Maltose formation by the maltose phosphorylase reversible reaction.

(*Ruminococcus flavefaciens*)

4-*O*-β-glucosyl-D-glucose + Pi ⇌ α-D-glucose 1-phosphate + D-glucose.
(cellobiose)

Cellobiose formation by the cellobiose phosphorylase reversible reaction.

$$\text{3-}O\text{-}\beta\text{-glucosyl-D-glucose (laminaribiose)} + \text{Pi} \overset{(\textit{Euglena gracilis})}{\rightleftharpoons} \alpha\text{-D-glucose 1-phosphate} + \text{D-glucose.}$$

Laminaribiose formation by the laminaribiose phosphorylase reversible reaction.

The number of enzymes responsible for the formation of polysaccharides from sucrose is also restricted. The two best known enzymes that form polysaccharides from sucrose are the dextran, synthesized by the microorganism *Leuconostoc mesenteroides* (*8*) and related organisms, and the levan, produced from the same substrate by *Acetobacter levanicum* (*9*) and other species.

Energy Relations of the Phosphorylated Sugar Substrates. From the thermodynamic point of view, nucleoside diphosphate sugars are superior donors for formation of complex saccharides because they have the highest negative free energy of hydrolysis of all known compounds containing glycosyl groups that can serve as a monosaccharide donor (*10*). Thus, the $\Delta G^{\circ\prime}$ of uridine phosphate (UDP)–D-glucose at pH 7.4 is -7600 cal mole^{-1} while that of α-D-glucose 1-phosphate at pH 8.5 is -4800 cal mole^{-1}. The value of $\Delta G^{\circ\prime}$ of the α-D-glucose-$(1 \rightarrow 4)$ linkage of glycogen which is produced from these substrates is -4300 cal mole^{-1}. Since the $\Delta G^{\circ\prime}$ of hydrolysis of UDP–D-glusose is -7600, the free energy change during the formation of glycogen from UDP–D-glucose can be calculated as -3300 mole^{-1}. This value corresponds to an equilibrium of about 250, which amounts to a practically quantitative conversion of the nucleotide bound glucose into glycogen.

Glycosyl–Enzyme Complex Intermediates in Biosynthesis of Complex Saccharides. The synthesis of nucleoside diphosphate sugars involves the transfer of a nucleotidyl group from a nucleoside triphosphate to a sugar 1-phosphate with the simultaneous release of pyrophosphate according to the following general reaction (*11*):

$$\text{Nucleoside triphosphate} + \text{sugar 1-phosphate} \underset{\text{pyrophosphorylase}}{\rightleftharpoons} \text{nucleoside diphosphate sugar} + \text{pyrophosphate}$$

Many nucleoside diphosphate sugars containing different bases and different sugar moieties were found to be synthesized by this enzymic process.

A similar reaction was observed (*12*) that leads to the synthesis of a nucleoside monophosphate sugar which occurs as follows:

$$\text{cytidine triphosphate} + N\text{-acetylneuraminic acid} \rightarrow \text{cytidine monophosphate } N\text{-acetylneuraminic acid} + \text{pyrophosphate.}$$

As previously mentioned, sucrose can be formed from α-D-glucose 1-P and D-fructose by a sucrose phosphorylase reversible reaction:

$$\text{Sucrose} + \text{Pi} \rightleftharpoons \alpha\text{-D-glucose-1-P} + \text{D-fructose.}$$

The glucosyl phosphate in this reaction does not seem to be an essential product or substrate of sucrose phosphorylase activity for the synthesis of disaccharides. This ester can be regarded as one of several glucose donors for the enzyme. The sucrose phosphorylase can act not only as a phosphorylase but also as a transglucosylase capable of mediating the transfer of the D-glucose portion of substrate to a variety of acceptors (*1*). The evidence for the double function of the enzyme is cited from the observation that when ^{32}P-labeled inorganic phosphate and nonradioactive α-D-glucose 1-P are added to sucrose phosphorylase preparations in the absence of ketose sugars, a rapid redistribution of the isotope occurs between the organic and inorganic fractions without the liberation of D-glucose. This observation led to the assumption that the enzyme combines reversibly with the D-glucose of α-D-glucose 1-P, forming a D-glucose–enzyme complex and releasing inorganic phosphate, according to the equation:

$$\alpha\text{-D-glucose–1-P} + \text{enzyme} \rightleftharpoons \text{D-glucose–enzyme} + \text{Pi}$$

The equilibrium reaction would require that the energy of the α-D-glucose-1-P linkage be preserved in the D-glucose–enzyme bond. The transfer of phosphate could not involve the formation of free D-glucose because if this occurred, about 4800 cal mole^{-1} would be released in the decomposition of the ester and would be required for its resynthesis. Since no external source of energy would be available for the resynthesis of the ester, it can be concluded that the original bond energy is conserved in the D-glucose–enzyme complex.

The enzyme is really a glucosyl transfer agent, as shown by the catalysis of an exchange of glycosidic bonds in the absence of phosphate (*13*):

$$\underset{\text{(sucrose)}}{\beta\text{-D-fructofuranosyl } \alpha\text{-D-glucopyranoside}} + \text{L-sorbose} \rightleftharpoons \text{D-glucopyranosyl L-sorboside} + \text{D-fructose.}$$

In a similar manner, sucrose can be prepared by a reaction between D-fructose and the corresponding disaccharide containing D-threo-pentulose (*14*).

Supporting evidence for the formation of a D-glucose–enzyme complex in reactions catalyzed by sucrose phosphorylase from *Pseudomonas saccharophila* was obtained by Voet and Abeles (*15*). They showed that when sucrose phosphorylase is denatured after exposure to uniformly labeled sucrose-^{14}C, the denatured protein contains firmly bound D-glucose or a compound derived from the D-glucose moiety of sucrose. The D-fructose moiety of sucrose is not bound to the protein. The molecular weight of the enzyme was shown (*16*) to be 80,000–100,000.

Recently, the formation of a covalent glycosyl–enzyme intermediate was also shown by Bell and Koshland (*17*) in another reaction. Evidence was presented that the mechanism of the enzyme, phosphoribosyl-adenosine triphosphate: pyrophosphate phosphoribosyl transferase, proceeds through a covalent phosphoribosyl-enzyme intermediate. The intermediate has been demonstrated after incubating the enzyme with ^{14}C-5-phosphoribosyl-1-pyrophosphate (PRPP) under native and denaturing conditions. The intermediate also forms from the reverse direction as shown when the enzyme is mixed with its product *N*-(5-phosphoribosyl-adenosine triphosphate (PR-ATP). These data give evidence for a covalent enzyme-substrate intermediate. The enzyme which catalyzes the overall reaction proceeds as follows:

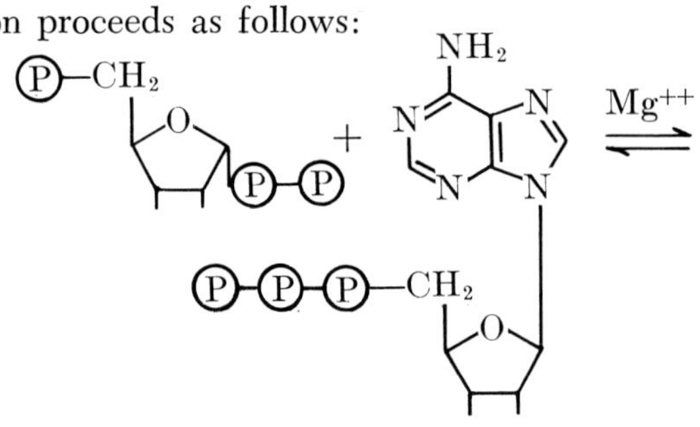

PRPP + ATP

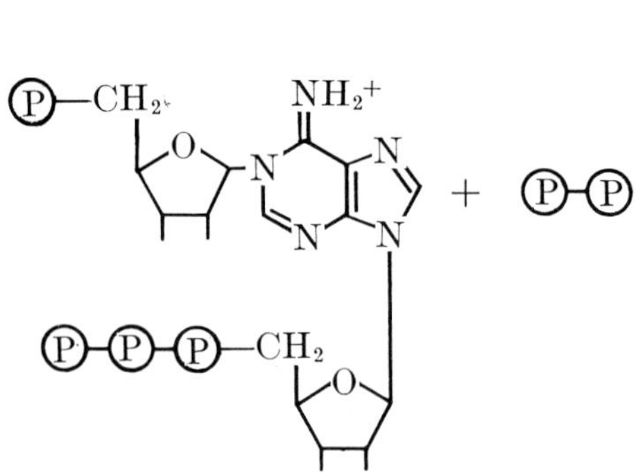

PR–ATP + PPi

Two glycosyl-enzyme intermediates have thus far been definitely shown to be formed, and only one involved in the formation of a disaccharide (sucrose) is known. However, it is now believed that all oligosaccharides and polysaccharides are synthesized by transglycosylation *via* glycosyl-enzyme intermediates. A search of the literature by Bell and Koshland (*18*) showed that during the last decade 58 cases have been recorded for which there is strong evidence for covalent enzyme–substrate intermediates involving various enzymes other than transglycosylases; these include phosphoryl-, acyl-, acetyl-enzyme, and other intermediates.

Enzymic transfer of a glycosyl moiety from a donor substrate to an acceptor may result in a product with the same configuration of the anomeric carbon atom as the original substrate or in an α–β inversion.

Natural complex saccharides found to be formed from a particular substrate with an inversion of configuration seem to be more prevalent than those without inversion.

Koshland (*19*) has postulated a theory of transglycosylation that accounts for both types:

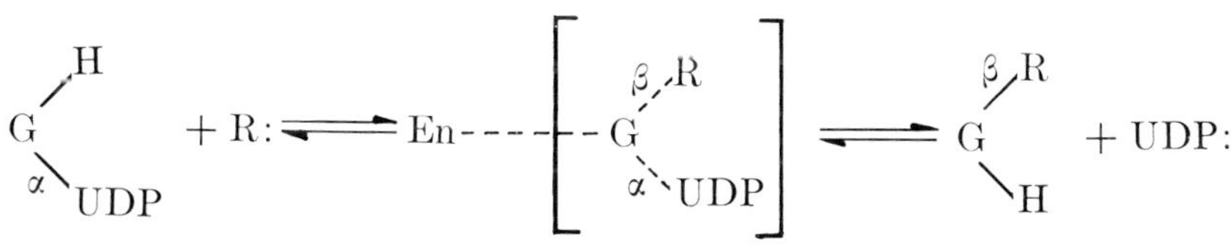

UDP-D-Glucose β-D-Glucoside

Single displacement resulting in inversion

Transglycosylation with inversion is considered to occur by a nucleophilic attack on the anomeric carbon atom by the acceptor substrate. The acceptor approaches the glycosyl moiety of the donor from one side as the group being displaced leaves from the other. The result is a Walden inversion, so that if the donor is an α-D-glycoside, the linkage in the product becomes β. The theory was originally formulated to account for the phosphorolysis of maltose, which gives β-D-glucose 1-phosphate and glucose. However, it can be applied to the synthesis of β-D-glycosides from sugar nucleotides.

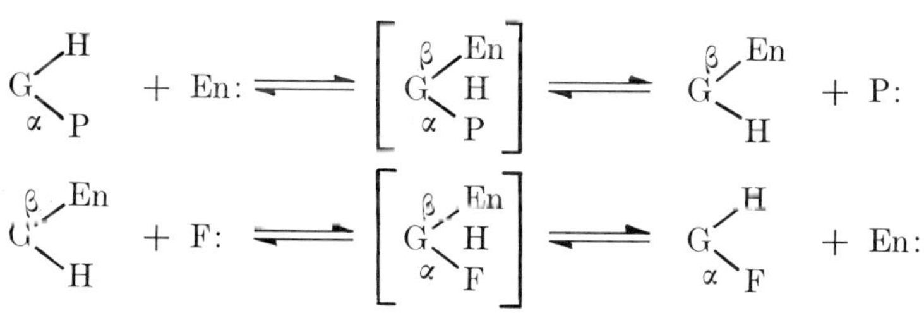

Double displacement resulting in retention of configuration

In cases where the configuration is retained, it is postulated that the glycosyl group is transferred twice, an inversion occurring at each transfer. For example, in the phosphorolysis of sucrose by sucrose phosphorylase, a reactive group on the enzyme would make a nucleophilic attack on the glycosyl carbon to yield a β-D-glucosyl-enzyme. A subsequent displacement of the enzyme group by a phosphate ion would yield a α-D-glucose 1-phosphate.

SPECIFICITY OF TRANSGLYCOSYLASES. The specificity of transglycosylases, regarding the particular base of the glycosyl nucleotide, varies considerably. Thus, the transglycosylase from plants which produces cellulose is specific for GDP-D-glucose and shows a high degree of specificity. This transglycosylase will not form cellulose from glucose nucleotide containing bases other than guanine—*i.e.*, uridine, adenine, cytosine or thymine (*20, 21*). Similarly the soluble transglycosylase found in milk will catalyze the formation of lactose only from UDP-D-galactose and dUDP-D-galactose (*22*).

The enzyme, glycogen synthetase, discovered by Leloir and Cardini (*23*), which synthesizes glycogen from UDP-D-glucose is not specific for this glycosyl nucleotide. This enzyme is also capable of synthesizing glycogen from ADP-D-glucose, but only at 50% of the rate obtained with UDP-D-glucose. However, starch synthetase, which also uses UDP-D-glucose as substrate, will form starch from ADP-D-glucose at a rate which is ten times as rapid as that from UDP-D-glucose. These transglycosylases are not strictly specific with regard to formation of polysaccharides containing α-D-glucose 1,4-linkages.

Lin (*24*) showed the presence of the following sugar diphosphate nucleosides in the marine brown alga, *Fucus gardneri:* GDP-D-mannose, GDP-D-mannuronic acid, GDP-L-guluronic acid, and GDP-L-galactose. He showed that the GDP-D-mannuronic acid is derived from GDP-D-mannose by a series of enzymic reactions while the GDP-L-galactose and the L-fucose polymer which is also present in this alga, are possible derivatives of D-mannose diphosphate nucleoside formed by known enzymic reactions. Whereas the seaweed polysaccharides (alginic acid and fucoidin) appear to be derived by a series of various reactions from the common precursor, GDP-D-mannose, most of the polysaccharide sugar constituents of higher plants, namely, D-galactose, D-glucuronic and D-galacturonic acids, D-xylose, and L-arabinose are derived from UDP-D-glucose.

BIOSYNTHESIS OF CELLULOSE AND OTHER GLUCANS. Elbein, Barber and Hassid (*20, 21*) obtained evidence based on radioactive data indicating that the direct precursor of cellulose is GDP-D-glucose. Using rapidly growing mung beans and other plants (cotton, string bean, squash, pea, and corn), the particulate enzymes were capable of using GDP-D-glucose-^{14}C as substrate for the formation of radioactive polysaccharide in which the D-glucose units are combined by β-(1 → 4) glucosyl linkages.

It has also been shown that mung beans, peas, and other plants contain a pyrophosphorylase which forms GDP-D-glucose from α-D-glucose 1-P and GTP. Based on the data obtained with enzymic plant preparations, we proposed the following mechanism for cellulose synthesis:

$$\text{Guanosine triphosphate} + \alpha\text{-D-glucose 1-phosphate} \xrightleftharpoons{\text{pyrophosphorylase}} \text{guanosine diphosphate D-glucose} + \text{pyrophosphate}$$

$$n(\text{Guanosine diphosphate D-glucose}) + \text{acceptor} \xrightarrow{\text{transferase}} \underset{\text{(cellulose)}}{\text{acceptor-}(\beta\text{-1,4-D-glucose})_n} + n(\text{guanosine diphosphate})$$

However, other workers claim that UDP-D-glucose may also be an effective donor for cellulose formation with preparations from higher plants, but we could not substantiate their results in our laboratory.

Subsequently, Ordin and Hall (*25, 26*) found that particulate preparations from oat coleoptiles could use UDP-D-glucose as substrate for polysaccharide formation. Upon degradation of the polysaccharide derived from UDP-D-glucose with impure cellulase, cellobiose, and to a lesser extent a substance identified as a trisaccharide containing mixed β-(1 → 4), β-(1 → 3) glucosyl linkages were obtained.

These results with enzyme preparations from oat coleoptiles and UDP-D-glucose as substrate could be substantiated by chemical methods (*27*). However, the question remained whether a single enzyme is involved in the synthesis of this oat coleoptile polysaccharide containing mixed linkages or whether two enzymes are present, each forming one of the two linkages of this polymer(s)—*i.e.*, one synthesizing β-(1 → 4) and the other the β-(1 → 3) linkage.

Concerning this problem, we have observed that when a 1×10^{-3} *M* UDP-D-glucose substrate concentration with particulate or digitonin solubilized enzyme preparations from oat coleoptiles was used, a β-(1 → 3) glucan is formed as the main product. Glucan produced from a UDP-D-glucose of 1×10^{-5} *M* or lower concentration contained practically only β-(1 → 4) glucosyl linkages. Also, a separation of the β-(1 → 4) and β-(1 → 3) glucan synthetase activities could be achieved at 1×10^{-3} *M* UDP-D-glucose concentration when the digitonin solubilized enzyme was adsorbed on a hydroxylapatite gel and then eluted with strong potassium phosphate buffer (*28*).

The results indicate that the particulate enzyme contains two enzymes; using UDP-D-glucose as substrate, one is capable of synthesizing β-(1→4) and another β-(1→3) linkages. The *Km* of the β-(1→4) synthesizing enzyme was found to be 1.2×10^{-5} *M* and that of the β-(1→3) 6×10^{-4} *M* (*29*).

Cellulose is the main component of the cell-wall which forms the insoluble skeletal framework of all higher plants. This polymer is associated with other polysaccharides, chiefly xylan, pectin, glucomannan, and hemicelluloses.

These polysaccharides have been synthesized *in vitro* from various sugar nucleotides with enzymes isolated from plant sources.

Xylan. Experiments *in vivo* by several investigators (*30*) indicated that xylan in plants originated from glucose by a series of reactions first leading to the formation of pentose polysaccharide. UDP-D-xylose has been isolated from plant seedlings and synthesized enzymically from UDP-D-glucose by the following sequence of reactions:

$$\text{UDP-D-glucose} \xrightarrow{\text{dehydrogenase}} \text{UDP-D-glucuronic acid} \xrightarrow{\text{decarboxylase}} \text{UDP-D-xylose}$$

Particulate preparations from corn shoots readily incorporate ^{14}C-labeled D-xylose from UDP-D-xylose-^{14}C into a polysaccharide in which the D-xylose residues are combined by β-1,4-D-xylosyl bonds (*31*). It was shown that this polysaccharide, similar to natural plant xylan, contains a small proportion of L-arabinose units which have the furanose configuration.

Pectins. The basic building unit of pectins is known to be α-1,4 linked D-galacturonic acid which forms the polygalacturonic acid chain. The carboxyl groups of the D-galacturonic acid in the chain are methylated to various degrees.

It has been shown that UDP-D-galacturonic acid is present in higher plants (*32*) and that they contain enzymes which lead to the formation of this uronic acid nucleotide, starting with UDP-D-glucose by the following pathway (*33, 34*):

$$\text{UDP-D-Glucose} \xrightarrow{\text{dehydrogenase}} \text{UDP-D-glucuronic acid} \xrightleftharpoons{\text{epimerase}} \text{UDP-D-galacturonic acid.}$$

UDP-D-galacturonic acid has also been isolated from mung beans.

A particulate preparation from mung beans was found to catalyze the polymerization of the D-galacturonic acid from the UDP-D-galacturonic acid, resulting in the formation of a polygalacturonic acid chain (*35*). The synthetic polygalacturonate could be hydrolyzed with *Penicillum chrys-*

ogenum polygalacturonase to D-galacturonic acid and with an exo-polygalacturonic acid transeliminase from *Clostridium multifermentans* to unsaturated 4,5-digalacturonic acid. The action of these enzymes is specific for degradation of the polygalacturonic acid chain (*36, 37*). The structure of the unsaturated digalacturonic acid was shown to be as follows (*38*):

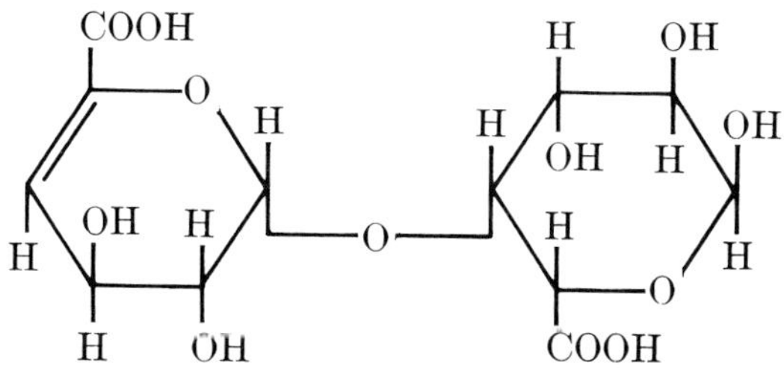

Kauss (*39*) has shown that the esterification of the carboxyl groups in the D-galacturonic acid chain takes place by a transfer of the methyl groups from S-adenosyl-L-methionine, analogous to the case in which the 4-methyl ether groups are transferred to D-glucuronic acid of hemicellulose (*40*).

Glucomannan. Elbein (*41, 42*) has shown that when GDP-D-mannose-^{14}C is used as substrate, a radioactive glucomannan is synthesized and that Mg^{2+} is required for its formation. The addition of unlabeled GDP-D-glucose to the reaction mixture containing GDP-D-mannose-^{14}C resulted in a marked inhibition of incorporation of radioactive D-mannose into an insoluble polysaccharide. The enzyme(s) involved in the synthesis of glucomannan apparently has a greater affinity for GDP-D-glucose than for GDP-D-mannose, which would account for the inhibition of mannose incorporation by GDP-D-glucose. Several of the oligosaccharides obtained from degradation of the glucomannan contained D-glucose and D-mannose in various proportions, indicating that the glucomannan is not a mixture of cellulose and mannan. The D-mannose units in this polysaccharide were shown to be linked by β-1,4 bonds.

The fact that glucomannan is obtained from GDP-D-mannose as substrate suggests that the particulate enzyme contains an epimerase which converts GDP-D-mannose to GDP-D-glucose. However, such an epimerase has not been observed. In this connection cellulose isolated from wood contains a considerable amount of D-mannose.

Hemicellulose. The hemicelluloses consisting of hexoses, pentoses, and uronic acids may be separated into two fractions, A and B. Hemicellulose B usually contains a higher proportion of uronic acid, mainly 4-methyl-D-glucuronic acid, than the A fraction. This methyl derivative of D-glucuronic acid is most frequently isolated in combined form as the

4-methyl aldobiuronic acid because it is hydrolyzed with difficulty with acid.

As previously mentioned, Kauss (*40*) has shown that the methyl donor for the formation of 4-methyl-D-glucuronic acid of hemicellulose B proved to be S-adenosyl-L-methionine, the same as in pectin. A particulate preparation from immature corn cobs containing hemicellulose B was found capable of transferring the ^{14}C-labeled methyl group from S-adenosyl-L-methionine to a macromolecular acceptor present in the particles. The radioactive product was shown to be hemicellulose B labeled in the 4-methyl-D-glucuronic acid residues. It was isolated chiefly as 4-methylglucuronosyl-(1 → 2)-D-xylose.

In another experiment with the particulate enzyme from the same plant he obtained 4-methyl D-glucuronosyl-D-galactose together with 4-methyl D-glucuronosyl-D-xylose. Kauss (*43*) also found that the particulate enzyme preparation from immature corn cobs contains, in addition to the methyl transferase, an enzyme which introduces the D-glucuronic acid group from UDP-D-glucuronic acid into hemicellulose B.

The observation that the methyl ether or ester groups of plant heteropolysaccharides are introduced at the macromolecular level is similar to the finding that C- and N-methyl groups of RNA and DNA are also introduced into preformed macromolecules.

While we are far from knowing all the biochemical and physiological details of the synthesis of these polymers, we obtained information pertaining to basic biochemical reactions involved in their synthesis.

Literature Cited

1. Hassid, W. F., "The Carbohydrates, Chemistry and Biochemistry," p. 301, Vol. IIA, W. Pigman and D. Horton, Eds., 2nd ed., Academic, 1970.
2. Cori, G. T., Cori, C. F., *J. Biol. Chem.* (1940) **135,** 733.
3. Hanes, C. S., *Proc. Roy. Soc. (London)* **B128,** 421; *ibid.* (1941) **B129,** 174.
4. Hassid, W. Z., Doudoroff, M., Barker, H. A., *J. Amer. Chem. Soc.* (1944) **66,** 1416.
5. Fitting, C., Doudoroff, M., *J. Biol. Chem.* (1952) **199,** 153.
6. Ayers, W. A., *J. Biol. Chem.* (1959) **234,** 2819.
7. Marechal, L. R., Goldenberg, S. H., *Biochem. Biophys. Res. Commun.* (1963) **13,** 106.
8. Stacey, M., Barker, S. A., "Polysaccharides of Microorganisms," p. 136, Oxford University, London and New York, 1960.
9. Hestrin, S., Feingold, D. S., Avigad, G., *Biochem. J.* (1956) **64,** 340.
10. Leloir, C. F., Cardini, C. E., Cabib, E., "Comparative Biochemistry," p. 97, M. Florkin and H. S. Mason, Eds., Vol. 2, Academic, New York, 1960.
11. Munch-Peterson, A., Kalckar, H. M., Culoto, E., Smith, E. E. B., *Nature* (1953) **172,** 1037.
12. Roseman, S., *Proc. Natl. Acad. Sci. U. S.* (1962) **48,** 437.
13. Hassid, W. Z., Doudoroff, M., *Advan. Enzymol.* (1950) **10,** 123.
14. Kalckar, H. M., "The Mechanism of Enzyme Action," p. 675, W. D. McElroy and G. Glass, Eds., Johns Hopkins, Baltimore, 1953.
15. Voet, J., Abeles, R. H., *J. Biol. Chem.* (1967) **242,** 1338.

16. Silverstein, R., Voet, J., Reed, D., Abeles, R. H., *J. Biol. Chem.* (1967) **242**, 1338.
17. Bell, R. M., Koshland, D. E., Jr., *Biochem. Biophys. Res. Commun.* (1970) **38**, 539.
18. Bell, R. M., Koshland, D. E., Jr., *Science* (1971) **172**, 1253.
19. Koshland, D. E., Jr., "Symposium on the Mechanism of Enzyme Action," p. 608, W. D. McElroy, and B. Glen, Eds., Johns Hopkins, Baltimore, 1954.
20. Elbein, A. D., Barber, G. A., Hassid, W. Z., *J. Amer. Chem. Soc.* (1964) **86**, 309.
21. Barber, G. A., Elbein, A. D., Hassid, W. Z., *J. Biol. Chem.* (1964) **239**, 4056.
22. Babad, H., Hassid, W. Z., *J. Biol. Chem.* (1966) **241**, 2672.
23. Leloir, L. F., Cardini, C. A., *J. Biol. Chem.* (1957) **79**, 6340.
24. Lin, T. S., Hassid, W. Z., *J. Biol. Chem.* (1966) **241**, 3282, 5284.
25. Ordin, L., Hall, M. A., *Plant Physiol.* (1967) **42**, 205.
26. Ordin, L., Hall, M. A., *Plant Physiol.* (1968) **43**, 473.
27. Flowers, N. H., Batra, K. K., Kemp, J., Hassid, W. Z., *Plant Physiol.* (1968) **43**, 1703.
28. Tsai, C. M., Hassid, W. Z., *Plant Physiol.* (1971) **47**, 740.
29. Tsai, C. M., Hassid, W. Z., in preparation.
30. Hassid, W. Z., Neufeld, E. F., Feingold, D. S., *Proc. Natl. Acad. Sci. U. S.* (1959) **45**, 905.
31. Bailey, R. W., Hassid, W. Z., *Proc. Natl. Acad. Sci. U. S.*, **56**, 1586.
32. Neufeld, E. F., Feingold, D. S., *Biochem. Biophys. Acta* (1961) **53**, 589.
33. Strominger, J. L., Mapson, L. W., *Biochem. J.* (1957) **66**, 567.
34. Feingold, D. S., Neufeld, E. F., Hassid, W. Z., *J. Biol. Chem.* (1960) **235**, 910.
35. Villemez, C. L., Lin, T. S., Hassid, W. Z., *Proc. Natl. Acad. Sci. U. S.* (1965) **54**, 1626.
36. Macmillan, J. D., Vaughn, R. H., *Biochemistry* (1964) **3**, 564.
37. Macmillan, J. D., Phaff, H. J., Vaughn, R. H., *Biochemistry* (1964) **3**, 564.
38. Hasegawa, S., Nagel, C. W., *J. Biol. Chem.* (1962) **237**, 619.
39. Kauss, H., Hassid, W. Z., *J. Biol. Chem.* (1967) **242**, 3449.
40. Kauss, H., Hassid, W. Z., *J. Biol. Chem.* (1967) **242**, 1680.
41. Elbein, A., Hassid, W. Z., *Biochem. Biophys. Res. Commun.* (1966) **23**, 311.
42. Elbein, A. D., *J. Biol. Chem.* (1969) **244**, 1608.
43. Kauss, H., *Biochim. Biophys. Acta* (1967) **148**, 572.

RECEIVED December 2, 1971.

20

Glucan Glucohydrolases: Action Mechanisms and Enzyme–Carbohydrate Complexes

JOHN H. PAZUR, DANIEL J. DROPKIN, and CHARLES E. HETZLER

Department of Biochemistry, Pennsylvania State University, University Park, Pa. 16802

Glucoamylase (α-1,4-glucan glucohydrolase) from Aspergillus niger *acts on α-glucans and α-glucosyl oligosaccharides by a multichain mechanism liberating glucose from the non-reducing end of these substrates. A β-1,4-glucan glucohydrolase observed in almond seeds acts by a similar mechanism on β-glucans and β-glucosyl oligosaccharides. Based on substrate specificities and rate data, multiple subsite models are suggested for the active enzyme–carbohydrate complexes for these hydrolases with a four subsite model for glucoamylase and a two subsite model for the new glucohydrolase. A glucosyl–enzyme complex is probably a common, terminal complex in the reaction sequences for both enzymes.*

Glucan hydrolases are present in virtually all types of living cells and are particularly important in the degradation of glucans of the starch, glycogen, and cellulose types. Well known glucan hydrolases are the α-amylases (α-1,4-glucan glucanosyl hydrolases, EC-3.2.1.1), the cellulases (β-1,4-glucan glucanosyl hydrolases, EC-3.2.1.4), the β-amylases (α-1,4-glucan maltohydrolases, EC-3.2.1.2), and glucoamylases (α-1,4-glucan glucohydrolase EC-3.2.1.3). Studies on the properties and on the mechanism of action of these glucan hydrolases (*1, 2, 3*) are extensively reviewed in a recent volume on enzymes. Most of the α-amylases and many of the cellulases are endo-amylases hydrolyzing interior α-1,4- or β-1,4-glucosidic bonds of the glucans in more or less random manner to low molecular weight reducing oligosaccharides. The β-amylases and the glucoamylases, however, are exo-amylases hydrolyzing the glucosidic bonds in a uniform manner, beginning at one end of the polymeric chain. Thus β-amylase removes maltosyl units from the non-reducing end of

α-1,4-glucans and progressively shortens the glucan chains while glucoamylases remove single glucose units from the non-reducing ends by a similar mechanism.

In nature the α-amylases are found in mammalian, plant, and microbial secretions or tissues, beta amylases in germinated plant seeds and glucoamylases and cellulases primarily in microorganisms. The latter enzymes are used by microorganisms to convert starch or cellulose to glucose which is taken up and used for metabolic purposes. In contrast to α- and β-amylases, glucoamylases are capable of hydrolyzing different types of α-glucosidic linkages (*4*) and accordingly convert quantitatively starch and related compounds to glucose (*5*). Because of the latter property, glucoamylases are technologically important in the commercial production of glucose from starch (*6*). Highly purified and crystalline preparations of glucoamylase have been obtained (*7, 8*), and such preparations while not essential for technological applications are used to study the action mechanism of the enzyme (*9*).

In this study the action mechanism of the glucoamylase on α-glucosyl oligosaccharides has been studied by use of an oligosaccharide mapping technique (*10*), and a new type of glucohydrolase activity has been observed in almond seeds. The new findings show that glucoamylase acts essentially by a multichain mechanism (*1*) on linear glucans in which single glucose units are removed from the non-reducing ends of the substrate chains and all chains are progressively shortened. The almond glucohydrolase effects a hydrolysis of β-1,4- and β-1,6-glucosyl oligosaccharides, as well as soluble β-glucans, directly to glucose; it is, therefore, a β-1,4-glucan glucohydrolase. The β-glucohydrolase activity may result from the well-known β-glucosidase in almond seeds, and this aspect is still being studied. In plant seeds the glucohydrolase activity of β-type may be important in the germination process for degrading the cellulose in the seed coat and allowing for growth and differentiation of the seedling. Since soluble β-glucans are also degraded by the almond glucohydrolase, enzymes of this type may have potential value for degrading soluble cellulose wastes.

Because of the similarity in action patterns of the fungal glucoamylase and the plant glucohydrolase, comparative studies have been made on the mechanism of action of the two types of glucohydrolases and on the nature of the enzyme–carbohydrate complexes. The mechanism of action of the two enzymes has been found to be similar—*i.e.*, a multichain mechanism, but the nature of the active complexes is quite different. The complexes of glucohydrolases with the substrates is best represented by a subsite model (*11, 12*) with a four subsite model for the fungal glucoamylase and a two subsite model for the almond glucohydrolase.

Experimental

Enzymes. Glucoamylase (α-1,4-glucan glucohydrolase) from *Aspergillus niger* has been prepared in pure form in our laboratory by using DEAE-cellulose chromatographic and alcohol fractionation procedures (*13, 14*). Solutions of the enzyme in 0.1*M* sodium acetate buffer of pH 5 and with protein concentrations ranging from 0.01–0.5% were used in the experiments. Solutions of the β-glucan glucohydrolase were prepared from almond emulsion obtained from Nutritional Biochemicals Corp., Cleveland, Ohio. Samples of the almond preparation were suspended in distilled water, stirred for 0.5 hour, and centrifuged to remove insoluble material. The supernatant solution was collected and constituted the enzyme samples. The protein content of these enzyme solutions ranged from 0.1–1%, and the pH was slightly acidic. Density gradient centrifugation of the almond glucohydrolase and reference proteins (bovine serum albumin and fungal glucoamylase) was conducted by procedures previously described (*15, 16*). The glucohydrolase were assayed on a substrate of cellobiose by a glucose oxidase procedure (*14*).

Substrates. Maltotriose, isomaltotriose, and the ^{14}C labeled malto-oligosaccharides were prepared by enzymic procedures using salivary amylase (*17*), glucosyl transferase (*18*), and macerans amylase (*19*) and appropriate substrates. Cellotriose and cellotetraose (*20*) were provided by G. L. Miller, Department of Microbiology, Jefferson Medical College, Philadelphia, Pa. Gentiotriose, gentiotetraose, and gentiopentaose were prepared from luteose (*21*) which was provided by J. Marshall, Department of Biochemistry, University of Miami, Fla. To prepare the latter oligosaccharides, a sample of 0.1 gram of luteose was dissolved in 3 ml of 0.1*N* hydrochloric acid and heated in a boiling water bath for 2 hours. The oligosaccharides in the hydrolysate were separated by paper chromatography and obtained in pure form by elution from the paper as described previously (*17*). Other carbohydrates used in this study were obtained from commercial suppliers. All oligosaccharides and glucosides used here yielded single spots when examined by paper chromatographic procedures by the multiple ascent technique in the solvent system of *n*-butyl alcohol–pyridine–water (6:4:3 by volume) (*22*) and stained with a copper sulfate–molybdic acid or a silver nitrate reagent (*22, 23*).

Methods. Samples of from 0.1–10 mg of the oligosaccharide were dissolved in 0.05–0.2 ml of water, mixed with an equal volume of enzyme solution, and incubated at room temperature. In the experiments the concentrations of the oligosaccharide and enzymes were selected in a ratio such that saturation levels of the substrates were present during the incubations (*8*). Samples of 5 μliters were removed from the various digests at specified time intervals, placed on paper chromatograms, and heated in an oven at 100°C for 5 minutes to inactivate the enzyme. Upon completion of the experiment, the chromatograms were developed in a solvent system of *n*-butyl alcohol–pyridine–water (6:4:3 by volume) by the multiple ascent technique (*22*). The finished chromatograms were then stained to reveal reducing sugars. The nature of hydrolytic products was established by comparing R_f values of these products with R_f values for standard compounds. Quantitative values were obtained by comparing intensities of the oligosaccharide spots to those for known concentrations of the compound. Since the incubations with the various oligosaccharides were performed at saturation levels of the substrates, the relative hydrolysis rates of these substrates could be calculated. The hydrolysis rate for the compound hydrolyzed with the fastest velocity has been assigned an arbitrary value of 100, and other rates have been calculated relative to this value.

The nature of the hydrolytic fragments and reaction rates for glucoamylase action on labeled maltooligosaccharides were also determined by the oligosaccharide mapping procedure (*10*). In this procedure the maltooligosaccharides-1-^{14}C were separated in one direction on a paper chromatogram, and the area of the chromatogram containing the oligosaccharide was then sprayed with a dilute solution (generally 0.01%) of the glucoamylase. After incubating for 15 minutes, the chromatogram was developed in a second direction in the same solvent system. The radioactive products on the finished chromatogram were detected by radioautography. Quantitative measurements of the hydrolytic products were obtained by radioactivity determinations. For the latter, the areas of the chromatogram containing the compound were cut from the chromatogram and counted in a liquid scintillation counter. From these data the relative rates of enzyme action on the various oligosaccharides were calculated. The non-radioactive products from the oligosaccharide were detected by staining duplicate chromatograms with the silver nitrate reagent (*23*).

Results

Photographs of chromatograms showing the products of glucoamylase action on maltotriose and isomaltotriose are reproduced in Figure 1 and Figure 2. In Figure 1 maltotriose, an α-1,4-glycosyl trisaccharide, can be seen to hydrolyze initially to glucose and to maltose, but as the concentration of maltose increases, the disaccharide is hydrolyzed to glucose by the glucoamylase. Results in Figure 2 show that isomaltotriose, an α-1,6-glucosyl trisaccharide, is readily hydrolyzed by the glucoamylase, yielding glucose and isomaltose. Chromatograms of other experiments showed that isomaltose is also slowly hydrolyzed to glucose by the glucoamylase. The above experiments were performed at comparable concentrations of enzyme and substrates, and the relative rate of hydrolysis of the various glucosidic bonds can be estimated by comparing the rates of appearance of hydrolytic products. Such comparisons reveal that the hydrolysis rate of an α-1,4-glucosidic bond of maltotriose is 20 to 30 times the rate for an α-1,6-glucosidic bond of isomaltotriose. Since maltose and isomaltose accumulate in the digests of the trisaccharides (Figures

1 and 2), the α-1,4- and the α-1,6-bonds of the disaccharides are hydrolyzed at slower rates than the corresponding bonds of the trisaccharides. The relative rate of hydrolysis of the bonds of the disaccharides (*8*) is in the same ratio as that observed for the trisaccharides.

From the above results it was not possible to find out which of the glucosidic bonds in the trisaccharides was initially hydrolyzed by the glucoamylase. Oligosaccharides labeled with ^{14}C in the reducing moiety have been used to study this aspect of the enzyme mechanism. The products of action of glucoamylase on pure maltotetraose-1-^{14}C are shown on the chromatogram and radioautogram reproduced in Figure 3. The chromatogram in this figure was sprayed with an excess of periodate–permangenate reagent (*24*), and some diffusion of the spots occurred. In the chromatogram maltotriose and glucose apparently are the initial products of enzyme action on the tetrasaccharide. On longer incubation the concentration of these products increases, and maltose is produced.

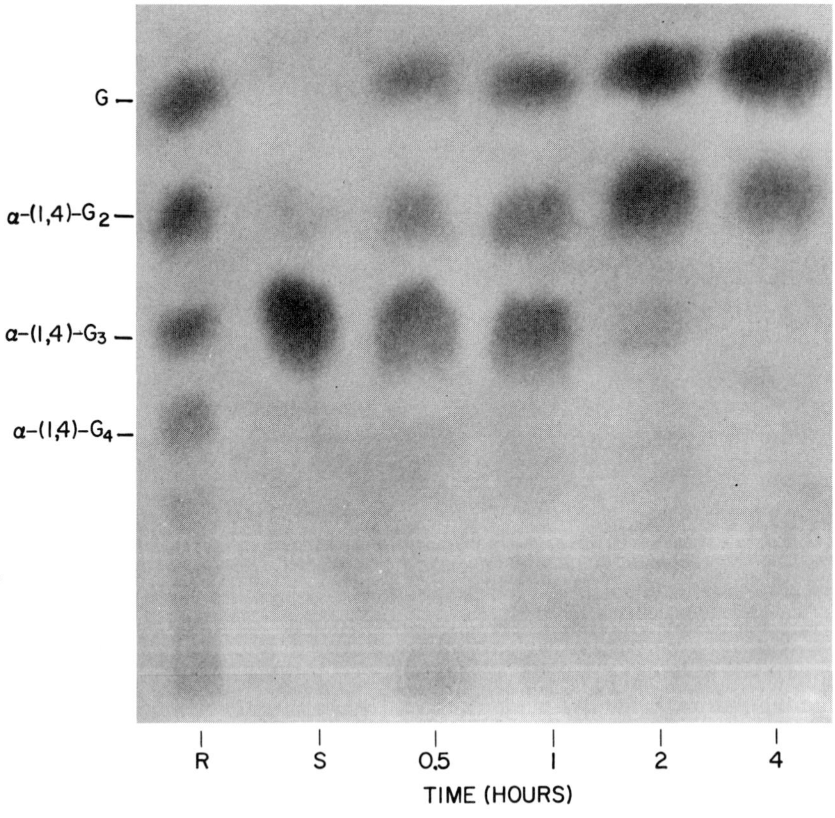

Figure 1. Photograph of a paper chromatogram of the products liberated from maltotriose by glucoamylase

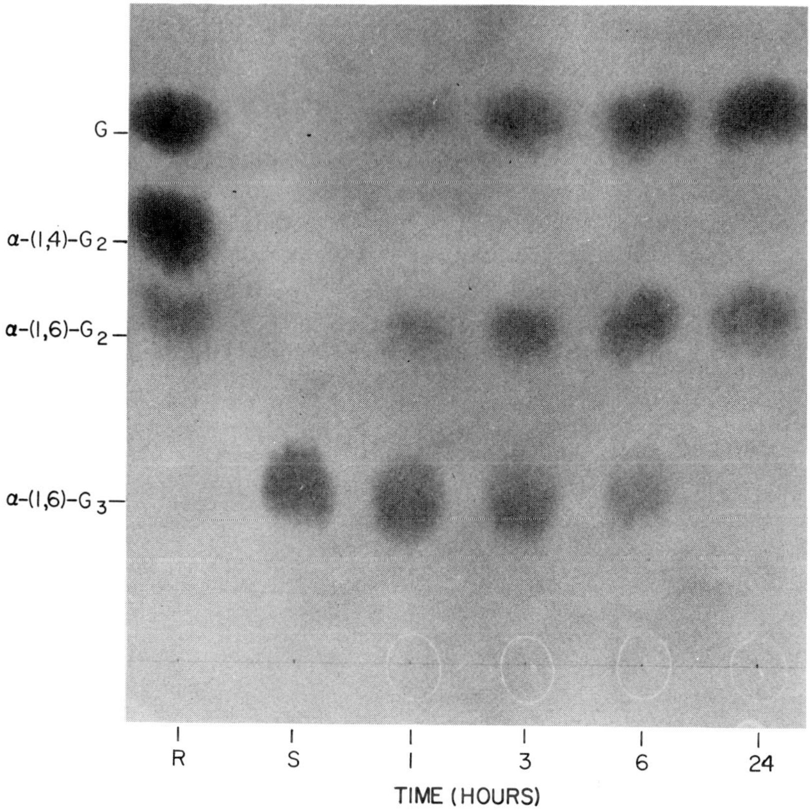

Figure 2. Photograph of a paper chromatogram of products liberated from isomaltotriose by glucoamylase

On the radioautogram of the hydrolytic products initially produced from the tetrasaccharide, the maltotriose was labeled, but the glucose was not. As the concentration of the maltotriose increased in the incubation mixture, a hydrolysis of the trisaccharide occurred to yield labeled maltose and non-labeled glucose. On prolonged incubation the tetrasaccharide was completely converted to glucose by the glucoamylase.

The results of the isotope experiment show that the glucoamylase hydrolyzed the bond at the non-reducing end of the maltotetraose-1-^{14}C, yielding non-labeled glucose and a labeled oligosaccharide with one less glucose residue. This mechanism of action has also been shown to be operative for the action of the enzyme on other oligosaccharides by using the oligosaccharide mapping technique. A radioautogram, showing the labeled products in digests of glucoamylase on maltooligosaccharides-1-^{14}C, is reproduced in Figure 4.

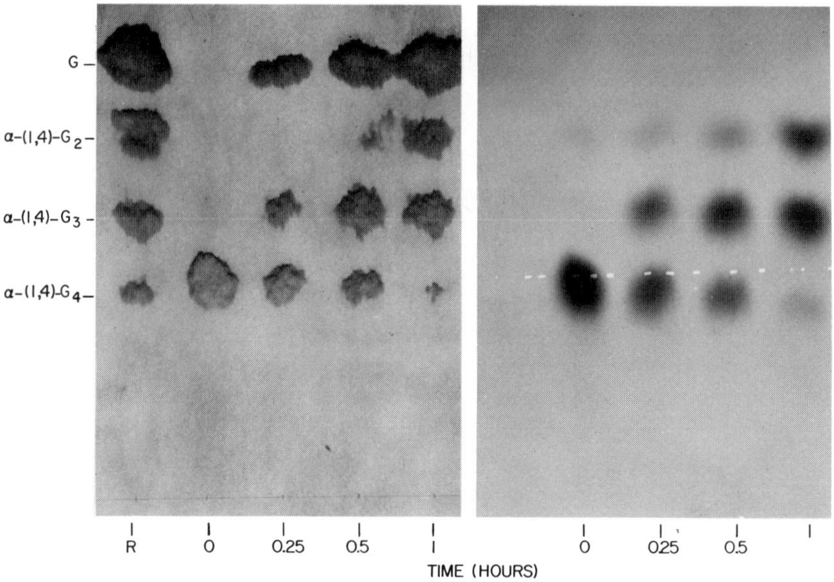

Figure 3. Photograph of a paper chromatogram and a radioautogram of the products liberated from maltotetraose-1-^{14}C by glucoamylase

In each case the major labeled hydrolytic product from these substrates was an oligosaccharide with one less glucose unit. Since the terminal glucosyl unit was not labeled, the glucose does not show up on the radioautogram. The labeled hydrolytic products from the higher molecular weight oligosaccharides were hydrolyzed by removing a second glucosyl unit from the non-reducing end of the substrate. This process is repeated until the oligosaccharides are completely hydrolyzed to glucose. Since maltose is slowly hydrolyzed by the enzyme, labeled glucose is produced from all of the labeled oligosaccharides when the hydrolysis is complete.

Quantitative data on the hydrolysis of the oligosaccharides were obtained by radioactivity measurements. On the basis of these data, the relative rates of hydrolysis of the oligosaccharides were calculated and are recorded in Table I. Also for comparison, results of earlier experiments (*8*) on the hydrolysis rates of various α-glucosides by the glucoamylase are presented in the table.

The data in Table I show that the enzyme acts with highest velocity on maltotetraose and oligosaccharides of higher molecular weight. The rate of action on maltotriose is one-half of the rate for the tetrasaccharide, and on maltose the rate is only one-tenth of the rate for the tetrasaccharide. Oligosaccharides in which the glucose units are joined by linkages other than α-1,4 are hydrolyzed at much slower rates than the 1,4 linked compounds. Apparently not only the number but the spatial orientation of the functional groups of the glucose residues are important in the formation of the enzyme–substrate complex.

In Table I there are significant differences in the hydrolysis rates of α-glucosides containing different aglycon moieties. Apparently alterations in the second residues of the substrates markedly influence the ability of the compound to form an enzyme substrate complex. The hydroxyl groups at positions 2, 3, and 6 of the aglycon seem to be specially important in the formation of the complex.

A chromatogram showing the action of the almond glucohydrolase solution on gentiopentaose is presented in Figure 5. In this figure the initial products of the enzyme action on the pentasaccharide are glucose and gentiotetraose; however, as the concentration of the tetrasaccharide increases, gentiotriose also appears with a concurrent increase in the glucose concentration. In other comparable experiments with gentiotriose as the substrate for the enzyme solution, the trisaccharide was hydrolyzed to glucose with an intermediate accumulation of gentiobiose.

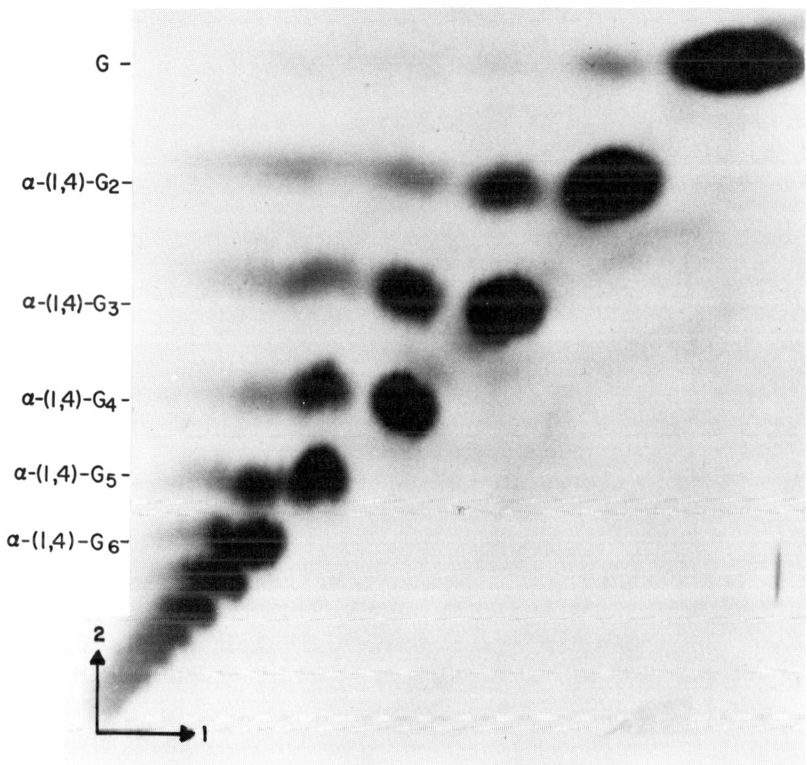

Figure 4. Photograph of a radioautogram of products from labeled malto-oligosaccharides-1-^{14}C liberated by glucoamylase

Table I. Relative Rates of Hydrolysis of Maltooligosaccharides-1-^{14}C and α-D-Glucosides by Fungal Glucoamylase

Compound	*Structure*	*Relative Rate*
Maltopentaose-1-^{14}C	G-(α-1,4)-G-(α-1,4)-G-(α-1,4)-G-(α-1,4)-G-*[a]	100
Maltotetraose-1-^{14}C	G-(α-1,4)-G-(α-1,4)-G-(α-1,4)-G-*	100
Maltotriose-1-^{14}C	G-(α-1,4)-G-(α-1,4)-G-*	50
Maltose-1-^{14}C	G-(α-1,4)-G-*	10
Maltobionic acid	G-(α-1,4)-gluconate	4
Isomaltose	G-(α-1,6)-G-	0.4
Arabinosyl glucoside	G-(α-1,3)-arabinose	0.2
Maltulose	G-(α-1,4)-fructose	0.2
Glycerol glucoside	G-(α-1,1)-glycerol	0.1
Methyl glucoside	G-(α)-OCH_3	0.03

[a] Asterisk indicates glucose residues labeled with ^{14}C.

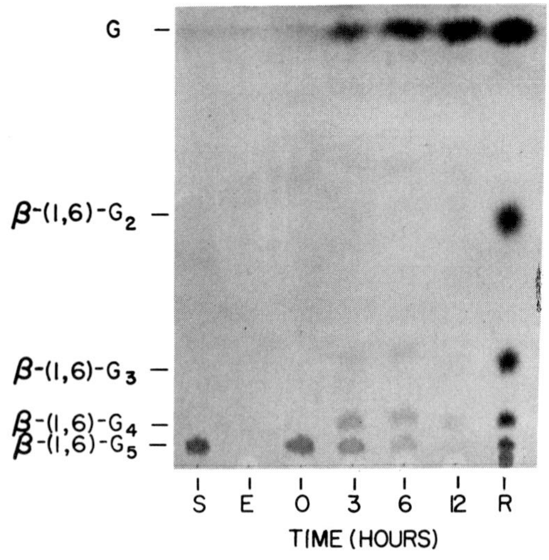

Figure 5. Photograph of a paper chromatogram of products liberated from gentiopentaose by almond glucohydrolase

The almond glucohydrolase hydrolyzed pure gentiotetraose, cellotetraose, and cellotriose by similar reaction sequences—*i.e.*, the removal of glucose units from the oligosaccharide and a concurrent production of a fragment with one less glucose unit. Furthermore the enzyme preparation rapidly effected the hydrolysis of cellobiose, gentiobiose, and methyl β-glucoside to glucose. Density gradient centrifugation of the preparation showed that the glucohydrolase activity on the various substrates were located in the same fractions of the column, indicating that a single enzyme may be responsible for the hydrolysis of the various substrates.

The sedimentation rate of the almond glucohydrolase on density gradient tests was slightly smaller than the rate for glucoamylase. Apparently the almond glucohydrolase is a somewhat smaller molecule.

The relative rates of action of almond glucohydrolase on the various oligosaccharides have been calculated, based on the rate of appearance of the hydrolytic products from the substrates. These values are recorded in Table II. The data in the table show that the enzyme acts most rapidly on the lower molecular weight glucosyl compounds with cellobiose and methyl β-glucoside being hydrolyzed at the fastest rate. The higher molecular weight oligosaccharides are hydrolyzed at slower rates, and a β-1,6-glucan (luteose) at a very slow rate.

Table II. Relative Rates of Hydrolysis of β-Glucosyl Oligosaccharides, Luteose and Methyl β-D-Glucoside, by Almond Glucohydrolase

Compound	*Structure*	*Relative Rate*
Cellobiose	G-(β-1,4)-G-	100
Gentiobiose	G-(β-1,6)-G-	50
Cellotriose	G-(β-1,4)-G-(β-1,4)-G-	40
Gentiotriose	G-(β-1,6)-G-(β-1,6)-G-	20
Cellotetraose	G-(β-1,4)-G-(β-1,4)-G-(β-1,4)-G-	20
Gentiotetraose	G-(β-1,6)-G-(β-1,6)-G-(β-1,6)-G-	10
Luteose	β-1,6-glucan	1
Methyl glucoside	G-(β)-OCH_3	100

Discussion

The results in Figures 1 and 2 show that under comparable reaction conditions, the fungal glucoamylase hydrolyzes the α 1,4- and the α 1,6-glucosidic bonds of maltotriose (α-1,4-glucosyl trisaccharide) and isomaltotriose (α-1,6-glucosyl trisaccharide) at significantly different rates, with the hydrolysis of the α-1,6-glucosyl compound proceeding at a much slower rate than for the α-1,4-glucosyl compound. The glucosidic bond at the non-reducing end of such oligosaccharides is shown to be hydrolyzed in the initial attack of the enzyme by the results of isotope experiments given in Figures 3 and 4. In Figure 3 the initial products from maltotetraose-1-^{14}C are labeled maltotriose and non-labeled glucose. In Figure 4 glucoamylase acts on other glucosyl oligosaccharides by a similar mechanism yielding a labeled oligosaccharide with one less glucose residue than the original substrate. The other hydrolytic fragment from the oligosaccharides was glucose which was detected on the chromatogram by staining procedures. Apparently when one glucosidic bond of the substrate is hydrolyzed, both hydrolytic fragments dissociate from the active site of the enzyme. The oligosaccharide fragment from such a reaction can then form an enzyme–substrate complex, and a second glucosidic bond of the original substrate can be cleaved. This pattern of action of glucoamylase on the oligosaccharides is a multichain mechanism.

The glucoamylase most likely acts by a similar mechanism on high molecular weight glucans—*e.g.*, starch and glycogen. However when an α-1,6-linked glucosyl residue is encountered in such glucans, a reduction in the hydrolysis rate of this specific chain occurs. Since the enzyme is capable of hydrolyzing α-1,6 linkages at a slow rate, such a bond is eventually cleaved, and the substrate is ultimately converted quantitatively to glucose.

The data in Table I show that the glucoamylase acts with the highest velocity on compounds containing four or more contiguous glucose units. Compounds with fewer glucose residues are hydrolyzed at much reduced rates. Also compounds with linkages other than α-1,4 are hydrolyzed at much reduced rates. These observations can best be interpreted on the basis of a multiple subsite model (*1*). Thus, it is suggested that the glucoamylase active site is composed of four consecutive subsites with each subsite capable of binding a single glucosyl moiety. The residue at the non-reducing end of the substrate occupies the subsite at one end of the total site, and the hydrolytic events occur at this subsite. Compounds with four or more residues can occupy the subsites equally well. However, compounds with less than four glucose residues may occupy any consecutive set of subsites according to the number of glucosyl units in the compounds. With such substrates several types of enzyme substrate complexes will be formed, and those in which the hydrolytic subsite is vacant will not contribute to the hydrolytic rates. The monomeric residues of the substrates are bound at the various subsites by interactions between the hydroxyl groups of the residues and appropriate functional groups of the amino acid residues at the active site. The glycosidic oxygen of the terminal glucosyl unit of the substrate contributes to the binding of this residue. However, other glycosidic oxygens of the substrate probably do not participate in binding the substrate to the enzyme.

For the almond glucohydrolase Figure 5 shows that this enzyme like glucoamylase removes glucosyl units from the non-reducing ends of β-1,6-linked glucosyl oligosaccharides. Thus from gentiopentaose the initial products are glucose and gentiotetraose. Then the gentiotetraose is hydrolyzed to glucose and gentiotriose, and repetition of the process leads to the quantitative conversion of the pentasaccharide to glucose. Similar studies with other compounds showed that a variety of β-1,4- and β-1,6-glucosyl oligosaccharides were hydrolyzed by the same type of mechanism. Thus the almond glucohydrolase acts by a multichain mechanism on the oligosaccharides. In Table II β-methylglucoside is hydrolyzed at rates comparable with hydrolysis of cellobiose. Since the enzymes which hydrolyzed methyl glucoside β-1,4-glucosyl oligosaccharides and β-1,6-glucosyl oligosaccharides sedimentated at the same rate on density gradient centrifugation, it is possible that a single enzyme is responsible for the hydrolysis of all substrates. Thus it would seem that the well-known β-glucosidase of almond extract like glucoamylase is a glucohydrolase with broad substrate specificity.

The rate data on the hydrolysis of different oligosaccharides by the almond glucohydrolase are shown in Table II. These data can be interpreted to indicate that the active site for the enzyme is composed of two subsites. Apparently the active site is so located on the enzyme that additional glucose residues in the substrate interfere with the binding of the substrate to the active site.

The hydrolysis mechanism of the glucosidic bond by the two glucohydrolases is similar. The initial step of the hydrolytic process involves the protonation of the glycosidic oxygen. Spectral evidence with two different glucan hydrolases (*26, 27*) indicates that the proton comes from the imidazole group of histidine. Resulting from this protonation is a shift of electrons to the glucosidic oxygen and carbon 1 becomes a nucleophilic center stabilized by electrons from the hydroxyl groups or carboxyl groups of other amino acids present at the active site of the enzyme (*28, 29*). The electrons between carbon 1 of the glucosyl unit and the protonated glucosidic oxygen become an integral part of the aglycon moiety of the substrate, and this moiety dissociates from the enzyme. Such a mechanism is consistent with the results of ^{18}O studies with several types of glucan hydrolases (*23*). The glucosyl unit of the substrate remains at the active site as a transient carbonium ion in the form of a glucosyl–enzyme complex which is stabilized by the interactions with other functional groups at the active center. A hydroxyl group from the solvent adds to the carbonium ion with or without inversion of the configuration at carbon 1. It is known that glucoamylase reactions proceed by inversion of configuration, and β-D-glucose is liberated from α-glucan (*30*). Whether inversion occurs with the almond glucohydrolase is not yet known. Finally the imidazole group is protonated by hydrogen ions from the solvent. The hydrolytic process can now be repeated, provided a collision of enzyme and substrate has occurred with proper orientation of the functional groups of both reactants for formation of an active enzyme–carbohydrate complex.

Acknowledgment

We express appreciation to Austra Cepure for assistance with several aspects of this work and to Paula McDermott for assistance with the preparation of the gentiooligosaccharides.

Literature Cited

1. Thoma, J. A., Spradlin, J. E., Dygert, S., "The Enzymes," pp. 115, Vol. V, P. D. Boyer, Ed., Academic, New York, 1972.

2. Takagi, T., Toda, H., Isemura, T., "The Enzymes," pp. 1–235, Vol. V, P. D. Boyer, Ed., Academic, New York, 1972.
3. Whitaker, D. R., "The Enzymes," pp. 273, Vol. V, P. D. Boyer, Ed., Academic, New York, 1972.
4. Pazur, J. H., Ando, T., *J. Biol. Chem.* (1960) **235**, 297.
5. Kerr, R. W., Cleveland, F. C., Katzbeck, W. J., *J. Amer. Chem. Soc.* (1951) **73**, 3916.
6. Kool, E. R., "Starch: Chemistry and Technology," Vol. II, R. L. Whistler and E. F. Paschall, Eds., Academic, New York, 1967.
7. Tsujisaka, Y., Fukujoto, J., Yamamoto, T., *Nature* (1958) **181**, 770.
8. Pazur, J. H., Kleppe, K., *J. Biol. Chem.* (1962) **237**, 1002.
9. Pazur, J. H., "Starch: Chemistry and Technology," pp. 1–133, Vol. I, R. L. Whistler and E. F. Paschall, Eds., Academic, New York, 1965.
10. Pazur, J. H., Okada, S., *J. Biol. Chem.* (1966) **241**, 4146.
11. Robyt, J., French, D., *Arch. Biochem. Biophys.* (1963) **100**, 451.
12. Thomas, J. A., Brothers, C., Spradlin, J. E., *Biochemistry* (1970) **9**, 1768.
13. Pazur, J. H., Ando, T., *J. Biol. Chem.* (1959) **234**, 1966.
14. Pazur, J. H., Knull, H. R., Cepure, A., *Carbohyd. Res.* (1971) **20**, 83.
15. Martin, R. G., Ames, B. M., *J. Biol. Chem.* (1961) **236**, 1372.
16. Pazur, J. H., Kleppe, K., Anderson, J. S., *Biochim. Biophys. Acta* (1962) **65**, 369.
17. Pazur, J. H., *J. Biol. Chem.* (1953) **205**, 75.
18. Pazur, J. H., French, D., *J. Biol. Chem.* (1952) **196**, 265.
19. Pazur, J. H., *J. Amer. Chem. Soc.* (1955) **77**, 1015.
20. Miller, G. L., Dean, J., Blum, R., *Arch. Biochem. Biophys.* (1960) **91**, 21.
21. Anderson, C. G., Haworth, W. N., Raistrick, H., Stacey, M., *Biochem. J.* (1939) **33**, 272.
22. French, D., Knapp, D. W., Pazur, J. H., *J. Amer. Chem. Soc.* (1950) **72**, 5151.
23. Mayer, F. C., Larner, J., *J. Amer. Chem. Soc.* (1959) **81**, 188.
24. Lemieux, R. U., Bauer, H. F., *Anal. Chem.* (1954) **26**, 920.
25. Whelan, W. J., *Biochem. J.* (1971) **242**, 4813.
26. Yamamoto, T., *Bull. Agr. Chem. Soc. (Japan)* (1955) **19**, 121.
27. Larner, J., Gillespie, R. E., *J. Biol. Chem.* (1956) **223**, 709.
28. French, D., *Bakers Digest* (1957) **32**, 50.
29. Ono, S., Hiromi, K., Yoshikawa, Y., *Bull. Chem. Soc. Jap.* (1958) **31**, 957.
30. Weill, C. E., Burch, R. J., Van Dyk, J. W., *Cereal Chem.* (1954) **31**, 150.

RECEIVED January 7, 1972.

21

Biological Mechanisms Involved in the Formation of Deoxy Sugars: Enzymatic Hydrogen Mediation

A Possible Example for the Evolutionary Process of Enzyme Catalysis

OTHMAR GABRIEL[a]

Department of Biochemistry, Georgetown University Schools of Medicine and Dentistry, Washington, D.C. 20007

For the purpose of orientation, an overall view of deoxyhexose biosynthesis is provided with examples of some of the well documented enzymatic reactions. Detailed studies of the reaction mechanism of deoxy sugar biosynthesis are reported with special emphasis upon enzymatic hydrogen transfer reactions, substrate conformation, and coenzyme participation. Extension of these studies to enzymes involved in various other sugar transformations reveal close similarities in the mechanism of hydrogen mediation for several enzymes. As a result, some unifying principles for enzyme catalyzed hydrogen mediation are established. The objective of this approach is to provide eventually a rational basis for an understanding of the close relation existing between substrate conformation and stereochemical changes of protein conformation at the active site during enzyme catalyzed transformations.

Deoxy sugars can be defined as sugars with one or more alcoholic hydroxy groups in a monosaccharide replaced by hydrogen(s). For example, the naturally-occurring pentose, D-ribose, is also found as its correponding deoxypentose, 2-deoxy-D-ribose. We are aware of the biological significance caused by this substitution and the fundamental difference in the metabolism and role of ribonucleic acid and deoxy-

[a] Present address: Harvard Medical School, Massachusetts General Hospital, Boston, Mass. 02114.

ribonucleic acid, respectively. It is, therefore, not surprising to find that a similar disparity of metabolic fate and function also exists between hexoses and deoxyhexoses. While the major function of naturally-occurring hexoses is to provide the necessary energy for the living processes of the cell, deoxyhexoses are not involved in this process. Deoxy sugars have been recognized as immunological determinants when attached to macromolecules and are responsible for the specificity of immune reactions. The attachment of various deoxyhexoses to strategic points on macromolecules imply far reaching biological consequences for our understanding of problems ranging from our natural defense mechanisms to cell surface and membrane interactions. Consequently, attention has been paid to the elucidation of pathways leading to the formation of these important sugar derivatives.

For the purpose of orientation, the metabolic pathways of some of the naturally-occurring deoxy sugars is presented. This is followed by a detailed study of the enzyme reaction mechanism with emphasis on hydrogen mediation and its relation to substrate conformation and stability. The picture emerging from these experiments is compared with studies carried out by other investigators on several different enzymatic systems not related to deoxyhexose biosynthesis. We hope to show similarities in the mechanism of hydrogen mediation common to several enzymatic reactions, thereby indicating a more general applicability of the reported findings.

Metabolic Pathways

6-Deoxyhexoses. The interconversions leading to the biosynthesis of 6-deoxyhexoses occur at the level of sugar nucleotides. The first conversion of this type, GDP-D-mannose to GDP-L-fucose (GDP-6-deoxy-L-galactose), was reported by Ginsburg (*1*, *2*). Similarly, TDP-D-glucose is transformed into TDP-L-rhamnose (*3*, *4*, *5*) (TDP-6-deoxy-L-mannose). The strategy for the biosynthesis of 6-deoxyhexoses is indicated in Figure 1. In each instance the first stage is the formation of a nucleotide-linked 4-keto-6-deoxy intermediate. The synthesis of the 4-keto derivative is common to all systems of biosynthesis of deoxyhexoses described thus far. The reaction is catalyzed by enzymes referred to as oxidoreductases and is irreversible. Once a hexose becomes converted to the nucleotide-4-keto-derivative, it is no longer available for use by the main metabolic pathways or for energy production. It is converted to a nucleotide linked deoxyhexose, which serves as a precursor for incorporation into such macromolecules as glycoproteins. It is advantageous for the following discussion to emphasize the key role of the 4-keto-intermediate as common to all deoxyhexose biosynthetic pathways. The formation of a 4-

ulose is an organic synthesis step chosen by nature to accomplish the necessary transformations. Epimerizations at carbons 3 and 5 of the hexose moiety followed by stereospecific reduction at carbon-4 can lead to a variety of deoxy sugars. Many of these biosynthetic systems have been elucidated, and some of the established pathways are listed in Table I. All take advantage of a similar principle and follow the scheme outlined above. Also, the same sugar may be conjugated with different purine or pyrimidine bases.

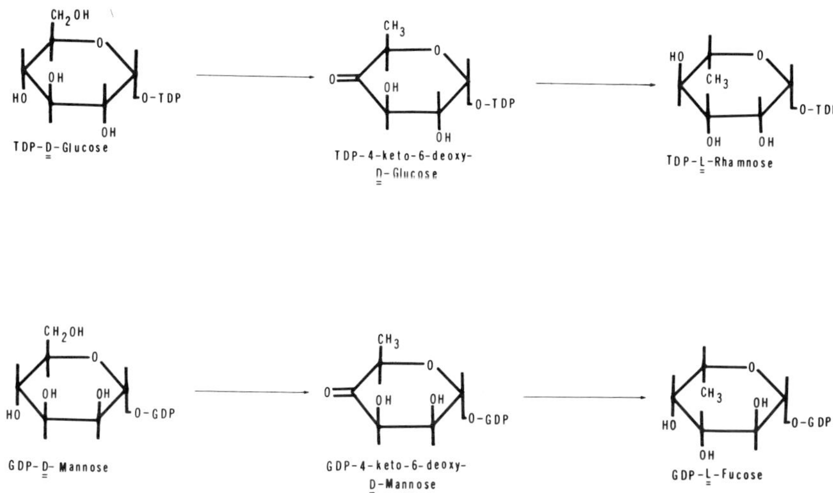

Figure 1. Biosynthesis of 6-deoxyhexoses

3,6-Dideoxyhexoses. Several bacterial antigenic determinants with the general structure of 3,6-dideoxyhexoses occur in the cell wall of Pasteurella and Salmonella strains. Most of the transformations reported so far occur as cytidine nucleotides (*see* Table I, References *15, 16, 17, 18, 19*). Here, again the first step is the transformation of the cytidine diphospho-linked glucose into its corresponding 4-keto derivative. By at least two distinct steps, requiring NADPH, reduction to several different 3,6-dideoxyhexoses have been reported. One 3,6-dideoxyhexose CDP-tyvelose (3,6-dideoxy-D-arabino hexose) is formed by a specific 2-epimerase from CDP-paratose (*24*).

Acetamidodeoxyhexoses. A further modification of the 4-keto-intermediate has been independently shown by Ashwell and by Strominger and associates (Table I, References *20, 21, 22, 23*). Transamination reactions with L-glutamate as the amino donor and pyridoxal phosphate as coenzyme led to formation of 3-amino 3,6-dideoxy- and 4-amino 4,6-dideoxyhexoses, respectively. Acetylation with acetyl coenzyme A yields the naturally-occurring *N*-acetyl amino sugar derivatives.

Detailed Reaction Mechanism of Deoxy Sugar Biosynthesis

TDPG-oxido Reductase. The enzyme initiating deoxyhexose biosynthesis in *E. coli*, TDPG-oxidoreductase, was selected as a model for a careful study of its reaction mechanism. As mentioned before, the molecular rearrangement leading to the formation of a 4-keto-intermediate is common to all 6-deoxyhexose pathways. It involves oxidation at carbon-4 and conversion of the primary alcoholic group at carbon-6 to a methyl group. The elucidation of the reaction mechanism was accomplished by using three different approaches: 1) a study of properties of model compounds to mimic the action of the enzyme, 2) preparation of selectively tritiated substrates to trace the fate of the tritium during the enzymatic reaction, and 3) isolation of the enzyme protein as a homogeneous component and studies with the pure protein.

Table I. Enzyme Systems for

Sugar Nucleotide Nucleoside	Oxidoreductase → Diphosphohexose	4-Keto Intermediate →
		6-Deoxyhexoses
Guanosine	D-Mannose	GDP-6-deoxy-D-lyxo-4-hexulose (4-keto-6-deoxy-D-mannose)
Uridine	D-Glucose	UDP-6-deoxy-D-xylo-4-hexulose (4-keto-6-deoxy-D-glucose)
Thymidine	D-Glucose	TDP-6-deoxy-D-xylo-4-hexulose (4-keto-6-deoxy-D-glucose)
Guanosine	D-Mannose	GDP-6-deoxy-D-lyxo-4-hexulose (4-keto-6-deoxy-mannose)
Guanosine	D-Mannose	GDP-6-deoxy-D-lyxo-4-hexulose (4-keto-6-deoxy-mannose)
Thymidine	D-Glucose	TDP-6-deoxy-D-xylo-4-hexulose (4-keto-6-deoxy-D-glucose)
		3, 6-Dideoxyhexoses
Guanosine	D-Mannose	GDP-6-deoxy-D-xylo-4-hexulose (4-keto-6-deoxy-glucose)
Cytidine	D-Glucose	CDP-6-deoxy-D-xylo-4-hexulose (4-keto-6-deoxy-glucose)
Cytidine	D-Glucose	CDP-6-deoxy-D-xylo-4-hexulose (4-keto-6-deoxy-glucose)
Cytidine	D-Glucose	CDP-6-deoxy-D-xylo-4-hexulose (4-keto-6-deoxy-glucose)
		6-Deoxy Amino Sugars
Thymidine	D-Glucose	TDP-6-deoxy-D-xylo-4-hexulose (4-keto-6-deoxy-glucose)
Thymidine	D-Glucose	TDP-6-deoxy-D-xylo-4-hexulose (4-keto-6-deoxy-glucose)

The major phases of our experiments are presented below. To facilitate the discussion, a summary of our findings is shown in Figure 2, indicating the mechanism involved.

TDP-D-glucose is initially attacked by enzyme-NAD$^+$ (enzyme protein containing one mole of firmly bound NAD$^+$) at carbon 4 to yield TDP-D-xylo-4-hexulose and accompanying formation of enzyme NADH. The 4-ulose rearranges by β-elimination of water between carbons 5 and 6 to form an unsaturated glucoseen. This unsaturated 5,6-glucoseen serves as hydrogen acceptor for enzyme-NADH to restore enzyme-NAD$^+$ and leads to the end product of the reaction TDP-6-deoxy-D-xylo-4-hexulose.

Returning to the first phase of our studies on the detailed reaction mechanism, we discuss some model experiments which were carried out

6-Deoxyhexose Biosynthesis

Epimerizations of 4-Keto Intermediate	NADPH ⟶ End Product
C3 and C5	GDP-6-deoxy-L-galactose (L-Fucose) (*1, 6*)
C3 and C5	UDP-6-deoxy-L-mannose (L-Rhamnose) (*7*)
C3 and C5	TDP-6-deoxy-L-mannose (L-Rhamnose) (*3, 4, 5, 8*)
None	GDP-6-deoxy-D-mannose (D-Rhamnose) (*9, 10, 11, 12, 13*)
None	GDP-6-deoxy-D-talose (*9, 12, 13*)
C3 and C5	TDP-6-deoxy-L-talose (*14*)
C3? and C5	GDP-3, 6-dideoxy-L-xylohexose (Colitose) *15, 16*)
None	CDP-3, 6-dideoxy-D-ribohexose (Paratose) (*17, 18, 19*)
None	CDP-3, 6-dideoxy-D-xylohexose (Abequose) (*17, 18, 19*)
C3? and C5	CDP-3, 6-dideoxy-L-arabinohexose (Ascarylose) (*17*)
None	TDP-3-acetamino-3, 6-dideoxy-D-galactose (*20*)
None	TDP-4-acetamido-4, 6-dideoxy-D-glucose (*21, 22, 23*)

to get insight into the stability of the 4-ulose derivative, a postulated intermediate of the enzymatic reaction.

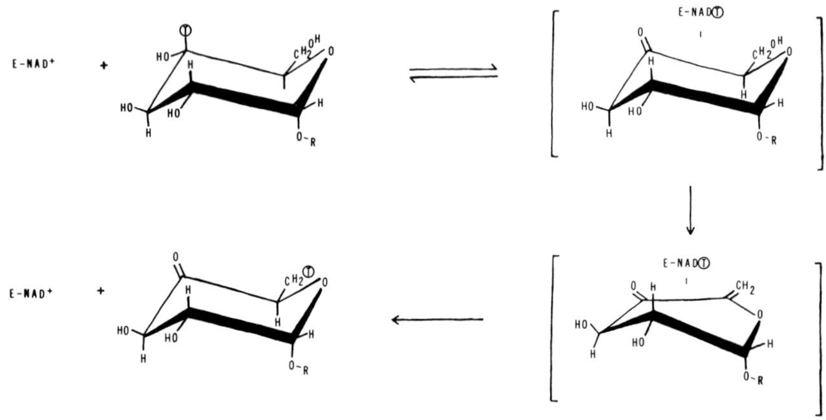

Figure 2. Reaction mechanism for TDPG oxidoreductase

It is well established that oxygen in the presence of platinum (Adams catalyst) can achieve specific oxidation of secondary alcohols by a preferential attack upon hydrogen in an equatorial position (*25*). Catalytic oxidation of methyl α- and β-D-galactopyranoside (*26*), followed by catalytic reduction with hydrogen, led to the formation of methyl α- and β-6-deoxy-D-galactopyranoside (D-fuco-pyranoside) in 15% and 35% yield, respectively. This oxidation–reduction sequence with selective oxidation at carbon 4 as the initial step is structurally closely related to the above described pathway for TDPG-oxidoreductase.

The isolation of D-fucose (6-deoxy-D-galactose) instead of galactose as a major product of the catalytic oxidation–reduction was considered to be the consequence of β-elimination of the initially formed 4-ulose-derivative. The removal of water resulted in the formation of a 4-keto-5,6-glucoseen which upon catalytic reduction yielded 6-deoxy-D-glactose. Thus, it seems that it is the intrinsic property of methyl-D-xylo-4-hexuloside to undergo molecular rearrangement spontaneously.

Considering this property of 4-uloses to be applicable to the postulated intermediate TDP-D-xylo-4-hexulose formed in the enzymatic reaction, we decided to verify this reaction mechanism by the use of specifically labeled substrate. For this purpose TDP-D-glucose-*4T* was prepared in the following way.

First, *4T*-D-glucose was synthesized starting from methyl-α-D-galactopyranoside (*27*). Taking advantage of the axial hydroxy group of galactose at carbon 4, reaction with benzoyl chloride resulted in formation of methyl-2,3,6-tri-*O*-benzoyl-α-D-galactopyranoside. Oxidation with dimethyl sulfoxide led to formation of methyl 2,3,6-tri-*O*-benzoyl-α-D-xylohexo-pyranoside-4-ulose. This step is followed by reduction with sodium borotritide and gave the expected 4-tritiated-D-gluco- and D-galacto-derivatives. Removal of the protective groups yields the selectively tritiated sugars, D-glucose-*4T* and D-galactose-*4T*. The radiopurity of D-glucose-*4T* was established by chemical degradation methods. The proper degradation products were isolated, and radioactivities were determined. Evaluation of the results indicated that more than 99% of the tritium activity was associated with carbon 4.

The formation of the sugar nucleotide TDPG-*4T* from D-glucose-*4T* by hexokinase resulted in isolation of D-glucose-6-phosphate-*4T*. Conversion to glucose-1-phosphate-*4T* by phosphoglucomutase and reaction with TDPG-pyrophosphorylase to yield TDP-D-glucose-*4T* was carried out in the second step. In this way the desired end product TDPG-*4T*, labeled uniquely at the 4 position of the hexose was obtained.

When the preparation of TDPG-*4T* was incubated with partially purified TDP-glucose-oxido reductase, the substrate was converted quantitatively to TDP-4-keto-6-deoxyglucose-*6T* (*28*). No exchange of tritium with the medium occurred during the reaction. The tritium originally present on carbon 4 of the glucose was transferred to carbon 6 of the 4-keto-derivative. These experimental findings are consistent with formation of a nucleotide-bound 4-keto-5,6-glucoseen intermediate as suggested by our model experiments (*see* Figure 2). An identical intramolecular hydrogen transfer was observed independently by Melo, Elliott, and Glaser (*29*) when TDPG-*4D* was used as the substrate, and similar results were obtained in the conversion of cytidine diphosphoglucose oxidoreductase in *Salmonella typhimurium*. Furthermore, along with the formation of the 4-keto-derivative, these authors (*29*) found incorporation of one atom of deuterium from the medium at carbon 5.

With the knowledge of an intramolecular hydrogen transfer from carbon 4 to carbon 6, further studies on the enzyme mechanism required the availability of a homogeneous enzyme preparation. For this purpose a large scale culture of *E. coli* was prepared. By conventional methods of protein purification a 1200 fold increase in purity was attained, and the enzyme was isolated in crystallized form (*30*). The preparation behaved like a single component in the ultracentrifuge and on disc gel electrophoresis. Determination of molecular weight by sedimentation equilibrium gave a value of 88,000. Fluorometric measurements established that one mole of NAD^+ was bound per mole of enzyme. Studies on this purified enzyme preparation were carried out using TDP-6-deoxy-D-glucose, a close structural analog of the natural substrate TDP-glucose (*31*). In contrast to the natural substrate, TDP-glucose, the substrate analog TDP-6-deoxy-D-glucose cannot undergo β-elimination of water and rearrangement. Consequently, there is no acceptor for enzyme-NADH with resultant accumulation of enzyme-NADH.

The reaction sequence catalyzed by TDPG oxido reductase is indicated below:

$$\text{TDP-6-deoxy-D-glucose} + \text{enzyme-NAD}^+ \rightleftharpoons$$

$$\text{TDP-4-keto-6-deoxy-glucose} + \text{enzyme-NADH} + \text{H}^+$$

The stability of enzyme-NADH was evidenced by its isolation after Sephadex chromatography and led to the direct spectrophotometric demonstration of enzyme-NADH. The formation of enzyme-NADH in this reaction was used to examine the stereospecificity of hydrogen transfer for acceptance from carbon 4 of the sugar nucleotide by enzyme-NAD^+ and donation of the same hydrogen back to carbon 6. For both steps β-stereospecificity of hydrogen transfer to the nicotinamide moiety of the pyridine nucleotide was established.

In a similar study on this enzyme, Glaser and co-workers (*32*) used the same substrate analog, TDP-6-deoxy-glucose, and provided indirect evidence that during the reaction yielding enzyme-NADH, the accumulating TDP-4-keto-6-deoxy-glucose is released very slowly from the enzyme. This is consistent with the well established fact that reaction intermediates are not released; hence, the final step, the donation of hydrogen from enzyme-NADH to restore enzyme-NAD^+, is essential to release products from the enzyme.

Another important property of TDPG-oxidoreductase is its reversible dissociation with enzyme bound-NAD^+ (*30, 31*). Treatment of the enzyme with *p*-chloromercuriphenylsulfonate results in the release of enzyme bound NAD^+ and formation of two protein subunits. Incubation of these inactive subunits with cysteine and excess NAD^+ leads to reassociation of the subunits and complete return of enzymatic activity. Conditions for the isolation of the subunits of molecular weight of about 40,000 were described by Zarkowsky *et al.* (*32*).

Subsequent Reactions Catalyzing 6-Deoxyhexose Formation: 3,5-Epimerase and Reductase. After the formation of the 4-keto intermediate, a step common to all deoxyhexose formations, at least two and possibly three additional enzymes are necessary for the biosynthesis of the end product. This sequence of transformations is illustrated in Figure 3. The first step is the conversion of the 4-keto-6-deoxy-hexose intermediate described above. The intermediates in brackets are postulated and are assumed to be bound to the enzyme. An enzyme (or enzymes), referred to below as 3,5-isomerase, catalyzes epimerizations at carbons 3 and 5, probably *via* the enediol form. The epimerizations are followed by a stereospecific reduction by an enzyme referred to as reductase. In this reaction a stoichiometric amount of NADPH is required to reduce the keto-group at carbon 4, resulting in release of enzyme bound intermediate as the final endproduct of the pathway. As indicated in Figure 3, two possible 4-epimers can be obtained; in *E. coli* B the only 6-deoxyhexose formed is 6-deoxy-L-mannose (L-rhamnose) (*3, 4, 5*). In *E. coli* 045, 6-deoxy-L-talose was reported to be replacing L-rhamnose in the O-antigen of the lipo-polysaccharide (*33*). We verify the pathway for 6-deoxy-L-talose biosynthesis (*14*) as an example of enzymatic conversion of the 4-keto-intermediate to the endproduct.

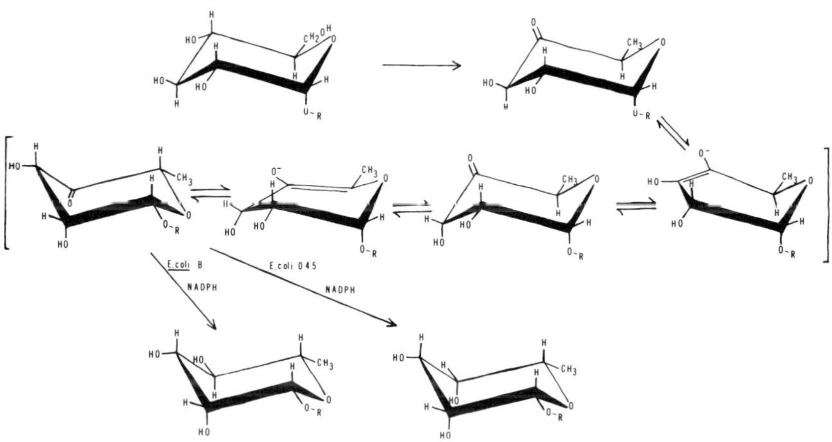

Figure 3. Pathway for 6-deoxyhexose synthesis

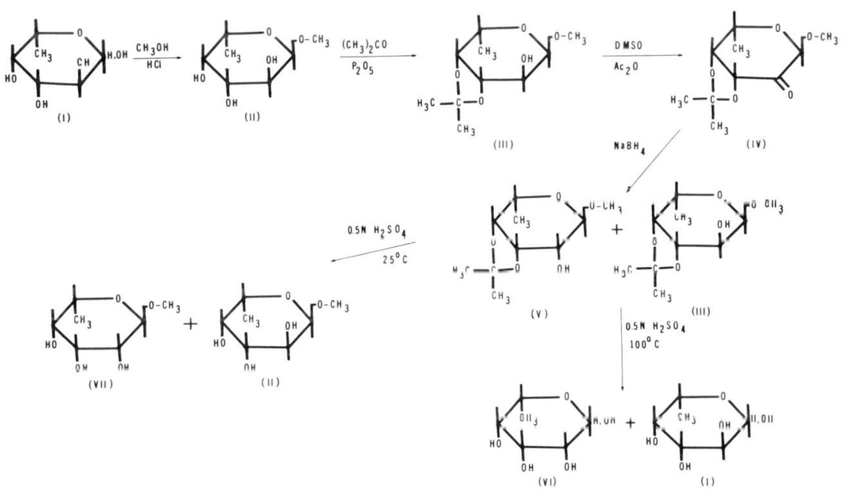

Figure 4. Synthesis of L-talomethylose (Overend)

To provide an authentic sample of 6-deoxy-L-talose for comparison in these studies, a synthetic procedure, following essentially that of Overend (*34*) was used with 6-deoxy-L-galactose as the starting material (Figure 4). Conversion to the methylglycoside was followed by preparation of the isopropylidene derivative. All carbons of the hexose except for carbon 2 were protected. Oxidation with dimethyl sulfoxide in acetic anhydride resulted in formation of the 2-keto derivative. Reduction with sodium borohydride yielded a mixture of 6-deoxy-L-talo- and 6-deoxy-L-galactose derivatives. Removal of protective groups leads to 6-deoxy-L-talose. With this sugar as a reference standard, enzymatic conversion with crude extracts of *E. coli* 045 were carried out. Incubation of TDPG-U^{14}C with the extracts in the presence of NADPH led to the quantitative conversion to a thymidine diphospho-6-deoxyhexose as the sole product. Upon release of the sugar from the nucleotide by mild acid hydrolysis, the sugar was identified as 6-deoxytalose. The L-configuration of the enzymatically formed sugar was established by a radioisotope dilution technique. All the analytical data obtained on the sugar nucleotide were consistent with the structure of TDP-6-deoxy-L-talose (Figure 5).

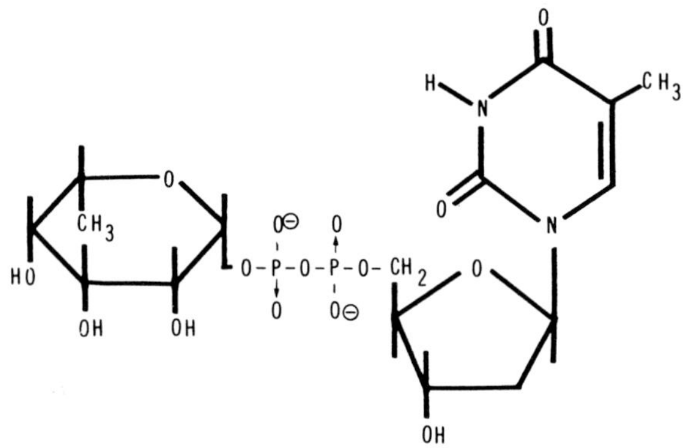

Figure 5. TDP-L-Talomethylose

The instability of TDP-6-deoxy-L-talose should be mentioned. When using the routine procedure for the isolation of TDP-rhamnose, a consistent loss of the 6-deoxy-L-talose nucleotide was observed. A systematic study revealed that in contrast to the rhamnose derivative, the 6-deoxy-L-talose nucleotide was extremely labile to alkali. Prolonged exposure to pH 8.0 at 37°C resulted in significant degradation. The degradation products are TMP and 6-deoxy-L-talose monophosphate. Careful degradation studies provided experimental evidence for its identification as cyclic-1,2-monophospho-6-deoxy-L-talose. The extreme alkali lability of TDP-6-deoxy-L-talose in contrast to its 4-epimer TDP-6-deoxy-L-mannose can be rationalized best by considering the conformation of the two sugars involved. In Figure 6 the 1-C conformation at the right is predominant. As can be seen, the axial hydroxy function at carbons 2 and 4 are at the same side of the ring, thereby facilitating stabilization by formation of the 1,2-cyclic monophosphate ester. The corresponding rhamnose conformation does not have this interaction—*i.e.*, the 4-epimer has an equatorial hydroxyl in the 1-C conformation and does not readily result in cyclic ester formation.

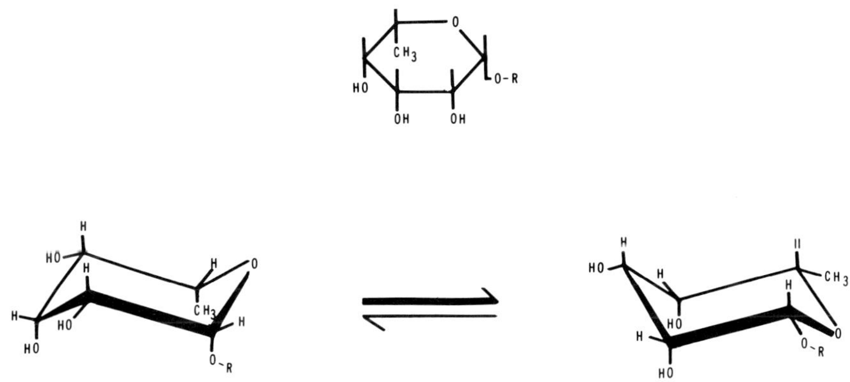

Figure 6. 6-Deoxy-L-talopyranoside

For more detailed studies of the overall pathway, a separation of the individual enzymes was carried out. For this purpose a single general procedure, effective in measuring 3,5-epimerase activities, was developed; incubation of TDP-4-keto-6-deoxyglucose-*3T* (*35, 36*) in the presence of the enzyme resulted in tritium exchange with the medium. After completion of the incubation, further tritium exchange with the medium was prevented by the addition of sodium borohydride (Figure 7).

Only in the presence of 3,5-epimerase could a significant exchange of tritium be observed. One line of evidence consistent with this interpretation came from the use of a mutant strain *E. coli* Y-10, incapable of producing 6-deoxyhexose. When extracts of *E. coli* Y-10 were incubated with TDP-4-keto-6-deoxyglucose-*3T*, no exchange with the medium was observed.

A large scale preparation of *E. coli* 045 was subjected to enzyme purification using the assay for 3,5-epimerase. Protamin sulfate precipitation, ammonium sulfate fractionation was followed by DEAE-chromatography. The fraction containing enzymatic activity, as measured by tritium exchange, was eluted from the DEAE column early. This fraction was incapable of producing any net synthesis of TDP-6-deoxy-L-

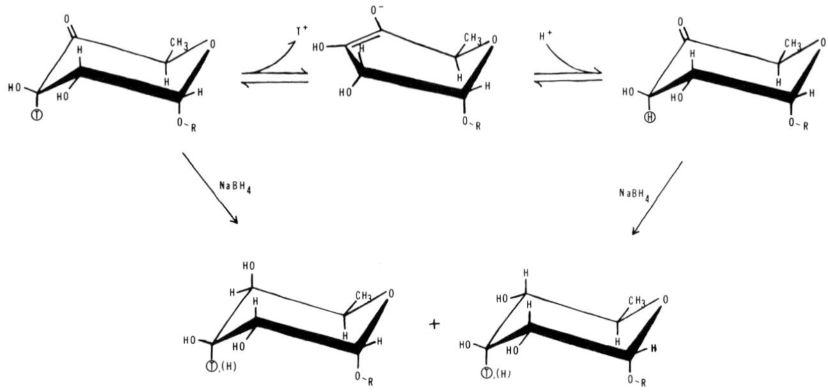

Figure 7. 3-Epimerase assay reactions

talose. However, combination of this fraction I with fraction II, eluted at a much higher ionic strength from the same column, resulted in synthesis of deoxyhexose. The progress of synthesis was followed spectrophotometrically by measuring disappearance of NADPH.

Conversely, fraction II alone did not catalyze net synthesis of 6-deoxyhexose. Thus, fraction I and II were not cross contaminated since no 6-deoxyhexose synthesis was observed in the absence of either. These observations are consistent with the following reaction sequence:

TDP-6-deoxy-D-xylo-4-hexulose $\xrightarrow[\text{(3,5-epimerase)}]{\text{fraction I}}$

[TDP-6-deoxy-L-lyxo-4-hexulose] $\xrightarrow[\text{(4-reductase)}]{\text{NADPH} \;\; \text{fraction II}}$

TDP-6-deoxy-L-talose

By varying the proportions of enzyme I to enzyme II it was possible to arrange reaction conditions for making either one of the enzymes rate limiting. Further studies on more purified preparations are necessary to establish the interrelation that must exist between these enzymes. Without enzyme II none of the postulated intermediates (Figure 3, structures in brackets) are released from the enzyme. Analogous results for the biosynthesis of TDP-6-deoxy-L-mannose (L-rhamnose) in *Pseudomonas aeroginosa* were previously reported by Melo and Glaser (37). In their system TDP-6-deoxy-D-xylo-4-ulose is epimerized at carbons 3,5 to the same hypothetical intermediate, TDP-6-deoxy-L-lyxo-4-hexulose, mentioned earlier. In contrast to the above system stereospecific reduction produces the 4-epimer of 6-deoxy-L-talose, TDP-6-deoxy-L-mannose (L-rhamnose).

From the two examples studied thus far, it seems that 3,5-epimerization of the 4-keto-intermediate proceeds by stereospecific reduction at carbon-4 to result in formation of a specific deoxysugar. The existence of the 4-keto intermediate is a rational, necessary prerequisite to facilitate these epimerizations *via* the corresponding enediols.

Formation of the identical sugars of the D-series, 6-deoxymannose (rhamnose) and 6-deoxytalose, seems to proceed by a different pathway. According to Winkler and Markowitz (*13*), GDP-6-deoxy-D-mannose is first converted to GDP-6-deoxy-D-lyxo-4-hexulose. This 4-keto intermediate is the direct precursor for the unspecific enzymatic reduction leading to GDP-6-deoxy-D-mannose and GDP-6-deoxy-D-talose. For a pyridine-nucleotide requiring enzyme, the transformation seems to be unusual because of its lack of stereospecificity. However, closer examination and evaluation of properties of the different 4-keto-intermediate reductases must await availability of more highly purified enzyme preparations.

Enzymes Not Directly Related to 6-Deoxyhexose Biosynthesis with Similar Reaction Mechanisms

UDP-Galactose-4-Epimerase. Our attention was directed to other enzymes—*e.g.*, UDP-galactose-4-epimerase—apparently unrelated to 6-deoxyhexose biosynthesis. One reason for our interest is that the postulated reaction mechanism also involves a 4-ulose intermediate, and the enzyme has been studied extensively (*38*). The general type of reaction catalyzed by this enzyme, shown in Figure 8, is the reversible epimerization of the center of asymmetry at carbon 4. Several modifications of the natural substrate UDP-glucose are possible without loss of ability to epimerize at carbon 4. For comparative studies with TDPG-oxidoreductase, an enzyme with a strict specificity requirement for the nucleotide moiety, we examined the ability of UDP-galactose-4-epimerase to accept TDP-glucose as a substrate. The following lines of evidence indicated that UDP galactose 4 epimerase from *E. coli* can catalyze the conversion of UDP-galactose as well as for the TDP-galactose.

(1) At different stages of enzyme purification the ratios of activities for substrates, UDP-galactose and TDP-galactose, remain unchanged.

(2) Upon kinetic analysis UDP-galactose is a competitive inhibitor for TDP-galactose epimerization.

(3) Mutant strains deficient in UDP-galactose-4-epimerase activity are essentially without activity for TDP-galactose.

(4) Cultures induced for UDP-galactose-4-epimerase display the same ratio of activity increase when TDP-galactose is used as a substrate.

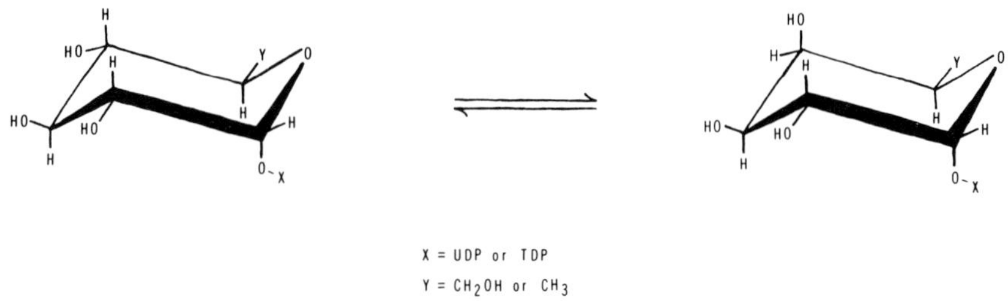

Figure 8. Reactions catalyzed by E. coli *UDP-galactose 4-epimerase*

Table II. Summary of Isotope Effect Data for TDPG-Oxidoreductase and UDP-Galactose-4-Epimerase (*40*)[a]

	Isotope Effect K_H/K_T	
Substrate	*TDPG-oxidoreductase*	*UDP-galactose-4-epimerase*
TDP-glucose-*3T*	0.9	1.0
TDP-glucose-*4T*	2.4	2.6

[a] The reactions were carried out under standard conditions using the specifically tritiated substrate mixed with uniformly ^{14}C labeled TDP-glucose. Samples were taken at different time intervals and the reaction products were isolated. ^{14}C activity was used to measure the reaction coordinate. The reactions were permitted to proceed until isotopic equilibrium was achieved.

The epimerization catalyzed by UDP-galactose-4-epimerase could occur *via* two different mechanisms, both involving participation of enzyme bound NAD^+. In the first mechanism to be considered, oxidation would occur at carbon 4 of the hexose leading to the corresponding 4-ulose derivative. This intermediate will then be reduced to the 4-epimer of the original substrate. A second possible mechanism would involve oxidation at carbon 3 followed by a keto–enediol transformation between carbons 3 and 4, causing epimerization at carbon 4. The final step of this sequence would be stereospecific reduction of the keto group at carbon 3. In this context it should be emphasized that lack of exchange of substrate hydrogen with the medium should not be considered as evidence for a hydride mechanism. Consequently, a reaction mechanism involving proton transfer cannot be ruled out (*39*). Having established that TDP-glucose can serve as a substrate for UDP-galactose-4-epimerase (*40*), we decided to distinguish between the above reaction mechanisms. For this purpose we used specifically labeled TDP-glucose-*3T* and TDP-glucose-*4T* as substrates to determine the rate limiting steps for UDP-galactose-4-epimerase and TDPG-oxidoreductase. A summary of these data is presented in Table II. In both enzymatic reactions a small but significant isotope effect is apparent when the 4-tritiated substrate is used. In contrast, TDP-glucose-*3T* does not seem to be rate limiting for these reactions. These data are interpreted to indicate that the initial attack by the enzyme involves removal of hydrogen at carbon 4 for both reactions; however, the numerical value for a tritium isotope effect is unusually small. It is surprising to find this low isotope effect for both reactions to be identical within experimental error.

Several independent lines of experimental evidence have been given (*41, 42*) to substantiate the proposed reaction mechanism for UDP-galactose-4-epimerase, shown in Figure 9. Recently, Kalckar and co-workers (*42*) used TDP-glucose-*4T* as a substrate. The initial attack occurs at carbon 4 and results in conversion to the enzyme-bound-4-ulose intermediate, accompanied by formation of enzyme-NADT. The 4-ulose intermediate serves as hydrogen acceptor to restore enzyme-NAD^+ and to release product.

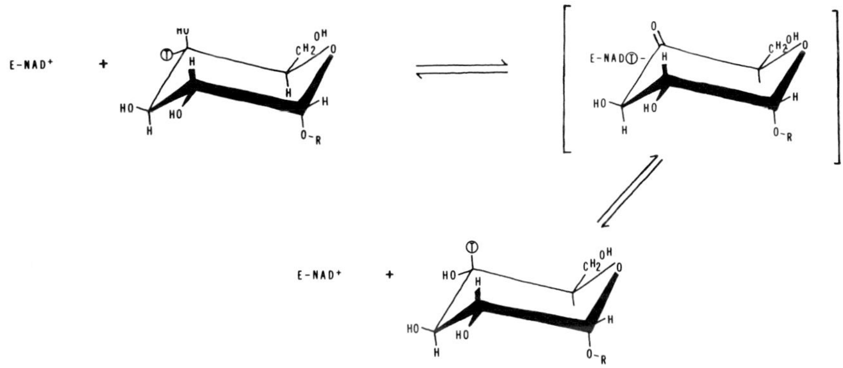

Figure 9. Proposed reaction mechanism for UDP-galactose-4-epimerase

When this mechanism is compared with the one for TDPG-oxido reductase (Figure 2), several similarities are apparent. The axial hydrogen at carbon 4 is the initial point of attack which leads to the enzyme bound 4-ulose intermediate. Release of product from the enzyme occurs only after reoxidation of enzyme-NADH to enzyme-NAD^+. A summary of available data to further substantiate our comparative study of TDPG-oxido reductase and UDP-galactose-4-epimerase is given in Table III.

UDP-Glucuronic Acid Carboxy-Lyase. After establishing some of the close similarities existing between the enzymes discussed above, we decided to examine the literature for other enzymes that may belong to this group. A well-documented example described by Feingold and co-workers (*44*) is the reaction sequence for UDP-glucuronic acid carboxy-lyase shown in Figure 10. As in previous examples, the initial hydrogen abstraction is the removal of the axial hydrogen at carbon 4 and accompanying formation of enzyme-NADH. Stabilization of the enzyme bound 4-ulose intermediate occurs by loss of CO_2 at carbon 5 which is followed by acceptance of one proton in this position. Inversion

Table III. Comparison of TDPG-Oxidoreductase (*28, 30, 31, 32*) and UDP-Galactose-4-Epimerase (*38, 40, 41, 42, 43*) from *E. Coli B*

I. SIMILARITIES

(a) Molecular weight (about 80,000)
(b) 1 mole of active enzyme contains 2 subunits
(c) 1 mole of active enzyme contains 1 mole of NAD^+ firmly bound
(d) Initial reaction step involves substrate oxidation at carbon 4 by removal of axial hydrogen and conversion to 4-ulose intermediate
(e) Reduction of enzyme-NAD^+ to enzyme-NADH
(f) Oxidoreductase mechanism—intramolecular hybride shift
(g) Enzyme-NADH does not release intermediate(s)
(h) Reoxidation of enzyme-NADH to enzyme-NAD^+ by hydrogen transfer to enzyme bound intermediate is last step resulting in product release
(i) Identical isotope effects for 4-tritiated substrates
(j) Similar or identical substrates

II. DIFFERENCES

TDPG-oxidoreductase	*UDP-galactose-4-epimerase*
(a) 4 → 6 hydride shift	4 → 4 hydride shift
(b) Rearrangement by β-elimination	No rearrangement

of configuration occurs at carbon 5 at this enzyme catalyzed proton addition (*44*). The final step is again the donation of hydrogen to carbon 4 and reoxidation of enzyme-NADH to enzyme-NAD^+. This is accomplished by release of UDP-xylose from the enzyme. As expected, rate studies with UDP-glucuronic acid, labeled specifically with tritium at carbon 3, carbon 4 or carbon 5, established an isotope effect of 2.4 only for the 4-tritiated substrate. This value agrees well with the isotope effects measured for TDPG-oxidoreductase and UDP-galactose-4-epimerase.

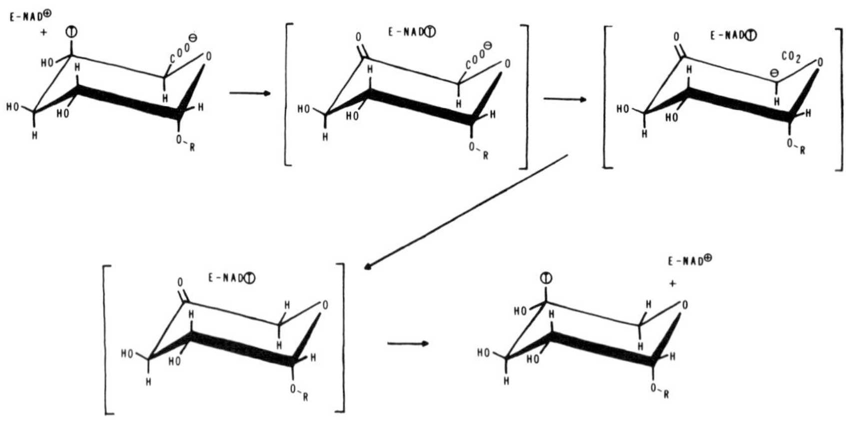

Figure 10. Reaction mechanism for UDP-GA carboxy-lyase (D. S. Feingold et al.)

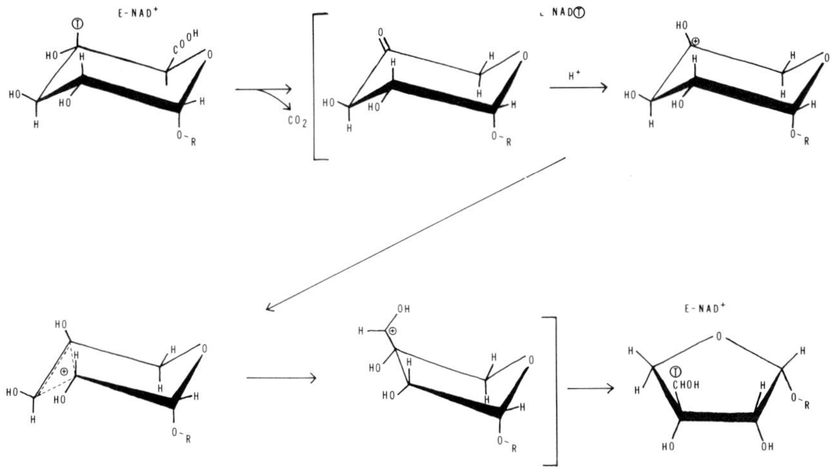

Figure 11. UDP apiose synthetase (H. Grisebach et al.)

UDP-D-Apiose Biosynthesis. Recently, another example for the same type of reaction, UDP-D-apiose synthetase (*45*), the biosynthesis of a branched sugar (3-*C*-hydroxy-methyl-D-erythro-furanose) came to our attention. The reaction mechanism is shown in Figure 11. Reaction of enzyme-NAD$^+$ and UDP-glucuronic acid leads to decarboxylation of carbon 6 and formation of a 4-ulose intermediate and enzyme-NADH. This is followed by a pinacol-type of rearrangement with the expulsion of carbon 3 of D-glucuronic acid and ring contraction, establishing a new linkage between carbons 2 and 4. Thus, C-3 of the glucuronic acid becomes the C-3′ branch carbon of apiose. As in the previously mentioned examples, the rearrangements occur as enzyme bound intermediates, the final step being donation of hydrogen to C-3′, reoxidation to enzyme-NAD$^+$, and release of UDP-D-apiose. When a UDP-glucuronic acid-*4T* was used in this reaction (*45*), a hydride shift from carbon 4 of the glucuronic acid to the branch carbon C 3′ of apiose was found experimentally consistent with the above reaction mechanism.

TDP-streptose Biosynthesis and 5-Dehydroquinic Acid Synthetase? The question mark in the above title should indicate that the next two examples are only a tempting, though hopefully constructive, speculation to demonstrate that TDP-streptose biosynthesis as well as 5-dehydroquinic synthetase are possible candidates for this group of enzymes with identical initiation of enzyme catalysis by hydrogen transfer to form enzyme-NADH.

For the example of streptose synthetase, we point out that no one has yet demonstrated a system capable of carrying out the biosynthesis of streptose. However, as early as 1961 Baddiley and co-workers proposed TDP-4-keto-6-deoxy-D-glucose as a precursor of streptose (*46*). In view of the recent work of Grisebach *et al.* (*45*) on the biosynthesis of apiose, it appears likely that an analogous reaction sequence such as the one shown in Figure 12 may be operative for TDP-streptose synthetase. All the features of this system are similar and consistent with the enzymes discussed above.

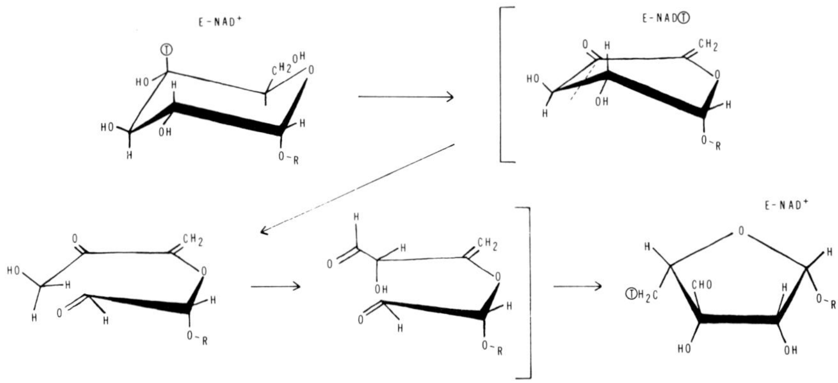

Figure 12. TDP-streptose biosynthesis (J. Baddiley et al.)

The last example we would like to present concerns 5-dehydroquinic acid synthetase. This complex system has been studied by Sprinson and co-workers (*47*). Because of the close similarities with the above enzymes we propose the reaction mechanism shown in Figure 13. The substrate, 3-deoxy-D-arabino-heptulosonic acid-7-phosphate in its chair conformation, is oxidized at carbon 5 by removing the axial hydrogen in a fashion identical to that described for all the other enzyme systems. The resulting 5-ulose derivative rearranges by eliminating phosphate. (The close resemblance to TDPG-oxidoreductase in which water is eliminated instead of phosphate should be noted). Final rearrangement leading to a homocyclic ring system and donation of hydrogen results in formation of 5-dehydroquinic acid.

Common Mechanistic Properties in the Enzymatic Catalysis for Enzyme-NAD⁺ Mediated Hydrogen Transfer Reactions—A Model System for the Study of the Evolutionary Process of Enzyme Catalysis?

A summary of some of the common features for the first three reactions described in this preparation, TDPG-oxidoreductase, UDP-galactose-4-epimerase, and UDP-glucuronic acid carboxylase, are shown in Figure 14. Our initial discussion is restricted to these three examples since we have more information about them.

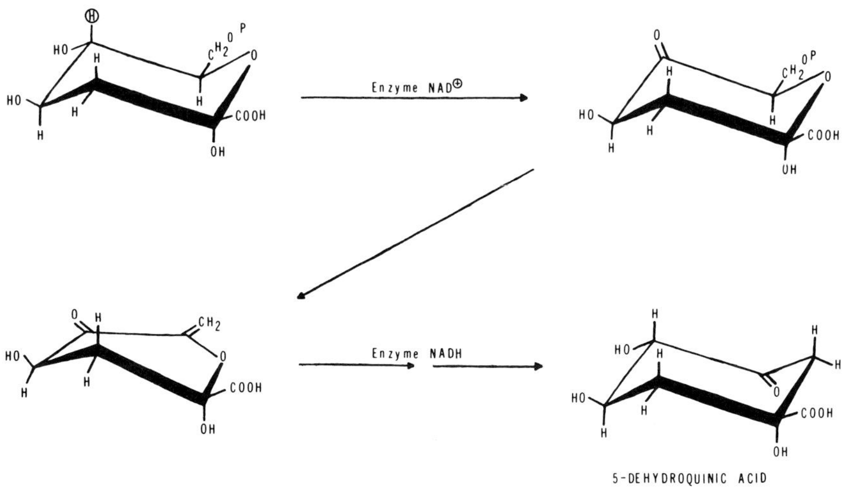

Figure 13. Proposed reaction mechanism for 5-dehydroquinate synthetase (D. B. Sprinson et al.)

In these systems the initial step of enzyme catalysis is identical. This includes some specific stereochemical requirements, which are abstraction of an axial hydrogen at carbon 4 and conversion to the corresponding 4-ulose derivative. In this process enzyme-NAD⁺ is converted to enzyme-NADH. There is increasing evidence that transformation of enzyme-NAD⁺ to enzyme-NADH must cause a conformational change of the enzyme protein since no intermediates are released from the enzyme once NADH is formed. The stereospecificity with respect to the nicotinamide moiety of the coenzyme has been examined for TDPG-oxidoreductase and for UDP-galactose-4-epimerase and was shown to have B specificity for the acceptance as well as for the donation of hydrogen. The most striking similarity for all three enzymes is the identical, small isotope effect of 2.5 observed when 4-tritiated substrates are used. With-

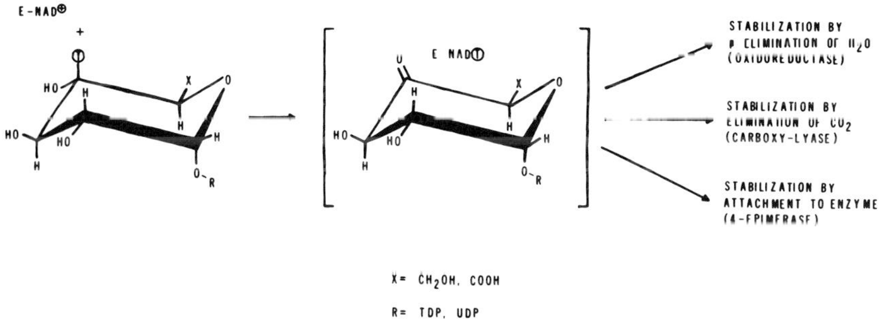

Figure 14. Common features of the TDPG-oxidoreductase, UDP-galactose-4-epimerase, and UDP-glucuronic acid carboxylase

out further data it is not possible to interpret this effect in mechanistic terms except to say that only hydrogen at carbon 4 is rate limiting. More important, however, is the similarity of the numerical value of this isotope effect for all three enzymes.

We interpret the effect to indicate that the three enzymes must have close resemblance of the transition state(s) for the initial part of the reaction sequence, leading to the 4-ulose intermediate. When we talk about the 4-ulose intermediate, we do not intend to imply that in each instance actual formation of a 4-keto group occurs. The intermediate to initiate subsequent rearrangements may well be the remaining carbonium ion after hydride abstraction. There is now no experimental evidence to distinguish these possibilities.

From a mechanistic viewpoint this group of enzymes is best described as oxidoreductases where intramolecular hydride transfer is mediated by enzyme bound NAD^+ (Table IV). Different enzymatic end products for these enzymes are a consequence of different ways of stabilization of the 4-ulose intermediates or the 4 carbonium ion. The stabilization of the 4-ulose derivative can occur in several ways as shown by the various examples mentioned. During the molecular rearrangement, initiated by the formation of the 4-ulose, the reaction intermediates are held bound to the enzyme until at the final step enzyme-NADH donates the hydrogen to the last intermediate and enzyme-NAD^+ releases the end product from the enzyme.

Table IV. Hydride Shift in Enzyme NAD^+ Mediated Reactions

	Hydride Shift		
Enzyme	*From Carbon*	*To Carbon*	*Reference*
UDP-galactose-4-epimerase	4	4	*40, 48*
UDP-glucuronic acid carboxy lyase	4	4	*44*
UDP-D-apiose synthetase	4	3′	*45*
TDPG-oxidoreductase	4	6	*28, 29*
TDP-streptose synthetase [a]	4	5	*46*
5-Dehydroquinic acid synthetase [a]	5	5	*47*

[a] Predicted, but not experimentally verified.

One important aspect of the catalytic principle of these oxidoreductases—*i.e.*, enzyme-NAD^+ as the vehicle for hydrogen transfer—is the obligatory presence of enzyme-NAD^+ for catalytic action; only enzyme-NAD^+ can serve as hydrogen acceptor to initiate catalysis. In contrast, enzyme-NADH is inactive and can not accept hydrogen from the substrate. Consequently, the ratio of enzyme-NAD^+ to enzyme-NADH is responsible for the net catalytic activity of a particular preparation. Kalckar and co-workers (*38*) were the first to recognize that preincubation of UDP-galactose-4-epimerase with UMP and a free sugar such as D-galactose leads to formation of enzyme-NADH and concomitant decrease in enzymatic activity. A different way for the formation of enzyme NADH—*i.e.*, with sodium borohydride—was reported by Nelsestuen and Kirkwood (*49*). The same authors reoxidized enzyme-NADH by adding TDP-4-keto-6-deoxy-D-glucose. An extensive report of this work was published during the preparation of this paper (*50*). The formation of enzyme-NADH by the addition of UMP and free sugar remains controversial (*51*). However, recently Seyama and Kalckar (*52*) reported that incubation of D-glucose-*1T* in presence of UMP resulted in tritium transfer to enzyme-NAD^+. This hydrogen transfer was accompanied by the accumulation of enzyme-NADT and oxidation of D-galactose-*1T* to D-galactonic acid, free of tritium label. The same investigators (*53*) showed that tritium of UDP-galactose-*1T* did not lead to enzyme-NADT formation.

To resolve some of the conflicting reports in the literature concerning the formation of enzyme-NADH, we started to investigate this problem for which a detailed report will be presented elsewhere (*42*). One surprising result of this effort was the recognition of adenosine diphosphoribose-(5), a degradation product of NAD^+, as an efficient donor of hydrogen for the formation of enzyme-NADH. In other words, adenosine diphosphoribose-(5) is a specific inhibitor for UDP-galactose-4-epimerase. An interesting consequence of this observation could be a regulatory function of adenosine diphosphoribose-(5) for this group of oxido reductase enzymes.

We have experimental evidence (*42*), indicating that the reductive inhibition caused by incubating UDP-galactose-4-epimerase with adenosine diphosphoribose-(5) is a result of hydrogen transfer from carbon 1 of the reducing end of ADPR-(5) to enzyme-NAD^+, leading to formation of enzyme-NADH. Examination of a molecular model of UDP-glucose-(1) reveals that the distance from the phosphate, attached to the anomeric carbon in UDP-glucose, to the axial hydrogen at carbon 4 (the hydrogen involved in the initiation of enzyme catalysis) is about 6A. Similarly, a model of ADPR-(5) shows a distance of about 6A between the phosphate linked to carbon 5 of the ribose moiety with the unsubstituted reducing end and the axial hydrogen at carbon 1 (β-ribo-furanose configuration). The same applies in the reaction between epimerase bound NAD^+ and L-arabinose in presence of UMP where the distance between the axial hydrogen at carbon 1 and the axial hydroxy group at carbon 4 is also 6A (β-pyranose configuration). Also, the topographical location of the hydrogen being accepted by the epimerase bound NAD^+ is almost identical with respect to the functional group in 6A distance in the reaction of epimerase with UDP-glucose-(1), ADP-ribose-(5) and L-arabinose in presence of UMP. We interpret these findings to indicate that the same site at the active site of the enzyme protein must be involved in all of these hydrogen transfer reactions.

From the rather arduous accumulation of isolated facts concerning mechanistic details of enzymes, several unifying principles seem to emerge. A group of enzymes, apparently metabolically unrelated, have an identical mechanism for the initiation of enzyme catalysis (Figure 15). We do not believe it is fortuitous that all these enzymes happen to have the same catalytic principle. We, therefore, propose the working hypothesis that the common properties of these enzymes are a consequence of the necessity of maintenance of catalytic principles in the evolutionary process of enzyme catalysis. We realize fully the necessity for further vigorous research to prove or disprove our hypothesis. However, it should be emphasized that, irrespective of the final decision leading to acceptance or rejection of the hypothesis, this comparative approach is an important asset to stimulate research in this area.

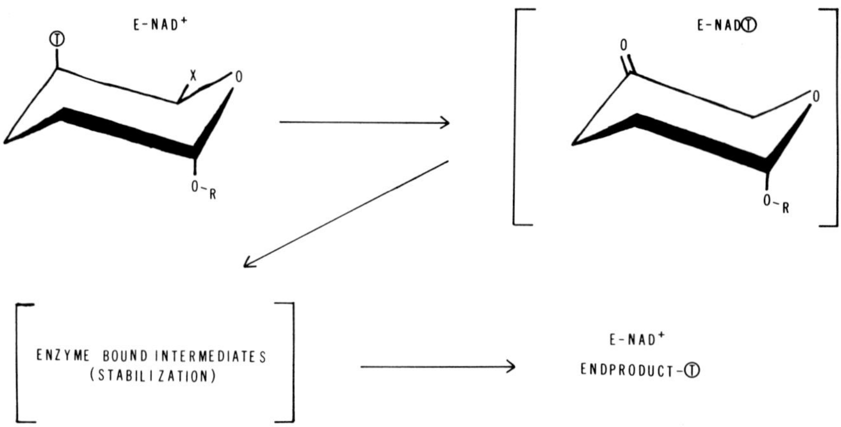

Figure 15. General scheme for 4-ulose intermediate formation and hydrogen transfer

Thus far, our approach has been that of gaining insight into the most likely conformation of the substrate, followed by a study of the change in its conformation during the enzymatic reaction to learn how to correlate properties and conformation of possible intermediates. More importantly, we hope to extend this approach by utilizing the complementary nature between substrate and enzyme protein to the active site. Consequently, conformational changes of the substrate and substrate intermediates during the enzymatic reaction must be accompanied by complementary changes on the active site of the enzyme protein. Thus, the studies concerning the mechanism of enzymatic reactions at the substrate level with emphasis upon conformation will eventually have some bearing on the catalytic and functional part of the enzyme. A look at the literature made us aware that several enzymes catalyzing different metabolic reactions have identical spatial requirements for their substrate and also have an identical reaction mechanism for the initiation of the catalytic reaction. Considering the complementary relationship between enzyme and substrate at the active site, it seems reasonable to assume that these similarities apply to the active site of the enzymes as well. For these reasons, we advance the working hypothesis for a common evolutionary origin of enzyme-NAD^+ mediated hydrogen transfer reactions.

Acknowledgment

The author wishes to express his deep appreciation for helpful discussions and constructive criticism provided by Gilbert Ashwell, Frank Eisenberg, Jr., William B. Jakoby, National Institutes of Health, Bethesda, Md.; Herman M. Kalckar, Harvard University Medical School, Boston, Mass. and Hans Grisebach, University of Freiburg, Germany. Finally the author would like to direct special thanks to Irwin A. Rose, Institute for Cancer Research, Philadelphia, Pa. for many valuable comments and suggestions. This research was supported by Grant AI-07241 from the National Institute of Allergy and Infectious Diseases, National Institutes of Health, United States Public Health Service.

Literature Cited

1. Ginsburg, V., *J. Biol. Chem.* (1960) **235,** 2196.
2. Ginsburg, V., *J. Biol. Chem.* (1961) **236,** 2389.
3. Glaser, L., Kornfeld, S., *J. Biol. Chem.* (1961) **236,** 1795.
4. Okazaki, R., Okazaki, T., Strominger, J. L., Michelson, M. A., *J. Biol. Chem.* (1962) **237,** 3014.
5. Pazur, J. H., Shuey, E. W., *J. Biol. Chem.* (1961) **236,** 1791.
6. Foster, D. W., Ginsburg, V., *Biochem Biophys. Acta* (1961) **54,** 376.
7. Barber, G. A., *Biochem. Biophys. Res. Comm.* (1962) **8,** 204.
8. Baddiley, J., Blumson, N. L., Girolamo, A. D., Girolamo, M. D., *Biochem. Biophys. Acta* (1961) **50,** 391.
9. Markowitz, A., *Biochem. Biophys. Res. Comm.* (1961) **6,** 250.
10. Markowitz, A., *Fed. Proc. Amer. Soc. Exp. Biol.* (1963) **22,** 464.
11. LeLoir, L. F., Cabib, E., *J. Amer. Chem. Soc.* (1953) **75,** 5445.
12. Markowitz, A., *J. Biol. Chem* (1962) **237,** 1767
13. Winkler, N. W., Markowitz, A., *J. Biol. Chem.* (1971) **246,** 5868.
14. Gaugler, R. W., Gabriel, O., *Fed Proc. Fed. Amer. Soc. Exp. Biol.* (1970) **29,** 337.
15. Elbein, A. D., Heath, E. C., *J. Biol. Chem.* (1965) **240,** 1920.
16. Heath, E. C., Elbein, A. D., *Proc. Natl. Acad. Sci.* (1962) **48,** 1209.
17. Nikaido, H., Nikaido, G., *J. Biol. Chem.* (1966) **242,** 1376.
18. Matsuhashi, S., Matsuhashi, M., Strominger, J. L., *J. Biol. Chem.* (1966) **241,** 4267.

19. Matsuhashi, S., Matsuhashi, M., Brown, J. G., Strominger, J. L., *J. Biol. Chem.* (1966) **241**, 4283.
20. Volk, W., Ashwell, G., *Biochem. Biophys. Res. Comm.* (1966) **24**, 113.
21. Matsuhashi, M., *Fed. Proc. Fed. Amer. Soc. Exp. Biol.* (1963) **22**, 465.
22. Matsuhashi, M., Strominger, J. L., *J. Biol. Chem.* (1964) **239**, 2454.
23. Gilbert, J. M., Matsuhashi, M., Strominger, J. L., *J. Biol Chem.* (1965) **240**, 1305.
24. Matsuhashi, S., *J. Biol. Chem.* (1966) **241**, 4275.
25. Heyns, K., Paulsen, H., *Advan. Carbohydrate Chem.* (1962) **17**, 169.
26. Gabriel, O., *Carbohyd. Res.* (1968) **6**, 111.
27. Gabriel, O., *Carbohyd. Res.* (1968) **6**, 319.
28. Gabriel, O., Lindquist, L. C., *J. Biol. Chem.* (1968) **243**, 1479.
29. Melo, A., Elliott, W. H., Glaser, L., *J. Biol. Chem.* (1968) **243**, 1467.
30. Wang, S. F., Gabriel, O., *J. Biol. Chem.* (1969) **244**, 3430.
31. Wang, S. F., Gabriel, O., *J. Biol. Chem.* (1970) **245**, 8.
32. Zarkowsky, H., Lipkin, E., Glaser, L., *J. Biol. Chem.* (1970) **245**, 6599.
33. Ørskov, F., Ørskov, I., Jann, B., Jann, K., Muller-Seitz, E., Westphal, O., *Acta Pathol. Microbiol. Scand.* (1967) **71**, 339.
34. Collins, P. M., Overend, W. G., *J. Chem. Soc.* (1965) 1912.
35. Gabriel, O., Ashwell, G., *J. Biol. Chem.* (1965) **240**, 4123.
36. Gabriel, O., Ashwell, G., *J. Biol. Chem.* (1965) **240**, 4128.
37. Melo, A., Glaser, L., *J. Biol. Chem.* (1968) **243**, 1475.
38. Kalckar, H. M., Bertland, II, A. U., Johansen, J. T., Ottensen, M., "The Role of Nucleotides for the Function and Conformation of Enzymes," p. 247, Academic, 1969.
39. Rose, I. A., *Ann. Rev. Biochem.* (1966) **25**, 23.
40. Adair, W. L., Gaugler, R. W., Gabriel, O., *Fed. Proc. Fed. Amer. Soc. Exp. Biol.* (1971) **30**, 376.
41. Maitra, U. S., Ankel, H., PNAS (1971) **68**, 2660.
42. Gabriel, O., Adair, W. L., Kalckar, H. M., Stathakos, D., *Fed. Proc. Fed. Amer. Soc. Exp. Biol.* (1972) **31**, 419.
43. Wilson, D. B., Hogness, D. S., *J. Biol. Chem.* (1964) **239**, 2469.
44. Schutzbach, J. S., Feingold, D. S., *J. Biol. Chem.* (1970) **245**, 2476.
45. Kelleher, W. J., Grisebach, H., *Eur. J. Biochem.* (1971) **23**, 136.
46. Blumson, N. L., Baddiley, J., *Biochem. J.* (1961) **81**, 114.
47. Rotenberg, S. L., Sprinson, D. B., *Proc. Natl. Acad. Sci.* (1970) **67**, 1669.
48. Kohn, B. D., Hoffmann, R. C., Kohn, P., *J. Biol. Chem.* (1963) **238**, 1193.
49. Nelsestuen, G. L., Kirkwood, S., *Fed. Proc. Fed. Amer. Soc. Exp. Biol.* (1970) **29**, 337.
50. Nelsestuen, G. L., Kirkwood, S., *J. Biol. Chem.* (1971) **246**, 7523.
51. David, L., Glaser, L., *Biochem. Biophys. Res. Comm.* (1971) **43**, 1429.
52. Seyama, Y., Kalckar, H. M., *Biochem.* (1972) **11**, 36.
53. Seyama, Y., Kalckar, H. M., *Biochem.* (1972) **11**, 40.

RECEIVED January 7, 1972.

INDEX

INDEX

H

S